COLLEGE CHEMISTRY

HARCOURT BRACE JOVANOVICH COLLEGE OUTLINE SERIES

COLLEGE CHEMISTRY

Harold Goldwhite and John R. Spielman

California State University, Los Angeles

Books for Professionals
Harcourt Brace Jovanovich, Publishers
San Diego   New York   London

Requests for permission to make copies of any part of the work should be mailed to:

Permissions
Harcourt Brace Jovanovich, Publishers
Orlando, Florida 32887

Printed in the United States of America

Library of Congress Cataloging in Publication Data

Goldwhite, Harold.
    College chemistry.

    (Books for professionals) (Harcourt Brace Jovanovich college outline series)
    Includes index.
    1. Chemistry.   I. Spielman, John R.   II. Title.
III. Series.   IV. Series: Harcourt Brace Jovanovich college outline series.
QD31.2.G653 1984   540   83-22616
ISBN 0-15-601561-7

First edition

    C   D   E

PREFACE

Do not *read* this Outline—**use** it. You can't learn chemistry simply by reading it: You have to *do* it. Solving specific, practical problems is the best way to master—and to demonstrate your mastery of—the theories, laws, and definitions upon which the science of chemistry is based. Outside the laboratory, you need three tools to do chemistry: a pencil, paper, and a calculator. Add a fourth tool, this Outline, and you're all set.

This HBJ College Outline has been designed as a tool to help you sharpen your problem-solving skills in chemistry. Each chapter covers a single topic, whose fundamental principles are broken down in outline form for easy reference. The outline text is heavily interspersed with worked-out examples, so you can see immediately how each new idea is applied in problem form. Each chapter also contains a Summary and a Raise Your Grades section, which (taken together) give you an opportunity to review the primary principles of a topic and the problem-solving techniques implicit in those principles.

Most important, this Outline gives you plenty of problems to practice on. Work the Solved Problems, and check yourself against the step-by-step solutions provided. Test your mastery of the material in each chapter by doing the Supplementary Exercises. (In the Supplementary Exercises, you're given answers only—the details of the solution are up to you.) Finally, you can review all the topics covered in the Outline by working the problems in the Semester and Final Exams. (The solution to each exam question is explained, so you can diagnose your own strengths and weaknesses.)

Having the tools is one thing; knowing how to use them is another. The solution to any problem in chemistry requires six procedures: (1) UNDERSTANDING, (2) ANALYZING, (3) PLANNING, (4) EXECUTING, (5) CHECKING, (6) REPORTING. Let's look at each of these procedures in more detail.

1. **UNDERSTANDING:** Read over the problem carefully and be sure you understand every part of it. If you have difficulty with any of the terms or ideas in the problem, reread the text material on which the problem is based. (In this Outline, important ideas, principles, laws, and terms are printed in boldface type, so they will be easy to find.) Make certain that you understand what kind of answer will be required. If the problem is quantitative, make an estimate of the magnitude of the answer.

2. **ANALYZING:** Break the problem down into its components. Ask yourself
   - What are the data?
   - What is (are) the unknown(s)?
   - What equation, law, or definition connects the data to the unknowns?

3. **PLANNING:** Trace a connection between the data and the unknowns as a series of discrete operations (steps). This often involves manipulating one or more mathematical or chemical expressions to isolate unknown quantities. Once you have a clear, stepwise path between data and solution, take note of any steps that require ancillary operations, such as balancing equations or converting units. (Keep a sharp watch on units—they are often useful clues.)

4. **EXECUTION:** Follow your plan and execute any mathematical operations. It helps to work with symbols whenever possible: Substituting data for variables should be the *last* thing you do. Make sure you've used the correct signs, exponents, and units.

5. **CHECKING:** Never consider a problem solved until you have checked your work. Does your answer
   - make sense?
   - have the right units?
   - answer the question?

   Is your math right?

6. REPORTING: Make sure you have shown your reasoning and method clearly, and that your answer is readable. (It can't hurt to write the word "Answer" in front of your answer. That way, you—and your instructor—can find it at a glance, saving time and trouble all 'round.)

The path from concept to bound book is a tortuous one and we have not traveled it alone. We thank our reviewers, especially Dr. Marinus P. Bardolph, Professor Emeritus, Southern Illinois University at Edwardsville, for their helpful advice and counsel. We also thank our families for their patience and support.

California State University,
*   Los Angeles*

HAROLD GOLDWHITE

JOHN R. SPIELMAN

CONTENTS

CHAPTER 1	**Matter and Measurement**	**1**
	1-1: Classifying Matter	1
	1-2: Physical Properties of Matter	2
	1-3: Measurements and Units	2
	1-4: Scientific Notation	4
	1-5: Significant Figures	4
	Solved Problems	*7*
CHAPTER 2	**Chemical Composition**	**13**
	2-1: Atoms	13
	2-2: Atomic Mass and Isotopes	14
	2-3: Molecules, Molecular Formula, and Molecular Mass	15
	2-4: Formulas and Formula Mass	16
	2-5: The Mole	17
	2-6: Nomenclature	17
	Solved Problems	*19*
CHAPTER 3	**Formulas**	**25**
	3-1: Empirical Formulas	25
	3-2: Molecular Formulas	25
	3-3: Determining Percentage Composition from a Formula	25
	3-4: Determining the Empirical Formula from a Composition	26
	3-5: Converting an Empirical Formula to a Molecular Formula	28
	Solved Problems	*29*
CHAPTER 4	**Equations and Stoichiometry**	**37**
	4-1: Chemical Equations	37
	4-2: Balancing Equations	37
	4-3: Molar Interpretation of Equations	38
	4-4: Stoichiometry: Mass Calculations from Equations	38
	4-5: Limiting Reagent	39
	4-6: Percentage Yield	40
	Solved Problems	*41*
CHAPTER 5	**Gases**	**53**
	5-1: General Properties of Gases	53
	5-2: The Gas Laws	55
	5-3: Kinetic Theory of Gases	60
	5-4: Real Gases	61
	Solved Problems	*62*
CHAPTER 6	**Intermolecular Forces and Liquids**	**74**
	6-1: Intermolecular Forces	74
	6-2: Liquids	75

	6-3: Surface Tension	76
	6-4: Viscosity	76
	Solved Problems	*77*

CHAPTER 7	**Solids**	**81**
	7-1: The Solid State	81
	7-2: Types of Solids	81
	7-3: X-Ray Diffraction	82
	7-4: Crystal Structure	82
	7-5: Common Structures of Salts	86
	Solved Problems	*87*

CHAPTER 8	**Thermochemistry**	**91**
	8-1: Temperature and Temperature Scales	91
	8-2: Heat Energy and Heat Capacity	91
	8-3: Enthalpy	92
	Solved Problems	*96*

CHAPTER 9	**Atomic Structure and the Periodic Table**	**103**
	9-1: Electromagnetic Radiation	103
	9-2: Quantum Theory of Radiation	104
	9-3: Orbitals	109
	9-4: Ionization Energies	111
	9-5: The Periodic Table	115
	Solved Problems	*116*

CHAPTER 10	**Chemical Bonding**	**123**
	10-1: Valence	123
	10-2: Bonding	124
	10-3: Molecular Structure	126
	10-4: Polarity and Electronegativity	129
	10-5: Bond Properties	131
	10-6: Molecular Orbitals	134
	Solved Problems	*138*

CHAPTER 11	**Solutions**	**147**
	11-1: Definitions	147
	11-2: Concentration Expressions	148
	11-3: Changes in Solubility	150
	11-4: Colligative Properties	151
	11-5: Electrolytes	153
	Solved Problems	*155*

| **SEMESTER EXAM** | | **163** |

CHAPTER 12	**Acids and Bases**	**175**
	12-1: Acid–Base Systems	175
	12-2: Acid–Base Strength	176
	12-3: Titration	178
	12-4: Stoichiometry	179
	Solved Problems	*181*

CHAPTER 13   Redox Reactions   **190**
13-1: Redox Reactions: Definitions   190
13-2: Oxidation Numbers   191
13-3: Balancing Redox Equations
by Half-Reactions   192
13-4: Balancing Redox Equations
by Oxidation Numbers   195
13-5: Analytical Uses of
Redox Reactions   197
Solved Problems   *197*

CHAPTER 14   Chemical Kinetics   **208**
14-1: Definition and Expression of
Reaction Rates   208
14-2: Reaction Rates and
Concentration Effects   209
14-3: Reaction Rates and
Temperature Effects:
The Arrhenius Equation   213
14-4: How Reactions Take Place:
Reaction Mechanisms   214
Solved Problems   *217*

CHAPTER 15   Equilibrium   **228**
15-1: The Equilibrium State   228
15-2: The Law of Chemical
Equilibrium   228
15-3: Equilibrium Changes   230
15-4: How to Solve
Equilibrium Problems   231
15-5: Heterogeneous Equilibria   233
Solved Problems   *234*

CHAPTER 16   Chemical Thermodynamics   **246**
16-1: Chemical Systems   246
16-2: The First Law   246
16-3: The Second Law and Entropy   247
16-4: Free Energy   249
Solved Problems   *252*

CHAPTER 17   Acid–Base Equilibria   **259**
17-1: Acids and Bases:
A Reminder   259
17-2: Finding the pH in Solutions
of Weak Acids   262
17-3: Finding the pH in Solutions
of Weak Bases   265
17-4: Salts of Weak Acids   267
17-5: Salts of Weak Bases   268
17-6: Buffers and Indicators   269
17-7: Titrations   272
17-8: Polyprotic Acids   275
Solved Problems   *277*

CHAPTER 18   **Heterogeneous Solution Equilibrium**   291
            18-1: Solubility Product   291
            18-2: The Common Ion Effect   293
            18-3: pH Effects   294
            18-4: Precipitation   294
            18-5: Solubility Rules   296
                  Solved Problems   *297*

CHAPTER 19   **Electrochemistry**   304
            19-1: Electrochemical Fundamentals   304
            19-2: Electrolytic Cells   305
            19-3: Galvanic (Voltaic) Cells   307
            19-4: Standard Electrode Potentials   308
            19-5: The Nernst Equation   312
                  Solved Problems   *314*

CHAPTER 20   **Radiochemistry**   323
            20-1: Radioactivity   323
            20-2: Nuclear Reactions   324
            20-3: Rates of Nuclear Reactions   325
            20-4: Mass/Energy Equivalence   327
            20-5: Fusion and Fission   329
                  Solved Problems   *330*

CHAPTER 21   **Coordination Chemistry**   339
            21-1: Coordinate Covalence   339
            21-2: Ligand Classification   341
            21-3: Geometry of Complexes   341
            21-4: Bonding in Complexes   344
            21-5: Stability of Complexes   347
                  Solved Problems   *349*

FINAL EXAM   355

INDEX   369

TEXT CORRELATION TABLE

1 MATTER AND MEASUREMENT

THIS CHAPTER IS ABOUT

☑ **Classifying Matter**
☑ **Physical Properties of Matter**
☑ **Measurements and Units**
☑ **Scientific Notation**
☑ **Significant Figures**

1-1. Classifying Matter

Matter is anything that occupies space and has mass. Chemists classify different kinds of matter in a scheme that leads from complex to simpler substances: mixtures, pure substances, compounds, and elements.

A. Mixtures

A mixture is an aggregate of two or more substances that retain their own properties and can be separated by physical means. Mixtures can be **homogeneous**, having the same physical and chemical properties throughout. For example, air is a homogeneous mixture of the gases nitrogen, oxygen, water vapor, argon, carbon dioxide, and traces of other gases; seawater is a homogeneous mixture of water and many dissolved salts, principally sodium chloride. The proportions of components in a homogeneous mixture may vary from sample to sample, but each sample is uniform in appearance.

Mixtures can be **heterogeneous**, having different physical and chemical properties from region to region in the mixture. For example, blood is a heterogeneous mixture of plasma, red blood cells, white blood corpuscles, platelets, and other factors that can be seen through a microscope. The components of a heterogeneous mixture remain separate and are visually distinguishable.

B. Pure substances

A **pure substance** has a definite fixed composition, and any sample will always have the same properties as any other sample when the external conditions (e.g., temperature and pressure) are the same. Pure substances include compounds and elements.

1. Compounds

Compounds are substances composed of two or more elements chemically united in fixed proportions. For example, sodium chloride (table salt) is a compound: All samples of NaCl will always contain 39.34% by mass of sodium and 60.66% by mass of chlorine. Water is also a compound: All samples of H_2O will always contain 88.81% by mass of oxygen and 11.19% by mass of hydrogen. The elements in a compound cannot be separated, except by chemical means.

2. Elements

Elements are the simplest forms of matter normally encountered. There are only just over a hundred different elements from which the millions of different substances you know are made. The known elements and some of their properties are listed on the inside front cover of this book.

1

1-2. Physical Properties of Matter

Physical properties are those characteristics that can be observed or measured in a sample of matter without producing a new substance. The physical properties of a piece of copper wire, for example, include its length, diameter, mass, density, color, melting point, electrical resistance, and hardness.

There are two kinds of physical properties:

1. **Extensive** physical properties depend on the *amount*, or *quantity*, of the sample and therefore can vary from sample to sample. The extensive properties of a piece of copper wire, for instance, include its length, diameter, mass, and electrical resistance.
2. **Intensive** physical properties depend on the *qualities* of the substance that do not involve the size or amount of the sample. The intensive properties of that piece of copper wire include its density, color, melting point, and hardness. Intensive properties are useful in distinguishing between different substances because they do not vary from sample to sample.

1-3. Measurements and Units

In chemistry properties are described as quantities, which can be measured and expressed as products of numbers and units.

A. Fundamental quantities

The most important **fundamental quantities** used in chemistry are *length, mass, time, amount of substance, temperature,* and *electric current.* (There's one more—*luminosity*—which will not be very important to you now.) Each of these quantities has its own irreducible unit.

B. Derived quantities

Derived quantities are physical quantities made up of combinations of fundamental quantities. For example, volume is a derived quantity.

EXAMPLE 1-1: The volume of a rectangular prism is calculated as length × breadth × height. In terms of fundamental quantities, the *dimensions* of the quantity volume are length × length × length, or length cubed. So the units of volume could be, for instance, in cubic meters (m^3) or cubic millimeters (mm^3).

C. Units

1. Systems of units

Two systems of units are commonly used in chemistry: the **cgs** (centimeter–gram–second) or metric system, and the **SI** (*Système Internationale d'Unités*), which is a revision of the metric system. In the cgs system the basic unit of length is the centimeter (cm); of mass, the gram (g); and of time, the second (s). In SI the basic unit of length is the meter (m); of mass, the kilogram (kg); and of time, the second (s). Both systems define individual **base units** for each of the fundamental quantities.

2. Unit prefixes

Prefixes are used in both systems to indicate decimal multiples or fractions of the base units. The common prefixes are

Multiple		Prefix	Abbreviation	Fraction		Prefix	Abbreviation
10	(10^1)	deka	da	0.1	(10^{-1})	deci	d
100	(10^2)	hecto	h	0.01	(10^{-2})	centi	c
1000	(10^3)	kilo	k	0.001	(10^{-3})	milli	m
1 000 000	(10^6)	mega	M	0.000 001	(10^{-6})	micro	μ
1 000 000 000	(10^9)	giga	G	0.000 000 001	(10^{-9})	nano	n

You can find a more complete table of prefixes in your textbook or in any standard reference.

3. Derived units

Derived physical quantities are measured in **derived units**. Although units used to measure derived physical quantities are actually derived from base units, they are often given special names for convenience. For example, volume is a very handy derived quantity in chemistry, so it is given a special unit—the **liter** (L). One liter is defined in SI as being equal to 1000 cubic centimeters (cm^3). So 1 milliliter (mL) is equal to 1 cm^3. Force and energy are also derived quantities: The derived unit of energy is the *erg* in cgs, and the *joule* in SI. Some derived units of force and energy in the two systems, and the relationships between them, follow:

Derived Units

	Force	Energy
Name of SI unit	newton	joule
Abbreviation	N	J
Definition in base units	$kg\,m\,s^{-2}$	$kg\,m^2\,s^{-2}$
Name of cgs unit	dyne	erg
Definition in base units	$g\,cm\,s^{-2}$	$g\,cm^2\,s^{-2}$
Conversion factors	1 N = 10^5 dynes	1 J = 10^7 ergs
	1 dyne = 10^{-5} N	1 erg = 10^{-7} J
Interconversion	1 erg = 1 dyne cm	

4. Conversion of units

There are several other useful relationships among cgs, SI, and other units that you should know—some of which you can figure out from their prefixes (and some of which you just have to memorize or look up).

Length: 1 m = 100 cm, 1 angstrom (Å) = 10^{-10} m
Mass: 1 kg = 1000 g
Volume: 1 m^3 = 1000 L
Pressure: 1 atmosphere (atm) = 760 torr = 101 325 pascals (Pa)

For complete tables of all units, you should consult a standard reference book.

It is often convenient to use the *"factor-cancel"* method to convert units into other units. Start with the original number with its units and then multiply by a succession of factors, each expressing a relation between units. Set up your multiplication in such a way that the units you want to get rid of are mathematically canceled. You will then end up with the units you want. For example, to convert 42.3 mi/h to ft/s, set it up as follows:

$$\frac{42.3 \text{ mi}}{h} \times \frac{5280 \text{ ft}}{\text{mi}} \times \frac{h}{60 \text{ min}} \times \frac{\text{min}}{60 \text{ s}} = \frac{42.3 \times 5280 \text{ ft}}{60 \times 60 \text{ s}} = \frac{62.0 \text{ ft}}{s}$$

[Note that the answer is in three digits. The numbers 5280 and 60 are absolute numbers and are not considered in determining the number of significant figures.]

EXAMPLE 1-2: Convert 59.32 km/h to cm/s.

Solution:

$$\frac{59.32 \text{ km}}{h} \times \frac{10^3 \text{ m}}{\text{km}} \times \frac{10^2 \text{ cm}}{\text{m}} \times \frac{h}{60 \text{ m}} \times \frac{\text{m}}{60 \text{ s}} = \frac{59.32 \times 10^3 \times 10^2 \text{ cm}}{60 \times 60 \text{ s}}$$

$$\frac{5.932 \times 10^6 \text{ cm}}{3.6 \times 10^3 \text{ s}} = \frac{1.648 \times 10^3 \text{ cm}}{s}$$

---

note: Base or derived, cgs or SI—all units are interconvertible and can be manipulated mathematically. You may not have to know specific conversion formulas (you can always look them up), but you do have to know how to handle units in an equation.

1-4. Scientific Notation

Scientific notation is a method of expressing large or small numbers as factors of the powers of 10. You can use exponents of 10 to make the expression of scientific measurements more compact, easier to understand, and simpler to manipulate. (See how 0.000 000 001 3 m compares to 1.3×10^{-9} m and how 7 500 000 g compares to 7.5×10^6 g.)

To express numbers in scientific notation, you use the form

$$a \times 10^b$$

where a is a decimal number between 1 and 10 (but not equal to 10), and b is a positive or negative integer or zero. (Remember, $10^0 = 1$.) Let's see how this form works for 7 500 000, a large number.

1. Set a equal to 7.5, which is a decimal number between 1 and 10.
2. To find b, count the places to the right of the decimal point in a to the original decimal point:

$$\overset{1\,2\,3\ 4\,5\,6}{7.500\,000_{\textstyle .}}$$

There are 6 places to the right ($+6$) from the decimal point in a to the original decimal point, so $b = 6$. The number is expressed as 7.5×10^6.

- For a large number, the exponent of 10 (b) will be a POSITIVE integer equal to the number of decimal places to the RIGHT from the decimal point in a to the original decimal point.

Now let's try the process for 0.000 000 001 3, a small number.

1. $a = 1.3$, which is between 1 and 10.
2. To find b, count the places to the left of the decimal point in a, finishing up at the original decimal point:

$$\overset{9\,8\,7\ 6\,5\,4\ 3\,2\,1}{0.000\,000\,001.3}$$

There are 9 places from the left (-9) of the decimal point in a to the original decimal point, so $b = -9$. The number is expressed as 1.3×10^{-9}.

- For a small number, the exponent of 10 will be a NEGATIVE integer equal to the number of decimal places to the LEFT from the decimal point in a to the original decimal point.

1-5. Significant Figures

The accuracy of a measurement depends on the quality of the instrument you use for measurement and on the carefulness of your measurement. When you report a measurement, you express it in the number of **significant figures** that best represents your own precision and that of your instrument. So when you write significant figures, you show the limits of accuracy and you show where uncertainty begins.

EXAMPLE 1-3: Measuring with an ordinary meter stick, you might report the length of an object as 1.4 m, which means you measured it as being longer than 1.35 m, but shorter than 1.45 m. The measurement 1.4 has *two* significant figures. If you had a better ruler, or were more careful, you might have reported the length as 1.42 m, which means that you measured the object as being longer than 1.415 m, but shorter than 1.425 m. The measurement 1.42 has *three* significant figures.

- The last digit in a significant figure is uncertain because it reflects the range of accuracy.

A. Significant zeros

You'll have to decide whether zeros are significant in three different situations.

1. *If the zeros precede the first nonzero digit, they are not significant.* Such zeros merely locate the decimal point; i.e., they define the magnitude of measurement. For example, 0.000 14 m has two significant figures, and 0.01 has one significant figure.

2. *If the zeros are between nonzero digits, they are significant.* For example, 103 307 kg has six significant figures while 0.044 03 has four significant figures.

3. *If the zeros follow nonzero digits, there is ambiguity if no decimal point is given.* If a force is given as 300 N, you have no way of knowing if the final two zeros are significant. But if the force is given as 300. N, you know that it has three significant figures; and if it is given as 300.0 N, it has four significant figures.

> *note:* You can avoid ambiguity by expressing your measurements in scientific notation. Then if you record final zeros in *a*, they are significant. So if you report "300 N" as 3×10^2 N, it has only one significant figure; 3.0×10^2 N has two significant figures; and 3.00×10^2 N has three significant figures.

B. Using significant figures in calculations

Accuracy in chemical calculation differs somewhat from arithmetic accuracy.

- a result can only be as accurate as the least significant measurement that goes into its calculation.

EXAMPLE 1-4: Express the result of adding 4.37 g and 1.002 g to the correct number of significant figures.

Solution:

$$\begin{array}{ll} 4.37 \text{ g} & \text{(significant to one-hundredth of a gram)} \\ + 1.002 \text{ g} & \text{(significant to one-thousandth of a gram)} \\ \hline 5.372 \text{ g} & \end{array}$$

This result cannot be significant beyond the hundredths of a gram. It must be reported as 5.37 g because its accuracy cannot exceed that of the least significant measurement.

EXAMPLE 1-5: Calculate the volume of a rectangular rod of aluminum that is 1.84 cm long (*l*), 0.413 cm wide (*w*), and 0.207 cm high (*h*). Report the result with the correct number of significant figures.

Solution:

$$\text{Volume} = l \times w \times h = (1.84 \text{ cm})(0.413 \text{ cm})(0.207 \text{ cm})$$
$$= 0.157\,303\,44 \text{ cm}^3$$

The numerical result of the multiplication on your calculator is 0.157 303 44. But since each measurement has only three significant figures, as a rule of thumb, the result should have only three significant figures. So you must report the result as 0.157 cm^3, or 1.57 cm $\times 10^{-1}$ cm^3 in scientific notation.

> *note:* An answer computed using multiplication or division should have approximately the same **percent relative error**, which is (uncertainty/actual value) $\times$ 100%, as the *least exact value* in the problem. For example,
>
> $$9.3 \text{ mL} \times 1.15 \text{ g/mL} = 10.7 \text{ g}$$
>
> Rounding off using the rule of thumb would give an answer of 11 g, which has 9.1% relative error. But the least exact value (9.3 mL) in the problem has only 1.1% relative error, so we go with the answer of 10.7 g (.93% relative error). This type of example happens rarely, but you should keep in mind that the rule of thumb is only an approximation.

You will use some numbers that do not have a specific number of significant figures. In a formula,

- integer whole numbers and constants do not alter your calculation of significant figures.

For example, the volume of a sphere is $V = \frac{4}{3}\pi r^3$. The 4 and 3 are exact whole numbers, while the constant π can be reported to any desired degree of accuracy (3.141 59. . .). The result for the volume will depend *only* on the accuracy of the measurement for the radius *r*. Other numbers without a specific number of significant figures occur in definitions. For example, definitions like

1 cm = 10^{-2} m and 1 atm = 760 torr don't contain experimentally determined quantities. Using these definitions has no effect on the number of significant figures in the answer to a problem.

EXAMPLE 1-6: Convert 234.17 torr to atm.

Solution: You see that

$$\frac{234.17 \text{ torr}}{760 \text{ torr/atm}} = 0.308\,12 \text{ atm}$$

You report the answer to five significant figures because 760 torr/atm is a defined or exact quantity.

EXAMPLE 1-7: Convert 0.377 m to mm.

Solution: Since 1 m = 10^3 mm,

$$0.377 \text{ m} \times \frac{10^3 \text{ mm}}{\text{m}} = 3.77 \times 10^2 \text{ mm}$$

C. Rounding off

We've already rounded off some results in some of our calculations. The rules are simple: If the digit following the last reportable digit is

4 or less, you drop it
6 or more, you increase the last reportable digit by one
5, you use the arbitrary odd–even rule: If the last reportable digit is even, you leave it unchanged; if the last reportable digit is odd, you increase it by one.

EXAMPLE 1-8: Round off 108.75 and 108.65 to four significant figures.

Solution: Since the fourth significant digit in 108.75 is 7, an odd number, you increase it by one. Report the answer as 108.8. Since the fourth significant digit in 108.65 is 6, an even number, you leave it unchanged. Report the answer as 108.6.

> *note:* You may wonder just when to round off. The answer is, round off when it's most convenient. In this age of calculators and computers, it's just as easy to carry six or seven digits as it is to carry three or four. So, for economy and accuracy, do your rounding off at the last step of a calculation.

SUMMARY

1. A sample of matter can be classified as either a mixture or a pure substance. Mixtures are either homogeneous or heterogeneous; pure substances are either compounds or elements.
2. Physical properties of matter can be used to distinguish substances.
3. Physical properties may be extensive or intensive. Extensive properties vary, but intensive properties do not vary from sample to sample.
4. Measurements are expressed as products of numbers and units.
5. Physical quantities are classified as fundamental or derived.
6. Systems of units used in chemistry include cgs and SI.
7. Units can be classified as base or derived.
8. Scientific notation is used to make the expression of results more compact and simpler to manipulate.
9. Significant figures express the accuracy with which a measurement or result can be stated and show the beginning of uncertainty.
10. A numerical result in chemistry is only as accurate as the least significant measurement.

RAISE YOUR GRADES

Can you define . . .?

☑ compound
☑ element
☑ base unit
☑ derived unit
☑ significant zero

Can you explain . . .?

☑ the difference between homogeneous and heterogeneous mixtures
☑ the difference between extensive and intensive physical properties
☑ what the fundamental physical properties are
☑ the difference between the cgs and SI unit systems
☑ what a significant figure is
☑ how to express numbers in scientific notation
☑ how to round off numbers to a given number of significant figures

SOLVED PROBLEMS

PROBLEM 1-1 Is orange juice a pure substance?

Solution: Even casual inspection shows you that orange juice is a heterogeneous mixture. It contains clear juice and pulp fragments that settle to the bottom of the container. Thus in the sense that chemists mean it, orange juice is *not* a pure substance.

PROBLEM 1-2 Which of the following are elements: niobium, wood, lead, brass, water, xenon?

Solution: The elements are listed inside the cover of this book. When you check that table, you'll find that niobium, lead, and xenon are included as elements.

PROBLEM 1-3 Classify the following properties of a plastic sheet as intensive or extensive: mass, color, thickness, density, transparency.

Solution: The extensive properties are those that describe a specific piece of plastic—its mass, thickness, and transparency. The intensive properties are those that are characteristic of the material of the plastic, namely color and density.

PROBLEM 1-4 Convert 1.2 cm to μm.

Solution: Recall that $1 \text{ cm} = 10^{-2} \text{ m}$ and $1 \text{ } \mu\text{m} = 10^{-6} \text{ m}$. Consequently

$$\frac{1 \text{ cm}}{1 \text{ } \mu\text{m}} = \frac{10^{-2} \text{ m}}{10^{-6} \text{ m}} = 10^4 \quad \text{or} \quad 1 \text{ cm} = 10^4 \text{ } \mu\text{m}$$

Carrying out the conversion, you have

$$1.2 \text{ cm} = 1.2 \text{ cm} \times \frac{10^4 \text{ } \mu\text{m}}{1 \text{ cm}} = 1.2 \times 10^4 \text{ } \mu\text{m}$$

PROBLEM 1-5 Convert 75 ng to mg.

Solution: The prefix nano (n) is 10^{-9}, so 1 ng = 10^{-9} g. And you know that 1 mg = 10^{-3} g. So

$$\frac{1 \text{ ng}}{1 \text{ mg}} = \frac{10^{-9} \text{ g}}{10^{-3} \text{ g}} = 10^{-6} \qquad \text{or} \qquad 1 \text{ ng} = 10^{-6} \text{ mg}$$

Thus

$$75 \text{ ng} = 75 \text{ ng} \times \frac{10^{-6} \text{ mg}}{1 \text{ ng}} = 7.5 \times 10^{-5} \text{ mg}$$

PROBLEM 1-6 How many significant figures are there in **(a)** 103 kg, **(b)** 0.003 g, **(c)** 80 450 N, **(d)** 0.0701 cm?

Solution:
(a) Three: Zeros between nonzero digits are significant.
(b) One: Zeros preceding the first nonzero digit are not significant.
(c) Ambiguous—four or five: The final zero may or may not be significant. The first four digits, including the first zero, are significant.
(d) Three.

PROBLEM 1-7 Use scientific notation to express **(a)** 103 kg, **(b)** 0.003 g, **(c)** 0.0701 cm, **(d)** 181 597 km.

Solution:
(a) 1.03×10^2 kg (Just count how many places to the left you have to shift the decimal to give a number between 1 and 10, and this gives the exponent.)
(b) 3×10^{-3} g (Just count how many places to the right you have to shift the decimal to give a number between 1 and 10, and this gives the negative exponent.)
(c) 7.01×10^{-2} cm
(d) $1.815 97 \times 10^5$ km

PROBLEM 1-8 Express the result of adding 945.5 g and 12.74 g to the correct number of significant figures.

Solution:

$$
\begin{array}{ll}
945.5 \text{ g} & \text{(significant to tenths of a gram)} \\
+ \ \ 12.74 \text{ g} & \text{(significant to hundredths of a gram)} \\
\hline
958.24 \text{ g} &
\end{array}
$$

The result is significant only to tenths of a gram, so you round it off to 958.2 g.

PROBLEM 1-9 Express the result of adding 0.102 cm and 11.6 mm to the correct number of significant figures.

Solution: You can't add apples to oranges—one of these measurements has to be converted to the units of the other. Since 1 cm = 10^{-2} m and 1 mm = 10^{-3} m, you get

$$\frac{1 \text{ mm}}{1 \text{ cm}} = \frac{10^{-3} \text{ m}}{10^{-2} \text{ m}} = 10^{-1}$$

and you can convert

$$11.6 \text{ mm} = 11.6 \text{ mm} \times \frac{10^{-1} \text{ cm}}{1 \text{ mm}} = 1.16 \text{ cm}$$

Now you can add the two quantities:

$$
\begin{array}{ll}
0.102 \text{ cm} & \text{(significant to a thousandth of a centimeter)} \\
1.16 \ \ \text{cm} & \text{(significant to a hundredth of a centimeter)} \\
\hline
1.262 \text{ cm} &
\end{array}
$$

The total can be significant only to a hundredth of a centimeter, so you must report it as 1.26 cm (or 1.26 cm $\times$ 10 mm/cm = 12.6 mm).

PROBLEM 1-10 Density is defined as the ratio of mass to volume. What is the density in g/cm^3 of a metal bar of mass 45.7 g and volume 9.3 cm^3? Express your answer to the correct number of significant figures.

Solution: The density (d) is

$$d = \frac{\text{mass}}{\text{volume}} = \frac{45.7 \text{ g}}{9.3 \text{ cm}^3}$$

When you enter this into your calculator, the display shows that this ratio equals 4.913 978 5 g/cm³. But the volume is given to only two significant figures, so the density can only have two significant figures. Your correct answer is therefore 4.9 g/cm³.

PROBLEM 1-11 The density of aluminum is 2.70 g cm⁻³. Find the volume of an aluminum block of mass 79.76 g.

Solution: Since $d = \text{mass/volume} = m/V$, it follows that $V = m/d$:

$$\frac{m}{d} = \frac{79.76 \text{ g}}{2.70 \text{ g cm}^{-3}}$$

Your calculator displays 29.5407. The mass is given to four significant figures, but the density is given to only three significant figures. So you can only report three significant figures in your answer:

$$V = 29.5 \text{ cm}^3$$

PROBLEM 1-12 A beaker containing sodium chloride was weighed on an analytical balance and its mass was found to be 57.3492 g. Some of the NaCl was transferred carefully to a flask. The mass of the beaker plus the remaining NaCl was then found to be 55.641 g. What mass of NaCl was transferred to the flask?

Solution: The mass transferred is just the difference between the initial and final masses of the beaker plus NaCl:

$$\begin{array}{r} 57.3492 \text{ g} \\ - 55.641 \text{ g} \\ \hline 1.7082 \text{ g} \end{array}$$

Since the second mass was only determined to the nearest milligram, you can only report the result to the nearest milligram, as 1.708 g.

PROBLEM 1-13 Round off each of the following to four significant figures: **(a)** 7.038 46 g, **(b)** 15 759 m, **(c)** 0.002 864 5 L, **(d)** 8.3100 × 10⁶ J.

Solution:
(a) The digit following the fourth significant digit is 4, so you drop it—and the 6 following it—and report the result as 7.038 g.
(b) The digit following the fourth significant digit is 9, so you increase the fourth digit by one to 6. If you report the result as 15 760 m, the trailing zero leads to ambiguity, so you should use scientific notation and report the result as 1.576 × 10⁴ m.
(c) The digit following the fourth significant digit is 5. Since the fourth significant digit is even (it is 4), you follow the odd–even rule and leave it unchanged, reporting the result as 0.002 864 L or as 2.864 × 10⁻³ L.
(d) The digit following the fourth significant digit is 0, so you drop it, reporting the result as 8.310 × 10⁶ J.

PROBLEM 1-14 The density of air at room temperature and pressure is 1.2 g/L. Calculate the mass in kilograms of the air in a room that is 5 m long, 4 m wide, and 3 m high.

Solution: Since $d = m/V$, you know that

$$m = V \times d$$

The volume of the room = (5 m)(4 m)(3 m) = 60 m³. Now 1 m³ = 10³ L and 1 kg = 10³ g. So

$$m = (60 \text{ m}^3)\left(\frac{10^3 \text{ L}}{\text{m}^3}\right)\left(\frac{1.2 \text{ g}}{\text{L}}\right)\left(\frac{1 \text{ kg}}{10^3 \text{ g}}\right) = 72 \text{ kg}$$

Since the room dimensions have only one significant figure, your answer can have only one significant figure. So, to report it without ambiguity, you should write it as 7 × 10¹ kg.

PROBLEM 1-15 Convert a pressure of 1.00 × 10⁶ Pa to **(a)** atm and **(b)** torr.

Solution: Recall that 1 atm = 760 torr = 101 325 Pa (or $1.013\,25 \times 10^5$ Pa). So

(a)
$$1.00 \times 10^6 \; \cancel{\text{Pa}} \times \frac{1 \text{ atm}}{1.013\,25 \times 10^5 \; \cancel{\text{Pa}}} = 9.87 \text{ atm}$$

(b)
$$1.00 \times 10^6 \; \cancel{\text{Pa}} \times \frac{760 \text{ torr}}{1.013\,25 \times 10^5 \; \cancel{\text{Pa}}} = 7.50 \times 10^3 \text{ torr}$$

PROBLEM 1-16   The density of gold is 19 g cm^{-3}. Which is heavier (has a larger mass): a 1-kg standard mass, or a cube of gold that measures 3.7 cm on each side?

Solution: The volume of the cube of gold $= (3.7 \text{ cm})^3 = 51 \text{ cm}^3$. The mass of the gold cube is then

$$(51 \text{ cm}^3)(19 \text{ g cm}^{-3}) = 9.7 \times 10^2 \text{ g} \qquad \text{(2 significant figures)}$$

By definition, a 1-kg standard mass contains exactly 1000 g, so it has the larger mass.

PROBLEM 1-17   The density of ice is 0.917 g/cm^3. If a cubic block of ice is 75 mm on each side, which of the following is its mass?

(a) 3.9×10^5 g   (d) 3.9×10^2 g
(b) 4.2×10^2 g   (e) 52 g
(c) 4.6×10^2 g

Solution: Before you start your calculations, think about how large this piece of ice is. A quick inspection of the dimensions should tell you that the cube must be about half a liter or less in volume. Since the density of ice is nearly 1 g/cm^3, the mass must be about 500 g. Obviously, answer (a) is much too large and answer (e) is much too small.
     The volume of a cube is $V = l^3$. It will be convenient to have the volume in cubic centimeters, so convert 75 mm to centimeters before computing the answer:

$$V = (7.5 \text{ cm})^3 = 422 \text{ cm}^3$$

Now use the density to determine the mass:

$$d = \frac{m}{V} = m = dV = \left(\frac{0.917 \text{ g}}{\cancel{\text{cm}^3}} \right)(422 \; \cancel{\text{cm}^3}) = 387 \text{ g}$$

Rounded off to two significant figures, the correct answer is 3.9×10^2 g, answer (d).
     You will obtain answer (a) if you don't convert the length from mm to cm. Answer (b) is the volume in cm^3, not the mass. In answer (c) the volume was divided by 0.917 g/cm^3 (pay attention to units!). In answer (e) the length was squared, not cubed, to obtain the volume.

d = m/V

Supplementary Exercises

PROBLEM 1-18   Which of the following would be classified as pure substances: milk, blood, salt, pepper, sugar, coffee?

Answer: salt and sugar

PROBLEM 1-19   Which of the following are elements: sulfur, chalk, dry ice, iron, nitrogen, ammonia?

Answer: sulfur, iron, and nitrogen

PROBLEM 1-20   Which of the following properties of a piece of gold foil are intensive: thickness, area, color, melting point, translucency, electrical resistance?

Answer: color and melting point

PROBLEM 1-21   Convert (a) 1.4 g to mg, (b) 1.4 mg to g, (c) 1.4 kg to μg, (d) 1.4 μg to mg.

Answer: (a) 1.4×10^3 mg   (b) 1.4×10^{-3} g   (c) 1.4×10^9 µg   (d) 1.4×10^{-3} mg

PROBLEM 1-22   Convert 6.75 m^3 to µL.

Answer: 6.75×10^9 µL

PROBLEM 1-23   Express the following in scientific notation: (a) 10 504 g, (b) 0.000 079 m, (c) 802 L, (d) 10.5 g cm^{-3}.

Answer: (a) 1.0504×10^4 g   (b) 7.9×10^{-5} m   (c) 8.02×10^2 L   (d) 1.05×10^1 g cm^{-3}

PROBLEM 1-24   How many significant figures are there in each of the following measurements: (a) 10 504 g, (b) 0.000 079 m, (c) 800 L, (d) 0.030 80 mL?

Answer: (a) 5   (b) 2   (c) ambiguous   (d) 4

PROBLEM 1-25   Round off each of the following to three significant figures: (a) 10 504 g, (b) 0.030 80 mL, (c) 8645 m, (d) 7.336 cm^2, (e) 1.675 s.

Answer: (a) 1.05×10^4 g   (b) 0.0308 mL   (c) 8.64×10^3 m   (d) 7.34 cm^2 (e) 1.68 s

PROBLEM 1-26   Calculate the total mass when 1.483 g of carbon is added to a flask of mass 58.7 g, expressing the answer to the correct number of significant figures.

Answer: 60.2 g

PROBLEM 1-27   Calculate the density of a rectangular block of metal whose length is 8.335 cm, width is 1.02 cm, height is 0.982 cm, and whose mass is 62.3538 g.

Answer: 7.47 g cm^{-3} (only 3 significant figures)

PROBLEM 1-28   What is the total volume produced when 1.037 L of water is mixed with 5.44 mL of water?

Answer: 1.042 L (or 1042 mL)

PROBLEM 1-29   Convert a pressure of 125 torr to atmospheres.

Answer: 0.164 atm

PROBLEM 1-30   The thickness of a sheet of iron is 0.887 mm. Its density is 7.86 g/cm^3. Its mass is 795 g. Calculate the area of the sheet of iron.

Answer: 1.14×10^3 cm^2

PROBLEM 1-31   The number 275 362 correctly rounded off to four significant figures is (a) 2754, (b) 2.754×10^5, (c) 28 000, (d) 275 362.0, (e) 2.7536×10^5.

Answer: (b)

PROBLEM 1-32   The answer to the expression 8.735 g + 2.3 g + 93.683 g, to the correct number of significant figures is (a) 104.7 g, (b) 104.718 g, (c) 1.05×10^2 g, (d) 100 g, (e) 104.6 g.

Answer: (a)

PROBLEM 1-33   For 2.05×10^2 Å, the equivalent expression in meters is (a) 2.05×10^{-12} m, (b) 2.05×10^{12} m, (c) 2.05×10^{-10} m, (d) 2.05×10^{-8} m, (e) 205 m.

Answer: (d)

PROBLEM 1-34   The speed of light is 2.9979×10^8 m s^{-1}. How far will light travel in 1.0 ns?

Answer: 0.30 m

PROBLEM 1-35 The element antimony has a density of 6.62 g cm^{-3}. Calculate the edge length in meters of a cube of antimony whose mass is 1.0×10^5 kg.

Answer: 2.5 m

PROBLEM 1-36 A 7.05-μL sample is taken from a vial holding 1.00 mL of blood. What volume of blood is left in the vial?

Answer: 0.99 mL

PROBLEM 1-37 Which is heavier: a 500.0-g weight or a rectangular slab of copper measuring 10.0 cm by 5.0 cm by 1.00 cm? (The density of copper is 8.92 g/cm^3.)

Answer: the 500.0-g weight

PROBLEM 1-38 Calculate the mass in kilograms of a cylinder of lead whose radius is 1.085 cm and height is 6.74 cm. The density of lead is 11.3 g/cm^3. (The volume of a cylinder of height h and radius r is $\pi r^2 h$.)

Answer: 0.282 kg

PROBLEM 1-39 The recommended daily intake of thiamine is 1.5 mg. A 100.0-g apple (a small apple) contains about 4×10^{-2} mg of thiamine. How many small apples should you eat daily to get the recommended daily intake of thiamine if this is your sole source of the vitamin (a very unlikely assumption!)?

Answer: about 40

PROBLEM 1-40 Determine the mass in kilograms of 1.00 km^3 of water. (The density of water is 1.00 g cm^{-3}.)

Answer: 1.00×10^{12} kg

2 CHEMICAL COMPOSITION

THIS CHAPTER IS ABOUT

☑ **Atoms**
☑ **Atomic Mass and Isotopes**
☑ **Molecules, Molecular Formula, and Molecular Mass**
☑ **Formulas and Formula Mass**
☑ **The Mole**
☑ **Nomenclature**

2-1. Atoms

An **atom** is the smallest electrically neutral component of an element that has all the chemical properties of the element. Elements and their atoms are represented by symbols, and you use the same symbol for the atom that you use for the element. The symbols for all the known atoms can be found in the periodic chart.

A. Fundamental subatomic particles

Atoms are composed of fundamental subatomic particles that are smaller than the smallest atom. Although atoms are electrically neutral, some subatomic particles carry charge. The subatomic particles that chemists use to explain the structure of the atom are the **proton**, the **neutron**, and the **electron**. For all practical purposes, you can think of the atom as having a very small core, the **nucleus**, which contains both protons and neutrons. Electrons form a cloud around the nucleus. This model is very simplistic (even a little misleading), but you will find it useful for solving certain problems in chemistry. (We'll cover atomic structure in detail in Chapter 9.)

The characteristics of the fundamental subatomic particles are listed here:

Particle	Mass (amu)	Charge
proton	1	+1
neutron	1	0
electron	effectively 0	−1

B. Atomic number

All atoms of a particular element have the same amount of positive charge on their nuclei. This positive charge, measured in **electron charge units**, is called the nuclear charge Z and is defined as the **atomic number** of that atom or element. For example, hydrogen is the simplest atom, having a nucleus whose charge is $+1$, so its atomic number is 1. The atomic number also represents the numerical position of an element in an arrangement of all the elements in order of increasing nuclear charge, as in the periodic table.

Finally, the atomic number equals the number of protons in the nucleus of an atom. (Values of Z for all the elements are listed on the inside cover of this book.) In an electrically neutral atom, then, the number of electrons is equal to the number of protons.

EXAMPLE 2-1: How many electrons are present in a neutral atom of lithium (Li); of manganese (Mn); of tin (Sn)?

Solution: You know that the number of electrons in a neutral atom equals Z, the charge on the nucleus. Looking up the atomic numbers of each element, you find that for Li $Z = 3$, and so it has 3 electrons; for Mn $Z = 25$, so it has 25 electrons; for Sn $Z = 50$, so it has 50 electrons.

C. Ions

Usually you think of an atom as being electrically neutral. However, it is possible to remove electrons from or add electrons to an atom, thus forming an electrically charged particle called an **ion**. Removing electrons from an atom forms a positively charged ion called a **cation**, and adding electrons forms a negatively charged ion called an **anion**.

EXAMPLE 2-2: What is the charge on a chlorine (Cl) atom if two electrons are removed? What is the charge on a Cl atom if one electron is added?

Solution: Since each electron has a -1 charge, removing two electrons gives the Cl atom a $+2$ charge, which is written as Cl^{2+}. Adding one electron gives the Cl atom a -1 charge, written as Cl^-.

EXAMPLE 2-3: What will the charge be on a beryllium (Be) atom if all of the electrons are removed?

Solution: Beryllium is element number 4, or $Z = 4$, so the neutral atom has four electrons. Removing all four electrons gives the Be atom a $+4$ charge, written as Be^{4+}.

2-2. Atomic Mass and Isotopes

A. Atomic mass units

Because the mass of an atom is very small when measured in SI or cgs units, there is a special unit for atomic mass. The **atomic mass unit** (amu), or **dalton**, is defined as $\frac{1}{12}$ of the mass of one atom of the carbon-12 isotope (^{12}C). Using this unit, we can characterize atoms in terms of their relative masses.

B. Isotopes

Atoms of the same element that differ in mass but not in atomic number are called **isotopes** of that element. Most carbon atoms, for example, have a mass of exactly 12 amu. A carbon atom that has this mass is the isotope carbon-12, shown in symbols as ^{12}C, where the left-hand superscript signifies the atomic mass. About 1% of the atoms of naturally occurring carbon have a mass of about 13 amu. These atoms are another isotope, carbon-13 (^{13}C).

note: Isotopes of an element have the same charge on the nuclei of their atoms and the same number of electrons; only the masses of their nuclei differ.

EXAMPLE 2-4: For all neutral atoms of tin (Sn), $Z = 50$, and there are 50 electrons. Ten isotopes of Sn occur in nature. Their masses and relative abundances are

Mass (amu)	Abundance (%)	Mass (amu)	Abundance (%)
112	0.95	118	24.01
114	0.65	119	8.58
115	0.34	120	32.97
116	14.24	122	4.71
117	7.57	124	5.98

C. Composition of nuclei

Chemists use a simple model—based on mass and charge—to characterize the composition of nuclei. Nuclei are made up of two different particles, the positively charged protons and the neutral neutrons. The proton's charge is $+1$, and its mass is very close to 1 amu. The neutron has no charge, and its mass is also very close to 1 amu. Given these facts, it's easy to account for isotopes.

- Isotopes of an element differ only in the number of neutrons present in the nuclei of the atoms.

For example, there are three known isotopes of hydrogen. Hydrogen atoms that have a single proton and no neutrons have a mass of about 1 amu. They are given the symbol ^1H. *Deuterium* atoms have nuclei composed of one proton and one neutron and have a mass of about 2 amu. They are given the symbol ^2H (or D). *Tritium* atoms have nuclei composed of one proton and two neutrons and have a mass of about 3 amu. They are given the symbol ^3H (or T). The numerical superscript to the atomic symbol gives the approximate mass of the nucleus in atomic mass units.

EXAMPLE 2-5: Using the chemist's model, analyze the composition of the nuclei of the isotopes ^{35}Cl and ^{37}Cl, the naturally occurring isotopes of chlorine.

Solution: For chlorine $Z = 17$, so the nucleus must have a charge of $+17$. You can account for this by assuming that there are 17 protons, each having a charge of $+1$, in the chlorine nucleus. These 17 protons have a total mass of about 17 amu. The rest of the nuclear mass is due to neutrons.

In ^{35}Cl the mass is about 35 amu, so 17 amu must come from the 17 protons and $(35 - 17)$ amu $= 18$ amu must come from 18 neutrons. In ^{37}Cl, whose mass is about 37 amu, $(37 - 17)$ amu $= 20$ amu must come from 20 neutrons.

---

note: In general, in a nucleus of mass A amu and charge Z, there are Z protons and $(A - Z)$ neutrons.

EXAMPLE 2-6: Find the number of neutrons for each isotope of tin listed in Example 2-4.

Solution: Since $Z = 50$ for Sn, you find the number of neutrons by subtracting 50 from the mass of each isotope.

Isotope:	^{112}Sn	^{114}Sn	^{115}Sn	^{116}Sn	^{117}Sn	^{118}Sn	^{119}Sn	^{120}Sn	^{122}Sn	^{124}Sn
Neutrons:	62	64	65	66	67	68	69	70	72	74

D. Average atomic mass

Most elements found in nature are mixtures of isotopes. The average mass for the atoms in an element is called the **atomic mass** or the **atomic weight** of the element. (The terms are often used interchangeably in this application. We'll use "atomic mass" predominantly throughout this book.) The atomic masses of the elements are listed on the inside cover of this book.

EXAMPLE 2-7: There are two isotopes of lithium found on earth: Isotope ^6Li (6.015 12 amu) accounts for 7.42% of the total; isotope ^7Li (7.016 00 amu) accounts for the remaining 92.58%. The average atomic mass of lithium is therefore

$$\left(\frac{7.42\%}{100\%}\right)(6.015\,12 \text{ amu}) + \left(\frac{92.58\%}{100\%}\right)(7.016\,00 \text{ amu}) = 6.942 \text{ amu}$$

2-3. Molecules, Molecular Formula, and Molecular Mass

In many substances groups of atoms are joined together by chemical bonds (see Chapter 10) to form molecules. A **molecule** is the smallest particle of a substance that retains any individual chemical character of the substance.

A. Molecular formula

You can express the composition of a molecule, its **molecular formula**, by writing the symbols of the atoms it contains, with numerical subscripts showing the number of that kind of atom present in the molecule. Where no subscript is shown, only one atom of that kind is present in the molecule.

EXAMPLE 2-8: The molecular formulas of water, butane, and isopropanol are

Compound:	water	butane	isopropanol
Molecular formula:	H_2O	C_4H_{10}	C_3H_8O

How many hydrogen atoms are there in molecules of water, butane, and isopropanol?

Solution: The numerical subscripts in the formulas show that there are two H atoms in a molecule of H_2O, ten H atoms in C_4H_{10}, and eight H atoms in C_3H_8O.

B. Molecular mass

By adding the atomic masses of all the atoms in a molecule you can obtain the **molecular mass** or **molecular weight** of the molecule (the terms are used interchangeably in this context). The usual units of molecular mass are atomic mass units. Molecular masses are *average* masses, just like the atomic masses they are derived from.

EXAMPLE 2-9: Determine the molecular mass of isopropanol (C_3H_8O).

Solution: You can obtain the atomic masses of C, H, and O from the periodic table. The molecular mass, often given the symbol M, is

$$\underset{\substack{\text{atomic mass} \\ \text{of 3 C atoms}}}{3(12.01 \text{ amu})} + \underset{\substack{\text{atomic mass} \\ \text{of 8 H atoms}}}{8(1.008 \text{ amu})} + \underset{\substack{\text{atomic mass} \\ \text{of 1 O atom}}}{1(16.00 \text{ amu})} = \underset{\substack{\text{molecular mass of} \\ C_3H_8O}}{60.09 \text{ amu}}$$

Since the atomic masses are significant to four places, you report that $M = 60.09$ amu.

2-4. Formulas and Formula Mass

Not all compounds are molecular (*please* see Chapter 3). If a compound is not molecular, or if you have no information about it, you can express its composition in a simple formula that expresses the smallest possible integer ratio of the different types of atoms present in the compound. Few inorganic solids are molecular, and their formulas are written in this way:

Compound:	salt	limestone	quartz
Formula:	NaCl	$CaCO_3$	SiO_2

You calculate the **formula mass (formula weight)** of compounds exactly as you calculate a molecular mass.

EXAMPLE 2-10: Calculate the formula mass of limestone ($CaCO_3$).

Solution:

$$\underset{\substack{\text{atomic mass} \\ \text{of 1 Ca atom}}}{1(40.08 \text{ amu})} + \underset{\substack{\text{atomic mass} \\ \text{of 1 C atom}}}{1(12.01 \text{ amu})} + \underset{\substack{\text{atomic mass} \\ \text{of 3 O atoms}}}{3(16.00 \text{ amu})} = \underset{\substack{\text{formula mass} \\ \text{of } CaCO_3}}{100.1 \text{ amu}}$$

(Round off to 100.1 amu from 100.09 amu because only four significant figures are justified.)

2-5. The Mole

The **mole** is the fundamental SI unit of amount of substance. It is defined as the amount of substance that contains *Avogadro's number* of *any* species. This species can be atoms, molecules, electrons, nuclei, formula units, wheelbarrows,.... The mole is abbreviated in calculations as mol.

A. Avogadro's number

Avogadro's number is defined as the number of atoms in exactly 0.012 kg (exactly 12 g) of ^{12}C. This number is about **6.02 × 10²³**.

B. Grams and moles

Because of the ways in which mole and Avogadro's number are defined, the mass of a mole of a pure substance is easy to determine. You just calculate the molecular or formula mass of the substance, and change the units from atomic mass units to grams. It's a good idea to specify the formula of the pure substance that is under consideration.

note: **Molar mass** is probably the most appropriate term to use for the mass of a mole. However, many chemists (and textbooks) use the terms molecular weight, formula weight, gram molecular weight, gram formula weight, or gram atomic weight (for an element). In this text we'll consistently use molar mass for the mass of a mole of any particle or group of particles, but you'll need to recognize these other terms.

EXAMPLE 2-11: Calculate the mass of 1 mol of (**a**) isopropanol (C_3H_8O) molecules and (**b**) limestone $(CaCO_3)$ formula units.

Solution:

(a) The molecular mass of C_3H_8O is 60.09 amu, so the mass of 1 mol of C_3H_8O molecules is 60.09 g.
(b) The formula mass of $CaCO_3$ is 100.1 amu, so the mass of 1 mol of $CaCO_3$ formula units is 100.1 g.

2-6. Nomenclature

Nomenclature is the term applied to the naming of chemical compounds, which can be a complex and confusing process. The names of chemical compounds can be divided into two main types, *trivial* and *systematic*.

A. Trivial names

Trivial names have historical origins and convey nothing about composition. You must simply memorize them. Many important simple compounds have trivial names, such as water for H_2O, ammonia for NH_3, and hydrazine for N_2H_4.

B. Systematic names for inorganic compounds

Systematic names convey information about the compounds they represent. The systematic names for simple inorganic compounds are made up of the names of the elements they contain. Prefixes (di- , tri- , tetra- , etc.) indicate the relative numbers of different kinds of atoms. If there is no prefix, you can assume that only one atom of that kind is present in the formula. By convention, the more electropositive element (see Chapter 10) is written first and the more electronegative element follows with an *-ide* suffix.

EXAMPLE 2-12: Name the following compounds: NCl_3, P_2F_4, CaO, B_2O_3, SO_2, CCl_4.

Solution:

NCl_3:	nitrogen trichloride	B_2O_3:	diboron trioxide
P_2F_4:	diphosphorus tetrafluoride	SO_2:	sulfur dioxide
CaO:	calcium oxide	CCl_4:	carbon tetrachloride

Certain groups, usually negative ions containing oxygen atoms, are identified by special names and often retain their identities in some chemical reactions. Thus, sodium sulfate (Na_2SO_4) will dissolve in water to give sodium ions Na^+ and sulfate ions SO_4^{2-}. The systematic nomenclature of these substances will not be discussed here, but you should begin to recognize the more common ones: *sulfate* SO_4^{2-}, *nitrate* NO_3^-, *carbonate* CO_3^{2-}, *chlorate* ClO_3^-. There is also one positive ion: *ammonium* NH_4^+.

It is important to note here that in a compound consisting of oppositely charged ions the sum of the charges carried by the ions is zero. That is why the formula for sodium sulfate is Na_2SO_4 and not $NaSO_4$. It takes two sodium ions, each bearing a $+1$ charge, to balance the -2 charge of the sulfate ion. Among the common positively charged ions are sodium Na^+, calcium Ca^{2+}, potassium K^+, and aluminum Al^{3+}. Note also that in the names of ionic compounds the prefixes di- , tri- , etc., are not used. The names are sodium sulfate (not disodium sulfate) and calcium chloride $CaCl_2$ (not calcium dichloride).

EXAMPLE 2-13: The formula for calcium carbonate is $CaCO_3$. What are the formulas of calcium sulfate and calcium nitrate?

Solution:

Calcium sulfate: $CaSO_4$

$Ca(NO_3)_2$. Note that Ca^{2+} requires *two* NO_3^- groups to make a neutral compound.

SUMMARY

1. Elements are made up of atoms; the same symbol is used for both.
2. Atoms contain positively charged nuclei and negatively charged electrons.
3. Ions are formed when atoms gain or lose electrons.
4. The magnitude of the charge on the nucleus is characteristic of the element and is called the atomic number (Z).
5. The unit of atomic mass (amu) equals $\frac{1}{12}$ of the mass of one atom of ^{12}C.
6. Isotopes are atoms that have the same atomic number but differ in mass.
7. The nucleus is composed of positively charged protons and uncharged (electrically neutral) neutrons.
8. The atomic mass (atomic weight) of an element is the average mass of the atoms of the naturally occurring element relative to $\frac{1}{12}$ the weight of an atom of ^{12}C.
9. Molecules are groups of atoms held together by bonds.
10. Molecular formulas express the composition of molecules.
11. A mole is Avogadro's number of any species.
12. Avogadro's number (approximately 6.02×10^{23}) is the number of ^{12}C atoms in exactly 12 g of ^{12}C.
13. Molar mass is the mass of a mole of any species.
14. Systematic names specify composition.

RAISE YOUR GRADES

Can you define . . . ?

- ☑ atomic number (Z)
- ☑ isotope
- ☑ atomic mass
- ☑ molecular mass
- ☑ formula mass
- ☑ Avogadro's number
- ☑ molar mass
- ☑ ion

Can you explain . . . ?

- ☑ the composition of an atom
- ☑ how isotopes of an element differ

☑ the composition of a nucleus, given charge and mass
☑ how to calculate a molecular mass or a formula mass
☑ how to determine the mass of a mole of a particular substance
☑ the meaning of Avogadro's number
☑ how to name simple binary inorganic compounds
☑ how to write formulas for simple inorganic compounds, given their systematic names

SOLVED PROBLEMS

PROBLEM 2-1   Give the symbols for the elements tungsten, potassium, bromine, and dysprosium.

Solution: The symbols of the atoms are listed on the inside cover of this book (and in the periodic table): tungsten W, potassium K, bromine Br, dysprosium Dy.

PROBLEM 2-2   What are the atomic numbers of the atoms of tungsten, potassium, bromine, and dsyprosium?

Solution: The atomic numbers Z are listed on the inside cover of this book. For W, $Z = 74$; for K, $Z = 19$; for Br, $Z = 35$; for Dy, $Z = 66$.

PROBLEM 2-3   What ncutral atom has a nuclear charge of $+35$ and a nuclear mass of 79 amu? How many neutrons does it have? How many electrons?

Solution: Since the nuclear charge, which equals the atomic number, is $+35$, the element is bromine, which has $Z = 35$. The 35 protons in its nucleus, which give the $+35$ charge, have a total mass of 35 amu; the rest of the nuclear mass, 79 amu $-$ 35 amu $=$ 44 amu, must be due to the presence of 44 neutrons, each of mass 1 amu. Since the atom is neutral, there must be 35 extranuclear electrons, each of charge -1, giving a total charge of -35, which neutralizes the nuclear charge of $+35$.

PROBLEM 2-4   The element chlorine contains two isotopes: ^{35}Cl, which has a mass of 34.98 amu, and ^{37}Cl, which has a mass of 36.98 amu. The atomic weight of chlorine is listed as 35.46 amu. Calculate the percentages of each of the isotopes in chlorine.

Solution: With problems of this kind, you can usually make a good start by thinking about moles, and their implications. Consider exactly 1 mol of chlorine atoms. Since the atomic weight of chlorine is 35.46, 1 mol of chlorine atoms weighs 35.46 g. Now let the fraction of the mole of chlorine atoms that is ^{35}Cl atoms ($\%\,^{35}Cl/100$) be the **mole fraction** α. Then the mass of α moles of ^{35}Cl atoms equals $(\alpha)(34.98 \text{ g})$. The mole fraction of ^{37}Cl atoms in chlorine must be $1 - \alpha$, since only the two isotopes are present and the sum of their mole fractions must equal exactly 1. The mass of $1 - \alpha$ moles of ^{37}Cl atoms equals $(1 - \alpha)(36.98 \text{ g})$. So the total mass m_T of the two isotopes is

$$m_T = (\alpha)(34.98 \text{ g}) + (1 - \alpha)(36.98 \text{ g})$$

But this is just 1 mol of chlorine atoms, and you know the mass of that is exactly 35.46 g. So

$$(\alpha)(34.98 \text{ g}) + (1 - \alpha)(36.98 \text{ g}) = 35.46 \text{ g}$$
$$34.98\alpha + 36.98 - 36.98\alpha = 35.46$$

Rearranging and collecting terms:

$$1.52 = 2.00\alpha$$

$$\alpha = \frac{1.52}{2.00} = 0.760$$

So the percentages are

$$0.760 \times 100\% = 76.0\% \text{ of } ^{35}Cl$$

and

$$100\% - 76.0\% = 24.0\% \text{ of } ^{37}Cl$$

PROBLEM 2-5   Calculate the molecular masses of (a) carbon tetrabromide (CBr_4) and (b) glycerol ($C_3H_5(OH)_3$).

Solution:

(a)

Atom	Atomic mass (amu)	Multiplier	Total mass (amu)
C	12.01	1	12.01
Br	79.91	4	319.64
			$M = \overline{331.65}$

(b) When a group of atoms is written in parentheses with a numerical subscript, like the $(OH)_3$ part of $C_3H_5(OH)_3$, that means that the whole group of atoms inside the parentheses occurs in the molecule or formula the number of times given by the subscript. Glycerol contains 3 OH groupings and so its molecular formula is $C_3H_5O_3H_3$ or $C_3H_8O_3$.

Atom	Atomic mass (amu)	Multiplier	Total mass (amu)
C	12.01	3	36.03
H	1.008	8	8.064
O		3	48.00
		$M =$	$\overline{92.09}$

Only 4 significant figures are justified.

PROBLEM 2-6   Calculate the formula masses of (a) potassium permanganate $KMnO_4$ and (b) cadmium nitrate $Cd(NO_3)_2$.

Solution:

(a)

Atom	Atomic mass (amu)	Multiplier	Total mass (amu)
K	39.10	1	39.10
Mn	54.94	1	54.94
O	16.00	4	64.00
		Formula mass =	$\overline{158.04}$

(b) As in Problem 2-5b, treat the parenthetic group in $Cd(NO_3)_2$ by multiplying it out. The formula of cadmium nitrate is CdN_2O_6:

Atom	Atomic mass (amu)	Multiplier	Total mass (amu)
Cd	112.4	1	112.4
N	14.01	2	28.02
O	16.00	6	96.00
		Formula mass =	$\overline{236.42}$

Only 4 significant figures are justified.

PROBLEM 2-7   What is the mass of one atom of ^{12}C?

Solution: One mole of ^{12}C contains Avogadro's number of atoms, 6.02×10^{23}, and has a mass of exactly 12 g. So

$$\text{mass of one atom} = \frac{12.00 \text{ g}}{6.02 \times 10^{23} \text{ atoms}} = 1.99 \times 10^{-23} \text{ g}$$

PROBLEM 2-8   Determine the mass of 1 mol of (a) glycerol molecules, and (b) cadmium nitrate formula units.

Solution: Since you have already determined the molecular mass of glycerol and the formula mass of cadmium nitrate (see Problems 2-5 and 2-6), you simply have to change units from atomic mass units to grams to get the solutions:

(a) $C_3H_5(OH)_3$,   92.09 g/mol
(b) $Cd(NO_3)_2$,   236.4 g/mol

note: In common chemical usage you would shorten the expressions used in this problem and simply refer to a "mole" of glycerol, or a "mole" of cadmium nitrate.

PROBLEM 2-9   For 1.000 g of the compound carbon disulfide (CS_2), **(a)** how many molecules are present? **(b)** How many atoms of sulfur are present?

Solution:

(a) You know that there are 6.02×10^{23} molecules of CS_2 in 1 mol of CS_2, and you can readily calculate the mass of 1 mol of CS_2:

Atom	Atomic mass (amu)	Multiplier	Total mass (amu)
C	12.01	1	12.01
S	32.06	2	64.12
			$M = \overline{76.13}$

So the molar mass of CS_2 is 76.13 g, and 1 mol contains 6.02×10^{23} molecules. The number of molecules in 1.000 g of CS_2 is

$$\frac{6.02 \times 10^{23} \text{ molecules}}{76.13 \text{ g CS}_2} \times 1.000 \text{ g CS}_2 = 7.91 \times 10^{21} \text{ molecules}$$

(b) Since each CS_2 molecule contains 2 atoms of sulfur, the number of sulfur atoms is

$$7.91 \times 10^{21} \text{ molecules CS}_2 \times \frac{2 \text{ atoms S}}{1 \text{ molecule CS}_2} = 1.58 \times 10^{22} \text{ atoms S}$$

PROBLEM 2-10   Metals are more electropositive than nonmetals. With this in mind, name the following compounds: **(a)** KCl, **(b)** MgO, **(c)** AlN.

Solution: The metal is named first, the nonmetal taking the *-ide* suffix. So the names are **(a)** potassium chloride, **(b)** magnesium oxide, **(c)** aluminum nitride.

PROBLEM 2-11   Write the formulas for the following compounds: **(a)** phosphorus trifluoride, **(b)** disilicon hexachloride, **(c)** germanium dioxide.

Solution: The prefix gives the numbers of atoms; when no prefix is given, only one atom of that kind is present in the formula (there are some exceptions to this general rule). So the formulas are **(a)** PF_3, **(b)** Si_2Cl_6, **(c)** GeO_2.

PROBLEM 2-12   Calculate the atomic mass of boron, which contains 18.83% of ^{10}B (mass 10.016 amu) and 81.17% of ^{11}B (mass 11.013 amu).

Solution: One mole of boron contains 0.1883 mol of ^{10}B, of mass (0.1883 mol)(10.016 g mol^{-1}), and 0.8117 mol of ^{11}B, of mass (0.8117 mol)(11.013 g mol^{-1}). The total mass of 1 mol of boron is therefore

$$(0.1883)(10.016 \text{ g}) + (0.8117)(11.013 \text{ g}) = 10.83 \text{ g}$$

and so the atomic mass of boron is 10.83 amu.

PROBLEM 2-13   For a 0.685-g sample of sulfur hexafluoride (SF_6) calculate **(a)** the number of moles, **(b)** the number of molecules, **(c)** the number of fluorine atoms.

Solution:

(a) The mass of 1 mol of SF_6 is calculated in the usual way:

Atom	Atomic mass (amu)	Multiplier	Total mass (amu)
S	32.06	1	32.06
F	19.00	6	114.00
			$M = \overline{146.06}$

So the molar mass of $SF_6 = 146.1$ g.

The number of moles of SF_6 in 0.685 g is

$$\frac{0.685\ g}{146.1\ g\,mol^{-1}} = 4.69 \times 10^{-3}\ mol$$

(b) The number of molecules in 4.69×10^{-3} mol SF_6 is

$$(6.02 \times 10^{23}\ molecules\ mol^{-1})(4.69 \times 10^{-3}\ mol) = 2.82 \times 10^{21}\ molecules$$

(c) The number of fluorine atoms in 2.82×10^{21} molecules of SF_6 is

$$\frac{6\ atoms\ F}{molecule\ SF_6}(2.82 \times 10^{21}\ molecules\ SF_6) = 1.69 \times 10^{22}\ atoms\ of\ fluorine$$

PROBLEM 2-14　It has been estimated that the average density of the universe is 1×10^{-30} g/cm³. Assuming that this density is due to hydrogen atoms, what is the average number of hydrogen atoms in 1 km³ of the universe?

Solution:　The mass of H in 1 km³ is

$$\left(\frac{1 \times 10^{-30}\ g}{cm^3}\right)\left(\frac{10^2\ cm}{m} \cdot \frac{10^3\ m}{km}\right)^3 = \left(\frac{1 \times 10^{-30}\ g}{cm^3}\right)\left(\frac{10^5\ cm}{km}\right)^3 = \left(\frac{1 \times 10^{-30}\ g}{cm^3}\right)\left(\frac{10^{15}\ cm^3}{km^3}\right)$$

$$= 1 \times 10^{-15}\ g/km^3$$

The mass of 1 mol of H atoms (6.02×10^{23} H atoms) is 1.00 g. So the number of H atoms per km³ is

$$\left(\frac{1 \times 10^{-15}\ g}{km^3}\right)\left(\frac{6.02 \times 10^{23}\ H\ atoms}{g}\right) = 6 \times 10^8\ H\ atoms\ per\ km^3$$

PROBLEM 2-15　Which of the following is the molecular mass (in amu) of ethanol (C_2H_5OH): **(a)** 46.1, **(b)** 29.0, **(c)** 20.0, **(d)** 34.0, **(e)** 45.1?

Solution:

Atom	Atomic mass (amu)	Multiplier	Total mass (amu)
C	12.01	2	24.02
H	1.008	6	6.05
O	16.00	1	16.00
			$M = 46.07$

The correct answer is **(a)**.

　note: In answer **(b)**, the multipliers were omitted; in answer **(c)**, atomic numbers were used; in answer **(d)**, the multiplier for carbon was omitted; in answer **(e)**, the multiplier 5 was used for H instead of 6.

PROBLEM 2-16　Which of the following is the number of boron atoms in 13.8 g of B_2H_6 molecules: **(a)** 2, **(b)** 1, **(c)** 6.02×10^{23}, **(d)** 10.8, **(e)** 12.0×10^{23}?

Solution:　The molar mass of B_2H_6 is calculated in the usual way:

Atom	Atomic mass (amu)	Multiplier	Total mass (amu)
B	10.81	2	21.62
H	1.008	6	6.05
			$M = 27.67$

The molar mass is 27.67 g mol⁻¹. The number of moles is

$$\frac{13.8\ g}{27.67\ g\,mol^{-1}} = 0.500\ mol\ B_2H_6$$

The number of B_2H_6 molecules is

$$(0.500\ mol)(6.02 \times 10^{23}\ molecule\ mol^{-1}) = 3.01 \times 10^{23}\ molecules$$

There are 2 B atoms in 1 B_2H_6 molecule, so

$$(3.01 \times 10^{23} \text{ molecules})\left(\frac{2 \text{ B atoms}}{\text{molecule}}\right) = 6.02 \times 10^{23} \text{ B atoms}$$

The correct answer is (c).

Answer (a) is the number of B atoms in 1 B_2H_6 molecule; answer (b) is the number of B atoms in $\frac{1}{2}B_2H_6$ molecule; answer (d) is the mass of 1 B atom in atomic mass units or the mass in grams of B in 13.8 g of B_2H_6; answer (e) is the number of B atoms in 1 mol of B_2H_6.

note: With practice you will be able to work multiple-choice problems without writing everything down, but be careful! Whenever a question asks for the number of atoms or molecules, expect the answer to be a large number. In this case answers (a), (b), and (d) are clearly unreasonable.

Supplementary Exercises

PROBLEM 2-17 How many atoms are present in 1.00 ng of gold?

Answer: 3.06×10^{12} atoms

PROBLEM 2-18 Calculate the molecular masses of (a) dinitrogen tetroxide N_2O_4, (b) trichloromethane (chloroform) $CHCl_3$, (c) sucrose (cane sugar) $C_{12}H_{22}O_{11}$.

Answer: (a) 92.02 amu   (b) 119.4 amu   (c) 342.3 amu

PROBLEM 2-19 Calculate the formula masses of (a) lithium sulfate Li_2SO_4, (b) sodium hydrogen carbonate $NaHCO_3$, (c) magnesium nitrate $Mg(NO_3)_2$.

Answer: (a) 109.9 amu   (b) 84.01 amu   (c) 148.3 amu

PROBLEM 2-20 Calculate the mass of 1 mol of (a) ozone O_3, (b) benzene C_6H_6, (c) potassium dichromate $K_2Cr_2O_7$.

Answer: (a) 48.00 g   (b) 78.11 g   (c) 294.2 g

PROBLEM 2-21 Copper (Cu: atomic mass 63.546 amu) contains the isotopes ^{63}Cu (mass 62.9298 amu) and ^{65}Cu (mass 64.9278). What percentage of Cu atoms is ^{65}Cu?

Answer: 30.8%

PROBLEM 2-22 How many hydrogen atoms are present in 2.55 mol of propane C_3H_8?

Answer: 1.23×10^{25} H atoms

PROBLEM 2-23 How many neutrons are present in the nuclei of (a) ^{12}C, (b) ^{14}C, (c) ^{31}P, (d) ^{235}U?

Answer: (a) 6   (b) 8   (c) 16   (d) 143

PROBLEM 2-24 How many moles of calcium phosphate $Ca_3(PO_4)_2$ will contain 0.100 mol of oxygen atoms?

Answer: 1.25×10^{-2} mol

PROBLEM 2-25 How many moles of each of the following are present in exactly 1 kg of the substance: (a) H_2O, (b) NaCl, (c) CI_4, (d) $Bi(IO_4)_3$? Give your answer correct to four significant figures.

Answer: (a) 55.51   (b) 17.11   (c) 1.925   (d) 1.279

PROBLEM 2-26   How many electrons (e), protons (p), and neutrons (n) are present in neutral atoms of the following: (a) ^{10}B, (b) ^{11}B, (c) ^{50}V, (d) ^{108}Ag?

Answer: (a) $5\,e, 5\,p, 5\,n$ (b) $5\,e, 5\,p, 6\,n$ (c) $23\,e, 23\,p, 27\,n$ (d) $47\,e, 47\,p, 61\,n$

PROBLEM 2-27   The formula of iron oxide is Fe_2O_3. How many grams of iron are there in 375 mol of Fe_2O_3?

Answer: 4.19×10^4 g

PROBLEM 2-28   The directions for an experiment tell you to use 9.35×10^{-2} mol of PCl_3. How many grams of PCl_3 will you use?

Answer: 12.8 g

PROBLEM 2-29   Calculate the mass in grams of 7.33×10^{22} molecules of carbon disulfide.

Answer: 9.27 g

3 FORMULAS

THIS CHAPTER IS ABOUT

☑ **Empirical Formulas**
☑ **Molecular Formulas**
☑ **Determining Percentage Composition from a Formula**
☑ **Determining the Empirical Formula from a Composition**
☑ **Converting an Empirical Formula to a Molecular Formula**

3-1. Empirical Formulas

An **empirical formula** is an experimentally determined formula that expresses the smallest possible integer ratio of the different kinds of atoms present in a compound. If a formula is not expressed in the simplest possible numerical terms, it is not an empirical formula. To determine an empirical formula, you only need information about *composition*. You can't tell whether or not a compound is molecular from an empirical formula.

EXAMPLE 3-1: Are the following formulas empirical: H_2O, H_2SO_4, C_3H_8, $C_{12}H_{22}O_{11}$, NaCl, K_2SO_3, H_2O_2, N_2H_4, $C_4H_8O_2$, C_3H_6?

Solution: H_2O, H_2SO_4, C_3H_8, $C_{12}H_{22}O_{11}$, NaCl, and K_2SO_3 are empirical because they represent the simplest possible ratio for those numbers of atoms. The formulas that *cannot* be empirical are H_2O_2, N_2H_4, $C_4H_8O_2$, and C_3H_6, whose simplest ratios would be HO, NH_2, C_2H_4O, and CH_2, respectively.

3-2. Molecular Formulas

A **molecular formula** is a formula that expresses the actual number of atoms joined by chemical bonds to form a molecule. You can only write molecular formulas for compounds you know to be molecular. You need to know *both* the composition of a compound and its molecular mass to write a molecular formula.

EXAMPLE 3-2: The molecular formulas of some compounds known to be molecular are H_2O (water), H_2O_2 (hydrogen peroxide), N_2H_4 (hydrazine), C_3H_6 (cyclopropane), $C_{12}H_{22}O_{11}$ (sucrose). Note that the molecular formulas for water and sucrose are in the simplest possible ratios. They are therefore also empirical. The formulas for H_2O_2, N_2H_4, and C_3H_6 can be simplified to HO, NH_2, and CH_2, respectively. The formulas given are therefore molecular only. The molecular formula is always a whole-number multiple (including 1) of the empirical formula. Thus it will sometimes happen (as in water and sucrose) that the empirical and molecular formulas are the same.

3-3. Determining Percentage Composition from a Formula

For certain chemical calculations you'll need to determine the **percentage composition** by mass of a compound, i.e., the mass of each element present expressed as a percentage of the total mass. Let's examine the procedure using sulfuric acid H_2SO_4 as our example.

25

1. *Determine the molar or formula mass of the compound.*
   The formula mass of H_2SO_4 is

$$2(1.008 \text{ g}) + 32.06 \text{ g} + 4(16.00 \text{ g}) = 98.08 \text{ g/mol of } H_2SO_4 \text{ formula units}$$

2. *Determine the mass of each element present in the molar or formula mass.*
   For H_2SO_4 you get

$$\text{mass of H present} = 2(1.008 \text{ g}) = 2.016 \text{ g}$$
$$\text{mass of S present} = 32.06 \text{ g}$$
$$\text{mass of O present} = 4(16.00) \text{ g} = 64.00 \text{ g}$$

3. *Determine the percentage by mass of each element present.*

$$\% \text{ H} = \frac{\text{mass of H in formula mass}}{\text{formula mass}} \times 100\%$$

$$= \frac{2.016 \text{ g}}{98.08 \text{ g}} \times 100\% = 2.055\%$$

$$\% \text{ S} = \frac{\text{mass of S in formula mass}}{\text{formula mass}} \times 100\%$$

$$= \frac{32.06 \text{ g}}{98.08 \text{ g}} \times 100\% = 32.69\%$$

$$\% \text{ O} = \frac{\text{mass of O in formula mass}}{\text{formula mass}} \times 100\%$$

$$= \frac{64.00 \text{ g}}{98.08 \text{ g}} \times 100\% = 65.25\%$$

4. *Check that the percentages add up to 100%.*

Check for H_2SO_4:

$$\begin{array}{rl} 2.055\% & \text{H} \\ 32.69\% & \text{S} \\ 62.25\% & \text{O} \\ \hline 100.00\% & \text{(4 significant figures)} \end{array}$$

3-4. Determining the Empirical Formula from a Composition

If you're given the elemental composition of a compound, you can easily calculate the compound's empirical formula. The procedure is as follows:

1. Determine the masses of each element for a particular sample of the compound (1 mol, or 100.0 g, or whatever is convenient, given the terms of the problem).
2. Convert these masses to numbers of moles of atoms of each element.
3. Express the relative numbers of moles of each element as integers.

EXAMPLE 3-3: The percentage composition of sulfuric acid is 2.055% H, 32.69% S, and 65.25% O. Determine the empirical formula of sulfuric acid.

Solution: Following our procedure:

1. *Determine the mass of each element*: Given the percentage composition—i.e., parts per 100, or grams per 100 g—of sample, you know that 100 g of sulfuric acid contains 2.055 g of H, 32.69 g of S, and 65.25 g of O.
2. *Convert mass to numbers of moles*: From the table of atomic masses, you find that 1 mol of H atoms has a mass of 1.008 g. Assuming 100 g of sulfuric acid and using n for the number of moles of atoms

of each kind, you write

$$n_H = \frac{2.055 \text{ g H}}{1.008 \text{ g H/mol H atoms}} = 2.039 \text{ mol H atoms}$$

$$n_S = \frac{32.69 \text{ g S}}{32.06 \text{ g S/mol S atoms}} = 1.020 \text{ mol S atoms}$$

$$n_O = \frac{65.25 \text{ g O}}{16.00 \text{ g O/mol O atoms}} = 4.078 \text{ mol O atoms}$$

3. *Express as integers*: Divide each of the relative numbers of moles of the different kinds of atoms by the smallest number found in step 2. For sulfuric acid, 1.020 mol of S is the smallest number, so

$$\text{relative number of moles of H} = \frac{2.039}{1.020} = 2.000 \qquad 2$$

$$\text{relative number of moles of S} = \frac{1.020}{1.020} = 1.000 \qquad 1$$

$$\text{relative number of moles of O} = \frac{4.078}{1.020} = 3.998 \qquad 4$$

Integer

So the empirical formula of sulfuric acid is H_2SO_4.

---

note: This method generates empirical formulas. You should be aware that round-off errors can occur, producing results that may not be exact integers, as for oxygen in this example.

EXAMPLE 3-4: Analysis of a 47.25-mg sample of aluminum chloride (which, as its name suggests, contains only aluminum and chlorine) showed that it contained 9.56 mg of aluminum. Determine the empirical formula of the compound.

Solution:

1. Since 9.56 mg of the sample is Al, the rest of the 47.25-mg sample must be Cl: Subtracting gives 47.25 mg − 9.56 mg = 37.69 mg of Cl.
2. Convert the mass to moles:

$$n_{Al} = \frac{9.56 \text{ mg Al}}{26.98 \text{ g Al/mol Al atoms}} = 0.354 \text{ mmol Al atoms}$$

$$n_{Cl} = \frac{37.69 \text{ mg Cl}}{35.45 \text{ g Cl/mol Cl atoms}} = 1.063 \text{ mmol Cl atoms}$$

3. Divide by 0.354, the smaller of the two n values. Because the formula is a ratio, it doesn't matter what kind of moles you use: You can use moles, millimoles, or kilomoles if you want.

$$\text{relative } n_{Al} = \frac{0.354 \times 10^{-3}}{0.354 \times 10^{-3}} = 1.00$$

$$\text{relative } n_{Cl} = \frac{1.063 \times 10^{-3}}{0.354 \times 10^{-3}} = 3.00$$

Expressing the relative numbers as integers, you have 1.00 = 1 and 3.00 = 3, so the empirical formula is $AlCl_3$.

EXAMPLE 3-5: A 1.723-g sample of aluminum oxide (which consists of aluminum and oxygen only) contains 0.912 g of Al. Determine the empirical formula of the compound.

Solution:

1. The amount of oxygen in the sample is 1.723 g − 0.912 g = 0.811 g.

2.
$$n_{Al} = \frac{0.912 \text{ g Al}}{26.98 \text{ g Al/mol Al}} = 0.0338 \text{ mol Al}$$

$$n_O = \frac{0.811 \text{ g O}}{16.0 \text{ g O/mol O}} = 0.0507 \text{ mol O}$$

This gives a tentative formula of $Al_{0.0338}O_{0.0507}$

3. Divide each by the smaller of the two numbers (so that one of them is 1):

$$\text{relative } n_{Al} = \frac{0.0338}{0.0338} = 1.0$$

$$\text{relative } n_O = \frac{0.0507}{0.0338} = 1.5$$

This gives the formula $AlO_{1.5}$ in which the number of O atoms is not an integer. So you have to multiply both relative numbers by the same number in order to get integers. In this case, you multiply each by 2

$$1.0 \times 2 = 2$$
$$1.5 \times 2 = 3$$

and get Al_2O_3 as the empirical formula.

3-5. Converting an Empirical Formula to a Molecular Formula

The molecular formula of a molecular compound is always an *integral* multiple m ($m = 1, 2, 3, \ldots$) of its empirical formula. To find the value of m, you need to know the molar or molecular mass of the compound. (These masses are obtained by experimental methods mentioned later in this outline or in your textbook.) Then

$$\frac{\text{molecular mass}}{\text{empirical formula mass}} = m$$

and

$$\text{molecular formula} = m(\text{empirical formula})$$

EXAMPLE 3-6: In the gas phase, aluminum chloride (empirical formula $AlCl_3$) has a molecular mass of 267 amu. What is its molecular formula?

Solution: The empirical formula mass of $AlCl_3$ is

$$26.98 \text{ amu} + 3(35.45 \text{ amu}) = 133.3 \text{ amu}$$

and

$$m = \frac{\text{molecular mass}}{\text{empirical formula mass}} = \frac{267 \text{ amu}}{133.3 \text{ amu}} = 2$$

(Remember: The multiple m MUST be an integer. If m does not come out as very nearly an integer in one of your calculations, check your calculations.)

For aluminum chloride

$$\text{molecular formula} = 2(\text{empirical formula}) = 2(AlCl_3)$$

$$= Al_2Cl_6$$

This means that, in the gas phase, two aluminum atoms and six chlorine atoms are held together by chemical bonds to form one molecule of Al_2Cl_6.

SUMMARY

1. An empirical formula gives the smallest possible integer ratio of the different kinds of atoms in a compound.
2. A molecular formula shows the actual numbers of atoms joined by chemical bonds in a molecule.
3. Percentage composition by mass can be calculated from either empirical or molecular formulas.
4. The empirical formula of a compound can be calculated from its elemental composition.
5. The molecular formula of a compound is always a whole number times the empirical formula.
6. The molecular formula of a compound can be determined from its empirical formula and molecular mass.

RAISE YOUR GRADES

Can you . . . ?

☑ calculate percentage composition, given the formula of a compound
☑ generate an empirical formula from percentage composition or from analytical mass data
☑ determine a molecular formula from an empirical formula and a molecular mass

SOLVED PROBLEMS

PROBLEM 3-1 Which of the following cannot be empirical formulas: (a) C_3H_8, (b) H_2SO_4, (c) C_2H_2, (d) $Na_2Cr_2O_7$, (e) $Na_2S_2O_4$, (f) C_3H_6?

Solution: An empirical formula must express the ratio of the numbers of different kinds of atoms in a formula in the simplest possible numerical terms. When you see a formula in which there is a common factor for these numbers—the subscripts that follow the atom symbols—you know it cannot be an empirical formula. So the following cannot be empirical formulas:

(c) C_2H_2, because there is a common factor of 2 for the subscripts; the empirical formula for this compound would be CH.
(e) $Na_2S_2O_4$, because there is a common factor of 2 for the subscripts; the empirical formula for this compound would be $NaSO_2$.
(f) C_3H_6, because there is a common factor of 3 for the subscripts; the empirical formula for this compound would be CH_2.

note: Nonempirical formulas provide information: The compound may be molecular (in this example, C_2H_2 and C_3H_6 are molecular), or there may be a known complexity of structure in the compound that justifies the nonempirical formula, as is the case for $Na_2S_2O_4$.

PROBLEM 3-2 Calculate the percentage composition of (a) MgS, (b) CaC_2, (c) $KClO_4$, (d) $C_{12}H_{22}O_{11}$.

Solution:
(a) Determine the formula mass of MgS:

$$24.31 \text{ g} + 32.06 \text{ g} = 56.37 \text{ g/mol MgS formula units}$$

The mass of Mg present in 1 mol of formula units is 24.31 g, so

$$\% \text{ Mg} = \frac{24.31 \text{ g}}{56.37 \text{ g}} \times 100\% = 43.13\%$$

The mass of S present in 1 mol of formula units is 32.06 g, so

$$\% \text{ S} = \frac{32.06 \text{ g}}{56.37 \text{ g}} \times 100\% = 56.87\%$$

Finally, check: 43.13% Mg + 56.87% S = 100.0%.

(b) The formula mass of CaC_2 is

$$40.08 \text{ g} + 2(12.01 \text{ g}) = 64.10 \text{ g/mol } CaC_2 \text{ formula units}$$

The mass of Ca per mole of formula units is 40.08 g, so

$$\% \text{ Ca} = \frac{40.08 \text{ g}}{64.10 \text{ g}} \times 100\% = 62.53\%$$

The mass of C per mole of formula units is $2 \times 12.01 \text{ g} = 24.02 \text{ g}$, so

$$\% \text{ C} = \frac{24.02 \text{ g}}{64.10 \text{ g}} \times 100\% = 37.47\%$$

Check: total $= 62.53\% + 37.47\% = 100.0\%$

(c) The formula mass of $KClO_4$ is

$$39.10 \text{ g} + 35.45 \text{ g} + 4(16.00 \text{ g}) = 138.55 \text{ g}$$

so

$$\% \text{ K} = \frac{39.10 \text{ g}}{138.55 \text{ g}} \times 100\% = 28.22\%$$

$$\% \text{ Cl} = \frac{35.45 \text{ g}}{138.55 \text{ g}} \times 100\% = 25.59\%$$

$$\% \text{ O} = \frac{4(16.00 \text{ g})}{138.55 \text{ g}} \times 100\% = 46.19\%$$

Check: total $= \overline{100.00\%}$

(d) The formula mass of $C_{12}H_{22}O_{11}$ is

$$12(12.01 \text{ g}) + 22(1.008 \text{ g}) + 11(16.00 \text{ g}) = 342.3 \text{ g}$$

$$\% \text{C} = \frac{12 \times 12.01 \text{ g}}{342.3 \text{ g}} \times 100\% = 42.10\%$$

$$\% \text{ H} = \frac{22 \times 1.008 \text{ g}}{342.3 \text{ g}} \times 100\% = 6.48\%$$

$$\% \text{ O} = \frac{11 \times 16.00 \text{ g}}{342.3 \text{ g}} \times 100\% = 51.42\%$$

Check: total $= \overline{100.00\%}$

PROBLEM 3-3   A compound of sulfur and chlorine is found to contain 47.4% of sulfur and 52.6% of chlorine. What is its empirical formula?

Solution: Percent can be interpreted as grams per 100 g, so 100.0 g of the compound contains 47.4 g of S and 52.6 g of Cl. Determine the number of moles of S and Cl atoms in 100.0 g of the compound:

$$n_S = \frac{47.4 \text{ g S}}{32.06 \text{ g S/mol S atoms}} = 1.48 \text{ mol S atoms}$$

$$n_{Cl} = \frac{52.6 \text{ g Cl}}{35.45 \text{ g Cl/mol Cl atoms}} = 1.48 \text{ mol Cl atoms}$$

The ratio of n_S to n_{Cl} is 1.48 mol : 1.48 mol $= 1 : 1$, so the empirical formula of this compound is SCl.

PROBLEM 3-4   A sample of an iron oxide (a compound containing only iron and oxygen) was analyzed and found to contain 69.9% iron. What is the empirical formula of the compound?

Solution: Because the compound contains only iron and oxygen, 100.0 g of the oxide contains 69.9 g Fe and $(100.0 \text{ g} - 69.9 \text{ g}) = 30.1 \text{ g O}$, so

$$n_{Fe} = \frac{69.9 \text{ g Fe}}{55.8 \text{ g Fe/mol Fe atoms}} = 1.25 \text{ mol Fe atoms}$$

$$n_O = \frac{30.1 \text{ g O}}{16.0 \text{ g O/mol O atoms}} = 1.88 \text{ mol O atoms}$$

$$\frac{n_O}{n_{Fe}} = \frac{1.88}{1.25} = \frac{1.50}{1.00}$$

Because we must express empirical formulas in whole-number terms, we note that 1.50/1.00 is the same as 3.00/2.00, and so the empirical formula is Fe_2O_3.

note: This last step is the one that may bother you. Just remember that the object is to write the formula with integral (whole number) multiples for the atomic symbols. If a ratio comes out as 1.50 to 1, you can express it as 3 to 2, a ratio of integers. If you calculate $n_{Fe}/n_O = 1.25/1.88 = 0.66/1.00$, you can see that 0.66/1.00 is very nearly $\frac{2}{3}$, so again you get an empirical formula of Fe_2O_3.

PROBLEM 3-5   Determine the empirical formula of a compound whose mass composition is 23.1% Al, 15.4% C, and 61.5% O.

Solution: You should start setting up these problems systematically, as follows:

Element	Mass in 100 g of sample	Number of moles of atoms in sample	Relative number of moles	Ratio
Al	23.1 g	$\dfrac{23.1 \text{ g Al}}{26.98 \text{ g Al/mol Al}}$	0.856	1.000
C	15.4 g	$\dfrac{15.4 \text{ g C}}{12.01 \text{ g C/mol C}}$	1.28	1.50
O	61.5 g	$\dfrac{61.5 \text{ g O}}{16.00 \text{ g O/mol O}}$	3.84	4.49

note: You obtain the ratio of the elements in the compound by dividing the number of moles of each element by the number of moles for the element present in the smallest amount. In this example, $0.856/0.856 = 1$, $1.28/0.856 = 1.50$, and $3.84/0.856 = 4.49$.

To obtain an integral ratio for $Al_{1.00}C_{1.50}O_{4.49}$, double it, and do a little rounding up for O, to get $Al_2C_3O_9$ as the empirical formula.

PROBLEM 3-6   PAN, a component of smog, contains 19.8% C, 2.50% H, and 11.6% N; the rest of the compound is oxygen. What is its empirical formula?

Solution:

$$\% \text{ O} = 100.0\% - (19.8 + 2.50 + 11.6)\% = 66.1\%$$

Now set up systematically:

Element	Mass in 100 g of sample	Number of moles of atoms	Relative number of moles	Ratio
C	19.8 g	$\dfrac{19.8 \text{ g}}{12.01 \text{ g/mol}}$	1.65	1.99
H	2.50 g	$\dfrac{2.50 \text{ g}}{1.008 \text{ g/mol}}$	2.48	3.00
N	11.6 g	$\dfrac{11.6 \text{ g}}{14.01 \text{ g/mol}}$	0.828	1.00
O	66.1 g	$\dfrac{66.1 \text{ g}}{16.00 \text{ g/mol}}$	4.13	4.99

Rounding off to the nearest integer, the empirical formula of PAN is $C_2H_3NO_5$.

PROBLEM 3-7   When a 10.00-g sample of barium (Ba) was heated in oxygen, 12.33 g of an unknown oxide was formed. Determine the empirical formula of the oxide.

Solution: The 12.33 g of the oxide must contain the original 10.00 g of Ba; so the extra mass of the oxide comes from combined oxygen, which amounts to 12.33 g − 10.00 g = 2.33 g. Now you can set up as usual:

Element	Mass in sample	Number of moles of atoms	Relative number of moles	Ratio
Ba	10.0 g	$\dfrac{10.0\text{ g}}{137.3\text{ g/mol}}$	0.0728	1.00
O	2.33 g	$\dfrac{2.33\text{ g}}{16.00\text{ g/mol}}$	0.145	1.99

Rounding off, you get the empirical formula BaO_2.

PROBLEM 3-8   Platinum (Pt) forms two different chlorides (compounds containing only Pt and Cl), one of which contains 26.7% Cl and the other, 42.1% Cl. Determine the empirical formulas of these two compounds.

Solution: The compound that contains 26.7% Cl must contain 100.0% − 26.7% = 73.3% Pt. Now set up systematically as usual:

Element	Mass in 100 g of sample	Number of moles of atoms	Relative number of moles	Ratio
Pt	73.3 g	$\dfrac{73.3\text{ g}}{195.1\text{ g/mol}}$	0.376	1.00
Cl	26.7 g	$\dfrac{26.7\text{ g}}{35.5\text{ g/mol}}$	0.752	2.00

The empirical formula of this compound is therefore $PtCl_2$.
  The compound with 42.1% Cl contains 100.0% − 42.1% = 57.9% Pt.

Element	Mass in 100 g of sample	Number of moles of atoms	Relative number of moles	Ratio
Pt	57.9 g	$\dfrac{57.9\text{ g}}{195.1\text{ g/mol}}$	0.297	1.00
Cl	42.1 g	$\dfrac{42.1\text{ g}}{35.5\text{ g/mol}}$	1.19	4.01

The empirical formula of this compound is therefore $PtCl_4$.

PROBLEM 3-9   Analysis of a 1.000-g sample of a compound showed that it contained 0.528 g of tin, 0.124 g of iron, 0.160 g of carbon, and 0.188 g of nitrogen. What is its empirical formula?

Solution: Set this up systematically as usual—with the 1.000 g as your sample.

Element	Mass in sample	Number of moles	Relative number of moles	Ratio
Sn	0.528 g	$\dfrac{0.528\text{ g}}{118.7\text{ g/mol}}$	4.45×10^{-3}	2.00
Fe	0.124 g	$\dfrac{0.124\text{ g}}{55.85\text{ g/mol}}$	2.22×10^{-3}	1.00
C	0.160 g	$\dfrac{0.160\text{ g}}{12.01\text{ g/mol}}$	1.33×10^{-2}	5.99
N	0.188 g	$\dfrac{0.188\text{ g}}{14.01\text{ g/mol}}$	1.34×10^{-2}	6.03

Rounding off to whole numbers as usual, we get $Sn_2FeC_6N_6$. (Watch the exponents on this one—and always be careful with them.)

PROBLEM 3-10   Explain how there can be literally hundreds of different hydrocarbons (compounds containing only carbon and hydrogen) with the same empirical formula CH_2.

Solution: The empirical formula CH_2 simply expresses the fact that the *ratio* of the numbers of carbon and hydrogen atoms in the molecules of these hydrocarbons is $1:2$. The *molecular* formulas of the compounds must all be of the type $(CH_2)_n$, where n is a whole number. Because of the many different values of n possible for hydrocarbons, there can be hundreds of compounds with molecular formulas $(CH_2)_n$—such as $(CH_2)_2$ or C_2H_4, $(CH_2)_3$ or C_3H_6, $(CH_2)_4$ or C_4H_8, etc.—all of which have the same empirical formula CH_2.

PROBLEM 3-11   An organic compound is analyzed and found to contain 54.6% C and 9.1% H, the remainder being oxygen; its molar mass was determined as 88 g/mol. What is the molecular formula of this compound?

Solution: The molecular formula is always an integral multiple of the empirical formula. The first step in answering this problem is to calculate the empirical formula in the usual way.

Element	Mass in 100 g of sample	Number of moles of atoms	Relative number of moles	Ratio
C	54.6 g	$\dfrac{54.6 \text{ g}}{12.01 \text{ g/mol}}$	4.55	2.00
H	9.1 g	$\dfrac{9.1 \text{ g}}{1.008 \text{ g/mol}}$	9.03	3.98
O	36.3 g	$\dfrac{36.3 \text{ g}}{16.0 \text{ g/mol}}$	2.27	1.00

Rounding off as usual, the empirical formula is C_2H_4O. The mass of a mole of empirical formula units is

$$2(12.01 \text{ g}) + 4(1.008 \text{ g}) + 16.0 \text{ g} = 44.0 \text{ g}$$

The molar mass of the compound is 88 g/mol. So there must be $88/44.0 = 2.0$ empirical formula units per molecule, so the molecular formula is $(C_2H_4O)_2$ or $C_4H_8O_2$.

PROBLEM 3-12   A hydrocarbon contains 93.75% C and 6.25% H. A rough determination of its molar mass gives a value of 120 ± 10 g/mol. Determine the molecular formula of the hydrocarbon and its molar mass, correct to four significant figures.

Solution: As usual, your first step is to determine the empirical formula:

Element	Mass in 100 g of sample	Number of moles of atoms	Relative number of moles	Ratio
C	93.75 g	$\dfrac{93.75 \text{ g}}{12.01 \text{ g/mol}}$	7.81	1.26
H	6.25 g	$\dfrac{6.25 \text{ g}}{1.008 \text{ g/mol}}$	6.20	1.00

A C to H ratio of $1.26:1$ is very close to $5:4$, so the empirical formula must be C_5H_4. The mass of a mole of empirical formula units is

$$5(12.01 \text{ g}) + 4(1.008 \text{ g}) = 64.1 \text{ g}$$

The molar mass is approximately 120 ± 10 g/mol. So the number of empirical formula units per molecule is $(120 \pm 10 \text{ g})/64.1 \text{ g} \cong 2$, the nearest integer. Remember that the molecular formula *must* be an integral multiple of the empirical formula. So the molecular formula is $(C_5H_4)_2$ or $C_{10}H_8$. The molar mass is

$$10(12.01 \text{ g}) + 8(1.008 \text{ g}) = 128.2 \text{ g}$$

PROBLEM 3-13   At 250°C gaseous elemental phosphorus (P) has a molar mass of 124 g/mol. Explain this observation.

Solution: Although it's true that the atomic mass of phosphorus is 30.97 amu, there's no reason why phosphorus shouldn't form a multiatom molecule in the gas phase. This must be what's happening here. The number of P atoms in this molecule is

$$\frac{124 \text{ g/mol}}{30.97 \text{ g/mol P atoms}} = 4 \text{ P atoms}$$

So we conclude that phosphorus is P_4 in the gas phase at 250°C.

PROBLEM 3-14 An oxide of nitrogen was analyzed and found to contain 30.43% N and 69.57% O. An experiment to determine the molar mass of the compound gave a value of 68 g/mol. Explain these observations.

Solution: As usual, first determine the empirical formula:

Element	Mass in 100 g	Number of moles of atoms	Relative number of moles	Ratio
N	30.43 g	$\dfrac{30.43 \text{ g}}{14.01 \text{ g/mol}}$	2.172	1.00
O	69.57 g	$\dfrac{69.57 \text{ g}}{16.00 \text{ g/mol}}$	4.348	2.00

So the empirical formula is NO_2. The mass of a mole of empirical formula units is

$$14.01 \text{ g} + 2(16.00 \text{ g}) = 46.01 \text{ g/mol}$$

So the number of empirical formula units in the molecule is

$$\frac{68 \text{ g/mol}}{46.01 \text{ g/mol}} = 1.5$$

Now this ratio *must* be an integer—that's what it says in this chapter and in your textbook. One explanation of these observations is that the molar mass experiment was not correctly done, and just gave a bad result. But look again. Because 1.5 is just about halfway between 1 and 2, perhaps this strange oxide of nitrogen is a *mixture* of the two compounds NO_2 and $(NO_2)_2$ and the molar mass experiment was done on the mixture. This may seem odd, but it happens to be the correct explanation! As you'll see later (Chapter 15, Equilibrium) chemical compounds sometimes do exist in two forms under the same conditions, and sometimes those forms can't be separated, so that you have to work with a mixture of them.

PROBLEM 3-15 A compound of carbon, hydrogen, nitrogen, and oxygen has the composition 42.9% C, 2.40% H, 16.7% N, and 38.1% O. Which is the correct empirical formula: **(a)** $C_{18}HN_7O_{16}$, **(b)** $C_2H_3N_6O_3$, **(c)** $C_6H_2NO_2$, **(d)** $C_3H_2NO_2$, **(e)** $C_4H_2NO_2$?

Solution: Set up the usual systematic scheme:

Element	Mass in 100 g of sample	Number of moles of atoms	Relative number of moles	Ratio
C	42.9 g	$\dfrac{42.9 \text{ g}}{12.01 \text{ g/mol}}$	3.57	3
H	2.40 g	$\dfrac{2.40 \text{ g}}{1.008 \text{ g/mol}}$	2.38	2
N	16.7 g	$\dfrac{16.7 \text{ g}}{14.01 \text{ g/mol}}$	1.19	1
O	38.1 g	$\dfrac{38.1 \text{ g}}{16.00 \text{ g/mol}}$	2.38	2

The correct answer is **(d)**, $C_3H_2NO_2$. In **(a)** the mass percentages were used without changing to moles of atoms. In **(b)** the moles of atoms were incorrectly determined: 1/mol was used. In **(c)** the molar masses of H_2, N_2, and O_2 were used instead of the atomic masses of the elements. In **(e)** an attempt was made to round off the number of moles of atoms to the nearest whole numbers: 4, 2, 1, and 2.

PROBLEM 3-16 When 1.135 g of scandium (Sc) is heated with excess chlorine gas (Cl_2), all of the Sc is consumed, producing 3.820 g of a compound of Sc and Cl. The empirical formula of this compound is (a) ScCl, (b) $ScCl_2$, (c) $ScCl_3$, (d) $Sc_2(Cl_2)_3$, (e) $ScCl_4$.

Solution: The 3.820-g sample of scandium chloride contains both the Sc (1.135 g) and the Cl that reacted. So the mass of Cl that combined with the Sc is (3.820 − 1.135)g = 2.685 g. Now set up the determination of the formula in the usual way:

Element	Mass in sample	Number of moles of atoms	Relative number of moles	Ratio
Sc	1.135 g	$\dfrac{1.135 \text{ g}}{44.96 \text{ g/mol}}$	0.0252	1
Cl	2.685 g	$\dfrac{2.685 \text{ g}}{35.45 \text{ g/mol}}$	0.0757	3

The correct answer is (c), $ScCl_3$. In answer (d) the ratio is the same, but this is NOT the correct way to write the empirical formula. Chlorine gas is diatomic (Cl_2), but there is no reason to leave the element in that form in the formula of the compound.

Supplementary Exercises

PROBLEM 3-17 Which of the following cannot be empirical formulas: MgO, H_2O_2, C_5H_{10}, $K_2Cr_2O_7$, $Cd(NO_3)_2$, $P_3N_3Cl_6$?

Answer: H_2O_2, C_5H_{10}, $P_3N_3Cl_6$

PROBLEM 3-18 Calculate the percentage composition of the following compounds: (a) MgO, (b) NaH_2PO_4, (c) C_2H_6O.

Answer: (a) Mg, 60.31%; O, 39.69%
(b) Na, 19.16%; H, 1.68%; P, 25.81%; O, 53.34%
(c) C, 52.14%; H, 13.13%; O, 34.73%

PROBLEM 3-19 An oxide of lead is found to contain 90.67% Pb and 9.33% O. What is its empirical formula?

Answer: Pb_3O_4

PROBLEM 3-20 An oxide of arsenic was analyzed and found to contain 75.74% As. What is its empirical formula?

Answer: As_2O_3

PROBLEM 3-21 What is the empirical formula of a compound that contains 31.29% Ca, 18.75% C, and 49.96% O?

Answer: CaC_2O_4

PROBLEM 3-22 A compound of carbon, hydrogen, and oxygen contains 47.35% C, 10.60% H, and the remainder O. Determine its empirical formula.

Answer: $C_3H_8O_2$

PROBLEM 3-23   When a 0.267-g sample of zirconium (Zr) was treated with fluorine (F), 0.489 g of a zirconium fluoride was produced. What is the empirical formula of this fluoride?

Answer: ZrF_4

PROBLEM 3-24   Antimony (Sb) forms two different chlorides, one of which contains 46.62% Cl and the other, 59.28% Cl. Determine the empirical formulas of these chlorides.

Answer: $SbCl_3$ and $SbCl_5$

PROBLEM 3-25   Analysis of a 100.0-g sample of a compound showed that it contained 2.25 g H, 34.75 g P, and 63.00 g O. Determine its empirical formula.

Answer: $H_4P_2O_7$

PROBLEM 3-26   Two hydrocarbons both have an empirical formula of CH, one having a molar mass of 26.04 g/mol and the other, a molar mass of 52.07 g/mol. What are their molecular formulas?

Answer: C_2H_2 and C_4H_4

PROBLEM 3-27   A compound of silicon and fluorine was found to contain 33.01% Si and 66.99% F; its molar mass was determined to be 170 ± 5 g/mol. Determine its empirical and molecular formulas.

Answer: SiF_3, Si_2F_6

PROBLEM 3-28   A compound was found to contain 53.31% C, 11.19% H, and 35.50% O. A rough determination of its molar mass gave a value of 88 ± 6 g/mol. Determine the molecular formula of the compound and its molar mass, correct to four significant figures.

Answer: $C_4H_{10}O_2$, 90.12 g/mol

PROBLEM 3-29   The element sulfur is found to have an apparent molar mass of 255 ± 5 g/mol in carbon disulfide solution. What is the formula for sulfur under these conditions?

Answer: S_8

PROBLEM 3-30   At high temperatures a compound of nitrogen and fluorine that contains 26.94% N is found to have an apparent molar mass of 76 ± 5 g/mol. Explain these observations.

Answer: The compound is an equilibrium mixture of NF_2 and N_2F_4

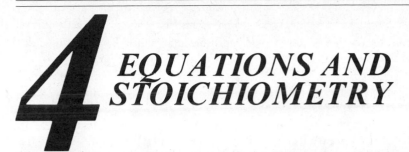

4 EQUATIONS AND STOICHIOMETRY

THIS CHAPTER IS ABOUT

☑ **Chemical Equations**
☑ **Balancing Equations**
☑ **Molar Interpretation of Equations**
☑ **Stoichiometry: Mass Calculations from Equations**
☑ **Limiting Reagent**
☑ **Percentage Yield**

4-1. Chemical Equations

The results of a chemical process are expressed by a chemical equation. The formulas of the materials reacting together—the **reactants** or **reagents**—are written on the left-hand side of the equation. The formulas of the materials that result from the process—the **products**—are written on the right-hand side of the equation. The physical states of the reactants or products can be indicated by writing a letter in parentheses after the formula it refers to: (*g*) for gas, (*l*) for liquid, (*s*) for solid, and (*c*) for crystalline solid.

Chemical equations, like equations in arithmetic or algebra, must *balance*; i.e., what's on the right side must equal what's on the left. In virtually all chemical processes (except those that involve nuclear changes) atoms of any particular kind are neither created nor destroyed. Thus the same numbers of atoms of each element must appear on the left- and the right-hand sides of each balanced equation.

EXAMPLE 4-1: Interpret these balanced equations at the atomic and molecular level.

$$I_2(s) \xrightleftharpoons{\Delta} 2I(g)$$

Interpretation: Solid diatomic iodine, when heated (Δ), gives two iodine atoms in the gas phase.

$$H_2(g) + Cl_2(g) \rightleftharpoons 2HCl(g)$$

Interpretation: One gaseous hydrogen molecule and one gaseous chlorine molecule react to give two gaseous hydrogen chloride molecules.

$$2Mg + O_2 \rightleftharpoons 2MgO$$

Interpretation: Two magnesium atoms react with one oxygen molecule to give two magnesium oxide molecules. No physical states are given.

$$6CO_2 + 6H_2O \rightleftharpoons C_6H_{12}O_6 + 6O_2$$

Interpretation: Six molecules of carbon dioxide and six molecules of water react to give a complex carbohydrate and six molecules of oxygen. No physical states are given. (You can also guess that the carbohydrate is molecular, since its empirical formula would be CH_2O, and a multiple of this is given.)

4-2. Balancing Equations

Balancing equations is a matter of simple arithmetic. Remember that atoms of any given kind are conserved.

EXAMPLE 4-2: Balance the equation for the reaction between molecular sulfur S_8 and molecular oxygen O_2 to yield sulfur trioxide SO_3.

Solution: The equation will be of the form

$$xS_8 + yO_2 \longrightarrow zSO_3$$

Because sulfur atoms must be conserved, *each* S_8 must yield 8 SO_3 molecules. We see that z must be an integral multiple of 8, i.e., 8, 16, 24, etc. Using trial and error, we'll try to solve this equation for the simplest case, where $x = 1$ and $z = 8$. Then

$$S_8 + yO_2 \longrightarrow 8SO_3$$

On the right-hand side we have 24 oxygen atoms; these can be provided from 12 molecules of O_2. So $y = 12$ and the balanced equation is

$$S_8 + 12O_2 \longrightarrow 8SO_3$$

EXAMPLE 4-3: Balance the equation for the reaction in which benzene C_6H_6 burns in O_2. The products are CO_2 and H_2O.

Solution: Let's take one molecule of C_6H_6 and follow the fate of its atoms separately. It's clear that

$$C_6 + 6O_2 \longrightarrow 6CO_2$$

and

$$H_6 + \tfrac{3}{2}O_2 \longrightarrow 3H_2O$$

Summing,

$$C_6H_6 + 6O_2 + \tfrac{3}{2}O_2 \longrightarrow 6CO_2 + 3H_2O$$

or

$$C_6H_6 + 7\tfrac{1}{2}O_2 \longrightarrow 6CO_2 + 3H_2O$$

By the usual chemical convention, you should include only *integral* numbers of molecules in balanced equations. To eliminate the fraction in the last equation, you simply multiply all the coefficients by 2 to get the final form:

$$2C_6H_6 + 15O_2 \longrightarrow 12CO_2 + 6H_2O \tag{4-1}$$

4-3. Molar Interpretation of Equations

You can interpret a balanced equation at both the molecular and the molar levels. Equation (4-1) means two things. At the molecular level it means that *2 molecules* of benzene react with *15 molecules* of oxygen to produce *12 molecules* of carbon dioxide and *6 molecules* of water. At the molar level it means that *2 moles* of benzene react with *15 moles* of oxygen to produce *12 moles* of carbon dioxide and *6 moles* of water. From the molar level, you can go directly to the mass relationships involved, recalling that

    1 mol of C_6H_6 has a mass of $6(12.01\text{ g}) + 6(1.008\text{ g}) = 78.11$ g
    1 mol of O_2 has a mass of $2(16.00\text{ g}) = 32.00$ g
    1 mol of CO_2 has a mass of $12.01\text{ g} + 2(16.00\text{ g}) = 44.01$ g
    1 mol of H_2O has a mass of $2(1.008\text{ g}) + 16.00\text{ g} = 18.02$ g

So 2×78.11 g of C_6H_6 reacts with 15×32.00 g of O_2 to produce 12×44.01 g of CO_2 and 6×18.02 g of H_2O.

4-4. Stoichiometry: Mass Calculations from Equations

Quantitative information about the mass relationships, i.e., the **stoichiometry**, of reactions allows you to calculate a wide range of other numerical results for reactions.

EXAMPLE 4-4: For the reaction in Equation (4-1), how much water (in grams) is produced by the reaction of 3.00 g of benzene with oxygen?

Solution: The direct approach is by proportions: Since 2×78.11 g of benzene gives 6×18.02 g of water, the mass of water is

$$\frac{6 \times 18.02 \text{ g H}_2\text{O}}{2 \times 78.11 \text{ g C}_6\text{H}_6} \times 3.00 \text{ g C}_6\text{H}_6 = 2.08 \text{ g H}_2\text{O}$$

A more generally useful approach—the factor-cancel method—starts with a tabulation of all the information contained in the initial equation:

Equation:	$2C_6H_6 +$	$15O_2 \longrightarrow$	$12CO_2 +$	$6H_2O$
Number of moles:	2	15	12	6
Mass per mole:	78.11 g	32.00 g	44.01 g	18.02 g

The mass of H_2O produced when 3.00 g of C_6H_6 reacts with oxygen is then calculated from an equation expressing all the factors and their units:

$$3.00 \text{ g C}_6\text{H}_6 \times \frac{\text{mol C}_6\text{H}_6}{78.11 \text{ g C}_6\text{H}_6} \times \frac{6 \text{ mol H}_2\text{O}}{2 \text{ mol C}_6\text{H}_6} \times \frac{18.02 \text{ g H}_2\text{O}}{\text{mol H}_2\text{O}} = 2.08 \text{ g H}_2\text{O}$$

Note that this equation is self-checking: Each unit cancels except the one you want, g H_2O.

4-5. Limiting Reagent

In real laboratory chemistry you don't always run reactions using the exact amounts of reagents that you would calculate from the stoichiometry. You might, for example, use an inexpensive reagent in excess of the stoichiometric amount. In that case the amount of another reagent—a **limiting reagent**—will limit the amount of product you get. Thus magnesium and chlorine react in a 1:1 molar ratio to produce magnesium chloride:

Equation:	$Mg +$	$Cl_2 \longrightarrow$	$MgCl_2$
Start:	1 mol	1 mol	0
Finish:	0	0	1 mol

If one mole of magnesium and two moles of molecular chlorine are mixed, the one mole of magnesium can react with only one mole of chlorine. Magnesium will be the limiting reagent. One mole of magnesium chloride will be produced and one mole of chlorine will be left unchanged:

Equation:	$Mg +$	$Cl_2 \longrightarrow$	$MgCl_2$
Start:	1 mol	2 mol	0
Finish:	0	1 mol	1 mol

If 0.63 mol of Mg is treated with 0.37 mol of Cl_2, chlorine is the limiting reagent: 0.37 mol of $MgCl_2$ is produced and 0.26 mol of Mg $(0.63 - 0.37)$ is unchanged.

EXAMPLE 4-5: Write a balanced equation for the reaction between molecular phosphorus (P_4) and molecular iodine (I_2) to produce phosphorus triiodide (PI_3). Use this equation to determine the limiting reagent and the amount of PI_3 produced when 6.00 g of P_4 is treated with 25.0 g of I_2.

Solution: Balance the equation and determine the molar relationships of the reactants:

Equation:	$P_4 +$	$6I_2 \longrightarrow$	$4PI_3$
g/mol:	$4 \times 31.0 = 124$	$2 \times 127 = 254$	$31 + 3(127) = 412$
Moles of reactant available:	$\frac{6.00}{124} = 0.0484$	$\frac{25.0}{254} = 0.0984$	

Determine how much I_2 would be required to react completely with the available amount of P_4:

$$0.0484 \text{ mol P}_4 \times \frac{6 \text{ mol I}_2}{1 \text{ mol P}_4} = 0.290 \text{ mol I}_2$$

Only 0.0984 mol of I_2 is available, so I_2 is the limiting reagent. Now determine how much PI_3 can be produced stoichiometrically from the limited amount of I_2:

$$0.0984 \text{ mol } I_2 \times \frac{4 \text{ mol } PI_3}{6 \text{ mol } I_2} \times \frac{412 \text{ g } PI_3}{1 \text{ mol } PI_3} = 27.0 \text{ g } PI_3$$

4-6. Percentage Yield

Percentage yield is defined as the actual yield of a product expressed as a percentage of the theoretical (stoichiometric) yield:

$$\% \text{ yield} = \frac{\text{yield of product actually obtained}}{\text{yield of product expected from stoichiometry}} \times 100\%$$

This concept arises because real reactions often do not proceed stoichiometrically.

EXAMPLE 4-6: Methane (CH_4) reacts with chlorine (Cl_2) to produce chloroform ($CHCl_3$) and hydrogen chloride (HCl). In an industrial process, 2.00×10^3 kg of CH_4 is treated with 2.00×10^4 kg of Cl_2 to produce 1.03×10^4 kg of $CHCl_3$. Determine the limiting reagent and the percentage yield of $CHCl_3$.

Solution: Balance the equation and determine the molar relationships of the reactants:

Equation: $CH_4 + 3Cl_2 \longrightarrow CHCl_3 + 3HCl$

g/mol: 16.0   70.9   119

Moles of CH_4 available: $\dfrac{(2.00 \times 10^3 \text{ kg})(10^3 \text{ g/kg})}{16.0 \text{ g/mol}} = 1.25 \times 10^5$

Moles of Cl_2 available: $\dfrac{(2.00 \times 10^4 \text{ kg})(10^3 \text{ g/kg})}{70.9 \text{ g/mol}} = 2.82 \times 10^5$

Determine how much Cl_2 would be required to react completely with the available amount of CH_4:

$$(1.25 \times 10^5 \text{ mol } CH_4)\left(\frac{3 \text{ mol } Cl_2}{1 \text{ mol } CH_4}\right) = 3.75 \times 10^5 \text{ mol } Cl_2$$

Only 2.82×10^5 mol of Cl_2 is available, so Cl_2 is the limiting reagent.

Now determine how much $CHCl_3$ can be produced stoichiometrically from the limited amount of Cl_2:

$$(2.82 \times 10^5 \text{ mol } Cl_2)\left(\frac{1 \text{ mol } CHCl_3}{3 \text{ mol } Cl_2}\right)\left(\frac{119 \text{ g } CHCl_3}{1 \text{ mol } CHCl_3}\right)\left(\frac{1 \text{ kg}}{10^3 \text{ g}}\right) = 1.12 \times 10^4 \text{ kg } CHCl_3$$

Given that only 1.03×10^4 kg of $CHCl_3$ is produced, the percentage yield is

$$\% \text{ yield} = \frac{\text{actual}}{\text{stoichiometric}} \times 100\% = \frac{1.03 \times 10^4 \text{ kg}}{1.12 \times 10^4 \text{ kg}} \times 100\% = 92.0\%$$

SUMMARY

1. Chemical equations express the results of a chemical process.
2. In chemical processes, atoms of each type are conserved.
3. Equations can be interpreted at the molecular level and at the molar level in terms of mass.
4. Stoichiometry is the calculation of mass relationships from equations.
5. A limiting reagent is one that limits the amount of product that can actually be produced in a reaction.
6. Percentage yield of a product is the actual yield obtained expressed as a percentage of the theoretical yield.

RAISE YOUR GRADES

Can you define...?

- ☑ reactants
- ☑ products
- ☑ stoichiometry
- ☑ limiting reagent
- ☑ percentage yield

Can you...?

- ☑ balance a simple reaction equation
- ☑ interpret an equation at the molecular level
- ☑ interpret an equation at the molar level
- ☑ use an equation to calculate mass relationships
- ☑ determine the limiting reagent in a reaction
- ☑ calculate the percentage yield of a product

SOLVED PROBLEMS

PROBLEM 4-1 Balance each of the following equations:

(a) $H_2(g) + O_2(g) \longrightarrow H_2O(l)$
(b) $NaOH + H_2SO_4 \longrightarrow Na_2SO_4 + H_2O$
(c) $Ca(OH)_2 + H_3PO_4 \longrightarrow Ca_3(PO_4)_2 + H_2O$
(d) $C_3H_6(g) + O_2(g) \longrightarrow CO_2(g) + H_2O(l)$
(e) $P_4 + Cl_2 \longrightarrow PCl_5$
(f) $NH_3 + O_2 \longrightarrow HNO_3 + H_2O$

Solution:

(a) You notice that the reactant oxygen is diatomic and that the product H_2O contains only one O atom per molecule. So your first partial answer shows one O_2 molecule giving two H_2O molecules or

$$H_2(g) + O_2(g) \longrightarrow 2H_2O(l)$$

You complete this partial answer by balancing the H atoms: Since four H atoms appear on the right in the $2H_2O$, you need four on the left. This is easily done by using two H_2 molecules. So the final equation is

$$2H_2(g) + O_2(g) \longrightarrow 2H_2O(l)$$

(b) The product Na_2SO_4 contains two atoms of Na, so you have to use two NaOH formula units in the reactants:

$$2NaOH + H_2SO_4 \longrightarrow Na_2SO_4 + H_2O$$

A count of atoms shows that you now have four H atoms in the reactants, and so you need two H_2O molecules in the products to balance them:

$$2NaOH + H_2SO_4 \longrightarrow Na_2SO_4 + 2H_2O$$

A quick check shows that the equation is balanced; the O atoms have taken care of themselves! You're done.

(c) You see that the product $Ca_3(PO_4)_2$ requires three $Ca(OH)_2$ and two H_3PO_4 species in the reactants:

$$3Ca(OH)_2 + 2H_3PO_4 \longrightarrow Ca_3(PO_4)_2 + H_2O$$

A check on H atoms shows twelve in the reactants, six in $3Ca(OH)_2$ and six in $2H_3PO_4$; so you need $6H_2O$ in the products:

$$3Ca(OH)_2 + 2H_3PO_4 \longrightarrow Ca_3(PO_4)_2 + 6H_2O$$

A final check on O atoms shows that this equation is balanced.

(d) As in Section 4-1, you tackle this by treating each kind of atom separately. You see that

$$C_3 + 3O_2 \longrightarrow 3CO_2 \quad \text{and} \quad H_6 + \tfrac{3}{2}O_2 \longrightarrow 3H_2O$$

You add these and get

$$C_3H_6(g) + 4\tfrac{1}{2}O_2(g) \longrightarrow 3CO_2(g) + 3H_2O(l)$$

Because convention specifies only integers in balanced equations, you double the coefficients to get the final, balanced equation:

$$2C_3H_6(g) + 9O_2(g) \longrightarrow 6CO_2(g) + 6H_2O(l)$$

(e) You realize that P_4 must end up in four PCl_5 molecules, requiring $4 \times 5 = 20$ atoms of Cl, derived from ten Cl_2 molecules. So, by inspection (all this thought occurring in your head much faster than you can write it down), you get

$$P_4 + 10Cl_2 \longrightarrow 4PCl_5$$

(f) First, concentrate on getting NH_3 to HNO_3; worry about the H_2O at the end:

$$NH_3 + \tfrac{3}{2}O_2 \longrightarrow HNO_3 + [H_2]$$

The bracketed $[H_2]$ simply means that at this stage of the balancing process you have an *intermediate* material that you'll have to convert to the actual final product H_2O. Now you add another $\tfrac{1}{2}O_2$ to end up with H_2O replacing the mysterious $[H_2]$:

$$NH_3 + \tfrac{3}{2}O_2 + \tfrac{1}{2}O_2 \longrightarrow HNO_3 + H_2O$$

Adding the fractional O_2 molecules, you tidy up:

$$NH_3 + 2O_2 \longrightarrow HNO_3 + H_2O$$

PROBLEM 4-2 Explain the meaning of the following equations at the molecular level:

(a) $3O_2(g) \longrightarrow 2O_3(g)$
(b) $SO_3(g) + H_2O(l) \longrightarrow H_2SO_4(l)$
(c) $2Sb(s) + 5F_2(g) \longrightarrow 2SbF_5(l)$

Solution:
(a) Three molecules of gaseous molecular oxygen yield two molecules of gaseous *ozone*.
(b) One molecule of gaseous sulfur trioxide reacts with one molecule of liquid water to give one molecule of liquid sulfuric acid.
(c) Two atoms of solid antimony react with five molecules of gaseous fluorine to give two molecules of liquid antimony pentafluoride.

PROBLEM 4-3 Give the molar and mass interpretations of the following equations:

(a) $SO_3 + H_2O \longrightarrow H_2SO_4$
(b) $2Sb + 5F_2 \longrightarrow 2SbF_5$
(c) $C_5H_{12} + 3Cl_2 \longrightarrow C_5H_9Cl_3 + 3HCl$

Solution: A convenient way for you to express all these relationships is to list them below the relevant material in the equation:

(a) Equation:	SO_3	+	H_2O	$\longrightarrow$	H_2SO_4
Atomic/molecular:	1 molecule		1 molecule		1 molecule
Molar:	1 mol		1 mol		1 mol
Mass:					2(1.008)
	32.06		16.00		+ 32.06
	+3(16.00)		+2(1.008)		+4(16.00)
	80.06 g		18.02 g		98.08 g

(b) Equation: **2Sb**   +   **5F$_2$**   **2SbF$_5$**

   Atomic/molecular: 2 atoms   5 molecules   2 molecules

   Molar: 2 mol   5 mol   2 mol

   Mass:

$$\frac{2[121.75]}{243.5 \text{ g}} \qquad \frac{5[2(19.00)]}{190.0 \text{ g}} \qquad \frac{2\left[\begin{array}{c}121.75\\ +5(19.00)\end{array}\right]}{433.5 \text{ g}}$$

(c) Equation: **C$_5$H$_{12}$** + **3Cl$_2$** $\longrightarrow$ **C$_5$H$_9$Cl$_3$** + **3HCl**

   Atomic/molecular: 1 molecule   3 molecules   1 molecule   3 molecules

   Molar: 1 mol   3 mol   1 mol   3 mol

   Mass:

$$\frac{\begin{array}{c}5(12.01)\\ +12(\;1.008)\end{array}}{72.15 \text{ g}} \quad \frac{3[2(35.45)]}{212.70 \text{ g}} \quad \frac{\begin{array}{c}5(12.01)\\ +9(\;1.008)\\ +3(35.45)\end{array}}{175.47 \text{ g}} \quad \frac{3\left[\begin{array}{c}1.008\\ +35.45\end{array}\right]}{109.37 \text{ g}}$$

PROBLEM 4-4   What mass of antimony pentafluoride SbF$_5$ is produced when 12.0 g of antimony is treated with excess fluorine?

Solution: You can solve this problem in a number of ways. At first sight the solution by proportion seems most direct. Looking at the solution to Problem 4-3b, you note that 243.5 g of Sb yields 433.5 g of SbF$_5$. Therefore 12.0 g of Sb will yield

$$12.0 \text{ g Sb} \times \frac{433.5 \text{ g SbF}_5}{243.5 \text{ g Sb}} = 21.4 \text{ g SbF}_5$$

A generally more powerful method, one that we *strongly* recommend, involves the molar ratios from the balanced equation. First, calculate the number of moles of Sb atoms in 12 g:

$$\frac{12.0 \text{ g Sb}}{121.75 \text{ g Sb/mol Sb}}$$

Then determine the number of moles of SbF$_5$ this would yield:

$$\frac{12.0 \text{ g Sb}}{121.75 \text{ g Sb/mol Sb}} \times \frac{2 \text{ mol SbF}_5}{2 \text{ mol Sb}} \qquad \text{(this ratio comes from the balanced equation)}$$

Finally, determine the mass of this number of moles of SbF$_5$:

$$\frac{12.0 \text{ g Sb}}{121.75 \text{ g Sb/mol Sb}} \times \frac{2 \text{ mol SbF}_5}{2 \text{ mol Sb}} \times \frac{216.75 \text{ g SbF}_5}{\text{mol SbF}_5} = 21.4 \text{ g SbF}_5$$

With practice, you'll find yourself writing these steps down directly in one line. The power of this method lies in its self-checking property. Notice how all the units cancel to leave you with only the units of the answer you expect.

PROBLEM 4-5   How many kilograms of sodium carbonate Na$_2$CO$_3$ can be produced by heating 1.75×10^4 kg of sodium hydrogen carbonate NaHCO$_3$? The other products of the reaction are CO$_2$ and H$_2$O.

Solution: First (as almost always) write the balanced equation with the molar interpretation and the mass of a mole of each of the significant compounds below. By inspection

$$2\text{NaHCO}_3 \xrightarrow{\;\Delta\;} \text{Na}_2\text{CO}_3 \; + \text{H}_2\text{O} + \text{CO}_2$$

   2 mol   1 mol

   84.0 g/mol   106.0 g/mol

Then go directly to your answer, using the factor-cancel method:

$$\frac{(1.75 \times 10^4 \text{ kg NaHCO}_3)(10^3 \text{ g/kg})}{84.0 \text{ g/mol NaHCO}_3} \times \frac{1 \text{ mol Na}_2\text{CO}_3}{2 \text{ mol NaHCO}_3} \times \frac{106.0 \text{ g/mol}}{10^3 \text{ g/kg Na}_2\text{CO}_3} = 1.10 \times 10^4 \text{ kg Na}_2\text{CO}_3$$

PROBLEM 4-6   Calcium carbide CaC$_2$ reacts with water to generate acetylene gas C$_2$H$_2$ and solid calcium hydroxide Ca(OH)$_2$. What mass of H$_2$O is needed to react completely with 25.0 g of CaC$_2$? What mass of C$_2$H$_2$ is produced in the reaction?

Solution: Balance and interpret the reaction equation:

$$CaC_2 \quad + \quad 2H_2O \qquad\qquad C_2H_2 \quad + \quad Ca(OH)_2$$

1 mol	2 mol	1 mol	1 mol
64.1 g/mol	18.02 g/mol	26.04 g/mol	—

(Since the problem has no answer that involves the mass of $Ca(OH)_2$, you don't need to determine its molar mass.)
Now you go directly to your two answers:

$$\text{mass of } H_2O = \frac{25.0 \text{ g } CaC_2}{64.1 \text{ g/mol } CaC_2} \times \frac{2 \text{ mol } H_2O}{1 \text{ mol } CaC_2} \times \frac{18.02 \text{ g}}{\text{mol } H_2O}$$

$$= 14.1 \text{ g } H_2O$$

$$\text{mass of } C_2H_2 = \frac{25.0 \text{ g } CaC_2}{64.1 \text{ g/mol } CaC_2} \times \frac{1 \text{ mol } C_2H_2}{1 \text{ mol } CaC_2} \times \frac{26.04 \text{ g}}{\text{mol } C_2H_2}$$

$$= 10.2 \text{ g } C_2H_2$$

PROBLEM 4-7   At a copper refinery metallic copper is extracted from copper sulfide CuS. During the refining process, the sulfur is converted into sulfur dioxide SO_2, which is trapped in a scrubber. How much metallic copper can be produced annually in a refinery that treats 3.7×10^4 kg of CuS daily, and how much SO_2 is trapped by the scrubbers annually?

Solution: You may be uneasy about this problem at first because you don't have enough information to write balanced equations. But don't give up! You can use the law of conservation of matter. Each atom of Cu starting in CuS must end up as metallic copper, and each atom of S must end up as SO_2, or the problem would be quite unsolvable. So you write the partial equation and interpret it as usual:

$$CuS \longrightarrow Cu \quad + \quad SO_2$$

1 mol	1 mol	1 mol
95.6 g/mol	63.54 g/mol	64.06 g/mol

Now you can go straight to the solutions—but just notice that the CuS treated is given as a *daily* amount and that the question asks about *annual* production, so

$$\text{amount of Cu} = \frac{3.7 \times 10^4 \text{ kg CuS/day}}{95.6 \text{ g/mol CuS}} \times \frac{1 \text{ mol Cu}}{1 \text{ mol CuS}} \times \frac{63.54 \text{ g}}{\text{mol Cu}} \times \frac{365 \text{ days}}{\text{year}}$$

$$= 9.0 \times 10^6 \text{ kg Cu/year}$$

$$\text{amount of } SO_2 = \frac{3.7 \times 10^4 \text{ kg CuS/day}}{95.6 \text{ g/mol CuS}} \times \frac{1 \text{ mol } SO_2}{1 \text{ mol CuS}} \times \frac{64.06 \text{ g}}{\text{mol } SO_2} \times \frac{365 \text{ day}}{\text{year}}$$

$$= 9.0 \times 10^6 \text{ kg } SO_2/\text{year}$$

PROBLEM 4-8   A 0.450-g sample of a tin chloride (containing tin and chlorine only) was added to water, and then excess silver nitrate was added. All the chlorine present in the sample was recovered as 0.680 g of silver chloride AgCl. What is the empirical formula of the tin chloride?

Solution: Since all the chlorine present was recovered as AgCl, you can immediately deduce that the number of moles of Cl in the original AgCl is

$$\frac{0.680 \text{ g AgCl}}{143.4 \text{ g/mol AgCl}} \times \frac{1 \text{ mol Cl}}{1 \text{ mol AgCl}} = 4.74 \times 10^{-3} \text{ mol Cl}$$

The mass of Cl in the AgCl is

$$4.74 \times 10^{-3} \text{ mol Cl} \times \frac{35.45 \text{ g}}{\text{mol Cl}} = 0.168 \text{ g}$$

which is also the mass of Cl in the original tin chloride sample. The mass of Sn in the tin chloride sample must have been $0.450 \text{ g} - 0.168 \text{ g} = 0.282 \text{ g}$. Thus

$$\text{Number of moles of Sn in the tin chloride} = \frac{0.282 \text{ g Sn}}{118.7 \text{ g/mol Sn}} = 2.38 \times 10^{-3} \text{ mol Sn}$$

$$\text{Molar ratio Cl/Sn in the sample} = \frac{4.74 \times 10^{-3} \text{ mol Cl}}{2.38 \times 10^{-3} \text{ mol Sn}} = 1.99$$

So the empirical formula of the tin chloride is $SnCl_2$.

[*Note*: This problem is representative of the way in which chemical analysis may be carried out: One element in a sample is determined by quantitative conversion into a compound of known formula—AgCl in this example—that can be weighed directly to give the number of moles of the element.]

PROBLEM 4-9   A 1.222-g sample of a hydrated form of barium chloride having the empirical formula $BaCl_2 \cdot yH_2O$ is heated until all the H_2O is driven off. A 1.042-g residue of dry $BaCl_2$ remains. What is the numerical value of y?

Solution: The number of moles of $BaCl_2$ left in the residue and present in the initial sample is

$$\frac{1.042 \text{ g BaCl}_2}{208.2 \text{ g/mol BaCl}_2} = 5.00 \times 10^{-3} \text{ mol BaCl}_2$$

The mass of H_2O driven off is 1.222 g − 1.042 g = 0.180 g. The number of moles of H_2O driven off (and present in the initial sample) is

$$\frac{0.180 \text{ g}}{18.0 \text{ g/mol}} = 0.0100 \text{ mol H}_2O$$

So the molar ratio is

$$\frac{H_2O}{BaCl_2} = \frac{0.0100}{5.00 \times 10^{-3}} = 2.00 \quad \text{in the initial sample}$$

Thus $y = 2$ and the original hydrate was $BaCl_2 \cdot 2H_2O$.

PROBLEM 4-10   A 1.00-g sample of hydrocarbon C_xH_y is burned in excess oxygen to yield 1.80 g of H_2O and 2.93 g of CO_2. What is the empirical formula of the hydrocarbon?

Solution: All the hydrogen in the sample is converted into H_2O, and all the carbon into CO_2. So start by determining the number of moles of C and H present in the sample:

$$\text{No. of moles of H} = \frac{1.80 \text{ g H}_2O}{18.0 \text{ g/mol H}_2O} \times \frac{2 \text{ mol H}}{1 \text{ mol H}_2O} = 0.200 \text{ mol H}$$

$$\text{No. of moles of C} = \frac{2.93 \text{ g CO}_2}{44.0 \text{ g/mol CO}_2} \times \frac{1 \text{ mol C}}{1 \text{ mol CO}_2} = 0.0666 \text{ mol C}$$

$$\frac{\text{mol H}}{\text{mol C}} = \frac{0.200}{0.0666} = 3.00$$

Hence the empirical formula of the hydrocarbon is CH_3.

PROBLEM 4-11   A 3.37-g sample of a mixture of MnO and Mn_2O_3 is treated with $H_2(g)$ under conditions in which only the Mn_2O_3 reacts, as follows:

$$Mn_2O_3(s) + H_2(g) \qquad 2MnO(s) + H_2O(l)$$

The reaction yields 0.165 g of H_2O. What is the percentage by mass of Mn_2O_3 in the mixture?

Solution: From the balanced equation, you can immediately calculate the mass of Mn_2O_3 in the mixture:

$$\text{mass Mn}_2O_3 = \frac{0.165 \text{ g H}_2O}{18.0 \text{ g/mol H}_2O} \times \frac{1 \text{ mol Mn}_2O_3}{1 \text{ mol H}_2O} \times \frac{157.9 \text{ g}}{\text{mol Mn}_2O_3}$$

$$= 1.45 \text{ g Mn}_2O_3$$

So

$$\% \text{ by mass of Mn}_2O_3 = \frac{1.45 \text{ g Mn}_2O_3}{3.37 \text{ g mixture}} \times 100\% = 43.0\%$$

PROBLEM 4-12   A 9.65-mg sample of a compound containing only carbon, hydrogen, and oxygen is burned in excess O_2; 22.9 mg of CO_2 and 5.47 mg of H_2O are obtained. What is the empirical formula of the compound?

Solution: You see that the data on CO_2 and H_2O lead you directly to the number of moles (and the mass) of carbon and hydrogen in the original sample; the oxygen will have to be determined by difference (which is the standard way in organic analysis).

$$\text{No. of moles of C in sample} = \frac{22.9 \times 10^{-3} \text{ g CO}_2}{44.0 \text{ g/mol CO}_2} \times \frac{1 \text{ mol C}}{1 \text{ mol CO}_2} = 5.20 \times 10^{-4} \text{ mol C}$$

$$\text{Mass of C in sample} = 5.20 \times 10^{-4} \text{ mol C} \times 12.0 \text{ g/mol C} = 6.25 \times 10^{-3} \text{ g}$$

$$\text{No. of moles of H in sample} = \frac{5.47 \times 10^{-3} \text{ g H}_2\text{O}}{18.0 \text{ g/mol H}_2\text{O}} \times \frac{2 \text{ mol H}}{1 \text{ mol H}_2\text{O}} = 6.08 \times 10^{-4} \text{ mol H}$$

$$\text{Mass of H in sample} = 6.08 \times 10^{-4} \text{ mol H} \times 1.008 \text{ g/mol H} = 6.13 \times 10^{-4} \text{ g}$$

$$\text{Mass of O in sample} = (9.65 - 6.25 - 0.613) \times 10^{-3} \text{ g} = 2.79 \times 10^{-3} \text{ g}$$

$$\text{No. of moles of O in sample} = \frac{2.79 \times 10^{-3} \text{ g O}}{16.0 \text{ g/mol O}} = 1.74 \times 10^{-4} \text{ mol O}$$

$$\text{Ratio of } n_C : n_H : n_O = 5.20 \times 10^{-4} : 6.08 \times 10^{-4} : 1.74 \times 10^{-4} = 2.99 : 3.50 : 1$$

$$= 6 : 7 : 2 \quad \text{(integer ratio)}$$

So the empirical formula is $C_6H_7O_2$.

PROBLEM 4-13   When $Cu(CN)_2$ is heated, C_2N_2 (cyanogen) and $CuCN$ are produced. What mass of $Cu(CN)_2$ would be needed to prepare 5.00 g of C_2N_2?

Solution: Write and interpret the balanced equation for the process:

$$2Cu(CN)_2 \xrightarrow{\Delta} C_2N_2 + 2CuCN$$
$$\begin{array}{ccc} 2 \text{ mol} & 1 \text{ mol} & 2 \text{ mol} \\ 115.6 \text{ g/mol} & 52.04 \text{ g/mol} & \end{array}$$

The mass of $Cu(CN)_2$ needed is

$$\frac{5.00 \text{ g C}_2\text{N}_2}{52.04 \text{ g/mol C}_2\text{N}_2} \times \frac{2 \text{ mol Cu(CN)}_2}{1 \text{ mol C}_2\text{N}_2} - \frac{115.6 \text{ g}}{\text{mol Cu(CN)}_2} = 22.2 \text{ g}$$

PROBLEM 4-14   A 0.504-g sample of a mixture of Na_2SO_4 and $MgSO_4$ is treated with an excess of $BaCl_2$ solution, which precipitates all the SO_4 in the mixture as $BaSO_4$. A mass of 0.962 g of $BaSO_4$ is obtained. Determine the percentage by mass of $MgSO_4$ in the original mixture.

Solution: The number of moles of $BaSO_4$ obtained equals the number of moles of sulfate in the mixture:

$$\frac{0.962 \text{ g BaSO}_4}{233.4 \text{ g/mol BaSO}_4} = 4.12 \times 10^{-3} \text{ mol}$$

Call the percentage of $MgSO_4$ in the mixture x; then the percentage of Na_2SO_4 is $100 - x$. The mass of $MgSO_4$ in the mixture is $(0.504 \text{ g})(x\%/100\%)$. So the number of moles of $MgSO_4$ in the mixture is

$$\frac{(0.504x) \text{ g}}{(100)(120.4 \text{ g/mol MgSO}_4)}$$

The number of moles of SO_4 from $MgSO_4$ in the mixture is

$$\frac{(0.504x) \text{ mol MgSO}_4}{(100)(120.4)} \times \frac{1 \text{ mol SO}_4}{1 \text{ mol MgSO}_4} = (4.19 \times 10^{-5})x$$

The number of moles of SO_4 from Na_2SO_4 in the mixture is

$$\frac{(100 - x)(0.504 \text{ g})}{100(142.0 \text{ g/mol Na}_2\text{SO}_4)} \times \frac{1 \text{ mol SO}_4}{1 \text{ mol Na}_2\text{SO}_4} = (3.55 \times 10^{-3}) - (3.55 \times 10^{-5})x$$

The total number of moles of SO_4 equals the number of moles of $BaSO_4$, so

$$(4.19 \times 10^{-5})x + (3.55 \times 10^{-3}) - (3.55 \times 10^{-5})x = 4.12 \times 10^{-3}$$
$$(0.64 \times 10^{-5})x = 0.57 \times 10^{-3}$$

$$x = \frac{0.57 \times 10^{-3}}{0.64 \times 10^{-5}} = 89$$

So 89% of the original mixture is $MgSO_4$. [Note: Only two significant figures are justified.]

PROBLEM 4-15   In a synthesis of bismuth trichloride $BiCl_3$, 10.0 g of bismuth is treated with 7.00 g of Cl_2. Determine the limiting reagent and the maximum yield of $BiCl_3$.

Solution: Write and interpret the balanced reaction:

$$2Bi \quad + \quad 3Cl_2 \quad \longrightarrow \quad 2BiCl_3$$

2 mol	3 mol	2 mol
209.0 g/mol	70.9 g/mol	315.4 g/mol

Now calculate the number of moles of one of the reactants, say bismuth:

$$\text{No. of moles of Bi} = \frac{10.0 \text{ g Bi}}{209.0 \text{ g/mol Bi}} = 0.0478 \text{ mol Bi}$$

For this much Bi we would need

$$0.0478 \text{ mol Bi} \times \frac{3 \text{ mol Cl}_2}{2 \text{ mol Bi}} \times \frac{70.9 \text{ g}}{\text{mol Cl}_2} = 5.09 \text{ g Cl}_2$$

But we have *more* Cl_2 than this (7.00 g), so Cl_2 is in excess and Bi is the limiting reagent. The maximum yield of $BiCl_3$ is

$$0.0478 \text{ mol Bi} \times \frac{2 \text{ mol BiCl}_3}{2 \text{ mol Bi}} \times \frac{315.4 \text{ g}}{\text{mol BiCl}_3} = 15.1 \text{ g BiCl}_3$$

PROBLEM 4-16   Antimony trichloride ($SbCl_3$) can be prepared by the following reaction:

$$Sb_2O_3 + 6NaCl + 3H_2SO_4 \longrightarrow 2SbCl_3 + 3Na_2SO_4 + 3H_2O$$

In a laboratory manual the suggested quantities of reagents are given as Sb_2O_3, 20.0 g; NaCl, 60.0 g; H_2SO_4, 150 g of a 50% by weight solution. Determine the maximum yield of $SbCl_3$ that can be made from these quantities of reagents.

Solution: First determine which is the limiting reagent. Interpret the equation:

$$Sb_2O_3 \quad + \quad 6NaCl \quad + \quad 3H_2SO_4 \quad \longrightarrow \quad 2SbCl_3 \quad + 3Na_2SO_4 + 3H_2O$$

1 mol	6 mol	3 mol	2 mol	3 mol	3 mol
291.5 g/mol	58.44 g/mol	98.08 g/mol	228.1 g/mol	—	

Since antimony oxide sounds more exotic (and expensive) than sodium chloride or sulfuric acid, you decide to work with it, first determining the number of moles available:

$$20.0 \text{ g of Sb}_2O_3 = \frac{20.0 \text{ g}}{291.5 \text{ g/mol}} = 0.0686 \text{ mol Sb}_2O_3.$$

The amount of NaCl needed is

$$0.0686 \text{ mol Sb}_2O_3 \times \frac{6 \text{ mol NaCl}}{1 \text{ mol Sb}_2O_3} \times \frac{58.44 \text{ g}}{\text{mol NaCl}} = 24.1 \text{ g NaCl}$$

You have more NaCl than this: 60.0 g, to be exact. So NaCl is present in excess. Now check H_2SO_4. The amount needed is

$$0.0686 \text{ mol Sb}_2O_3 \times \frac{3 \text{ mol H}_2SO_4}{1 \text{ mol Sb}_2O_3} \times \frac{98.08 \text{ g}}{\text{mol H}_2SO_4} = 20.2 \text{ g H}_2SO_4$$

You have 150 g of a 50% by weight H_2SO_4 solution, i.e., 150 g $\times$ 50/100 = 75 g H_2SO_4; again, a large excess. So your first impressions were justified: Sb_2O_3 is the limiting reagent. The maximum yield of $SbCl_3$ is

$$0.0686 \text{ mol Sb}_2O_3 \times \frac{2 \text{ mol SbCl}_3}{1 \text{ mol Sb}_2O_3} \times \frac{228 \text{ g}}{\text{mol SbCl}_3} = 31.3 \text{ g}$$

PROBLEM 4-17   When mercury (Hg) is heated with excess sulfur, the product is mercuric sulfide (HgS). But because mercury is volatile, a little of it always escapes during the heating. In an experiment of this kind, 15.00 g of Hg yielded 17.08 g of HgS. What was the percentage yield in this experiment?

Solution: Percentage yield is (actual yield/theoretical yield) $\times$ 100%. So first the *theoretical yield* must be calculated:

$$\frac{15.00 \text{ g Hg}}{200.6 \text{ g/mol Hg}} \times \frac{1 \text{ mol HgS}}{1 \text{ mol Hg}} \times \frac{232.7 \text{ g HgS}}{1 \text{ mol HgS}} = 17.40 \text{ g HgS}$$

The percentage yield is

$$\frac{17.08 \text{ g HgS}}{17.40 \text{ g HgS}} \times 100\% = 98.16\% \quad \text{(4 sig. figs)}$$

PROBLEM 4-18   In a blast furnace, iron oxide (Fe_2O_3) reacts with carbon to give metallic iron (Fe) and carbon monoxide (CO). If the blast furnace reaction gives a 94% yield, how much iron would be obtained from 3.95×10^5 kg of Fe_2O_3?

Solution: Since percentage yield is (actual yield/theoretical yield) × 100%, you can rearrange this to show that actual yield = (theoretical yield) × (% yield/100%). The balanced equation is

$$\begin{array}{ccccccc}
Fe_2O_3 & + & 3C & \longrightarrow & 2Fe & + & 3CO \\
1 \text{ mol} & & & & 2 \text{ mol} & & - \\
159.7 \text{ g/mol} & & & & 55.85 \text{ g/mol} & & -
\end{array}$$

The theoretical yield of Fe is

$$\frac{3.95 \times 10^5 \text{ kg } Fe_2O_3}{159.7 \text{ g/mol } Fe_2O_3} \times \frac{2 \text{ mol Fe}}{1 \text{ mol } Fe_2O_3} \times \frac{55.85 \text{ g}}{1 \text{ mol Fe}} = 2.76 \times 10^5 \text{ kg Fe}$$

and the actual yield is

$$2.76 \times 10^5 \text{ kg Fe} \times \frac{94\%}{100\%} = 2.6 \times 10^5 \text{ kg Fe} \quad \text{(2 sig. figs.)}$$

PROBLEM 4-19   A batch of potassium permanganate ($KMnO_4$) was accidentally contaminated with potassium chloride (KCl). A 0.586-g sample of the impure $KMnO_4$ was dissolved in base and treated with oxalic acid. The Mn in the sample was quantitatively (completely) converted into manganese dioxide (MnO_2), of which 0.297 g was obtained. What percentage of the sample was $KMnO_4$?

Solution: Since

$$\begin{array}{ccc}
KMnO_4 & \Longrightarrow & MnO_2 \\
1 \text{ mol} & & 1 \text{ mol} \qquad \text{(because Mn is conserved)} \\
158.04 \text{ g} & & 86.94 \text{ g}
\end{array}$$

the mass of $KMnO_4$ in the sample was

$$\frac{0.297 \text{ g } MnO_2}{86.94 \text{ g/mol } MnO_2} \times \frac{1 \text{ mol } KMnO_4}{1 \text{ mol } MnO_2} \times \frac{158.04 \text{ g } KMnO_4}{1 \text{ mol } KMnO_4} = 0.540 \text{ g}$$

So the percentage of $KMnO_4$ in the sample was

$$\frac{0.540 \text{ g } KMnO_4}{0.586 \text{ g sample}} \times 100\% = 92.1\%$$

PROBLEM 4-20   Salicylic acid is converted into aspirin by the following reaction:

$$\underset{\text{salicylic acid}}{HOC_6H_4CO_2H} + \underset{\text{acetic anhydride}}{(CH_3CO)_2O} \qquad \underset{\text{aspirin}}{CH_3COOC_6H_4CO_2H} + \underset{\text{acetic acid}}{CH_3CO_2H}$$

In a particular laboratory preparation, 7.50 g of salicylic acid was treated with 10.0 g of acetic anhydride, and 8.85 g of aspirin was obtained. Determine the limiting reagent and the percentage yield of aspirin.

Solution: Combine the atoms in the formulas given and follow the usual procedure:

$$\begin{array}{ccccccc}
C_7H_6O_3 & + & C_4H_6O_3 & \longrightarrow & C_9H_8O_4 & + & CH_3CO_2H \\
1 \text{ mol} & & 1 \text{ mol} & & 1 \text{ mol} & & - \\
138.1 \text{ g/mol} & & 102.1 \text{ g/mol} & & 180.2 \text{ g/mol} & & -
\end{array}$$

For 10.0 g of acetic anhydride, the mass of salicylic acid needed is

$$\frac{10.0 \text{ g } C_4H_6O_3}{102.1 \text{ g/mol } C_4H_6O_3} \times \frac{1 \text{ mol } C_7H_6O_3}{1 \text{ mol } C_4H_6O_3} \times \frac{138.1 \text{ g}}{\text{mol } C_7H_6O_3} = 13.5 \text{ g}$$

Since the preparation specifies only 7.50 g, salicylic acid must be the limiting reagent. The theoretical yield of aspirin must therefore be based on the limiting reagent, $C_7H_6O_3$, which gives a theoretical yield of

$$\frac{7.50 \text{ g C}_7\text{H}_6\text{O}_3}{138.1 \text{ g/mol C}_7\text{H}_6\text{O}_3} \times \frac{1 \text{ mol C}_9\text{H}_8\text{O}_4}{1 \text{ mol C}_7\text{H}_6\text{O}_3} \times \frac{180.2 \text{ g}}{\text{mol C}_9\text{H}_8\text{O}_4} = 9.79 \text{ g}$$

So the percentage yield is

$$\frac{\text{actual}}{\text{theoretical}} \times 100\% = \frac{8.85 \text{ g}}{9.79 \text{ g}} \times 100\% = 90.4\%$$

PROBLEM 4-21   Copper reacts with nitric acid (HNO_3) as follows:

$$8HNO_3 + 3Cu \longrightarrow 2NO + 3Cu(NO_3)_2 + 4H_2O$$

What is the minimum number of grams of Cu required to make 0.0372 mol of NO? (a) 2.36 g, (b) 1.67 g, (c) 1.11 g, (d) 0.0558 g, (e) 3.55 g.

Solution: The equation is already balanced. All you need to do is use it to determine how many moles of Cu are required:

$$\frac{(0.0372 \text{ mol NO})(3 \text{ mole Cu})}{(2 \text{ mol NO})} = 0.0558 \text{ mol Cu}$$

Now convert moles of Cu to grams:

$$(0.0558 \text{ mol Cu})(63.54 \text{ g/mol}) = 3.55 \text{ g Cu}$$

The correct answer is (e), 3.55 g. If you do not multiply moles of NO by $\frac{3}{2}$, you obtain answer (a). Answer (c) is the *mass* of NO: Note that you do not need to know the mass of NO. Answer (b) is the mass of NO $\times \frac{3}{2}$. The number of *moles*, not mass, should be multiplied by $\frac{3}{2}$. Answer (d) is moles of Cu. You must remember to complete the problem.

PROBLEM 4-22   Hydrogen fluoride (HF) is made from the reaction of CaF_2 (fluorspar) and excess H_2SO_4:

$$CaF_2 + H_2SO_4 \longrightarrow 2HF + CaSO_4$$

When 1.64×10^3 g of CaF_2 was used in this reaction, the yield of HF was 693 g. What was the percentage yield? (a) 82.5%, (b) 60.6%, (c) 42.3%, (d) 84.5%, (e) 51.2%.

Solution: The equation is already balanced. Since there is excess H_2SO_4, the limiting reagent is CaF_2. The stoichiometric yield is

$$\frac{1.64 \times 10^3 \text{ g CaF}_2}{78.1 \text{ g CaF}_2/\text{mol CaF}_2} \times \frac{2 \text{ mol HF}}{1 \text{ mol Ca}_2} \times \frac{20.01 \text{ g HF}}{\text{mol HF}} = 840 \text{ g HF}$$

$$\% \text{ yield} = \frac{\text{actual yield}}{\text{stoichiometric yield}} \times 100\% = \frac{693 \text{ g}}{840 \text{ g}} \times 100\% = 82.5\%$$

The correct answer is (a), 82.5%. If you do not multiply moles of CaF_2 by 2 and then invert (actual yield)/(stoichiometric yield), you obtain answer (b). Answer (c) is the (actual yield)/(mass of CaF_2 used) $\times$ 100%. Answer (d) is wrong answer (c) multiplied by 2. Answer (e) is 2(molar mass HF)/(molar mass CaF_2) $\times$ 100%.

Supplementary Exercises

PROBLEM 4-23   For each of the following reactions calculate how many grams of the second reactant react with exactly 1 mol of the first reactant. Give your answer to four significant figures.

(a) $2Sb + 3F_2 \longrightarrow 2SbF_3$
(b) $O_2 + 2Mg \longrightarrow 2MgO$
(c) $C_3H_6 + Cl_2 \longrightarrow C_3H_6Cl_2$

Answer: (a) 57.00 g F_2 (b) 48.62 g Mg   (c) 70.91 g Cl_2

PROBLEM 4-24   What mass of antimony trifluoride (SbF_3) is produced when 25.00 g of fluorine (F_2) is treated with excess antimony?

Answer: 78.39 g SbF_3

PROBLEM 4-25   How many grams of O_2 can be produced by heating 100.00 g of barium peroxide (BaO_2)? The other product of the reaction is barium oxide (BaO).

Answer: 9.4490 g O_2

PROBLEM 4-26   Sulfuric acid (H_2SO_4) reacts with sodium chloride (NaCl) to produce hydrogen chloride gas (HCl) and sodium sulfate (Na_2SO_4). What mass of H_2SO_4 is needed to react completely with 15.0 g NaCl; what mass of HCl will be produced?

Answer: 12.6 g H_2SO_4; 9.36 g HCl

PROBLEM 4-27   In a blast furnace, iron oxide (Fe_2O_3) reacts with carbon to produce metallic iron (Fe) and carbon monoxide (CO). How many kilograms of C and Fe_2O_3 will be needed each hour if the smelter wants to produce 2.75×10^4 kg of Fe a day? What is the daily production of CO in this process?

Answer: 3.70×10^2 kg C and 1.64×10^3 kg Fe_2O_3 each hour; 2.07×10^4 kg CO daily.

PROBLEM 4-28   A 0.625-g sample of a bromide of iron (containing Fe and Br only) was added to water, and then excess silver nitrate was added. All the Br^- present in the sample was recovered as 1.191 g of AgBr. What is the empirical formula of the iron bromide?

Answer: $FeBr_3$

PROBLEM 4-29   A 0.500-g sample of hydrated copper nitrate $Cu(NO_3)_2 \cdot 4H_2O$ was heated and decomposed, leaving a residue of copper oxide (CuO). All the copper in the nitrate was retained in the residue. What mass of CuO was obtained?

Answer: 0.153 g CuO

PROBLEM 4-30   A 1.000-g sample of a hydrate of sodium sulfate $Na_2SO_4 \cdot xH_2O$ was heated until all the H_2O was driven off and 0.441 g of anhydrous Na_2SO_4 remained. What is the formula of the original hydrate?

Answer: $Na_2SO_4 \cdot 10H_2O$

PROBLEM 4-31   A sample of a hydrocarbon (containing C and H only) weighing 0.153 g was burned in excess O_2 and yielded 0.237 g of H_2O and 0.464 g of CO_2. Determine the empirical formula of the hydrocarbon.

Answer: C_2H_5

PROBLEM 4-32   A mixture of $PtCl_2$ and $PtCl_4$ weighing 1.549 g was treated with hydrogen gas to give 1.071 g of pure metallic Pt. What percentage of $PtCl_2$ was present in the mixture?

Answer: 73% $PtCl_2$

PROBLEM 4-33   When a 7.626-mg sample of a compound containing only C, H, and O was burned in excess O_2, it yielded 13.59 mg of CO_2 and 5.564 mg of H_2O. Determine the empirical formula of the compound.

Answer: $C_3H_6O_2$

PROBLEM 4-34   A 1.244-g sample of a mixture of NaBr and $CdBr_2$ is treated with an excess of $AgNO_3$ solution, which precipitates all the Br^- as AgBr. A mass of 1.910 g of AgBr is obtained. What is the percentage by mass of $CdBr_2$ in the sample?

Answer: 65% $CdBr_2$

PROBLEM 4-35 In a synthesis of aluminum chloride ($AlCl_3$) 25.0 g of Al was treated with 60.0 g of Cl_2. Determine the limiting reagent and the maximum yield of $AlCl_3$.

Answer: Cl_2 limiting; 75.2 g of $AlCl_3$

PROBLEM 4-36 Sodium perborate ($NaBO_3$) is made from borax ($Na_2B_4O_7$) by the following reaction:

$$Na_2B_4O_7 + 2NaOH + 4H_2O_2 \longrightarrow 4NaBO_3 + 5H_2O$$

In a laboratory preparation, 10.0 g of $Na_2B_4O_7$ is treated with 5.00 g of NaOH and 60.0 g of 20% H_2O_2 solution. What is the maximum yield of $NaBO_3$ that can be expected from these quantities of reactants?

Answer: 16.3 g $NaBO_3$

PROBLEM 4-37 When potassium iodide (KI) is heated with H_2SO_4, its iodide is converted into iodine (I_2). In a particular experiment 7.50 g of KI was heated with excess H_2SO_4 and 5.05 g of I_2 was collected. What was the percentage yield of I_2 in this experiment?

Answer: 88%

PROBLEM 4-38 Aluminum oxide (Al_2O_3) is reduced by the action of electric current to metallic aluminum Al. The yield of the process is 97.5%. How much aluminum is obtained from 8.00×10^4 kg of Al_2O_3?

Answer: 4.13×10^4 kg of Al

PROBLEM 4-39 A sample of potassium dichromate ($K_2Cr_2O_7$) was accidentally contaminated with potassium sulfate (K_2SO_4). When 0.443 g of the impure dichromate was dissolved in base, its chromium was quantitatively converted into Cr_2O_3, of which 0.221 g was obtained. What percentage of the sample was $K_2Cr_2O_7$?

Answer: 96.6%

PROBLEM 4-40 Phenol (C_6H_5OH) is converted into phenyl benzoate ($C_6H_5OCOC_6H_5$) by the following reaction:

$$\underset{\text{phenol}}{C_6H_5OH} + \underset{\substack{\text{benzoyl} \\ \text{chloride}}}{C_6H_5COCl} + \underset{\substack{\text{sodium} \\ \text{hydroxide}}}{NaOH} \longrightarrow \underset{\substack{\text{phenyl} \\ \text{benzoate}}}{C_6H_5OCOC_6H_5} + NaCl$$

In a particular preparation 6.00 g of phenol was treated with 10.0 g of benzoyl chloride and 200 g of 10% by weight NaOH solution, and 10.55 g of phenyl benzoate was obtained. Determine the limiting reagent and the percentage yield of phenyl benzoate.

Answer: Phenol limiting; 83.5% yield

PROBLEM 4-41 The following reaction proceeds as shown:

$$Fe_2O_3 + 3H_2 \longrightarrow 2Fe + 3H_2O$$

A mixture of 10.0 g Fe_2O_3 and 1.00 g H_2 is allowed to react according to this equation. Determine the limiting reagent and the mass of Fe produced.

Answer: Fe_2O_3 limiting; 6.99 g Fe

PROBLEM 4-42 Sulfur dioxide (SO_2) can be made by heating FeS_2 (pyrite or "fool's gold") in air according to the following reaction:

$$3FeS_2 + 8O_2 \longrightarrow Fe_3O_4 + 6SO_2$$

What mass of SO_2 can be made from 10.0 kg of FeS_2?

Answer: 10.7 kg

PROBLEM 4-43   A 0.133-g sample of metallic Zr was treated with gaseous F_2 to give 0.245 g of a solid zirconium fluoride. Write a balanced equation for the reaction involved.

Answer: $Zr + 2F_2 \longrightarrow ZrF_4$

PROBLEM 4-44   A 0.611-g sample of a hydrate of barium chloride ($BaCl_2 \cdot xH_2O$) was heated to drive off all the bound water, and 0.521 g of anhydrous $BaCl_2$ was produced. What is the value of x?

Answer: $x = 2.0$

PROBLEM 4-45   A 0.250-g sample of a metal sulfate of formula M_2SO_4 was treated with excess $BaCl_2$ solution and gave 0.161 g of $BaSO_4$. What is the identity of M?

Answer: M is Cs

PROBLEM 4-46   Silica (SiO_2) reacts with HF to give H_2O and a compound of molar mass 104 g/mol containing 73.0% F and 27.0% Si. Determine the molecular formula of the compound produced, and a balanced equation for the reaction.

Answer: SiF_4; $SiO_2 + 4HF \longrightarrow SiF_4 + 2H_2O$

PROBLEM 4-47   When Al_2O_3 is reduced in an electrolytic cell, the products are metallic aluminum and carbon monoxide (from the carbon electrodes). In a commercial cell that consumes 68.0 kg of Al_2O_3 per hour, the yield of aluminum is 33.5 kg per hour. What is the percentage yield in the process?

Answer: 93.1%

PROBLEM 4-48   The iron in a sample of Mohr's salt $Fe(NH_4)_2(SO_4)_2 \cdot 6H_2O$ was converted quantitatively to Fe_2O_3. A 0.3794-g sample of the salt gave 0.0636 g of Fe_2O_3. Calculate the percentage purity of the salt sample.

Answer: 82.3%

5 GASES

THIS CHAPTER IS ABOUT

- ☑ **General Properties of Gases**
- ☑ **The Gas Laws**
- ☑ **Kinetic Theory of Gases**
- ☑ **Real Gases**

5-1. General Properties of Gases

A. Definition of a gas

A **gas** is a particular state of matter in which the particles (molecules or atoms) that make up the sample of matter are small compared with the distances between them. In a gas, the particles are in constant motion, bombarding the walls of the container. This bombardment gives rise to the pressure of a gas.

B. State of a gas

The **state** of a gas is specified by its *volume, pressure, temperature*, and *composition*. Before you can use these quantities, you have to know what they are and how they are expressed or measured.

1. Volume

A gas expands uniformly to fill the whole of any container. Thus the **volume** of a gas equals the volume of its container.

2. Pressure

Pressure is the force exerted by the gas on each unit area of surface. In SI, the unit of pressure is a derived unit called the **pascal** (Pa), which is measured in newtons per square meter (N/m^2 or $N\,m^{-2}$).

EXAMPLE 5-1: Express the pascal in the basic SI units of kilograms, meters, and seconds.

Solution: Pressure (in Pa) = force/area = N/m^2. You know that $N = kg\,m\,s^{-2}$. Thus

$$Pa = \frac{kg\,m\,s^{-2}}{m^2} = kg\,m^{-1}\,s^{-2}$$

Non-SI units of pressure still used in chemistry are the **standard atmosphere** (atm), now defined as precisely 101 325 Pa, and the **torr** (sometimes called the millimeter of mercury, mm Hg), defined as $101\,325/760 = 133.3$ Pa (1 atm = exactly 760 torr).

To convert one pressure unit into another, you can use conversion factors:

	Pressure conversion factors	
1 Pa	$= 7.50 \times 10^{-3}$ torr	$= 9.87 \times 10^{-6}$ atm
1 atm	$= 101\,325$ Pa	$= 760$ torr
1 torr (mm Hg)	$= 133.3$ Pa	$= 1.316 \times 10^{-3}$ atm

EXAMPLE 5-2:

(1) Express a pressure of 5.00×10^4 Pa in (a) atm and (b) torr.

Solution:

(a)
$$(5.00 \times 10^4 \text{ Pa}) \times \frac{1 \text{ atm}}{1.01325 \times 10^5 \text{ Pa}} = \frac{5.00 \times 10^4}{1.01325 \times 10^5} = 0.493 \text{ atm}$$

(b)
$$(5.00 \times 10^4 \text{ Pa}) \times \frac{7.6 \times 10^2 \text{ torr}}{1.01325 \times 10^5 \text{ Pa}} = \frac{(5.0 \times 7.6) \times 10^6}{1.01325 \times 10^5} = 375 \text{ torr}$$

(2) Express a pressure of 28.3 torr in Pa.

Solution:

$$28.3 \text{ torr} \times \frac{1.01325 \times 10^5 \text{ Pa}}{7.60 \times 10^2 \text{ torr}} = \frac{(28.3 \times 1.01325) \times 10^5}{7.60 \times 10^2} = 3.77 \times 10^3 \text{ Pa} \qquad \text{(or 3.77 kPa)}$$

3. Temperature

Temperature is a measure of kinetic energy. The SI unit for temperature is the kelvin degree (K). (Note that the degree sign is not used with this unit.) On the Kelvin scale, the point at which all molecules possess zero kinetic energy is termed **absolute zero** (0 K).

Water freezes at 273.15 K and boils at 373.15 K (the decimal part of these numbers is often deleted in calculations). Another commonly used temperature scale is the Celsius scale (°C), which defines the freezing point of water as 0°C and its boiling point as 100°C. Since both scales contain 100 degrees between the freezing and boiling points, they can be interconverted as follows:

TEMPERATURE CONVERSION
$$K = °C + 273.15$$
$$°C = K - 273.15$$
(5-1)

EXAMPLE 5-3: Express the following Celsius temperatures in kelvin degrees: (a) 993°C, (b) 29°C, (c) −93°C.

Solution: Rounding off the conversion factor to 273, we get

(a) $K = 993°C + 273 = 1266$ K
(b) $K = 29°C + 273 = 302$ K
(c) $K = -93°C + 273 = 180$ K

4. Composition

The **composition** of a gas is described in terms of the relative amounts of each gas present in a mixture. In a mixture of gases (in which the gases do not react with one another) the amounts of the constituent gases can be expressed in several ways:

(a) **number of moles** n of each gas
(b) **mole fraction** of each gas X_i: the number of moles of one of the gases in a mixture divided by the total number of moles of all gases in the mixture. Thus, if we have n_A moles of gas A, n_B moles of gas B, and n_C moles of gas C, the mole fraction of gas A (X_A) is computed as

MOLE FRACTION
$$X_A = \frac{n_A}{n_A + n_B + n_C}$$
(5-2)

EXAMPLE 5-4: A scuba tank is charged with a mixture of 1.0×10^3 mol of oxygen (O_2) and 4.5×10^3 mol of helium (He). Determine the mole fraction of each gas in the mixture.

Solution: Given that

$$n_{He} = 4.5 \times 10^3 \text{ mol} \quad \text{and} \quad n_{O_2} = 1.0 \times 10^3 \text{ mol}$$

we calculate the mole fraction as

$$X_{He} = \frac{4.5 \times 10^3 \text{ mol}}{(4.5 \times 10^3 \text{ mol}) + (1.0 \times 10^3 \text{ mol})} = \frac{4.5 \times 10^3}{5.5 \times 10^3} = 0.82$$

$$X_{O_2} = \frac{1.0 \times 10^3 \text{ mol}}{(4.5 \times 10^3 \text{ mol}) + (1.0 \times 10^3 \text{ mol})} = \frac{1.0 \times 10^3}{5.5 \times 10^3} = 0.18$$

note: $X_{O_2} + X_{He} = 0.18 + 0.82 = 1.00$. The sum of all the mole fractions of all the components of a mixture must be 1.

(c) **partial pressure** of each gas, p_i: the pressure a particular gas would exert if it were the only gas present in the volume occupied by the mixture. **Dalton's law of partial pressures** states that in a gas mixture, the total pressure is the sum of the partial pressures:

DALTON'S LAW OF PARTIAL PRESSURES $\qquad P_{total} = p_1 + p_2 + p_3 + \cdots + p_n \qquad$ (5-3)

EXAMPLE 5-5: The mole fractions of the three most abundant gases in the atmosphere are nitrogen (N_2), 0.781; oxygen (O_2), 0.209; and argon (Ar), 0.009. Determine the partial pressure, in torr, of each gas in an average sample of air collected at sea level.

Solution: The average atmospheric pressure at sea level is 1 atm (760 torr), so

$$p_{N_2} = 0.781 \times 760 \text{ torr} = 594 \text{ torr}$$
$$p_{O_2} = 0.209 \times 760 \text{ torr} = 159 \text{ torr}$$
$$p_{Ar} = 0.009 \times 760 \text{ torr} = 7 \text{ torr}$$

C. Standard temperature and pressure (STP)

To compare the quantities of gases whose volumes and densities vary with pressure and temperature, the conditions under which the measurements are made must be stated. Conditions of exactly 1 atm and 0°C (273 K) are termed **standard temperature and pressure (STP)** for gases.

5-2. The Gas Laws

A. Boyle's law: Pressure and volume

Boyle's law states that for a given mass of gas, whether pure or a mixture, at constant temperature, the volume occupied by the gas is inversely proportional to the pressure applied to the gas. This relationship may be expressed as $V = (1/P) \times$ constant or $PV =$ constant. A more useful form for comparing changes in volume and pressure is given by

BOYLE'S LAW $\qquad\qquad P_1 V_1 = P_2 V_2 \qquad\qquad$ (5-4)

where the subscripts 1 and 2 refer, respectively, to the initial and final states of the gas.

EXAMPLE 5-6: A sample of neon (Ne) occupies 4.00 L at a pressure of 5.00×10^4 Pa and a temperature of 273 K. Determine the volume of the sample at STP.

Solution: The initial conditions are

$$P_1 = 5.00 \times 10^4 \text{ Pa}, \qquad V_1 = 4.00 \text{ L}$$

The final conditions are

$$P_2 = 101\,325 \text{ Pa}, \qquad V_2 = \text{unknown}$$

Rearrange $P_1 V_1 = P_2 V_2$ to solve for V_2:

$$V_2 = \frac{P_1 V_1}{P_2}$$

$$= \frac{(5.00 \times 10^4 \text{ Pa})(4.00 \text{ L})}{(1.01325 \times 10^5 \text{ Pa})} = \frac{2.00 \times 10^5}{1.01325 \times 10^5} \text{ L}$$

$$= 1.97 \text{ L}$$

B. Charles' law: Volume and temperature

Charles' law states that for a given mass of gas, whether pure or a mixture, at constant pressure, the volume occupied by the gas is directly proportional to its *absolute* (Kelvin) temperature. Stated mathematically, $V = T \times$ constant or $V/T =$ constant. The more useful form, using the same symbols as in Section 5-2A, is

CHARLES' LAW $\qquad\qquad \dfrac{V_1}{T_1} = \dfrac{V_2}{T_2}$ (5-5)

EXAMPLE 5-7: At STP, a sample of nitrogen occupies 40 mL. Find its volume at 27°C at a pressure of 760 torr.

Solution: The initial conditions are

$$V_1 = 40 \text{ mL}, \qquad T_1 = 273 \text{ K}$$

The final conditions are

$$V_2 = \text{unknown}, \qquad T_2 = (27°C + 273) = 300 \text{ K}$$

(Note that the pressure remains constant and that absolute temperature must be used.)
Rearrange $V_1/T_1 = V_2/T_2$ to solve for V_2:

$$V_2 = \frac{V_1 \times T_2}{T_1}$$

$$= \frac{(40 \text{ mL})(300 \text{ K})}{273 \text{ K}} = \frac{12 \times 10^3}{2.73 \times 10^2} \text{ mL}$$

$$= 44 \text{ mL}$$

C. Combined gas law: Pressure, volume, and temperature

Boyle's and Charles' laws may be combined algebraically to yield the combined gas law. For a given mass of gas,

COMBINED GAS LAW $\qquad \dfrac{P_1 V_1}{T_1} = \dfrac{P_2 V_2}{T_2} =$ constant (5-6)

Temperature must be stated in degrees kelvin.

EXAMPLE 5-8: A 2.0-L sample of carbon dioxide (CO_2) at 77°C and 2500 torr is heated to 123°C at a pressure of 1500 torr. Calculate the volume of the sample at the new conditions.

Solution: The initial conditions are

$$V_1 = 2.0 \text{ L}, \qquad P_1 = 2500 \text{ torr}, \qquad T_1 = 350 \text{ K}$$

The final conditions are

$$V_2 = \text{unknown}, \qquad P_2 = 1500 \text{ torr}, \qquad T_2 = 396 \text{ K}$$

Noting that the temperature must be expressed in kelvins, we solve the combined gas law for V_2:

$$V_2 = V_1 \left(\frac{T_2}{T_1}\right)\left(\frac{P_1}{P_2}\right)$$

$$= (2.0 \text{ L})\left(\frac{396 \text{ K}}{350 \text{ K}}\right)\left(\frac{2500 \text{ torr}}{1500 \text{ torr}}\right) = \frac{2.0(396)(2.5 \times 10^3)}{(3.5 \times 10^2)(1.5 \times 10^3)} \text{ L}$$

$$= 3.8 \text{ L}$$

D. More gas laws: Avogadro's and others

1. Avogadro's law: Equal volumes

Avogadro's law states that at the same pressure and temperature, equal volumes of gases contain equal numbers of molecules (or atoms for the noble gases). In other words, under the same conditions of temperature and pressure, the volume of a gas is directly proportional to the number of molecules of the gas. Stated mathematically,

AVOGADRO'S LAW $\qquad V = n \times \text{constant} \qquad$ or $\qquad \dfrac{V_1}{n_1} = \dfrac{V_2}{n_2}$ $\qquad$ (5-7)

where the subscripts 1 and 2 refer to the two gases and n represents the number of moles of each gas. One mole of any gas contains 6.02×10^{23} gas particles. This quantity is called **Avogadro's number**.

EXAMPLE 5-9: What volume of O_2, measured at 1.86×10^6 Pa and 375 K, is needed to react completely with 1.78 L of H_2, measured at the same pressure and temperature, to give H_2O?

Solution: First write a balanced equation for the reaction:

$$2H_2 + O_2 \longrightarrow 2H_2O$$
$$\text{2 mol} \quad \text{1 mol} \qquad —$$

You see from the equation that the molar amount of O_2 needed is one-half the molar amount of H_2. By Avogadro's law n is proportional to V at the same pressure and temperature. Therefore the volume of O_2 needed is

$$1.78 \text{ L } H_2\left(\frac{1 \text{ L } O_2}{2 \text{ L } H_2}\right) = 0.890 \text{ L } O_2$$

Note that the pressure and temperature values stated in the problem were not needed in the calculation.

---

2. Molar volume

It follows from Avogadro's law that the volume of a mole of any ideal gas at STP occupies the same volume as a mole of any other ideal gas at STP. For ideal gases, this molar volume is 22.4 L at STP.

note: Real gases don't necessarily exhibit ideal behavior, so the standard molar volume of 22.4 L must be considered an *average*.

EXAMPLE 5-10: How many molecules of O_2 are present in 1.00 L of O_2 at STP?

Solution: You have 1 mol of O_2 in the standard volume of 22.4 L, so

$$\frac{1.00 \text{ L (STP)}}{22.4 \text{ L mol}^{-1} \text{ (STP)}} = 0.0446 \text{ mol } O_2$$

Since the number of O_2 molecules per mole must equal Avogadro's number, the number of O_2 molecules in 0.0446 mol is

$$(6.02 \times 10^{23} \text{ molecules mol}^{-1})(0.0446 \text{ mol}) = 2.69 \times 10^{22} \text{ molecules}$$

---

3. Density

Avogadro's law can be used to determine **density** (d), the mass per unit volume of a gas (SI units: kg/m^3 or g/L). Gas density is inversely proportional to volume, so

DENSITY $$d = \frac{1}{V} \times \text{constant} \qquad \text{or} \qquad \frac{d_1}{d_2} = \frac{V_2}{V_1} \qquad \text{(5-8)}$$

Stated in terms of the combined gas law (Eq. 5-6), density proportionalities can be expressed as

$$\frac{d_1 T_1}{P_1} = \frac{d_2 T_2}{P_2}$$

or

$$d_2 = d_1 \left(\frac{T_1}{T_2}\right)\left(\frac{P_2}{P_1}\right) \qquad \text{(5-9)}$$

Note that since density is inversely proportional to volume, the temperature and pressure terms are inverted.

EXAMPLE 5-11: The density of oxygen at STP is 1.43 g/L. Calculate the density of oxygen at 400 torr and 30°C.

Solution: The initial conditions are

$$d_1 = 1.43 \text{ g/L}, \qquad T_1 = 273 \text{ K}, \qquad P_1 = 760 \text{ torr}$$

The final conditions are

$$d_2 = \text{unknown}, \qquad T_2 = 303 \text{ K}, \qquad P_2 = 400 \text{ torr}$$

So

$$d_2 = d_1 \left(\frac{T_1}{T_2}\right)\left(\frac{P_2}{P_1}\right)$$

$$= 1.43 \left(\frac{273}{303}\right)\left(\frac{400}{760}\right) \text{ g/L}$$

$$= 0.678 \text{ g/L}$$

E. Ideal gas law

Using the proportionalities contained in Boyle's, Charles', and Avogadro's laws,

$$V = \text{constant} \times \frac{1}{P} \qquad \text{(fixed mass and temperature)}$$

$$V = \text{constant} \times T \qquad \text{(fixed mass and pressure)}$$

$$V = \text{constant} \times n \qquad \text{(fixed pressure and temperature)}$$

we can derive a single equation containing all the terms and a combined constant: $V = \text{constant} \times (1/P) \times T \times n$. In its more common form this single equation is known as the **ideal gas law**:

IDEAL GAS LAW $$PV = nRT \qquad \text{(5-10)}$$

where R is the **ideal** (or **universal**) **gas constant**.

EXAMPLE 5-12: Calculate the value of R and determine its units, using liters, atmospheres, and degrees kelvin.

Solution: Rearranging the ideal gas law to isolate R, you get

$$R = \frac{PV}{nT}$$

You know that 1 mol of any ideal gas occupies 22.4 L at 1 atm and 273 K. Solving for R,

$$R = \frac{(1 \text{ atm})(22.4 \text{ L})}{(1 \text{ mol})(273 \text{ K})}$$

$$= 0.0821 \text{ L atm/mol K}$$

The value of R depends on the unit system chosen. Commonly used values for R include

$$R = 0.0821 \text{ L atm/mol K} \quad \text{(or L atm mol}^{-1}\text{K}^{-1}\text{)}$$
$$R = 8.314 \text{ J/mol K} \quad \text{(or J mol}^{-1}\text{K}^{-1}\text{)}$$
$$R = 62.36 \text{ L torr/mol K} \quad \text{(or L torr mol}^{-1}\text{K}^{-1}\text{)}$$

EXAMPLE 5-13: A sample of gas weighs 1.25 g. At 28°C its volume is 2.50×10^2 mL and its pressure is 715 torr. What is the molar mass?

Solution: Remembering to use the correct units for R and to convert °C to K, set up the known conditions:

$$P = 715 \text{ torr}$$
$$V = 2.50 \times 10^2 \text{ mL} = 0.250 \text{ L}$$
$$R = 62.36 \text{ L torr/mol K}$$
$$T = 28 + 273 = 301 \text{ K}$$

Rearrange the ideal gas law to isolate n:

$$n = \frac{PV}{RT} = \frac{(715 \text{ torr})(0.250 \text{ L})}{(62.36 \text{ L torr/mol K})(301 \text{ K})} = 9.52 \times 10^{-3} \text{ mol}$$

Since the molar mass (M) is the mass (m) divided by the number of moles (n), you get

$$\text{molar mass} = M = \frac{m}{n} = \frac{1.25 \text{ g}}{9.52 \times 10^{-3} \text{ mol}} = 131 \text{ g/mol}$$

EXAMPLE 5-14: What is the density of chloromethane gas (CH_3Cl) at 20°C and 0.973 atm pressure? The molecular mass of chloromethane is 50.5 amu.

Solution: The molar mass of CH_3Cl is 50.5 g mol^{-1}. Using $m/M = n$, rewrite the ideal gas law:

$$PV = \frac{m}{M} RT$$

Rearrange the equation to give density $d = m/V = PM/RT$ and substitute the correct numerical values for P, M, R, and T ($T = 20 + 273 = 293$ K):

$$d = \frac{PM}{RT} = \frac{(0.973 \text{ atm})(50.5 \text{ g mol}^{-1})}{(0.0821 \text{ L atm mol}^{-1} \text{ K}^{-1})(293 \text{ K})} = 2.04 \text{ g L}^{-1}$$

(You could convert this answer to 2.04×10^{-3} g/cm^3, but the densities of gases are usually given in grams per liter.)

EXAMPLE 5-15: Calculate the partial pressures (p) of each component and the total pressure (P) of a mixture of 2.75 mol of carbon dioxide (CO_2) and 1.62 mol of carbon monoxide (CO) in a 5.00-L cylinder at 302 K (see Eq. 5-3).

Solution: For each gas the partial pressure is $p = nRT/V$, so

$$p_{CO_2} = (2.75 \text{ mol})(0.0821 \text{ L atm K}^{-1} \text{ mol}^{-1}) \frac{302 \text{ K}}{5.00 \text{ L}} = 13.6 \text{ atm}$$

$$p_{CO} = (1.62 \text{ mol})(0.0821 \text{ L atm K}^{-1} \text{ mol}^{-1}) \frac{302 \text{ K}}{5.00 \text{ L}} = 8.03 \text{ atm}$$

$$P = p_{CO_2} + p_{CO} = (13.6 + 8.03) \text{ atm} = 21.6 \text{ atm}$$

5-3. Kinetic Theory of Gases

The behavior of gases can be predicted by using the postulates of the **kinetic theory of gases**. The theory states that the molecules of an ideal gas

- are in continuous random motion in a straight line, occupy negligible volume, and have point masses;
- are perfectly elastic; i.e., no loss in the total kinetic energy of the molecules is observed during collisions with surfaces or other molecules, and no forces of repulsion or attraction occur;
- have an average kinetic energy independent of the type of gas and directly proportional to the temperature (in kelvins).

A. Kinetic theory and the ideal gas law

1. Kinetic energy

The average kinetic energy of a gas can be calculated by writing the ideal gas law in the following form:

KINETIC ENERGY RELATIONSHIPS
$$E_k = \frac{3}{2} RT \quad \text{or} \quad \bar{\varepsilon} = \frac{3}{2} \frac{R}{N_A} T = \frac{3}{2} kT \quad \text{(5-11)}$$

where E_k is the sum of the kinetic energies of one mole of ideal gas molecules, $\bar{\varepsilon}$ is the average kinetic energy of one gas molecule, N_A is Avogadro's number, and $k = R/N_A$ is **Boltzmann's constant**. It is important to remember that the average kinetic energy is independent of the identity of the gas; it depends only on the absolute temperature.

2. Molecular speed

Another important relationship that can be derived from kinetic theory involves molecular speed (v) and molecular mass (M). The average molecular speed is proportional to the square root of the absolute temperature and inversely proportional to the square root of the molecular mass. You can express this proportionality by comparing the average speeds of two different gases at the same temperature.

$$\frac{\bar{v}_A}{\bar{v}_B} = \frac{\sqrt{M_B}}{\sqrt{M_A}} \quad \text{(5-12)}$$

where $\bar{v}_A$ and M_A are, respectively, the average speed and molecular mass of gas A and $\bar{v}_B$ and M_B are the same for gas B.

note: Molecular speed may be averaged in different ways, the results of which are not necessarily equal:

$$\text{number average:} \quad \bar{v} = \frac{1}{N}(v_1 + v_2 + v_3 + \cdots)$$

$$\text{root mean square:} \quad v_{rms} = \left\{ \frac{1}{N}(v_1^2 + v_2^2 + v_3^2 + \cdots) \right\}^{1/2}$$

Since only ratios or proportionalities are considered here, you will not need to worry about the distinction.

B. Graham's law: Diffusion and effusion

Kinetic theory provides a rationale for another gas law accounting for the movement of gases. Gases move throughout a container or through a porous barrier by a process called **diffusion**. The rate (r) of diffusion for a gas is proportional to its average molecular speed. This proportionality is expressed in

Graham's law: For two gases A and B at the same temperature and pressure, the rates of diffusion r_A and r_B are inversely proportional to the square roots of their molecular weights (or densities). Stated mathematically,

GRAHAM'S LAW
$$\frac{r_A}{r_B} = \frac{\bar{v}_A}{\bar{v}_B} = \frac{\sqrt{M_B}}{\sqrt{M_A}} \tag{5-13}$$

Since the density d of a gas is proportional to the molecular mass at constant temperature and pressure (see Example 5-14), Graham's law may also be written as

$$\frac{r_A}{r_B} = \frac{\sqrt{M_B}}{\sqrt{M_A}} = \frac{\sqrt{d_B}}{\sqrt{d_A}}$$

Effusion is the process whereby a gas escapes through a small hole (of molecular dimensions) in its container. Graham's law applies to effusion as well.

5-4. Real Gases

An ideal gas, one that obeys the gas laws *exactly*, is a hypothetical concept. The behavior of real gases deviates from the exact gas laws:

1. When cooled, real gases liquefy, and finally solidify, because their molecules attract each other. Their volume does *not* become zero as T gets close to absolute zero.
2. At very high pressures, the molecules of real gases are squeezed close together and resist further compression.

Van der Waals equation is often applied to describe the behavior of real gases:

VAN DER WAALS EQUATION
$$\left(P + \frac{a}{V^2}\right)(V - b) = nRT \tag{5-14}$$

where a and b are constants for a particular gas (each gas has its own a and b constants, which you can look up). The a/V^2 term accounts for intermolecular attractions; the b term accounts for the volume actually occupied by the gas molecules. For example, the van der Waals constants for CO_2 are $a = 3.59 \text{ L}^2 \text{ atm mol}^{-2}$ and $b = 0.0427 \text{ L mol}^{-1}$. [*Note*: To make van der Waals equation (5.14) consistent in its units, a and b are expressed as extensive quantities; the number of moles of gas per unit of volume must be taken into account when using this equation.]

SUMMARY

1. The state of a gas may be specified by its volume, pressure, temperature, and composition.
2. The laws that relate these quantities include

   Boyle's law: $PV = \text{constant}$ or $P_1 V_1 = P_2 V_2$

   Charles' law: $\dfrac{V}{T} = \text{constant}$ or $\dfrac{V_1}{T_1} = \dfrac{V_2}{T_2}$

   Combined gas law: $\dfrac{P_1 V_1}{T_1} = \dfrac{P_2 V_2}{T_2}$

   Ideal gas law: $PV = nRT$

3. Avogadro's law states that equal volumes of all gases, under the same conditions of temperature and pressure, contain the same number of particles.
4. Dalton's law states that the total pressure of a mixture of gases equals the sum of the partial pressures of all the components.
5. The kinetic molecular theory, a model of gas behavior, can be used to derive the gas laws from simple assumptions about the nature of gas molecules.
6. Graham's law states that the rates of diffusion or effusion of gases through a barrier, under equal conditions of P and T, are inversely proportional to the square roots of their molecular masses.
7. Real gases deviate from the ideal gas law because their molecules have finite sizes and attract each other. The van der Waals equation can be used to describe real gases.

RAISE YOUR GRADES

Can you state...?

- ☑ Boyle's law
  Charles' law
- ☑ Avogadro's law
- ☑ Dalton's law of partial pressures
- ☑ the combined gas law
- ☑ Graham's law of diffusion
- ☑ van der Waals equation

Can you define...?

- ☑ STP
- ☑ Avogadro's number
- ☑ density
- ☑ diffusion
- ☑ effusion

Can you explain...?

- ☑ what the gaseous state is and how it is characterized
- ☑ how pressure is defined and measured
- ☑ how the Kelvin and Celsius temperature scales are related
- ☑ how real and ideal gases differ
- ☑ the kinetic theory of gases

SOLVED PROBLEMS

PROBLEM 5-1   Convert the following Celsius temperatures into degrees kelvin: **(a)** 0°C, **(b)** 25.3°C, **(c)** 174°C, **(d)** −82°C.

Solution:
(a) The conversion formula is

$$K = °C + 273.15$$

If °C = 0,

$$K = 273.15 + 0 = 273.15 \text{ K}$$

[*Note*: There's an uncertainty in significant figures here that you'll frequently meet in temperature conversions, but we don't usually discuss it. In practical terms, if "0°C" means that the measured temperature is $0 \pm 0.5°C$, then you would give the answer to the nearest degree kelvin, that is, as 273 K. Let's take this commonsense approach from here on.]

(b) At 25.3°C

$$K = (25.3 + 273.15) = 298.45$$
$$= 298.4 \text{ K}$$

(c) At 174°C

$$K = (174 + 273.15) = 447.15$$
$$= 447 \text{ K}$$

(d) At −82°C

$$K = (-82 + 273.15) = 191.15$$
$$= 191 \text{ K}$$

PROBLEM 5-2   Determine the mole fractions of each gas in a mixture of 3.7 mol He, 1.3 mol Ne, and 4.5 mol O_2.

Solution: The definition of mole fraction is

$$X_A = \frac{n_A}{n_A + n_B + n_C}$$

so n_{He} = 3.7 mol, n_{Ne} = 1.3 mol, and n_{O_2} = 4.5 mol. Here we have three gases, so the total number of moles of gas is

$$n_{He} + n_{Ne} + n_{O_2} = (3.7 + 1.3 + 4.5) \text{ mol} = 9.5 \text{ mol}$$

and the mole fraction of each gas is

$$X_{He} = \frac{n_{He}}{9.5 \text{ mol}} = \frac{3.7 \text{ mol}}{9.5 \text{ mol}} = 0.39 \quad \text{(2 sig. figs.)}$$

$$X_{Ne} = \frac{n_{Ne}}{9.5 \text{ mol}} = \frac{1.3 \text{ mol}}{9.5 \text{ mol}} = 0.14$$

$$X_{O_2} = \frac{n_{O_2}}{9.5 \text{ mol}} = \frac{4.5 \text{ mol}}{9.5 \text{ mol}} = 0.47$$

Note that the sum of *all* the mole fractions of *all* the components is (0.39 + 0.14 + 0.47) = 1.00. This is always true and provides a useful check on the accuracy of your calculations.

PROBLEM 5-3   A sample of gaseous oxygen originally held in a 275-L tank at a pressure of 155 atm is allowed to expand at a constant temperature of 298 K until its final pressure is 1.00 atm. What volume will it occupy?

Solution: Boyle's law says that

$$P_1 V_1 = P_2 V_2$$

Calling the initial conditions 1 and the final conditions 2, P_1 = 155 atm, V_1 = the volume of the tank = 275 L, and P_2 = 1.00 atm. V_2 is what you have to calculate:

$$(155 \text{ atm})(275 \text{ L}) = (1.00 \text{ atm})(V_2)$$

$$V_2 = \frac{(155 \text{ atm})(275 \text{ L})}{1.00 \text{ atm}}$$

$$= 4.26 \times 10^4 \text{ L}$$

PROBLEM 5-4   A 3.75-L sample of nitrogen gas at a pressure of 3.00×10^5 kPa and a constant temperature of 300 K is allowed to expand until its volume is 10.00 L. What is its new pressure?

Solution: Call the initial conditions 1 and the new conditions 2. According to Boyle's law,

$$P_1 V_1 = P_2 V_2$$
$$(3.00 \times 10^5 \text{ kPa})(3.75 \text{ L}) = (P_2)(10.00 \text{ L})$$

$$P_2 = (3.00 \times 10^5 \text{ kPa})\left(\frac{3.75 \text{ L}}{10.00 \text{ L}}\right)$$

$$= 1.13 \times 10^5 \text{ kPa}$$

PROBLEM 5-5   A 53.0-mL sample of N_2 at 298 K and 735 torr is cooled at constant pressure until its volume is 30.0 mL. What is its new temperature?

Solution: Charles' law states that

$$\frac{V_1}{T_1} = \frac{V_2}{T_2} \quad \text{(P being constant)}$$

Label the inital conditions 1, the final conditions 2. Since V_1 is given as 53.0 mL, T_1 as 298 K, and V_2 as 30.0 mL, you want to find T_2:

$$\frac{53.0 \text{ mL}}{298 \text{ K}} = \frac{30.0 \text{ mL}}{T_2}$$

so

$$T_2 = \left(\frac{30.0 \text{ mL}}{53.0 \text{ mL}}\right)(298 \text{ K})$$

$$= 169 \text{ K}$$

PROBLEM 5-6 A balloon of starting volume 3.7×10^5 L travels from Michigan, where the temperature is 25°C, to Alaska, where the temperature is $-45°C$. What is its new volume, assuming that there has been no pressure change?

Solution: From Charles' law, $V_1/T_1 = V_2/T_2$ when P is constant. Set up the conditions, remembering that temperatures for this equation must be in kelvins (K). $V_1 = 3.7 \times 10^5$ L, $T_1 = (25 + 273) = 298$ K, $T_2 = (-45 + 273) = 228$ K:

$$\frac{3.7 \times 10^5 \text{ L}}{298 \text{ K}} = \frac{V_2}{228 \text{ K}}$$

$$V_2 = \frac{228 \text{ K}}{298 \text{ K}} (3.7 \times 10^5 \text{ L}) = 2.8 \times 10^5 \text{ L}$$

PROBLEM 5-7 How many moles of hydrogen sulfide (H_2S) are there in a cylinder of volume 1.35 L at 302 K, in which the pressure of H_2S is 69 atm?

Solution: Use the ideal gas law, which relates pressure, volume, and temperature to the number of moles (n):

$$PV = nRT \qquad \text{or} \qquad n = \frac{PV}{RT}$$

P, V, and T are given in the data, and R is a constant. Choose the value of R dimensionally consistent with the data, i.e., liters, atmospheres, and degrees kelvin. So

$$n = \frac{PV}{RT} = \frac{(69 \text{ atm})(1.35 \text{ L})}{(0.0821 \text{ L atm K}^{-1} \text{mol}^{-1})(302 \text{ K})}$$

$$= 3.8 \text{ mol}$$

Note that, because of Avogadro's law, the *nature* of the gas has no influence on the answer; 3.8 mol of any gas in that cylinder at the given conditions would exert the same pressure.

PROBLEM 5-8 A 1.3-L cylinder containing H_2S at a pressure of 69 atm and a temperature of 302 K is heated to 175°C. What is the resulting pressure of H_2S?

Solution: The law relating pressure, temperature, and volume is the combined gas law:

$$\frac{P_1 V_1}{T_1} = \frac{P_2 V_2}{T_2}$$

In this example V is kept constant ($V_1 = V_2$), so we can simplify the law:

$$\frac{P_1}{T_1} = \frac{P_2}{T_2}$$

Given that $P_1 = 69$ atm, $T_1 = 302$ K, and $T_2 = (273 + 175) = 448$ K, we substitute

$$\frac{69 \text{ atm}}{302 \text{ K}} = \frac{P_2}{448 \text{ K}}$$

to get

$$P_2 = (69 \text{ atm})\left(\frac{448 \text{ K}}{302 \text{ K}}\right) = 1.0 \times 10^2 \text{ atm}$$

Only two significant figures are justified.

PROBLEM 5-9 What pressure, measured in torr, will a 1.56-g sample of methane (CH_4) exert in a flask of volume 3.36 L at a temperature of 273 K? (The boiling point of CH_4 is 111 K.)

Solution: Use the ideal gas law, $PV = nRT$, or $P = nRT/V$. Now we need the number of moles (n) of CH_4, so we calculate the molar mass

$$12.01 + 4(1.008) = 16.04 \text{ g mol}^{-1}$$

and use the given mass to get n:

$$n = \frac{1.56 \text{ g}}{16.04 \text{ g mol}^{-1}}$$

Picking the appropriate value of R, we get

$$P = \frac{1.56\,g}{16.04\,g\,mol^{-1}}\frac{(62.4\,L\,torr\,K^{-1}\,mol^{-1})(273\,K)}{3.36\,L} = 493\ torr$$

PROBLEM 5-10 A 0.856-g sample of a gaseous compound of sulfur and oxygen exerts a pressure of 644 torr in a 387-mL bulb at a temperature of 299 K. What is the molar mass of this compound?

Solution: We use the ideal gas law, $PV = nRT$, because molar and mass relationships are involved. Now the number of moles of a gas $n = m/M$, where m is the mass of the gas sample and M is the molar mass. Now we have a useful general relationship:

$$\frac{m}{M} = \frac{PV}{RT} \quad \text{or} \quad M = \frac{mRT}{PV}$$

For this compound, $m = 0.856$ g, $T = 299$ K, $P = 644$ torr, $V = 387$ mL $= 0.387$ L, and $R = 62.4$ L torr $K^{-1}\,mol^{-1}$ (dimensionally consistent with the data). So

$$M = \frac{(0.856\ g)(62.4\,L\,torr\,K^{-1}\,mol^{-1})(299\,K)}{(644\,torr)(0.387\,L)} = 64.1\ g\,mol^{-1}$$

PROBLEM 5-11 What is the total pressure, in atmospheres, inside a 2.45-L tank that holds 21.3 mol of helium (He) and 127 g of oxygen (O_2) at 35°C?

Solution: Dalton's law says that the total pressure equals the sum of the partial pressures of all components. Partial pressure is the pressure each gas would exert if it alone occupied the container. If p_{He} is partial pressure of He, and p_{O_2} partial pressure of O_2, then

$$P_{total} = p_{He} + p_{O_2}$$

$$p_{He} = n_{He}\frac{RT}{V} \quad \text{and} \quad p_{O_2} = n_{O_2}\frac{RT}{V}$$

so

$$P_{total} = n_{He}\frac{RT}{V} + n_{O_2}\frac{RT}{V} = (n_{He} + n_{O_2})\frac{RT}{V}$$

Then

$$n_{He} = 21.3\ mol \quad \text{and} \quad n_{O_2} = \frac{m_{O_2}}{M_{O_2}} = \frac{127\ g}{32.0\ g\,mol^{-1}} = 3.97\ mol$$

so

$$n_{He} + n_{O_2} = 21.3 + 3.97 = 25.3\ mol$$
$$T = (35 + 273)\ K = 308\ K$$

Thus,

$$P_{total} = \frac{(25.3\,mol)(0.0821\,L\,atm\,mol^{-1}\,K^{-1})(308\,K)}{2.45\,L} = 261\ atm$$

PROBLEM 5-12 When a 2.35-g sample of calcium carbonate ($CaCO_3$) is heated, it decomposes completely to yield solid calcium oxide (CaO) and carbon dioxide gas (CO_2). If the CO_2 is collected at 25°C at a pressure of 725 torr, what volume of CO_2 is produced?

Solution: First write a balanced equation for the reaction, so you know its stoichiometry. By inspection,

$$CaCO_3 \longrightarrow CaO(s) + CO_2(g)$$
$$1\ mol \qquad\qquad 1\ mol$$

The molar mass of $CaCO_3$ is

$$40.08 + 12.01 + 3(16.00) = 100.09\ g$$

So n, the number of moles of $CO_2(g)$ produced, is

$$\frac{2.35\ g\ CaCO_3}{100.09\ g\ CaCO_3/1\ mol\ CaCO_3} \times \frac{1\ mol\ CO_2}{1\ mol\ CaCO_3} = 0.0235\ mol$$

From the ideal gas law, $V = nRT/P$. So convert the temperature to degrees kelvin, $T = (25 + 273)\ \text{K} = 298\ \text{K}$, and substitute:

$$V = \frac{(0.0235\ \text{mol CO}_2)(62.4\ \text{L torr K}^{-1}\ \text{mol}^{-1})(298\ \text{K})}{725\ \text{torr}}$$

$$= 0.603\ \text{L}$$

PROBLEM 5-13   When potassium chlorate ($KClO_3$) is heated, oxygen gas (O_2) and solid potassium chloride (KCl), are produced. A 1.50-g sample of a mixture of $KClO_3$ and KCl is heated until all the chlorate has decomposed; the O_2 produced has a volume of 185 mL at 743 torr and 26°C. What is the percentage by mass of $KClO_3$ in the mixture?

Solution: First balance the decomposition equation and establish the stoichiometry. It is clear that since O_2 contains an even number of oxygen atoms, and $KClO_3$ an odd number, the balanced equation must involve 2 mol of $KClO_3$:

$$2KClO_3 \longrightarrow 2KCl + 3O_2$$
$$\ \ 2\ \text{mol} \qquad\qquad 3\ \text{mol}$$

Next, find the number of moles of O_2 produced in the decomposition:

$$n_{O_2} = \frac{PV}{RT} = \frac{(743\ \text{torr})(185\ \text{mL})(10^{-3}\ \text{L/mL})}{(62.4\ \text{L torr K}^{-1}\ \text{mol}^{-1})(26 + 273)\ \text{K}}$$

$$= 7.37 \times 10^{-3}\ \text{mol}$$

Then find the number of moles of $KClO_3$ that produced this amount of O_2:

$$n_{KClO_3} = (7.37 \times 10^{-3}\ \text{mol O}_2)\frac{(2\ \text{mol KClO}_3)}{(3\ \text{mol O}_2)}$$

$$= 4.91 \times 10^{-3}\ \text{mol}$$

The molar mass of $KClO_3$ is

$$39.10 + 35.45 + 3(16.00)\ \text{g} = 122.55\ \text{g}$$

so the mass of $KClO_3$ that produced the oxygen is

$$4.91 \times 10^{-3}\ \text{mol} \times 122.55\ \text{g mol}^{-1} = 0.602\ \text{g}$$

and the percentage by mass of $KClO_3$ in the original mixture is

$$\frac{0.602\ \text{g KClO}_3}{1.50\ \text{g mixture}} \times 100\% = 40.1\%$$

PROBLEM 5-14   How many moles of CH_4 are present in a 151-mL sample stored over water at a total pressure of 742 torr at 25°C? [*Hint*: Look up the vapor pressure of water in any standard reference.]

Solution: At 25°C water has an appreciable vapor pressure. The water vapor will be mixed with the CH_4 and contribute its vapor pressure to the total pressure of the gases in the sample. So Dalton's law of partial pressures can be applied here:

$$P_{\text{total}} = p_{CH_4} + p_{H_2O}$$

The vapor pressure of the water is a function of temperature: As shown in your reference table, at 25°C, $p_{H_2O} = 24$ torr. So

$$p_{CH_4} = P_{\text{total}} - p_{H_2O} = (742 - 24)\ \text{torr} = 718\ \text{torr}$$

Now use the ideal gas law to calculate the number of moles of CH_4:

$$n_{CH_4} = \frac{PV}{RT} = \frac{(718\ \text{torr})(151\ \text{mL})(10^{-3}\ \text{L mL}^{-1})}{(62.4\ \text{L torr K}^{-1}\ \text{mol}^{-1})(25 + 273)\ \text{K}}$$

$$= 5.83 \times 10^{-3}\ \text{mol}$$

PROBLEM 5-15   The density of a gas at a pressure of 1.34 atm and a temperature of 303 K is found to be 1.77 g/L. What is the molar mass of this gas?

Solution: Recall the ideal gas equation in the form $M = mRT/PV$ (see Problem 5-10). In this example you are given the gas density m/V, so

$$M = \frac{mRT}{VP} = \frac{(1.77 \text{ g L}^{-1})(0.0821 \text{ L atm K}^{-1} \text{ mol}^{-1})(303 \text{ K})}{1.34 \text{ atm}}$$

$$= 32.9 \text{ g mol}^{-1}$$

PROBLEM 5-16 The mole fraction of Ne in air is 1.8×10^{-5}. Calculate the partial pressure in torr of Ne in an air sample at STP.

Solution: At STP the total pressure of the sample is exactly 1 atm, or 760 torr. Now consider n moles of the air sample, which contains n times Avogadro's number nN_A of molecules of the components of air in a volume V. So

$$n_{total} = P_{total}\left(\frac{V}{RT}\right) \quad \text{and} \quad n_{Ne} = p_{Ne}\left(\frac{V}{RT}\right)$$

or

$$p_{Ne} = \frac{n_{Ne}}{n_{total}} P_{total}$$

You know that the mole fraction of Ne is $X_{Ne} = n_{Ne}/n_{total}$, so

$$p_{Ne} = X_{Ne}P_{total}$$

This is a useful general conclusion: *The partial pressure of any component of a gas mixture equals its mole fraction times the total pressure.* The answer to the question is now simple:

$$p_{Ne} = X_{Ne}P_{total} = (1.8 \times 10^{-5})(760 \text{ torr})$$
$$= 1.4 \times 10^{-2} \text{ torr}$$

PROBLEM 5-17 What volume of O_2 measured at STP, is needed to react completely with 1.75 L of ethylene (C_2H_4), also measured at STP, to yield CO_2 and H_2O?

Solution: First balance the equation to determine the stoichiometry:

$$\begin{array}{ccc} C_2H_4 & + & 3O_2 & \longrightarrow & 2CO_2 + 2H_2O \\ 1 \text{ mol} & & 3 \text{ mol} \\ 1 \text{ molecule} & & 3 \text{ molecules} \end{array}$$

Both the O_2 and the C_2H_4 are at the same temperature and pressure (STP). From Avogadro's law you know that 1.75 L of O_2 contains the same number of molecules as 1.75 L of C_2H_4. From the stoichiometry of the balanced equation, three times as many O_2 molecules as C_2H_4 molecules are required. The volume of O_2 required is $3 \times 1.75 \text{ L} = 5.25 \text{ L}$.

PROBLEM 5-18 If the average velocity of protons (H^+) in the outer atmosphere of the sun is $1.2 \times 10^3 \text{ m s}^{-1}$, what is the average velocity of ^{12}C atoms in the sun's outer atmosphere?

Solution: According to kinetic theory, the velocity of a gas particle is inversely proportional to the square root of its mass. So we use Eq. (5-12) to express that proportionality:

$$\frac{\bar{v}_A}{\bar{v}_B} = \frac{\sqrt{M_B}}{\sqrt{M_A}}$$

Remembering that what works on a molar level will also work on any other level in proportionalities, and that the masses of ^{12}C atoms and H^+ particles are their respective atomic masses, we can write

$$\frac{\bar{v}_{H^+}}{\bar{v}_{12C}} = \frac{\sqrt{m_{12C}}}{\sqrt{m_{H^+}}} = \frac{1.2 \times 10^3 \text{ m s}^{-1}}{\bar{v}_{12C}} = \frac{\sqrt{12}}{\sqrt{1}}$$

Thus

$$\bar{v}_{12C} = \frac{1.2 \times 10^3 \text{ m s}^{-1}}{\sqrt{12}}$$

$$= 3.5 \times 10^2 \text{ m s}^{-1}$$

PROBLEM 5-19   Gaseous neon, which is monatomic, is observed to diffuse through a ceramic disk at a rate of 1.90 cm^3/min. An unknown gas X is observed to diffuse through the same disk, under identical conditions of pressure and temperature, at a rate of 1.05 cm^3/min. What is the molar mass of X?

Solution: Graham's law states that

$$\frac{r_1}{r_2} = \frac{\sqrt{M_2}}{\sqrt{M_1}} \quad \text{or} \quad \left(\frac{r_1}{r_2}\right)^2 = \frac{M_2}{M_1}$$

Let M_2 be X and let M_1 be Ne, whose molar mass is 20.2 g mol^{-1}. Then

$$M_2 = M_1 \left(\frac{r_1}{r_2}\right)^2$$

$$= 20.2 \text{ g mol}^{-1} \left(\frac{1.90 \text{ cm}^3 \text{ min}^{-1}}{1.05 \text{ cm}^3 \text{ min}^{-1}}\right)^2$$

$$= 66.1 \text{ g mol}^{-1}$$

PROBLEM 5-20   Gaseous diffusion of uranium hexafluoride (UF_6) is used to separate the isotopes ^{235}U and ^{238}U. Calculate the enrichment factor (the ratio of the diffusion rates of $^{235}UF_6$ and $^{238}UF_6$) for this method.

Solution: From Graham's law, the enrichment factor is

$$\frac{r_1}{r_2} = \frac{\sqrt{M_2}}{\sqrt{M_1}} = \frac{\sqrt{\text{molar mass } ^{238}UF_6}}{\sqrt{\text{molar mass } ^{235}UF_6}}$$

$$= \frac{\sqrt{352 \text{ g mol}^{-1}}}{\sqrt{349 \text{ g mol}^{-1}}} = 1.004$$

This is a very small enrichment, and the process must be repeated many times to get a substantial enrichment in ^{235}U.

PROBLEM 5-21   How does kinetic theory account for the fact that (a) gases are very easily compressed, and (b) that they diffuse rapidly?

Solution: The kinetic theory model includes the following ideas:

(a) Gas molecules are small. Therefore most of the "volume" of a gas is actually empty space, and so gases are easily compressed.
(b) Gas molecules are in continual motion. Therefore they move through an empty space or through another gas very rapidly, and so gases diffuse rapidly.

PROBLEM 5-22   Which of the following are characteristic of a real gas rather than an ideal gas? Explain your answers.

(a) At low temperatures the gas liquefies.
(b) The pressure of the gas is caused by the force exerted by gas molecules colliding with the container walls.

Solution:
(a) This is characteristic of a real gas in which there are weak attractive forces between the gas molecules. When the kinetic energy of the molecules is sufficiently reduced by lowering the gas temperature, these weak forces hold the molecules together in the liquid state.
(b) This is characteristic of both real and ideal gases.

PROBLEM 5-23 Which of the following statements should be true for a real gas at a constant temperature? As the pressure increases, the PV product (a) increases, (b) stays constant, (c) decreases.

Solution: Statement (a) should be true. At high pressures the distance between gas molecules decreases and intermolecular repulsions become strong. It becomes harder to compress the gas and so a value of P greater than predicted from the PV product is needed to achieve the expected volume. Hence PV increases as P increases for real gases.

PROBLEM 5-24 What pressure is needed at 298 K to maintain 1.00 mol of CO_2 in a volume of **(a)** 1.00 L and **(b)** 0.050 L? Calculate your answer treating CO_2 (1) as an ideal gas, (2) as a real gas [*Hint*: Use van der Waals equation].

Solution:
(a) (1) Ideal gas: $PV = nRT$, so

$$P = \frac{nRT}{V} = \frac{(1.00 \text{ mol})(0.0821 \text{ L atm mol}^{-1}\text{K}^{-1})(298 \text{ K})}{1.00 \text{ L}}$$

$$= 24.5 \text{ atm}$$

(2) Real gas: Recall that the van der Waals equation is

$$\left(P + \frac{a}{V^2}\right)(V - b) = nRT$$

Looking up the van der Waals constants for CO_2 (see any standard reference), we have $a = 3.59 \text{ L}^2 \text{ atm mol}^{-2}$ and $b = 0.0427 \text{ L mol}^{-1}$. Since we have 1.00 mol CO_2 in 1.00 L, we know that $n = 1.00$ mol and $V = 1.00 \text{ L mol}^{-1}$. So

$$\left(P + \frac{3.59 \text{ L}^2 \text{ atm mol}^{-2}}{(1.00 \text{ L mol}^{-1})^2}\right)(1.00 \text{ L mol}^{-1} - 0.0427 \text{ L mol}^{-1}) = (1.00 \text{ mol})(0.821 \text{ L atm mol}^{-1}\text{K}^{-1})(298 \text{ K})$$

Simplifying,

$$(P + 3.59 \text{ atm})(0.957 \text{ L mol}^{-1}) = 24.5 \text{ L atm mol}^{-1}$$

$$P = 22.0 \text{ atm}$$

(b) (1) Ideal gas:

$$P = \frac{nRT}{V} = \frac{(1.00 \text{ mol})(0.0821 \text{ L atm mol}^{-1}\text{K}^{-1})(298 \text{ K})}{(0.050 \text{ L})}$$

$$= 4.9 \times 10^2 \text{ atm}$$

(2) Real gas:

$$\left(P + \frac{3.59}{(0.050)^2} \text{ atm}\right)(0.050 - 0.0427 \text{ L mol}^{-1}) = 24.5 \text{ L atm mol}^{-1}$$

$$(P + 1.4 \times 10^3 \text{ atm})(0.0073 \text{ L mol}^{-1}) = 24.5 \text{ L atm mol}^{-1}$$

$$P = 2.0 \times 10^3 \text{ atm}$$

(At high pressures, P must be greater than the calculated ideal pressure, as discussed in Problem 5-23.)

PROBLEM 5-25 What is the pressure in torr of the gas in each of the vessels shown in Figure 5-1? The atmospheric pressure in each case is 735 torr. The liquid in the vessels is mercury.

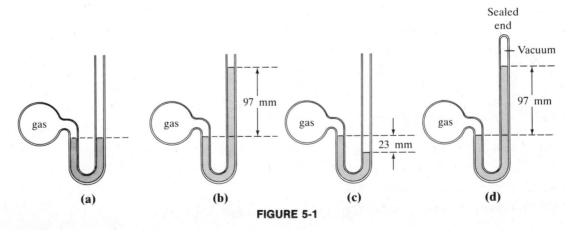

FIGURE 5-1

Solution:
(a) Because the levels of mercury on either sider of the U-tube are equal, the pressure in the vessel (P_{gas}) is equal to the atmospheric pressure (P_{atm}), which is given as 735 torr:

$$P_{gas} = P_{atm} = 735 \text{ torr}$$

(b) Here the level of Hg is higher on the right side. Remembering that 1 mm Hg = 1 torr, add the difference in levels to the atmospheric pressure:

$$P_{gas} = P_{atm} + 97 \text{ torr}$$
$$= (97 + 735) \text{ torr}$$
$$= 832 \text{ torr}$$

(c)
$$P_{gas} = P_{atm} - 23 \text{ torr}$$
$$= (735 - 23) \text{ torr}$$
$$= 712 \text{ torr}$$

(d)
$$P_{gas} = P_{atm} + 97 \text{ torr}$$
$$= (0 + 97) \text{ torr}$$
$$= 97 \text{ torr}$$

(Remember that pressure in an absolute vacuum is zero.)

PROBLEM 5-26 The mole fraction of N_2 in a mixture of gases containing 2.30×10^{-2} mol PH_3, 3.75×10^{-3} mol N_2, and 5.68×10^{-2} mol H_2 is (a) 0.320, (b) 4.70×10^{-2}, (c) 4.49, (d) 4.49×10^{-2}, (e) 0.275.

Solution: From the definition of mole fraction,

$$X_{N_2} = \frac{n_{N_2}}{n_{PH_3} + n_{N_2} + n_{H_2}} = \frac{3.75 \times 10^{-3} \text{ mol}}{(2.30 \times 10^{-2} + 3.75 \times 10^{-3} + 5.6 \times 10^{-2}) \text{ mol}}$$

$$= 4.49 \times 10^{-2}$$

The correct answer is (d).

You would obtain answer (a) if you add 3.75×10^{-3} as 3.75×10^{-2}; (b) if you leave the moles of N_2 out of the denominator; (c) if you calculate the mole percent—this answer is 100 times too large; and (e) if you calculate the mole fraction of PH_3.

PROBLEM 5-27 A sample of gaseous chlorine originally held in a 426-L container at 1.55 atm pressure is compressed until the volume is 50.0 L. The final pressure in kilopascals is (a) 13.2, (b) 18.4, (c) 1.34×10^3, (d) 1.37×10^4, (e) 1.34×10^6.

Solution: Apply Boyle's law to the conditions given:

$$\frac{P_1}{V_1} = \frac{P_2}{V_2} = \frac{1.55 \text{ atm}}{426 \text{ L}} = \frac{P_2}{50.0 \text{ L}}$$

$$P_2 = \frac{(1.55 \text{ atm})(426 \text{ L})}{50.0 \text{ L}} = 13.2 \text{ atm}$$

Convert atmospheres to kilopascals:

$$1 \text{ atm} = 101.325 \text{ kPa/atm} = 1.01325 \times 10^2 \text{ kPa/atm}$$

$$13.2 \text{ atm} \frac{(1.01325 \times 10^2 \text{ kPa})}{(1 \text{ atm})} = 1.34 \times 10^3 \text{ kPa}$$

The correct answer is (c).

You would obtain answer (a) if you fail to convert atmospheres to kilopascals; (b) and (d) if you perform the algebraic manipulations of Boyle's law incorrectly; and (e) if you convert atmospheres to pascals.

PROBLEM 5-28 The temperature in kelvins of 1.00 L of an ideal gas is doubled and its pressure is tripled. The volume in liters will then be (a) $\frac{1}{6}$, (b) $\frac{2}{3}$, (c) $\frac{3}{2}$, (d) 6, (e) $\frac{1}{3}$.

Solution: Use the combined gas law, solving for V_2:

$$V_2 = V_1 \left(\frac{P_1}{P_2}\right)\left(\frac{T_2}{T_1}\right)$$

The temperature is doubled, so T_2/T_1 is 2; the pressure is tripled, so P_2/P_1 is 3 and P_1/P_2 is $\frac{1}{3}$. Thus

$$V_2 = (1.00 \text{ L})(\tfrac{1}{3})(2) = \tfrac{2}{3} \text{ L}$$

The correct answer is (b).

PROBLEM 5-29 How many molecules of CO_2 gas are there in a 1.00-L container at 35.0°C and 0.987 atm pressure? (a) 2.35×10^{22}, (b) 6.02×10^{23}, (c) 3.91×10^{-2}, (d) 2.07×10^{23}, (e) 1.72.

Solution: In this multiple-choice problem, three of the five choices are obviously incorrect: (c) and (e) are far too small—usually, any question that asks for the number of molecules will have a large value for the answer; (b) is possible only for 1 mol of gas—the amount of gas in this problem is considerably less.

To work this problem, apply the ideal gas law to determine the number of moles:

$$n = \frac{PV}{RT} = \frac{(0.987 \text{ atm})(1.00 \text{ L})}{(0.0821 \text{ L atm mol}^{-1}\text{K}^{-1})(35.0 + 273) \text{ K}}$$

$$= 0.0390 \text{ mol}$$

The number of molecules is the number of moles times Avogadro's number:

$$N = nN_A = (0.0390 \text{ mol})(6.02 \times 10^{23} \text{ mol}^{-1}) = 2.35 \times 10^{22}$$

and the correct answer is (a). You would have obtained answer (d) if you had failed to convert Celsius degrees to Kelvin degrees.

PROBLEM 5-30 A 1.00-L vessel at STP is filled with oxygen gas while a vessel of equal size is filled with neon gas at STP. If N is the number of oxygen molecules in the first container, which of the following is true? (a) $N = 6.02 \times 10^{23}$, (b) the number of neon molecules is $\frac{20}{16}N$, (c) the number of neon molecules is $\frac{16}{20}N$, (d) the number of neon molecules is equal to N, (e) there is no relationship between the number of neon molecules and the number of oxygen molecules.

Solution: This problem tests your understanding of Avogadro's law. Both of the 1.00-L vessels are at the same temperature and pressure (in this case, STP), so they must contain the same number of molecules. The correct answer is (d). Answer (a) is true only if the vessels were 22.4 L. Answers (b) and (c) are incorrect because the number of molecules does not depend on the atomic or molar mass of the element.

Supplementary Exercises

PROBLEM 5-31 If 2.85 L of air at a pressure of 1.54 atm and a temperature of 273 K is cooled to 175 K at a new pressure of 3.64 atm, what is the new volume of the air?

Answer: 0.773 L

PROBLEM 5-32 Calculate the volume occupied by 2.00 kg of propane (C_3H_8) at a pressure of 1.00 atm and a temperature of 273 K.

Answer: 1.02×10^3 L

PROBLEM 5-33 Which has the greater density: gaseous O_2 or gaseous N_2, both measured at STP?

Answer: O_2

PROBLEM 5-34 Calculate the density of gaseous SF_6 at STP in grams per liter.

Answer: 6.52 g/L

PROBLEM 5-35 When limestone (calcium carbonate, $CaCO_3$) is heated, it decomposes to gaseous CO_2 and lime (calcium oxide CaO). Calculate the volume, in liters, of CO_2 produced when 2.0×10^3 kg of limestone is heated and the CO_2 collected at 10.0 atm and 298 K.

Answer: 4.89×10^4 L

PROBLEM 5-36 Calculate the ratio of the rates of diffusion through a porous plug, under identical conditions of temperature and pressure, of the gases $^{12}CH_4$ and $^{13}CH_4$.

Answer: 1.03

PROBLEM 5-37 Arrange the following in order of increasing mass: 1.00 mol of CH_4; 1.25 L of H_2 at STP; 6.0 L of SO_3 at 1.25 atm and 325 K.

Answer: H_2, CH_4, SO_3

PROBLEM 5-38 Which of the following contains the larger number of gas molecules: 18.0 g of HCl or 6.75 L of butene (C_4H_8) measured at 310 K and 0.978 atm?

Answer: HCl

PROBLEM 5-39 A good vacuum pump can produce a pressure of 1×10^{-8} torr at 300 K. How many gas molecules remain in each cubic centimeter of this "vacuum"?

Answer: 3×10^8 molecules

PROBLEM 5-40 A sample of an unknown gas diffuses through a plug at a rate of 1.05 cm^3/min under the same temperature and pressure conditions in which a sample of CO_2 diffuses at a rate of 0.64 cm^3/min. What is the molar mass of the unknown gas?

Answer: 16 g/mol

PROBLEM 5-41 Calculate the mass of CO_2 needed to charge a 5.5-L fire extinguisher with the gas at a pressure of 20.0 atm and a temperature of 28°C.

Answer: 196 g

PROBLEM 5-42 A 325-mL bulb is filled with a gas to a pressure of 414 torr at 24°C. The mass of the gas (determined by weighing the empty bulb and then weighing the gas-filled bulb) is 0.687 g. Calculate the molar mass of the gas.

Answer: 94.6 g mol^{-1}

PROBLEM 5-43 An 8.26-L flask containing helium at a pressure of 148 torr is connected to a 2.29-L flask containing argon at a pressure of 544 torr, the whole apparatus being held at 25°C throughout. Calculate the final partial pressures of each gas in torr, and the final total pressure in torr.

Answer: $p_{He} = 116$ torr, $p_{Ar} = 118$ torr, $P_{total} = 234$ torr

PROBLEM 5-44 What volume of O_2 at STP is needed to convert 15.0 g of sulfur to SO_3?

Answer: 15.8 L

PROBLEM 5-45 Calculate the density of gaseous CH_4 at 298 K and a pressure of 745 torr.

Answer: 0.641 g/L

PROBLEM 5-46 A mixture of 0.44 mol of Ne and 0.37 mol of Kr was put into a 5.25-L flask at 185 K. Calculate the partial pressure of each gas and the total pressure in the flask, both in atmospheres.

Answer: $p_{Ne} = 1.3$ atm, $p_{Kr} = 1.1$ atm, $P_{total} = 2.4$ atm

PROBLEM 5-47 A 256-mL flask contains 1.675 g of a gas at 373 K and 0.961 atm. What is the molar mass of the gas?

Answer: 208 g/mol

PROBLEM 5-48 A 50.0-mL sample of H_2 measured at STP is brought to a pressure of 731 torr at 25°C. What volume will it occupy?

Answer: 56.8 mL

PROBLEM 5-49 What volume of gaseous Cl_2, measured at 27°C and 730 torr, will react with 14.0 g of metallic sodium according to the following equation?

$$2Na + Cl_2 \longrightarrow 2NaCl$$

Answer: 7.80 L

PROBLEM 5-50 Oxygen is supplied to hospitals in 180-L tanks at a pressure of 150 atm at 290 K. What mass of O_2 in kilograms is contained in each tank?

Answer: 36 kg

PROBLEM 5-51 There are two naturally occurring isotopes of nitrogen, ^{14}N and ^{15}N, but only one of fluorine, ^{19}F. Calculate the ratio of the rates of diffusion of $^{14}NF_3$ and $^{15}NF_3$ under the same temperature and pressure conditions.

Answer: 1.007

PROBLEM 5-52 A cylinder holds 1.05 L of helium at a pressure of 97 atm and a temperature of 10°C. What volume would the gas occupy at STP?

Answer: 98 L

PROBLEM 5-53 A piece of zinc was allowed to react with excess aqueous acid to give 86.4 mL of H_2 measured at STP:

$$Zn(s) + 2H^+(aq) \longrightarrow H_2(g) + Zn^{2+}(aq)$$

What was the mass of the piece of zinc?

Answer: 0.252 g

PROBLEM 5-54 What is the ratio of the root mean square velocity of an H_2 molecule at 585 K to that of a ClO_2 molecule at the same temperature?

Answer: 5.8

PROBLEM 5-55 Urea (CH_4N_2O), molar mass 60.0 g/mol, reacts with sodium hypochlorite to give a mixture of gaseous CO_2 and N_2:

$$CH_4N_2O + 3NaOCl \longrightarrow CO_2 + N_2 + 2H_2O + 3NaCl$$

A sample of urine gave 2.70 mL of gas, measured at 25°C and 745 torr when treated with excess NaOCl. What mass of urea was present in the urine sample?

Answer: 3.25 mg

6 INTERMOLECULAR FORCES AND LIQUIDS

THIS CHAPTER IS ABOUT

- ☑ **Intermolecular Forces**
- ☑ **Liquids**
- ☑ **Surface Tension**
- ☑ **Viscosity**

6-1. Intermolecular Forces

Intermolecular forces act between molecules. Although these short-range forces are weak in comparison with chemical bonds (see Chapter 10), they are responsible for several important attractions between *electrically neutral* molecules. There are three principal intermolecular forces:

1. *Dipole–dipole forces*: Polar molecules are neutral molecules in which the charge is not distributed evenly, so that a permanent charge separation is maintained. These polar molecules align themselves so that the positive end of one molecule attracts the negative end of another, and so on. These attractions are called **dipole–dipole forces**.

2. *London forces*: Because electrons are in constant motion, the charge distribution of a neutral molecule fluctuates. At any given moment a molecule may act as an instantaneous dipole, *inducing* a dipole effect in a nearby molecule and thus generating an attraction between the instantaneous dipole and the induced dipole. These attractions are called **London forces** and can occur in all atoms and molecules, both polar and nonpolar. London forces are weaker than dipole–dipole forces; their strength increases as the number of electrons in the atom or molecule increases.

3. *Hydrogen bonding*: When a hydrogen atom is covalently bonded to a small, strongly electronegative atom, a highly polar molecule is formed. There is sufficient positive charge on the hydrogen end of such a molecule to attract and form a **hydrogen bond** with a nearby electronegative atom. The strongest hydrogen bonds form between H and F, N, or O. Hydrogen bonding is usually stronger than dipole–dipole forces.

Dipole–dipole forces and London forces are known as **van der Waals forces**.

note: Although your textbook may exclude one of these forces in the discussion of van der Waals forces, you should remember that *all* of these forces act on electrically neutral molecules to generate attractive forces between molecules.

At very short intermolecular distances all molecules *repel* each other. This repulsive effect is balanced with the van der Waals forces, and the kinetic energy of the molecules due to temperature. This balance is the actual equilibrium state of the molecules. At lower temperatures molecules of a substance have less kinetic energy and the van der Waals forces are dominant, so the substance may exist in a *liquid* or a *solid* state. These two states of matter are known as the **condensed states** of matter.

EXAMPLE 6-1: The boiling points of sulfur dioxide, xenon, and water at 1 atm are $-10°C$, $-107°C$, and $100°C$, respectively. Account qualitatively for the differences in these boiling points in terms of intermolecular forces.

Solution: Sulfur dioxide (SO_2) molecules are polar: They attract each other through dipole–dipole forces. Up to $-10°C$, the van der Waals forces are dominant and SO_2 remains a liquid.

Xenon atoms are spherical and nonpolar. They attract each other only through the relatively weak London forces, over which kinetic energy is easily dominant at higher temperatures. So Xe is a liquid only at temperatures below $-107°C$.

Water (H_2O) molecules are polar and attract each other not only through van der Waals forces, but also through hydrogen bonding. Water is a liquid up to $100°C$.

6-2. Liquids

Matter in the liquid state is fluid. Like gases, liquids have no fixed or definite shape: The molecules are not ordered, and they move about at random. A specified mass of liquid, unlike gases, has a definite volume at a given temperature. This property is also exhibited by solids.

A. Vapor pressure

When a liquid is placed in a closed vessel at a constant temperature, some of its molecules escape into the space above the liquid as vapor; i.e., some of the liquid evaporates and goes into the gaseous state. Eventually, the rate at which molecules escape from the liquid into the vapor is exactly balanced by the rate at which they return to the liquid from the vapor. This balance of rates is known as a state of **dynamic equilibrium**. The molecules of the vapor exert a **vapor pressure** that is characteristic of the liquid and the temperature. For most liquids the variation of vapor pressure P_{vap} (in torr) with absolute temperature T obeys the equation

VAPOR PRESSURE
$$\log P_{vap} = \frac{-a}{T} + b \qquad (6\text{-}1)$$

where a and b are constants given for a specific liquid. (These constants can be found in standard reference tables.)

EXAMPLE 6-2: The vapor pressure of water is 24 torr at $25°C$, 42 torr at $35°C$, and 72 torr at $45°C$. Account qualitatively for this rise in vapor pressure.

Solution: As the temperature increases, the water molecules have increased kinetic energy, which increases their tendency to escape from the liquid into the vapor phase and thus increases the vapor pressure.

EXAMPLE 6-3: For carbon tetrachloride (CCl_4) at $4.3°C$, $a = 1771$ and $b = 8.004$. What is the vapor pressure of CCl_4 at this temperature?

Solution: Substituting directly into the vapor pressure equation (6-1),

$$\log P_{vap} = \frac{-1771}{277.5} + 8.004 = 1.622$$

$$P_{vap} = \text{antilog}(1.622) = 41.9 \text{ torr}$$

B. Boiling point

A liquid *boils* when vapor bubbles form in the liquid and leave it. Boiling occurs when the vapor pressure of the liquid equals the external pressure. The **boiling point** of a liquid is the temperature at which its vapor pressure equals the external pressure at a specified external pressure. When the external pressure is exactly 1 standard atmosphere, the boiling point is called the **normal boiling point**. Each liquid has its own characteristic normal boiling point.

EXAMPLE 6-4: Here are some normal boiling point values: ammonia (NH_3) $-34°C$, bromine (Br_2) $58°C$, carbon tetrachloride (CCl_4) $77°C$, gold (Au) $2966°C$, oxygen (O_2) $-183°C$. Which of these liquids has the highest vapor pressure at room temperature (298 K)? Which has the lowest?

Solution: The higher the normal boiling point, the lower the vapor pressure, so O_2 (bp $-183°C$) has the highest vapor pressure and Au (bp $2966°C$) has the lowest.

C. Distillation

Distillation is a method of purifying a liquid by evaporation and condensation. The liquid is heated to boiling in one container in a closed system that allows the resulting vapor to be collected. The collected vapor is then cooled, so that it **condenses** (reverts to liquid form), and the condensed liquid is recovered in a second container (still in the closed system). Impurities (dissolved liquids and solids) remain in the first container. (Remember that boiling points are characteristic: As long as the temperature remains the same, a single substance is involved.) Distillation can be carried out at atmospheric pressure or, for liquids with a high normal boiling point, at a reduced pressure provided by a vacuum pump.

note: Some compounds can't be purified by distillation because they form constant-boiling *mixtures* called **azeotropes**.

6-3. Surface Tension

A liquid behaves as if there is a force pulling its surface together. It takes energy to increase the surface area of a liquid by, for instance, making many smaller drops from one large drop. The amount of energy needed to increase the surface area of a liquid is called the **surface tension** of the liquid. Surface tension is measured in SI units of newtons per meter ($N\,m^{-1}$). Liquids with large intermolecular forces have large surface tensions.

One way to quantify surface tension is to determine how high a liquid will rise in a capillary tube. If a liquid of density d rises a height h in a tube of radius r above the surface of the liquid, the surface tension of the liquid γ is given by

SURFACE TENSION $$\gamma = \tfrac{1}{2}rhdg \qquad\qquad (6\text{-}2)$$

where g is the acceleration due to gravity, $9.81\ m\,s^{-2}$.

EXAMPLE 6-5:   The surface tension of water is given as $7.26 \times 10^{-2}\ N\,m^{-1}$. How high will water rise in a capillary tube of radius 0.100 mm?

Solution: Rearrange the surface tension equation (6-2):

$$h = \frac{2\gamma}{rdg}$$

Set up your conditions, remembering to use consistent SI units:

$$r = (0.100\ \text{mm})\left(\frac{1\ \text{m}}{10^3\ \text{mm}}\right) = 1.00 \times 10^{-4}\ \text{m}$$

$$d = (1.00\ \text{g cm}^{-3})\left(\frac{1\ \text{kg}}{10^3\ \text{g}}\right)\left(\frac{100\ \text{cm}}{1\ \text{m}}\right)^3 = 1.00 \times 10^3\ \text{kg m}^{-3}$$

Change newtons to base units

$$1\ \text{N} = 1\ \text{m kg s}^{-2}$$

and solve for h:

$$h = \frac{2(7.26 \times 10^{-2}\ \text{m kg s}^{-2}\text{m}^{-1})}{(1.00 \times 10^{-4}\ \text{m})(1.00 \times 10^3\ \text{kg m}^{-3})(9.81\ \text{m s}^{-2})}$$

$$= 0.148\ \text{m} \qquad \text{(or 148 mm)}$$

Quite a rise—but the capillary tube is extremely narrow.

6-4. Viscosity

Viscosity, or the resistance to flow, is defined as the force needed to give a liquid a unit velocity gradient per unit area. In SI the units of viscosity are kilograms per meter per second ($kg\ m^{-1}\,s^{-1}$). In general, the greater the intermolecular forces and the lower the kinetic energy in a liquid, the higher its viscosity.

Lubricating oil for automobiles, for example, is available in different viscosities, labeled as 20 W, 30 W, etc. The higher numbers indicate higher-viscosity oils, which are useful at higher temperatures. Lower-viscosity oils would be used at lower temperatures.

SUMMARY

1. For neutral molecules in the liquid state the intermolecular forces that attract molecules to each other are the dipole–dipole forces, the London forces, and hydrogen bonding.
2. Liquids are fluid, possessing definite volumes at fixed temperatures, but no definite shape.
3. At any given temperature, liquids and their vapors are in dynamic equilibrium with each other if the rate of evaporation equals the rate of condensation.
4. Vapor pressure is the pressure exerted by a vapor in dynamic equilibrium with a liquid at a given temperature: $\log P_{vap} = (-a/T) + b$.
5. A liquid boils when its vapor pressure equals the external pressure.
6. The normal boiling point of a liquid is the temperature at which its vapor pressure equals one standard atmosphere (760 torr).
7. Surface tension is a measure of the amount of work needed to increase the surface area of a liquid: $\gamma = \frac{1}{2}rhdg$.
8. Viscosity is a measure of the resistance to flow of a liquid.

RAISE YOUR GRADES

Can you define . . . ?

☑ equilibrium
☑ evaporation
☑ the liquid state
☑ boiling point
☑ normal boiling point

☑ condensation
☑ distillation
☑ surface tension
☑ viscosity

Can you . . . ?

☑ specify the forces that act between neutral molecules
☑ explain vapor pressure
☑ calculate how vapor pressure varies with temperature

SOLVED PROBLEMS

PROBLEM 6-1   Arrange the following in order of increasing intermolecular force (hence, increasing boiling point): F_2 (nonpolar), Ne, HF, ClF (polar).

Solution: Neon atoms are spherical and can only have London forces acting between the atoms. The molecule F_2 is not polar and hence can only have London forces acting between the molecules. Since F_2 has more electrons than Ne, the London forces in F_2 should be stronger than those in Ne. The ClF molecule is polar and so the dipole–dipole forces in ClF should be stronger than the London forces in F_2. Finally, HF should be hydrogen-bonded, and so should have the strongest intermolecular forces. Consequently, we have the following predicted order of intermolecular force and boiling point: Ne < F_2 < ClF < HF.

The normal boiling points of these species should be in the same order—as we see when we look up the actual boiling points:

$$\text{Ne} -246°C, F_2 -188°C, \text{ClF} -101°C, \text{HF } 19°C$$

PROBLEM 6-2   Calculate the vapor pressure of NH_3 at $-90°C$ and at $-80°C$, given the constant values $a = 1724$ and $b = 9.9974$.

Solution: From the vapor pressure equation (6-1), we have

$$\log P_{vap} = \frac{-a}{T} + b$$

Using the given values for a and b (and remembering that P must be in torrs and temperature must be in kelvins), we calculate the vapor pressure at $-90°C$ by writing

$$\log P_{vap} = \frac{-1724}{(273 - 90)} + 9.9974$$

$$= -9.42 + 10.00$$
$$= 0.58 \quad \text{(3 sig. figs.)}$$

Then

$$P_{vap} = \text{antilog } 0.58$$
$$= 3.8 \text{ torr}$$

We use the same procedure to calculate the vapor pressure at $-80°C$:

$$\log P_{vap} = \frac{-1724}{(273 - 80)} + 9.997 = 1.06$$

$$P_{vap} = \text{antilog } 1.06 = 11.5 \text{ torr}$$

PROBLEM 6-3   What is the vapor pressure of Br_2 at its normal boiling point (58°C)?

Solution: By definition, the normal boiling point of any liquid is the temperature at which its vapor pressure equals 1 atm, or 760 torr. So at its normal boiling point, Br_2 must have a vapor pressure of 760 torr.

PROBLEM 6-4   A liquid of density 0.865 $g\,cm^{-3}$ is observed to rise 10.4 mm in a tube of radius 0.50 mm. Calculate the surface tension of the liquid.

Solution: From the surface tension equation (6-2), we have

$$\gamma = \tfrac{1}{2} rhdg$$

Keeping our units consistent (and in SI), we set up:

$$r = 0.50 \text{ mm} \times \frac{1 \text{ m}}{10^3 \text{ mm}} = 5.0 \times 10^{-4} \text{ m}$$

$$h = 10.4 \text{ mm} \times \frac{1 \text{ m}}{10^3 \text{ mm}} = 1.04 \times 10^{-2} \text{ m}$$

$$d = (0.865 \text{ g}\,cm^{-3}) \left(\frac{1 \text{ kg}}{10^3 \text{ g}}\right) \left(\frac{100 \text{ cm}}{m}\right)^3 = 8.65 \times 10^2 \text{ kg}\,m^{-3}$$

$$g = 9.81 \text{ m}\,s^{-2}$$

So

$$\gamma = \tfrac{1}{2}(5.0 \times 10^{-4} \text{ m})(1.04 \times 10^{-2} \text{ m})(8.65 \times 10^2 \text{ kg}\,m^{-3})(9.81 \text{ m}\,s^{-2})$$
$$= 2.2 \times 10^{-2} \text{ kg}\,s^{-2}$$

or, in the more descriptive units, $\gamma = 2.2 \times 10^{-2} \text{ N m}^{-1}$, since $N \equiv m\,kg\,s^{-2}$ and $N\,m^{-1} \equiv kg\,s^{-2}$.

PROBLEM 6-5   Why do foods cook more rapidly in a pressure cooker than in an ordinary pot?

Solution: In an ordinary pot, at around 1 atm pressure, water boils at 100°C, its normal boiling point, and so food in that pot is cooking at a temperature of 100°C. A pressure cooker maintains a pressure greater than 1 atm inside; when water boils in a pressure cooker, its boiling point is greater than 100°C. For instance, if the pressure is around 2 atm inside the pressure cooker, water in it will boil at around 120°C. Consequently, the food in the pressure cooker, subjected to the higher temperature of the water in the cooker, cooks more quickly.

PROBLEM 6-6   Why is glycerol, which has a molecular formula $HOCH_2(CHOH)CH_2OH$, much more viscous than ethyl alcohol, which has a molecular formula CH_3CH_2OH?

Solution: Generally, the stronger the intermolecular forces, the more viscous the liquid. Glycerol, with its three OH groups per molecule, can form many more hydrogen bonds than can ethyl alcohol, with its one OH group per molecule—and the hydrogen bond is one of the strongest intermolecular forces. So glycerol is more viscous than ethyl alcohol (and, incidentally, has a much higher boiling point for similar reasons).

PROBLEM 6-7   Which of the following compounds should have the greater surface tension?

$$CH_3OCH_2CH_3 \qquad CH_3CH_2CH_2OH$$
methyl ethyl ether   *n*-propanol

Solution: Stronger intermolecular forces usually lead to greater surface tension. Now *n*-propanol will have hydrogen bonds, since the molecule has an OH group, but methyl ethyl ether has no hydrogen bonds. Since hydrogen bonding is the strongest kind of intermolecular force you have encountered, you predict that *n*-propanol will have stronger intermolecular forces—and thus greater surface tension—than methyl ethyl ether.

PROBLEM 6-8   At 40°C the vapor pressure of acetone is 495 torr, that of water is 55 torr, and that of dichloromethane is 750 torr. Arrange the compounds in order of increasing normal boiling point.

Solution: The higher the vapor pressure of a liquid at a given temperature, the lower its normal boiling point. Since the vapor pressure order at 40°C is dichloromethane > acetone > water, the normal boiling point order should be dichloromethane < acetone < water.

PROBLEM 6-9   Calculate the normal boiling point, in degrees Celsius, of boron tribromide (BBr_3). For this compound the values of the vapor pressure constants are $a = 1740.3$ and $b = 7.655$.

Solution: Since we need the *normal* boiling point, we know we need the vapor pressure equation (6-1):

$$\log P_{vap} = \frac{-a}{T} + b$$

where T is in degrees kelvin and P is in torrs. We also know that $P = 760$ torr, because (by definition) the normal boiling point is the temperature at which the vapor pressure equals 1 atm, or 760 torr. Using the given values of a and b, we substitute and rearrange to get T:

$$\log 760 = 2.8808 = \frac{-1740.3}{T} + 7.655$$

so

$$T = \frac{1740.3}{4.774} = 365 \text{ K}$$

Converting to degrees Celsius, the boiling point is

$$365 - 273 = 92°C$$

PROBLEM 6-10   You live in Washington D.C., and you like your breakfast eggs boiled for $3\frac{1}{2}$ min. You're on vacation, camping in the Rockies at an elevation of over 3 km. You boil your eggs for $3\frac{1}{2}$ min, crack one open, and find it's a runny mess. What has happened?

Solution: At a height of 3000 m, the pressure of the atmosphere is around 500 torr. The water in your pot will boil when its vapor pressure equals 500 torr, i.e., at a temperature of around 89°C. But at this temperature the protein in egg white cooks much more slowly than at the 100°C, the temperature of boiling water in Washington D.C. (at sea level, where the atmospheric pressure is around 760 torr). So to cook your eggs properly in the Rockies, you must either use a pressure cooker, or allow more cooking time.

PROBLEM 6-11   The time required for 2.00 mL of ethyl alcohol to flow through a section of capillary tubing is 40 s. It takes only 10 s for 2.00 mL of acetone to flow through the same capillary tube. Which of the two liquids has the greater viscosity?

Solution: Viscosity is the *resistance* of a liquid to flow. Since it takes the alcohol a longer time to flow through the tube, the alcohol must be the more resistant to flow. So ethyl alcohol has the greater viscosity.

PROBLEM 6-12   "Many smaller drops have a larger surface area than one large drop of equal volume." Show that this is true for a spherical drop of volume 1.00 cm³ that is divided into eight spherical drops, each of which has a volume of 0.125 cm³.

Solution: Recall that the volume V of a sphere of radius r is given by $V = 4\pi r^3/3$, and that the surface area A of a sphere of radius r is given by $A = 4\pi r^2$. So, by rearranging and substituting, we find the radius for a single drop of $V = 1.00$ cm^3:

$$r = \left(\frac{3V}{4\pi}\right)^{1/3} = \left(\frac{3(1.00 \text{ cm}^3)}{4\pi}\right)^{1/3} = 0.620 \text{ cm}$$

So the surface area of this drop is

$$A = 4\pi r^2 = 4\pi(0.620 \text{ cm})^2 = 4.83 \text{ cm}^2$$

For the smaller drops ($V = 0.125$ cm^3)

$$r = \left(\frac{3(0.125 \text{ cm}^3)}{4\pi}\right)^{1/3} = 0.310 \text{ cm}$$

and

$$A = 4\pi(0.310 \text{ cm})^2 = 1.21 \text{ cm}^2$$

so the total surface area of eight 0.125-cm^3 drops is 8(1.21 cm^2) = 9.68 cm^2. Thus dividing one large drop into eight small drops doubles the surface area.

Supplementary Exercises

PROBLEM 6-13 Arrange the following in order of increasing intermolecular forces and boiling points: NH_3, He, Cl_2.

Answer: He $<$ Cl_2 $<$ NH_3

PROBLEM 6-14 Calculate the vapor pressure of NH_3 at 200 K ($a = 1724$, $b = 9.9974$).

Answer: 24 torr

PROBLEM 6-15 Calculate the normal boiling point of benzoyl chloride ($a = 2372$, $b = 7.9245$).

Answer: 470 K or 197°C

PROBLEM 6-16 A liquid of density 1.742 g cm^{-3} is observed to show a capillary rise of 7.3 mm in a tube of radius 0.65 mm. Calculate the surface tension of the liquid.

Answer: 4.1×10^{-2} N m^{-1}

PROBLEM 6-17 Which of the following should be the most viscous? Why?

$$CH_3CH_2OH \qquad CH_3OCH_3 \qquad HOCH_2CH_2OH$$

Answer: $HOCH_2CH_2OH$; most hydrogen-bonded (see Problem 6-6)

PROBLEM 6-18 At 30°C ethanol (C_2H_5OH) has a vapor pressure of about 90 torr, acetic acid (CH_3CO_2H) has a vapor pressure of about 30 torr, and ethyl iodide (C_2H_5I) has a vapor pressure of about 200 torr. Arrange these compounds in order of increasing normal boiling point.

Answer: C_2H_5I $<$ C_2H_5OH $<$ CH_3CO_2H

7 SOLIDS

THIS CHAPTER IS ABOUT

☑ **The Solid State**
☑ **Types of Solids**
☑ **X-Ray Diffraction**
☑ **Crystal Structure**
☑ **Common Structures of Salts**

7-1. The Solid State

A solid—matter that is in the solid state—has a definite shape that is resistant to change. At a given temperature a solid has a fixed volume and density. Solids don't flow.

7-2. Types of Solids

Solids are characterized by the nature of their structural units and by the pattern in which their structural units are arranged. There are two types of solids—amorphous and crystalline.

A. Amorphous solids

The structural units of an **amorphous solid**, whether they are atoms, molecules, or ions (charged species), occur at random positions. As in liquids, there is no ordered pattern to the arrangement of an amorphous solid. Glass and tar are familiar examples of amorphous solids.

B. Crystalline solids

The structural units of a **crystalline solid** have a characteristic repetitive pattern. There are four important classes of crystalline solids:

1. Metallic solids

The fundamental units of a pure **metallic solid** are identical metal atoms. Metallic crystals are opaque with reflective surfaces. They are *ductile*; that is, they can be shaped if sufficient pressure is applied. Metals are good conductors of heat and electricity, and they usually have high melting points. Copper, silver, aluminum, and iron are familiar examples of metals.

2. Molecular solids

The fundamental unit of a **molecular solid** is the molecule. Such solids are common among organic compounds and simple inorganic compounds. Molecular crystals are usually transparent, brittle, and break easily when stressed. They are usually poor conductors of heat and electricity and usually have low melting points. Familiar molecular crystalline solids include sugar, aspirin, and dry ice (solid CO_2).

3. Ionic solids

The fundamental units of an **ionic solid** are positive and negative ions. Crystalline ionic solids are usually transparent, brittle, and poor conductors of electricity, although molten crystals may be good conductors. They usually have high melting points. Some of the more familiar ionic solids are table salt (NaCl), saltpeter (KNO_3), blackboard chalk ($CaSO_4$), and washing soda (Na_2CO_3).

4. Covalent network solids

In a **covalent network solid** the whole crystal is one giant molecule. The fundamental units are atoms *covalently* bonded to their neighbors (see Chapter 10). Their crystals are usually brittle, poor conductors, and have high melting points. Examples of covalent network solids include quartz (SiO_2) and diamond (a form of carbon).

7-3. X-Ray Diffraction

Chemists investigate the internal structure of crystals with x rays. When an x-ray beam of fixed single wavelength λ (lambda) strikes a crystal, the beam is bent, or *diffracted*, by the regularly spaced rows of molecules, ions, or atoms in the crystal and produces a distinctive display of light and dark areas on a sensitive plate. This display is called a **diffraction pattern**.

The angles at which diffraction occurs can be calculated by **Bragg's law**:

BRAGG'S LAW $$n\lambda = 2d \sin \theta \qquad\qquad (7\text{-}1)$$

where n is an integer, λ the x-ray wavelength, d the distance between planes of units in the crystal, and θ (theta) the angle of diffraction.

note: Amorphous solids give only diffuse x-ray scattering and no regular diffraction pattern.

EXAMPLE 7-1: The NaCl crystal is made up of sets of planes that are 397 pm apart. If an x-ray beam with a wavelength of 154 pm strikes this set of planes, at what angle will a diffraction pattern be observed?

Solution: Rearrange Bragg's law to give $\sin \theta$. Use $n = 1$ to make the calculations easier:

$$\sin \theta = \frac{n\lambda}{2d}$$

$$= \frac{(1)(154 \text{ pm})}{(2)(397 \text{ pm})} = 0.1940$$

(Notice that the units of length must be the same and cancel.) Find arc sin 0.1940, using a table of trig functions or the inverse sine function on a calculator:

$$\theta = \text{arc sin } 0.1940 = 11.18°$$

So we see diffraction at an angle of 11°, which is close enough for our purposes.

Diffraction should also be observed at $n = 2$:

$$\sin \theta = \frac{(2)(154 \text{ pm})}{(2)(397 \text{ pm})} = 0.388, \qquad \theta = \text{arc sin } 0.388 = 23°$$

7-4. Crystal Structure

The regular and symmetrical external form of crystals suggests that there is a regular internal arrangement of the building blocks of a crystalline solid. Crystals have plane faces and often have regular geometry. For example, if you examine common table salt under the microscope, you'll see that its crystals are nicely regular cubes. When a crystalline solid breaks, it *cleaves* between planes of particles (atoms, molecules, or ions), producing fragments with planar faces. An amorphous solid like glass breaks every which way, leaving irregular pieces with curved surfaces.

A. Characterizing crystals: Definitions

- A **space lattice** is a pattern of points that describes the arrangement of particles in a crystal.
- A **unit cell** is the smallest portion of the space lattice that includes enough information to allow you to infer the whole crystal structure or shape—by repeating the unit cell in three dimensions. The stoichiometry, density, and spatial arrangement of the unit cell are equivalent to those of the whole crystal.

- A **primitive cell** is occupied by particles only at its corners. Primitive unit cells include cubic, tetragonal, and hexagonal patterns.
- The **crystal coordination number** is the number of nearest (particle) neighbors a particle can have in a particular crystal structure. This number reflects the way in which unit cells are arranged, or *packed*, within a crystal structure.

B. Types of crystal structures

Crystal structures can occur in several geometric shapes, which can be packed in several ways. Each shape has a characteristic number of particles in its unit cell, and each packing arrangement has a characteristic coordination number. (We'll discuss only a few of the simplest crystal structures here.)

1. Primitive cubic structure

A **primitive cubic structure** is a shape whose corners are occupied by particles and whose edges are all of equal length.

To determine how many particles there are in each unit cell, we have to think in three dimensions. If we look at a cubic unit cell (see Figure 7-1a), we can infer that two adjacent cubes meet at a face, four adjacent cubes meet along an edge, and *eight* adjacent cubes meet at a corner.

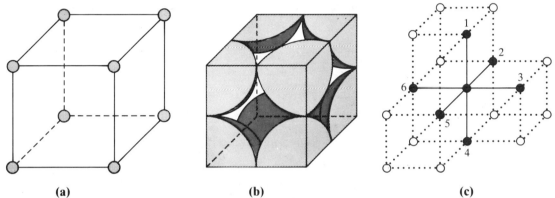

(a)   **(b)**   **(c)**

FIGURE 7-1. (a) Primitive cubic unit cell; (b) primitive cubic unit cell, expanded; (c) nearest particle neighbors in primitive cubic structure; crystal coordination number = 6.

Now, because each particle is at a corner, only *one-eighth* of a particle is at each corner in any one unit cell (see Figure 7-1b). Thus we calculate the total number of particles in a primitive unit cell by multiplying the number of corners by the fraction of particles per corner:

$$(8 \text{ corners}) \times \tfrac{1}{8} \text{ particle/corner} = 1 \text{ particle}$$

The crystal coordination number of each particle in a primitive cubic structure is 6, as each particle can have six nearest particle neighbors—four in its own plane and one each above and below (see Figure 7-1c).

2. Face-centered cubic structures

The **face-centered cubic structure** (*fcc*), whose unit cell is shown in Figure 7-2, has particles at the corners of a cube and also at the center of each face of the cube. The total content of the unit cell is

$$
\begin{array}{l}
8 \text{ corners} \times \tfrac{1}{8} \text{ particle/corner} \\
+\ 6 \text{ faces} \times \tfrac{1}{2} \text{ particle/face} \\
\hline
4 \text{ particles in a unit cell}
\end{array}
$$

As we can see in Figure 7-3, the *fcc* structure, whose coordination number is 12, is one way of packing a given space as tightly as possible with equal spheres. This arrangement is called **cubic closest packing**. Seventy-four percent of the available space is occupied.

note: **Hexagonal closest packing** is another arrangement that has a coordination number of 12 and occupies 74% of the available space. As you might guess, the units are arranged in a hexagonal pattern. About two-thirds of all metals are cubic or hexagonal closest-packed.

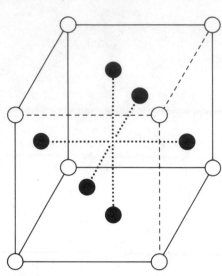

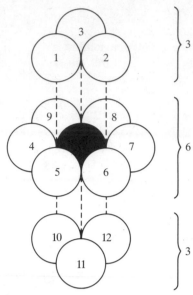

FIGURE 7-2.
Face-centered cubic unit cell. (The face-centered atoms are shaded.)

FIGURE 7-3.
Cubic closest packing in a face-centered cubic structure; coordination number = 12.

EXAMPLE 7-2: Silver crystallizes in an *fcc* structure like the one shown in Figure 7-2. A single face is isolated in Figure 7-4. Notice that the atoms touch along the *face diagonal* (d_f) so that its length is four times the radius of the silver atom, i.e., $d_f = 4r$. If the radius of a silver atom is 144.5 pm, what is the length of the face diagonal and the edge length a of the unit cell?

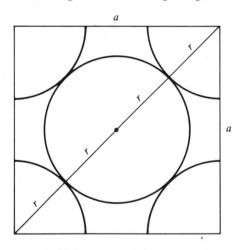

FIGURE 7-4.
One face of a face-centered cubic unit cell; a = edge length, $d_f = 4r$.

Solution: The face diagonal $d_f = 4r = 4(144.5 \text{ pm}) = 578.0 \text{ pm}$. The edges a of the unit cell are the legs of a right triangle and d_f is the hypotenuse, so (remembering that all the edges in a cube are of equal length) $a^2 + a^2 = d_f^2$ and

$$a = \frac{d_f}{\sqrt{2}} = \frac{578.0 \text{ pm}}{\sqrt{2}}$$

$$= 408.7 \text{ pm}$$

3. Body-centered cubic structure

The **body-centered cubic structure** (*bcc*), shown in Figure 7-5, has its particles at the corners of a cube with one additional particle at the center of the cube. The total content of the unit cell is

$$8 \text{ corners} \times \tfrac{1}{8}\text{particle/corner}$$
$$+ 1 \text{ particle/cube center}$$
$$\overline{2 \text{ particles in a unit cell}}$$

The coordination number of each *bcc* particle is 8.

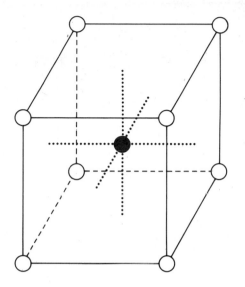

FIGURE 7-5.
Body-centered cubic unit cell. (The body-centered atom is shaded.)

EXAMPLE 7-3: Lithium crystallizes in a *bcc* structure like that shown in Figure 7-5. The atoms that lie along the *body diagonal* (d_b: the line between opposite corners of the cube) are touching. Thus the body diagonal is four times the radius of the Li atom, i.e., $d_b = 4r$. If the edge length of the unit cell is 353 pm, what is the radius of the Li atom?

Solution: Although the relationship between the edge and body diagonal is a bit more complicated than that between the edge and the face diagonal in the face-centered cube, the body diagonal is also the hypotenuse of a right triangle (see Figure 7-6). One leg of the triangle is an edge of the cube a and the other is d_f. Since $d_f = a\sqrt{2}$, then

$$d_b^2 = (a\sqrt{2})^2 + a^2$$
$$= 2a^2 + a^2$$
$$= 3a^2$$

Taking the square root, $d_b = a\sqrt{3}$. Since $d_b = 4r$,

$$r = \frac{d_b}{4}$$
$$= \frac{a\sqrt{3}}{4}$$
$$= \frac{(353 \text{ pm})\sqrt{3}}{4}$$
$$= 153 \text{ pm}$$

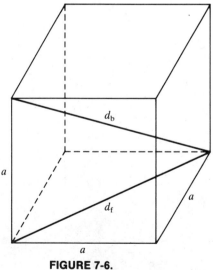

FIGURE 7-6.
Relationship of d_b to d_f.

C. Density and crystal structure

The ordinary bulk density of a solid must be the same as the density of a single unit cell: $d = m/v$, where m is the mass of the contents of the unit cell and v is the unit cell volume.

EXAMPLE 7-4: Use the data of Example 7-2 together with the value of the atomic mass of silver to calculate the density of metallic silver in grams per cubic centimeter.

Solution: In Example 7-2, we calculated that the edge length of a unit cell of the *fcc* structure of silver is 408.7 pm. Thus the volume of the unit cell (in cm^3) is

$$\left[\frac{(408.7 \text{ pm})(1 \text{ cm})}{10^{10} \text{ pm}}\right]^3 = 6.827 \times 10^{-23} \text{ cm}^3$$

An *fcc* unit cell contains 4 particles (Section 7-4B2), so the mass of the unit cell contents is the mass of 4 Ag atoms, which (in g) is

$$\frac{4(107.9 \text{ amu/atom})}{6.022 \times 10^{23} \text{ amu/g}} = 7.167 \times 10^{-22} \text{ g}$$

So

$$d = \frac{m}{v} = \frac{7.167 \times 10^{-22} \text{ g}}{6.827 \times 10^{-23} \text{ cm}^3} = 10.50 \text{ g/cm}^3$$

(This is exactly equal to the reported value of the density of silver, as you can check in a handbook of chemical constants—but don't be too impressed; in fact, the radius of the silver atom is calculated from the measured density of silver together with x-ray diffraction data.)

7-5. Common Structures of Salts

Common salts are ionic crystals composed of two different ions of opposite charge, so packing is less straightforward than in metallic crystals. In addition, electrical neutrality must be maintained. The structures of NaCl, ZnS, and CaF$_2$ are shown in Figure 7-7. Many other salts have similar structures. Notice in the illustration of NaCl structure that the chloride ions (open circles) are in the *fcc* arrangement. The sodium ions (shaded circles) are located at the center of the cube and the middle of each edge of the cube. Ions of like charge are relatively distant from one another and each ion is surrounded by ions of opposite charge, thus increasing the stability of the crystal.

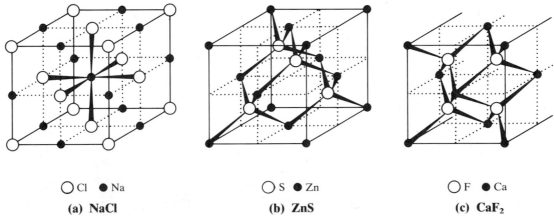

    ○ Cl ● Na   ○ S ● Zn   ○ F ● Ca

    (a) NaCl   **(b) ZnS**   **(c) CaF$_2$**

FIGURE 7-7. Common structures of salts: (**a**) NaCl (also LiF, NaF, KCl, LiI, etc.); (**b**) ZnS (also AgI, ZnO, HgS, etc.); (**c**) CaF$_2$ (also SrF$_2$, BaCl$_2$, ZrO$_2$, etc.)

EXAMPLE 7-5: What fractions of the sodium ions on the edges of the NaCl cube are within the unit cell? How many ions of each kind are there in the unit cell?

Solution: Four adjacent cubes meet at each edge, so each sodium edge ion is $\frac{1}{4}$ within the cube. The sodium ion at the center of the cube is entirely within the unit cell and there are 12 edges, so there are $1 + \frac{1}{4}(12) = 4$ sodium ions within the unit cell. The chloride ions are in the *fcc* arrangement, so there must be four of them also. Thus the electrical neutrality of the unit cell, and of the salt, is confirmed.

SUMMARY

1. The solid state is nonfluid.
2. The structural components of an amorphous solid are not ordered, whereas crystalline solids have characteristic repeating patterns.
3. Crystalline solids are of four types: metallic, molecular, ionic, and covalent network.
4. Crystalline solids cleave along planes and produce regular x-ray diffraction patterns; amorphous solids don't.
5. Bragg's law ($n\lambda = 2d \sin \theta$) can be used to find the angle of diffraction or the distance between the planes in a crystalline solid.
6. A unit cell is the smallest portion of the space lattice representation of a crystal that can be used to reproduce the entire crystal.
7. Important cubic structural types include simple, face-centered, and body-centered.
8. Common salts are electrically neutral ionic crystals whose ions form characteristic patterns.

RAISE YOUR GRADES

Can you...?

☑ explain the difference between crystalline and amorphous solids
☑ give examples of the four important types of crystalline solids
☑ apply Bragg's Law
☑ sketch simple cubic, body-centered cubic, and face-centered cubic unit cells
☑ use unit cell structures and geometric relationships to calculate densities, unit-cell dimensions, and atomic radii

SOLVED PROBLEMS

PROBLEM 7-1 Which of the following is false? A solid (**a**) has a definite shape, (**b**) always fills the entire container, (**c**) does not flow, (**d**) has a fixed volume and density at a given temperature, (**e**) does not deform to take the shape of its container.

Solution: The correct answer is (**b**). Only gases always fill the entire container. This is not the property of a solid.

PROBLEM 7-2 What is the difference between an amorphous solid and a crystalline solid? Give an example of an amorphous solid.

Solution: In crystalline solids the atoms, ions, or molecules are arranged in an ordered pattern. In an amorphous solid the components are not in an ordered pattern. Examples of amorphous solids include most solid plastics, obsidian, glass, and tar.

PROBLEM 7-3 Give an example of each of the following classes of crystalline solids: (**a**) metallic, (**b**) molecular, (**c**) ionic, (**d**) covalent network.

Solution: Look in your textbook for some examples besides those listed here. (**a**) sodium, (**b**) naphthalene, (**c**) magnesium sulfate, (**d**) silicon.

PROBLEM 7-4 A piece of glass may be cut into a cube and polished so that it appears identical to a large, clear sodium chloride crystal. What difference in behavior between the glass and sodium chloride would you expect to observe if you broke off a piece of either cube?

Solution: The sodium chloride (an ionic solid) would break along flat surfaces parallel to the faces of the cube because the planes of its component ions are parallel to the faces of the crystalline cube. The glass (an amorphous solid) would break irregularly, usually in curved shapes, because its component molecules are not arranged in an ordered pattern.

PROBLEM 7-5   The x-ray diffraction pattern of a particular crystal shows an angle of diffraction at 17°. The x-ray wavelength is 139 pm. Calculate the distance between the layers of atoms assuming $n = 1$.

Solution: Given the angle of diffraction ($\theta = 17°$) and the wavelength ($\lambda = 139$ pm), we can rearrange Bragg's law ($n\lambda = 2d \sin \theta$) to calculate the distance d between layers:

$$d = \frac{n\lambda}{2 \sin \theta} = \frac{(1)(139 \text{ pm})}{2(0.292)} = 238 \text{ pm}$$

PROBLEM 7-6   A very small cube of NaCl measures 562 pm on each edge and contains exactly 4 Na ions and 4 Cl ions. Calculate the bulk density of NaCl in $g \, cm^{-3}$.

Solution: Recognizing that the conditions described represent a unit cell of NaCl—whose density is the same as the bulk density of NaCl—we set up the known conditions in the desired units (g and cm^3):
The edge length a is

$$a = (562 \text{ pm})(10^{-10} \text{ cm/pm}) = 5.62 \times 10^{-8} \text{ cm}$$

So the volume V is

$$V = a^3 = (5.62 \times 10^{-8} \text{ cm})^3 = 177.5 \times 10^{-24} \text{ cm}$$

The mass m of 4 NaCl formula units is calculated (using Avogadro's number) as

$$m = \frac{4(23.0 + 35.5) \text{ amu}}{6.02 \times 10^{23} \text{ amu g}^{-1}} = 38.9 \times 10^{-23} \text{ g}$$

Now we can use the definition of density to find d:

$$d = \frac{m}{V} = \frac{38.9 \times 10^{-23} \text{ g}}{177.5 \times 10^{-24} \text{ cm}^3} = 2.19 \text{ g cm}^{-3}$$

(As a double-check, we look up the experimentally determined density of NaCl, which is 2.17 $g \, cm^{-3}$—fairly close!)

PROBLEM 7-7   Solid gold has a face-centered cubic (*fcc*) structure. The length of the unit cell edge is 408 pm. Calculate (**a**) the radius of the Au atom, (**b**) the volume of the unit cell in cm^3, and (**c**) the density of solid Au in $g \, cm^{-3}$.

Solution: Since Au is *fcc*, we know that there are four atoms in a unit cell whose edge lengths are all equal.

(**a**) To calculate the radius, we use the edge length ($a = 408$ pm) to find the face diagonal d_f by the Pythagorean theorem (remembering that d_f is the hypotenuse of a right triangle: $d_f^2 = 2a^2$):

$$d_f = a\sqrt{2} = (408 \text{ pm})\sqrt{2} = 577 \text{ pm}$$

Then, since the face diagonal must be 4 (atoms) times the radius,

$$r = \frac{d_f}{4} = \frac{577 \text{ pm}}{4} = 144 \text{ pm}$$

(**b**) The volume is simply the edge length cubed: $V = a^3$. But the answer must be in cm^3, so we convert first (which is often easier):

$$a = (408 \text{ pm})(10^{-10} \text{ cm/pm}) = 4.08 \times 10^{-8} \text{ cm}$$
$$V = a^3 = (4.08 \times 10^{-8} \text{ cm})^3 = 67.9 \times 10^{-24} \text{ cm}^3 = 6.79 \times 10^{-23} \text{ cm}^3$$

(**c**) The density of the unit cell is also the density of solid Au, so we calculate the mass of the unit cell in grams:

$$m = \frac{(4 \text{ atoms})(197.0 \text{ g mol}^{-1})}{6.02 \times 10^{23} \text{ atoms mol}^{-1}} = 1.31 \times 10^{-21} \text{ g}$$

Now the density is

$$d = \frac{m}{V} = \frac{1.31 \times 10^{-21} \text{ g}}{6.79 \times 10^{-23} \text{ cm}^3} = 19.3 \text{ g cm}^{-3}$$

PROBLEM 7-8 Solid iron has a *bcc* structure, and its density is 7.86 g/cm^3. Calculate **(a)** the edge length of the unit cell and **(b)** the radius of the Fe atom in picometers.

Solution: The *bcc* structure of Fe contains 2 atoms per unit cell. Therefore the mass of Fe in the unit cell is

$$m = \frac{(2 \text{ atoms})(55.8 \text{ g/mol})}{6.02 \times 10^{23} \text{ atoms/mol}} = 1.85 \times 10^{-22} \text{ g}$$

Because the density of Fe is also the density of the unit cell, the volume of the unit cell is

$$V = \frac{m}{d} = \frac{1.85 \times 10^{-22} \text{ g}}{7.86 \text{ g/cm}^3} = 2.36 \times 10^{-23} \text{ cm}^3$$

(a) The edge length of this cube is the cube root of the volume:

$$a = \sqrt[3]{V} = (23.6 \times 10^{-24} \text{ cm}^3)^{1/3}$$
$$= 2.87 \times 10^{-8} \text{ cm} = (2.87 \times 10^{-8} \text{ cm})(10^{10} \text{ pm/cm}) = 287 \text{ pm}$$

(b) The body diagonal is four times the radius and equals $a\sqrt{3}$ (see Example 7-3). Therefore

$$r = \frac{a\sqrt{3}}{4} = \frac{\sqrt{3}(287 \text{ pm})}{4} = 124 \text{ pm}$$

PROBLEM 7-9 The density of silicon is 2.33 g cm^{-3}. The unit cell is cubic and the edge length is 543 pm. The number of atoms in the unit cell is **(a)** 1, **(b)** 2, **(c)** 4, **(d)** 6, **(e)** 8.

Solution: Determine the mass of the unit cell and then divide it by the mass of one Si atom. The volume of the unit cell is the edge length cubed, so (changing pm to cm)

$$a = (543 \text{ pm})(10^{-10} \text{ cm/pm}) = 5.43 \times 10^{-8} \text{ cm}$$
$$V = a^3 = (5.43 \times 10^{-8} \text{ cm})^3 = 1.60 \times 10^{-22} \text{ cm}^3$$

The mass of the unit cell is

$$m = dV = (2.33 \text{ g cm}^{-3})(1.60 \times 10^{-22} \text{ cm}^3) = 3.73 \times 10^{-22} \text{ g}$$

The mass of one Si atom is

$$\frac{28.1 \text{ g/mol}}{6.02 \times 10^{23} \text{ atoms/mol}} = 4.67 \times 10^{-23} \text{ g}$$

so the number n of Si atoms in one unit cell must be

$$n = \frac{3.73 \times 10^{-22} \text{ g}}{4.67 \times 10^{-23} \text{ g}} = 8$$

The correct answer is **(e)**. (The structure of silicon is the same as that of diamond.)

PROBLEM 7-10 Cerium phosphide (CeP) crystallizes in a cubic structure identical to NaCl. The edge length of the unit cell is 591 pm. Calculate the density of CeP.

Solution: Determine the mass of the unit cell and its volume. The total mass includes the mass of four Ce atoms and four P atoms:

$$m = \frac{(4 \text{ atoms})(140.1 + 31.0) \text{ g/mol}}{6.02 \times 10^{23} \text{ atoms/mol}} = 1.14 \times 10^{-21} \text{ g}$$

The volume of the unit cell is $V = a^3$ and $a = (591 \text{ pm})(10^{-10} \text{ cm/pm}) = 5.91 \times 10^{-8} \text{ cm}$, so

$$V = (5.91 \times 10^{-8} \text{ cm})^3 = 2.06 \times 10^{-22} \text{ cm}^3$$

and

$$d = \frac{m}{V} = \frac{1.14 \times 10^{-21} \text{ g}}{2.06 \times 10^{-22} \text{ cm}^3} = 5.53 \text{ g/cm}^3$$

Supplementary Exercises

PROBLEM 7-11   A very small cube of lead contains four Pb atoms. The volume of the cube is 1.206×10^{-22} cm^3. Calculate (**a**) the edge length of the cube in pm and (**b**) the density of Pb in g/cm^3.

Answer: (**a**) 494 pm   (**b**) 11.4 g/cm^3

PROBLEM 7-12   Solid copper has an *fcc* structure. Its density is 8.96 g cm^{-3}. Calculate (**a**) the volume of the unit cell, (**b**) the edge length of the unit cell, and (**c**) the radius of the Cu atom.

Answer: (**a**) 4.71×10^{-23} cm^3 (**b**) 361 pm   (**c**) 128 pm

PROBLEM 7-13   Solid sodium metal has a *bcc* structure. The edge length of the unit cell is 429 pm. Calculate (**a**) the radius of the Na atom and (**b**) the density of solid Na.

Answer: (**a**) 186 pm   (**b**) 0.966 g cm^{-3}

PROBLEM 7-14   When titanium is heated above 1200 K, it changes structure from hexagonal to a cubic form. If the density above 1200 K is 4.33 g cm^{-3} and the edge length of the cubic unit cell is 333 pm, how many atoms are there in the unit cell and what is the type of cubic structure?

Answer: 2 atoms/unit cell; it is probably *bcc*

PROBLEM 7-15   MgO crystallizes in the same structure as NaCl. The edge length of the cubic unit cell is 411 pm. Calculate the density of MgO.

Answer: 3.86 g cm^{-3}

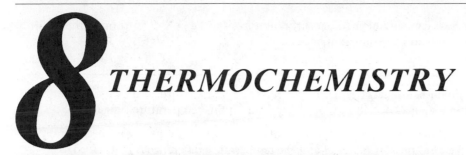

8 THERMOCHEMISTRY

THIS CHAPTER IS ABOUT

☑ **Temperature and Temperature Scales**
☑ **Heat Energy and Heat Capacity**
☑ **Enthalpy**

8-1. Temperature and Temperature Scales

Temperature measures the intensity with which heat energy is concentrated in a system. When two objects at different temperatures are put into contact, heat energy passes from the object at higher temperature to the one at lower temperature until their temperatures reach **thermal equilibrium**, i.e., become equal. You already know from the discussion of the kinetic molecular theory (Chapter 5) that temperature is a direct measure of the kinetic energy of the molecules of a gas, or the particles of any material.

You measure temperature with a *thermometer*. Common thermometers use the expansion of a liquid (mercury, alcohol, kerosene) as the basis of the measurement of temperature: The higher the temperature, the larger the volume of a fixed amount of liquid. You can use many other physical properties to measure temperature: the expansion of a gas, the generation of a voltage at a junction of dissimilar metals, a thermocouple, or the resistance of a wire to the passage of an electric current. You'll encounter two scales of temperature in chemistry: the absolute or Kelvin scale (K) and the Celsius or centigrade scale (°C). You met them earlier in the discussion of measurement (see Chapter 1) and of gases (see Chapter 5). As a reminder, the size of the degree interval is the same for these two scales, but the numerical values of the scales are different. The relationship between them is $K = °C + 273$. You'll still see the Fahrenheit scale (°F) in everyday use in cooking, weather forecasting, and so on. Its relationship to the Celsius scale is $°F = \frac{9}{5}(°C) + 32$.

8-2. Heat Energy and Heat Capacity

A. Heat energy

Heat energy is the kinetic energy of the particles that make up matter. In SI, heat energy, like all forms of energy, is measured in *joules* (J), where $J = N\,m = kg\,m^2\,s^{-2}$. In cgs, heat energy is measured in calories, where 1 calorie (cal), now defined in terms of SI units, equals exactly 4.184 J. Approximately 1 cal is required to increase the temperature of exactly 1 g of water by 1 K or 1°C degree at ordinary room temperature (around 17°C, 290 K).

note: The calorie counted by diet-watchers equals 1000 cal or 1 kilocalorie (kcal), sometimes abbreviated Cal.

B. Heat capacity

Heat capacity (C) is the amount of heat energy needed to raise the temperature of a certain amount of material by 1 K or 1°C. When the amount of material is given in moles, we use *molar* heat capacity. You may see the term *specific heat* used to indicate the amount of heat required to raise the temperature of 1 g of substance by 1°C. Because the heat capacity of any substance varies only slowly with temperature, you can often ignore the experimental temperature in work that is not of the very highest precision. Constant pressure or volume are indicated by a subscript on C (C_p or C_v). For example, you might say that the heat capacity of water at room temperature (near 25°C) is

$4.18 \, J\, g^{-1} \, K^{-1}$; the heat capacity of metallic chromium is $0.488 \, J\, g^{-1} \, K^{-1}$ at room temperature. Atmospheric pressure would be assumed in both cases.

EXAMPLE 8-1: How much heat energy is needed to raise the temperature of 58.4 g of water from 25.0°C to "blood heat" 36.9°C, if $C = 4.18 \, J\, g^{-1} K^{-1}$ for water in this temperature range?

Solution: It's clear from the definition of heat capacity that to isolate heat energy q, you need to multiply the heat capacity C by the amount n of material and the temperature difference ΔT involved, or

HEAT ENERGY $$q = nC \, \Delta T$$

where n is in moles or grams, depending on the units of C. For this problem

$$q = (58.4 \, g \, H_2O)\left(\frac{4.18 \, J}{g \, K}\right)(36.9° - 25.0°)$$

$$= 2.90 \times 10^3 \, J \quad \text{or} \quad 2.90 \, kJ$$

Note that the temperature *difference* is the same, whether expressed in kelvins or degrees Celsius, because the size of the degree interval is the same in the two scales.

It's interesting that the molar heat capacity of many metallic elements at room temperature is around $26 \, J\, mol^{-1} K^{-1}$ ($6.2 \, cal \, mol^{-1} K^{-1}$). Thus the molar mass times the heat capacity of a metal equals $\sim 26 \, J\, mol^{-1} K^{-1}$. This phenomenon is known as the **law of Dulong and Petit**.

EXAMPLE 8-2: Does metallic chromium obey the law of Dulong and Petit? (The heat capacity of metallic Cr is $0.448 \, J\, g^{-1} K^{-1}$.)

Solution: The mass of 1 mol of Cr atoms is 52.0 g, so the molar heat capacity of Cr is

$$\frac{0.448 \, J}{g \, K} \frac{52.0 \, g}{mol} = 23.3 \, J\, mol^{-1} K^{-1}$$

Because the law is only an approximation—it says heat capacities of many metallic elements are *around* $26 \, J\, mol^{-1} K^{-1}$—and because $23.3 \, J\, mol^{-1} K^{-1}$ is within 10% of the value given, you could say that metallic Cr does obey the law of Dulong and Petit.

8-3. Enthalpy

Most chemical reactions in the laboratory are carried out at constant pressure, usually the pressure of the atmosphere. The *change* in heat energy of a chemical system held at constant pressure is an **enthalpy** change, symbolized by ΔH. You'll often see the capital Greek Δ (delta) used in scientific writing as a symbol for a change (calculated by subtracting the initial value from the final value) in a property. For a system in which a chemical reaction is occurring, $\Delta H = H_2 - H_1$, where H_2 is the final enthalpy of the system, and H_1 is the initial enthalpy of the system.

A. Kinds of enthalpy changes

Some special kinds of enthalpy changes that you'll meet in your work in chemistry include

1. **Enthalpy of fusion, ΔH_m:** The enthalpy change that occurs when a substance is fused or melted. Water, for instance, has an enthalpy of fusion of $6.02 \, kJ \, mol^{-1}$.
2. **Enthalpy of vaporization, ΔH_{vap}:** The enthalpy change that occurs when a substance is boiled or vaporized. The enthalpy of vaporization varies with pressure but is usually expressed at a pressure of 1 atm. For water, ΔH_{vap} is $41.1 \, kJ \, mol^{-1}$.
3. **Standard enthalpy of formation, ΔH_f°:** The enthalpy change that occurs when 1 mol of a substance in its most stable state at 298 K is formed from its constituent elements in their most common states at 298 K. The degree sign (°) superscript on H indicates a *standard* enthalpy change.

note: You may see ΔH_m° and ΔH_{vap}° referred to as the *standard* enthalpy of fusion and vaporization, respectively. The degree sign signifies that the measurements were obtained at the normal melting or boiling point of the material and at a pressure of 1 atm. The degree sign on ΔH_f° means that the elements are in their most stable states at 298 K and that they are at a pressure of 1 atm. Do not confuse these standard enthalpies with STP, standard temperature and pressure (see Chapter 5).

The standard enthalpy of formation of liquid water is -286 kJ mol^{-1} for the reaction

$$H_2(g) + \tfrac{1}{2}O_2(g) \longrightarrow H_2O(l) \qquad \Delta H_f^{\circ} = -286 \text{ kJ mol}^{-1}$$

The standard enthalpies of formation of the elements in their common forms at 298 K are, by convention, assigned values of zero. Values of ΔH_f° for selected compounds are listed in Table 8-1.

TABLE 8-1: Standard Enthalpies of Formation ΔH_f° in kJ mol^{-1}

Substance and state	ΔH_f°	Substance and state	ΔH_f°
$BCl_3(l)$	-418	$NH_3(g)$	-46
$B_2H_6(g)$	31	$N_2H_4(l)$	50
$CaH_2(s)$	-189	$HNO_3(l)$	-173
$Ca(OH)_2(s)$	-987	$PH_3(g)$	9
$CO(g)$	-111	$CO_2(g)$	-394
$CF_4(g)$	-680	$NaCl(s)$	-411
$HCl(g)$	-92	$NaOH(s)$	-427
$HF(g)$	-269	$Na_2CO_3(s)$	-1131
$H_2O(l)$	-286	$SO_2(g)$	-297
$H_2O(g)$	-242	$PCl_3(l)$	-339
$H_2O_2(l)$	-188	$SO_3(g)$	-395
$N_2O(g)$	82	$H_2S(g)$	-20
$NO(g)$	90	$H_2SO_4(l)$	-811
$NO_2(g)$	34	$CH_4(g)$	-75

B. Exothermic and endothermic reactions

In an **exothermic** reaction, heat energy is given out by the reacting system to its surroundings. The burning of natural gas is a familiar example. Because the reacting system is losing heat energy, it will have a negative enthalpy change, indicated by a negative value for ΔH. In an **endothermic** reaction, heat energy is taken up by the reacting system from its surroundings. Endothermic reactions have positive enthalpy changes, indicated by positive values of ΔH.

EXAMPLE 8-3: Is the following reaction exothermic or endothermic?

$$C(graphite) + O_2(g) \longrightarrow CO_2(g) \qquad \Delta H = -393.5 \text{ kJ mol}^{-1}$$

Solution: The negative sign means that the process is exothermic; the system loses heat energy to the surroundings.

C. Hess' law

Hess' law states that for a process or reaction that is the result of a set of other reactions or processes, the total enthalpy change will equal the algebraic sum of the enthalpy changes of those other reactions or processes. Stated another way, the total enthalpy change is independent of the numbers of intermediate reactions.

EXAMPLE 8-4: Calculate the standard enthalpy change for the following process:

$$H_2O(s) \longrightarrow H_2O(g)$$

Solution: The conversion of ice to steam (solid water to gaseous water) is the sum of the two processes of melting ice to liquid water and then vaporizing the resulting liquid. The enthalpy changes for these processes were given in Section 8-3A. List the component reactions and add:

$$\begin{array}{ll} H_2O(s) \longrightarrow H_2O(l) & \Delta H_m = 6.02 \text{ kJ mol}^{-1} \\ H_2O(l) \longrightarrow H_2O(g) & \Delta H_{vap} = 41.1 \text{ kJ mol}^{-1} \\ \hline H_2O(s) + H_2O(l) \longrightarrow H_2O(l) + H_2O(g) & \Delta H = (6.02 + 41.1) \text{ kJ mol}^{-1} \end{array}$$

Remove the $H_2O(l)$ term common to both sides of the reaction equation:

$$H_2O(s) \longrightarrow H_2O(g) \qquad \Delta H = 47.1 \text{ kJ mol}^{-1}$$

As a consequence of Hess' law and the definition of the enthalpy of formation, we can define a straightforward relationship between the **standard enthalpy of reaction** ΔH_r° and the standard enthalpies of formation ΔH_f° of all the substances taking part in the reaction. If we use the symbol $\sum$ (capital Greek sigma) to mean *the sum of*, we can write

**STANDARD
ENTHALPY OF
REACTION**
$$\Delta H_r^\circ = \sum(\Delta H_f^\circ \text{ products}) - \sum(\Delta H_f^\circ \text{ reactants})$$

note: Because ΔH_f° values are expressed in kilojoules or kilocalories *per mole*, each reactant and product that takes part in a reaction must be multiplied by the number of moles consumed or produced. Remember that ΔH_f° for an *element* in the standard state is zero.

EXAMPLE 8-5: Calculate ΔH_r° for the following reaction:

$$CH_4(g) + 4F_2(g) \longrightarrow CF_4(g) + 4HF(g)$$

Solution: Using the definition of the enthalpy of reaction,

$$\begin{aligned} \Delta H_r^\circ &= \sum(\Delta H_f^\circ \text{ products}) - \sum(\Delta H_f^\circ \text{ reactants}) \\ &= [\Delta H_f^\circ(CF_4(g))] + 4[\Delta H_f^\circ(HF(g))] - [\Delta H_f^\circ(CH_4(g))] - 4[\Delta H_f^\circ(F_2(g))] \end{aligned}$$

Substitute in the ΔH_f° values from Table 8-1 (remembering that ΔH_f° for an element is 0):

$$\begin{aligned} \Delta H_r^\circ &= -680 + 4(-269) - (-75) - 4(0) \text{ kJ} \\ &= -1681 \text{ kJ} \end{aligned}$$

D. First law of thermodynamics

The first law of thermodynamics expresses the commonly observed and well-known observation that energy (or its equivalent, work) can't be created or destroyed: Energy is conserved in all processes. But energy can be transformed in a process. Thus if you do work on a chemical system, that work must appear as an increase in the energy of the system in one form or another. Conversely, if you use chemical energy to do work (as in burning gasoline in a car engine), that energy must be completely accounted for by the work done and by other forms of energy produced. The first law of thermodynamics is an *empirical* (based on experience) expression. It follows from the first law that the energy or enthalpy change for the reverse of any given process or reaction is equivalent to the enthalpy change for the forward process or reaction, but the signs of the changes are opposite.

EXAMPLE 8-6: Given that $\Delta H_{vap} = 41.1 \text{ kJ mol}^{-1}$, determine ΔH for the change $H_2O(g) \to H_2O(l)$.

Solution: For $H_2O(l) \to H_2O(g)$, $\Delta H = \Delta H_{vap} = +41.1 \text{ kJ mol}^{-1}$ (an *endo*thermic change). Thus for $H_2O(g) \to H_2O(l)$, $\Delta H = -\Delta H_{vap} = -41.1 \text{ kJ mol}^{-1}$. This change is *exo*thermic.

E. Bond energies

You can make good estimates of the enthalpy changes in many chemical reactions by using the bond-energy approach. Each type of chemical bond has an average **bond energy**, defined as the average amount of energy required to *break* one mole of that type of bond. From the first law of thermodynamics you can see that the *negative* of the bond energy is, on average, the amount of energy *released* by the formation of 1 mol of that type of bond. The values of bond energies for many common bond types are given in Table 8-2.

TABLE 8-2: Average Thermochemical Bond Energies

Bond type	Energy (kJ/mol)	Bond type	Energy (kJ/mol)
H—H	436	C—O	351
H—C	414	C—N	293
H—N	389	C—F	485
H—O	464	C—Cl	331
H—F	565	N—Cl	201
H—Cl	431	Cl—Cl	243
C—C	347	F—F	153

EXAMPLE 8-7: Using the values given in Table 8-2, calculate the enthalpy change for the following reaction. Is it exo- or endothermic?

$$NH_3(g) + Cl_2(g) \longrightarrow ClNH_2(g) + HCl(g)$$

Solution: Look carefully at the reaction! You can see that

Bonds broken:	N—H and Cl—Cl	
Energy input:	389 + 243 =	632 kJ/mol
New bonds formed:	N—Cl and H—Cl	
Energy obtained:	(−201) + (−431) =	−632 kJ/mol
Net energy change:	632 + (−632) =	0 kJ/mol

The reaction is *thermoneutral*, that is, neither exo- nor endothermic.

SUMMARY

1. Temperature measures the intensity with which heat energy is concentrated in a system.
2. Heat capacity expresses the amount of heat energy needed to raise the temperature of a fixed amount of a substance by 1 K or 1°C.
3. The law of Dulong and Petit states that the heat capacities of many metals are near 26 J mol^{-1} K^{-1} at room temperature.
4. Enthalpy change (ΔH) is the change in heat energy measured at constant pressure.
5. In exothermic reactions, systems lose heat energy, $\Delta H < 0$. In endothermic reactions, systems gain heat energy, $\Delta H > 0$.
6. Hess' law states that enthalpies for different processes can be added algebraically.
7. The standard enthalpy of reaction ΔH_r° is the sum of the standard enthalpies of formation of the products minus the sum of the standard enthalpies of formation of the reactants:

$$\Delta H_r^\circ = \sum(\Delta H_f^\circ \text{ products}) - \sum(\Delta H_f^\circ \text{ reactants})$$

8. The first law of thermodynamics states that energy can't be created or destroyed.
9. Bond energies measure the average energy needed to break or form a particular kind of bond.

RAISE YOUR GRADES

Can you...?

☑ calculate heat capacities
☑ use the law of Dulong and Petit to determine atomic mass
☑ determine whether a process is exothermic, endothermic, or thermoneutral
☑ use Hess' law to calculate ΔH for a reaction
☑ calculate ΔH_r°, given ΔH_f° for reactants and products
☑ use average bond energies to calculate ΔH for a reaction

SOLVED PROBLEMS

PROBLEM 8-1 The heat capacity of graphite is $8.6 \text{ J mol}^{-1} \text{ K}^{-1}$. How much heat energy is needed to heat 7.4 g of graphite from $28°C$ to $375°C$?

Solution: Using the equation for heat energy

$$q = nC \Delta T$$

set up the conditions in units consistent with the given heat capacity (C). Graphite is simply carbon, so n (in moles) is $(7.4 \text{ g})/(12 \text{ g mol}^{-1})$. The degree interval is the same in the Kelvin and Celsius scales, so ΔT is $375°C - 28°C = 347°C = 347 \text{ K}$. Now

$$q = \left(\frac{7.4 \text{ g}}{12 \text{ g mol}^{-1}}\right)\left(\frac{8.6 \text{ J}}{\text{mol K}}\right)(347 \text{ K}) = 1.8 \times 10^3 \text{ J} \quad \text{or} \quad 1.8 \text{ kJ}$$

PROBLEM 8-2 A silver-gray metal has a heat capacity of 0.23 J/g K and forms a white chloride that contains 38.68% chlorine. Determine the atomic mass of the metal and the formula of the chloride.

Solution: Start by calculating an *approximate* atomic mass of the metal (we'll call the metal M) using the law of Dulong and Petit: Assume that the mass of a mole of M atoms multiplied by the heat capacity will be $\sim 26 \text{ J/mol K}$ (see Example 8-2), so the approximate atomic mass must be

$$\left(\frac{26 \text{ J}}{\text{mol K}}\right) \div \left(\frac{0.23 \text{ J}}{\text{g K}}\right) = \left(\frac{26 \text{ J}}{\text{mol K}}\right)\left(\frac{\text{g K}}{0.23 \text{ J}}\right) = 1.1 \times 10^2 \text{ g/mol}$$

Now we can use the approximate atomic mass in the calculations necessary for working out the empirical formula of the metal chloride:

Element	% (g/100 g)	Number of moles	Integral ratio
M	61.32 (100 − 38.68)	$\dfrac{61.32 \text{ g}}{1.1 \times 10^2 \text{ g/mol}} = 0.56$	1
Cl	38.68	$\dfrac{38.68 \text{ g}}{35.45 \text{ g/mol}} = 1.091$	2

Thus the chloride is MCl_2.

Now we can determine the *accurate* atomic mass of M. We know that 61.32 g of M reacts with 1.091 mol of Cl atoms to form MCl_2. So the number of moles of M must be

$$\left(\frac{1.091 \text{ mol Cl atoms}}{2 \text{ mol Cl/mol } MCl_2}\right)\left(\frac{1 \text{ mol M}}{1 \text{ mol } MCl_2}\right) = 0.5456 \text{ mol M}$$

and the atomic mass of M is

$$\frac{61.32 \text{ g}}{0.5456 \text{ mol}} = 112.4 \text{ g/mol} \quad \text{(or amu/atom)}$$

We conclude that M must be the metal cadmium, which has an atomic mass of 112 amu and forms a chloride $CdCl_2$.

PROBLEM 8-3 A 5.8-g sample of a new mineral X at a temperature of 99.6°C is rapidly put into 49.0 g of H_2O at 24.4°C in a calorimeter. The final temperature reached by the system is 29.2°C. Calculate the heat capacity C of X in $J\,g^{-1}\,K^{-1}$. (Neglect any heat energy taken up by the calorimeter itself.)

Solution: In this typical problem of calorimetry the standard substance is H_2O, which has a heat capacity of $4.18\,J\,g^{-1}\,K^{-1}$. The heat energy gained by H_2O in warming from 24.4°C to 29.2°C is

$$q_{H_2O} = nC\,\Delta T = (49\text{ g})(4.18\,J\,g^{-1}\,K^{-1})(4.8\text{ K}) = 9.8 \times 10^2\text{ J}$$

The heat energy lost by X in cooling from 99.6°C to 29.2°C is

$$q_X = (5.8\text{ g})(C)(70.4\text{ K}) = (408\text{ g K})C$$

where C is our unknown. Then by the first law of thermodynamics, the heat energy gained by H_2O must equal the heat energy lost by X, so

$$q_{H_2O} = q_X = 9.8 \times 10^2\text{ J} = (408\text{ g K})C$$

and

$$C = \frac{9.8 \times 10^2\text{ J}}{408\text{ g K}} = 2.4\,J\,g^{-1}\,K^{-1}$$

PROBLEM 8-4 An 83-g sample of Hg $(C = 27.8\,J\,mol^{-1}\,K^{-1})$ at 215°C is poured into 100.0 g of H_2O at 18°C. Calculate the final temperature of the mixture, assuming no heat energy is lost.

Solution: Common sense tells us that the final temperature, which we'll call T_f, will be between 18°C and 215°C. Using this commonsense approximation to keep us in bounds, we proceed systematically:

$$q_{Hg} = nC\,\Delta T = \left(\frac{83\text{ g}}{200.6\text{ g mol}^{-1}}\right)\left(\frac{27.8\text{ J}}{\text{mol K}}\right)(215 - T_f)\text{ K}$$

$$q_{H_2O} = nC\,\Delta T = (100\text{ g})(4.18\,J\,g^{-1}\,K^{-1})(T_f - 18)\text{ K}$$

Since we know that q_{Hg} must equal q_{H_2O}, we can solve this equation for one unknown (T_f):

$$q_{Hg} = q_{H_2O}$$

$$\frac{(83)(27.8)(215 - T_f)}{200.6} = (100.0)(4.18)(T_f - 18)$$

$$2473 - 11.5T_f = 418T_f - 7524$$

$$9997 = 430T_f$$

So

$$T_f = \frac{9997}{430} = 23°C$$

Thus the final temperature is 23°C—which *is* between 18°C and 215°C. (Note the small change in the water temperature. This is a consequence of the large heat capacity of water.)

PROBLEM 8-5 Determine the total enthalpy change when 1.00 g of ice at 0°C is converted into 1.00 g of steam at 1 atm and 100°C.

Solution: Break this problem down into its three stages: First the ice melts; then the water is heated from 0°C to 100°C; finally the water is vaporized at 100°C. The relevant ΔH and heat capacity data are given in the text.

Melting:

$$\Delta H_1 = (6.02\text{ kJ mol}^{-1})\left(\frac{1.00\text{ g ice}}{18.0\text{ g mol}^{-1}}\right) = 0.334\text{ kJ}$$

Heating:

$$\Delta H_2 = (1.00\text{ g }H_2O)\left(\frac{4.18\text{ J}}{\text{g K}}\right)(100 - 0)\text{ K}\left(\frac{1\text{ kJ}}{10^3\text{ J}}\right) = 0.418\text{ kJ}$$

Vaporization:

$$\Delta H_3 = (41.1\text{ kJ mol}^{-1})\left(\frac{1.00\text{ g }H_2O}{18.0\text{ g mol}^{-1}}\right) = 2.28\text{ kJ}$$

So the total enthalpy change is

$$\Delta H_1 + \Delta H_2 + \Delta H_3 = (0.334 + 0.418 + 2.28)\,\text{kJ}$$
$$= 3.04\,\text{kJ}$$

PROBLEM 8-6   Consider the reaction

$$2C(\text{graphite}) + O_2(g) \longrightarrow 2CO(g)$$

If the standard enthalpy of reaction $\Delta H_r^\circ = -222\,\text{kJ}$, what is the value of H_f° for $CO(g)$?

Solution: The standard enthalpy of formation of a compound (ΔH_f°) is defined as the enthalpy change that occurs when 1 mol of a substance is formed from its elements in their standard states at 298 K. The reaction given forms exactly 2 mol of CO from the element C and compound O_2 in their standard states at 298 K. Thus

$$\Delta H_r^\circ = 2[\Delta H_f^\circ(CO(g))] \quad \text{and} \quad \Delta H_f^\circ(CO(g)) = \frac{\Delta H_r^\circ}{2} = \frac{-222\,\text{kJ}}{2} = -111\,\text{kJ/mol}$$

PROBLEM 8-7   Arrange the following reactions in order of increasing exothermicity:

		ΔH° (kJ)
(a)	$C(\text{graphite}) \longrightarrow C(\text{diamond})$	1.88
(b)	$CaCO_3(s) \longrightarrow CaO(s) + CO_2(g)$	178
(c)	$3C(\text{graphite}) + 4H_2(g) \longrightarrow C_3H_8(g)$	−104
(d)	$2C(\text{graphite}) + H_2(g) \longrightarrow C_2H_2(g)$	227
(e)	$2H_2(g) + O_2(g) \longrightarrow 2H_2O(l)$	−572

Solution: Remember to associate signs with exothermic (negative) or endothermic (positive) changes. The *most* exothermic reaction is **(e)**, where $\Delta H^\circ = -572\,\text{kJ}$; i.e., 572 kJ is released to the surroundings by this reaction. The *least* exothermic reaction is the most endothermic reaction, which is **(d)**. This reaction takes in 227 kJ from the surroundings. The required order is therefore

$$\begin{array}{ccccccccc} \textbf{(d)} & < & \textbf{(b)} & < & \textbf{(a)} & < & \textbf{(c)} & < & \textbf{(e)} \\ 227 & & 178 & & 1.88 & & -104 & & -572 \end{array}$$

PROBLEM 8-8   Calculate the enthalpy change for the following reaction:

$$2H_2O_2(l) \longrightarrow 2H_2O(l) + O_2(g)$$

Is it exo- or endothermic?

Solution: Use the rule from Hess' law

$$\Delta H_r^\circ = \sum(\Delta H_f^\circ \text{ products}) - \sum(\Delta H_f^\circ \text{ reactants})$$

and the ΔH_f° data from Table 8-1 (or any standard reference):

$$\Delta H_f^\circ(H_2O_2(l)) = -188\,\text{kJ mol}^{-1}$$
$$\Delta H_f^\circ(H_2O(l)) = -286\,\text{kJ mol}^{-1}$$
$$\Delta H_f^\circ(O_2(g)) = 0$$

(Remember that ΔH_f° for an element is *always* zero.) Now we have

$$\Delta H_r^\circ = 2[\Delta H_f^\circ(H_2O(l))] + \Delta H_f^\circ(O_2(g)) - 2[\Delta H_f^\circ(H_2O_2(l))]$$
$$= 2(-286) + 0 - 2(-188) = -196\,\text{kJ}$$

Because the sign is negative, the reaction is exothermic.

PROBLEM 8-9   Calculate the enthalpy change for the following reaction (which cannot be carried out directly):

$$2BCl_3(l) + 6H_2(g) \longrightarrow B_2H_6(g) + 6HCl(g)$$

Is the reaction exo- or endothermic?

Solution: Apply the rule—watch the signs—and realize that $H_2(g)$ is an element in its standard state. Consequently

$$\Delta H_r^\circ = \sum(\Delta H_f^\circ \text{ products}) - \sum(\Delta H_f^\circ \text{ reactants})$$
$$= \Delta H_f^\circ(B_2H_6(g)) + 6[\Delta H_f^\circ(HCl(g))] - 2[\Delta H_f^\circ(BCl_3(l))] - 6[\Delta H_f^\circ(H_2(g))]$$
$$= 31 + 6(-92) - 2(-418) - 6(0) = 315 \text{ kJ}$$

The reaction is endothermic.

PROBLEM 8-10 Calculate ΔH_f° for $CS_2(l)$, given that the standard enthalpy of combustion of $CS_2(l)$, i.e., ΔH_r° for the following reaction, is -1864 kJ:

$$CS_2(l) + 3O_2(g) = 3CO_2(g) + 2SO_2(g)$$

Solution: Using the equation defining ΔH_r° from Hess' law and ΔH_f° data from Table 8-1, we can set up so that $\Delta H_f^\circ(CS_2(l))$ is the only unknown:

$$\Delta H_r^\circ = 3[\Delta H_f^\circ(CO_2(g))] + 2[\Delta H_f^\circ(SO_2(g))] - [\Delta H_f^\circ(CS_2(l))] - 3[\Delta H_f^\circ(O_2(g))]$$
$$-1864 \text{ kJ} = 3(-394) + 2(-297) - [\Delta H_f^\circ(CS_2(l))] - 3(0)$$
$$= -1776 - \Delta H_f^\circ(CS_2(l))$$
$$\Delta H_f^\circ(CS_2(l)) = (-1776 + 1864) \text{ kJ} = 88 \text{ kJ}$$

PROBLEM 8-11 Which will be the better (more exothermic) fuel on a per-gram basis: (a) $CH_4(g)$ burning to $CO_2(g)$ and $H_2O(l)$ or (b) $N_2H_4(l)$ burning to $N_2(g)$ and $H_2O(l)$?

Solution: We have to solve this sort of problem in steps: First balance each equation. Then calculate ΔH_r° for each. Finally calculate ΔH per gram and compare.

Balance (by inspection):

(a) $CH_4(g) + 2O_2(g) \longrightarrow CO_2(g) + 2H_2O(l)$
(b) $N_2H_4(l) + O_2(g) \longrightarrow N_2(g) + 2H_2O(l)$

Calculate ΔH_r°:

(a) $\Delta H_r^\circ = \Delta H_f^\circ(CO_2(g)) + 2[\Delta H_f^\circ(H_2O(l))] - \Delta H_f^\circ(CH_4(g)) - 2[\Delta H_f^\circ(O_2(g))]$
$\quad = -394 + 2(-286) - (-75) - 2(0) = -891$ kJ
(b) $\Delta H_r^\circ = \Delta H_f^\circ(N_2(g)) + 2[\Delta H_f^\circ(H_2O(l))] - \Delta H_f^\circ(N_2H_4(l)) - \Delta H_f^\circ(O_2(g))$
$\quad = 0 + 2(-286) - 50 - 0 = -622$ kJ

Calculate $\Delta H/g$:

(a) Molar mass of $CH_4 = 16.04$ g, so

$$\frac{\Delta H}{g} = \frac{-891 \text{ kJ/mol}}{16.0 \text{ g/mol}} = -55.7 \text{ kJ/g}$$

(b) Molar mass of $N_2H_4 = 32.05$ g, so

$$\frac{\Delta H}{g} = \frac{-622 \text{ kJ/mol}}{32.05 \text{ g/mol}} = -19.4 \text{ kJ/g}$$

Thus CH_4 is a much better fuel than N_2H_4 on a per-gram basis.

PROBLEM 8-12 Picric acid $C_6H_2(NO_2)_3(OH)(s)$ is an explosive. Its enthalpy of combustion to $CO_2(g)$, $H_2O(l)$, and $N_2(g)$, is -2.560×10^3 kJ/mol. Calculate ΔH° for the detonation of 1 mol of picric acid, with no added oxygen, assuming the detonation products are as follows:

$$2C_6H_2(NO_2)_3(OH)(s) \longrightarrow 3N_2(g) + 3H_2O(g) + 11CO(g) + C(\text{graphite})$$

Solution: Solve this problem by calculating ΔH_f° for picric acid from the combustion reaction, then use the calculated ΔH_f° in the detonation reaction.

First balance the combustion reaction for $C_6H_2(NO_2)_3(OH)$ (or $C_6H_3N_3O_7$) as follows:

$$2C_6H_3N_3O_7(s) + 6\tfrac{1}{2}O_2(g) \longrightarrow 12CO_2(g) + 3H_2O(l) + 3N_2(g)$$

Set up the standard enthalpy of reaction equation in one unknown (remembering that we're dealing with 2 mol of picric acid and that we can eliminate elements because they have ΔH_f° of zero):

$$2\Delta H_r^\circ = 12[\Delta H_f^\circ(CO_2(g))] + 3[\Delta H_f^\circ(H_2O(l))] - 2[\Delta H_f^\circ(C_6H_3N_3O_7(s))]$$
$$-5120 \text{ kJ} = 12(-394) + 3(-286) - 2[\Delta H_f^\circ(C_6H_3N_3O_7(s))]$$
$$= -5586 - 2[\Delta H_f^\circ(C_6H_3N_3O_7(s))]$$

$$\Delta H_f^\circ(C_6H_3N_3O_7(s)) = \frac{-5586 + 5120}{2} = -233 \text{ kJ/mol}$$

And now for the detonation

$$2C_6H_3N_3O_7(s) \longrightarrow 3N_2(g) + 3H_2O(g) + 11CO(g) + C(\text{graphite})$$
$$\Delta H_r^\circ = 3[\Delta H_f^\circ(H_2O(g))] + 11[\Delta H_f^\circ(CO(g))] - 2[\Delta H_f^\circ(C_6H_3N_3O_7(s))]$$
$$= 3(-242) + 11(-111) - 2(-233) = -1481 \text{ kJ}$$

Like the combustion reaction, the detonation reaction involves 2 mol of picric acid. Since we've been asked to give ΔH° for 1 mol, we have $(-1481 \text{ kJ})/(2 \text{ mol}) = -740.5 \text{ kJ/mol}$. (Yes, the detonation is exothermic; but what makes picric acid an explosive is the large volume of hot gas produced by the detonation. Note that 2 mol of solid gives rise to 17 mol of gaseous products in this detonation.)

PROBLEM 8-13 If energy can be neither created nor destroyed, what happens to the kinetic energy of an automobile when you stop it?

Solution: The kinetic energy of the automobile is converted into heat energy, mostly in the brakes, of course (which is why brakepads are made of heat-resistant materials).

PROBLEM 8-14 Using the data given in Table 8-2, calculate the enthalpy change for the following reaction by the bond-energy method:

$$CH_4 + 2F_2 \longrightarrow CH_2F_2 + 2HF$$

Solution: First tabulate the bonds broken, and see how much heat energy is needed to break them (breaking bonds will always be an endothermic process). Then tabulate the bonds formed, and see how much energy is released in forming them (bond formation is always an exothermic process).

Reactant	Bonds broken	Energy required (kJ/mol)	Product	Bonds formed	Energy released (kJ/mol)
CH_4	2C—H	2(414) = 828	CH_2F_2	2C—F	2(-485) = - 970
$2F_2$	2F—F	2(153) = 306	2HF	2H—F	2(-565) = -1130
		Total $\overline{1134}$			Total $\overline{-2100}$

So the enthalpy change for the whole reaction is $(1134 - 2100) = -966 \text{ kJ}$. (Remember, this is only an estimate of the enthalpy change, because you've used average bond energy values in the calculation.)

PROBLEM 8-15 Calculate the enthalpy change for the following reaction by the bond-energy method:

$$CH_3CH_2CH_2CH_2Cl + (C_2H_5)_2NH \longrightarrow CH_3CH_2CH_2CH_2N(C_2H_5)_2 + HCl$$

Solution: This may look formidable at first, but on careful inspection we see that most of the bonds in the reactants are preserved in the products. In fact, the only bonds broken are

1 C—Cl in $CH_3CH_2CH_2CH_2Cl$	331 kJ
1 N—H in $(C_2H_5)_2NH$	389 kJ
heat energy needed:	720 kJ

The only bonds made are

1 C—N in $CH_3CH_2CH_2CH_2N(C_2H_5)_2$	-293 kJ
1 H—Cl in HCl	-431 kJ
heat energy released:	-724 kJ

So the overall enthalpy change is $720 - 724 = -4 \text{ kJ}$.

Supplementary Exercises

PROBLEM 8-16 Calculate the heat energy needed to heat a copper frying pan of mass 1.30 kg from 25°C to 190°C. The heat capacity of Cu is 24.5 J/mol K.

Answer: 82.7 kJ

PROBLEM 8-17 A metallic element forms an oxide that contains 74.39% of the metal. The heat capacity of the metal is 0.38 J/g K. Determine the formula of the metal oxide and the identity of the metal.

Answer: M_2O_3, Ga

PROBLEM 8-18 A 12.3-g sample of a polymer heated to 137°C was dropped into 65.0 g of carbon tetrachloride (which has a heat capacity of 0.86 J/g K) at 22.0°C. The final temperature of the system was 31.6°C. Calculate the heat capacity of the polymer.

Answer: 0.41 J/g K

PROBLEM 8-19 A 25.5-g portion of acetone (heat capacity 2.21 J/g K) at 24.0°C is mixed with 36.0 g of benzene (heat capacity 1.70 J/g K) at 47.5°C. Calculate the final temperature of the mixture.

Answer: 36.2°C

PROBLEM 8-20 What is the total enthalpy change when 25.0 g of metallic potassium is heated from 273 K to 373 K. Potassium melts at 335 K. The heat capacity of K(s) is 29.4 J/mol K; of K(l), 32.7 J/mol K. The enthalpy of fusion of potassium is 2.57 kJ/mol.

Answer: 3.60 kJ

PROBLEM 8-21 Calculate the enthalpy change for the following reaction:

$$H_2O_2(l) + SO_2(g) \longrightarrow H_2SO_4(l)$$

Is the reaction endo- or exothermic?

Answer: -326 kJ; exothermic

PROBLEM 8-22 Calculate the enthalpy change for the following reaction:

$$CaH_2(s) + 2H_2O(l) \longrightarrow Ca(OH)_2(s) + 2H_2(g)$$

Answer: -226 kJ

PROBLEM 8-23 The standard enthalpy of combustion of formaldehyde $CH_2O(g)$ to $CO_2(g)$ and $H_2O(l)$ is -561 kJ/mol. Calculate the standard enthalpy of formation (ΔH_f°) of $CH_2O(g)$.

Answer: -119 kJ/mol

PROBLEM 8-24 Which is the more exothermic reaction per gram of reactant?

(a) $2H_2O_2(l) \longrightarrow 2H_2O(l) + O_2(g)$
(b) $3N_2H_4(l) \longrightarrow 4NH_3(g) + N_2(g)$

Answer: reaction (b)

PROBLEM 8-25 The enthalpy of combustion of C(graphite) to $CO_2(g)$ is -394.0 kJ/mol. The enthalpy of combustion of C(diamond) to $CO_2(g)$ is -395.9 kJ/mol. Calculate the enthalpy change when 1.0 mol of graphite is converted to diamond. Is the change exo- or endothermic?

Answer: 1.9 kJ; endothermic

PROBLEM 8-26   Use the bond-energy method to calculate the enthalpy change for the following reaction:

$$CH_4 + CCl_4 \longrightarrow 2CH_2Cl_2$$

Answer: 0 kJ

PROBLEM 8-27   Use the bond-energy method to determine the enthalpy change for the following reaction:

$$2(C_2H_5)_2NCl + CH_2O \longrightarrow 2(C_2H_5)NH + Cl_2CO$$

Answer: -210 kJ

9 ATOMIC STRUCTURE AND THE PERIODIC TABLE

THIS CHAPTER IS ABOUT

☑ **Electromagnetic Radiation**
☑ **Quantum Theory of Radiation**
☑ **Orbitals**
☑ **Ionization Energies**
☑ **The Periodic Table**

9-1. Electromagnetic Radiation

Visible light is the most familiar kind of **electromagnetic radiation**, i.e., energy traveling through space as a series of waves. (We'll examine the concept of radiation as a stream of particles in Section 9-2.) There are several regions of electromagnetic radiation, which are classified according to frequency and wavelength. These regions are given special names: *Radio waves* are low-energy waves with low frequencies and long wavelengths; *gamma rays* are high-energy waves with high frequencies and short wavelengths; *visible light* falls about midway between these two extremes.

note: **Wavelength** is the distance a wave travels in one cycle (i.e., the distance between two points at the same relative height in two adjacent waves). **Frequency** is the number of cycles per unit of time (i.e., the number of wavelengths that pass a given point in a unit of time). The **wave number** is the number of cycles per unit of distance traveled (i.e., the reciprocal of the wavelength).

All electromagnetic radiation travels at the velocity of light c. In a vacuum, $c = 2.998 \times 10^8 \text{ m s}^{-1}$. This constant is handy because it is related to frequency v and wavelength λ, which are the quantities used to describe all electromagnetic radiation:

VELOCITY OF LIGHT
$$c = \lambda v = 2.998 \times 10^8 \text{ m s}^{-1}$$

One more quantity is used to describe electromagnetic radiation—the **wave number** $\bar{v}$, which depends inversely on the wavelength:

WAVE NUMBER
$$\bar{v} = \frac{1}{\lambda}$$

EXAMPLE 9-1: Light from the sun has a maximum intensity at a wavelength of about 6.0×10^{-7} m. What is the frequency of sunlight?

Solution: To determine the frequency, rearrange the formula for the velocity of light and substitute. From $c = v\lambda$ we get

$$v = \frac{c}{\lambda} = \frac{3.00 \times 10^8 \text{ m s}^{-1}}{6.0 \times 10^{-7} \text{ m}} = 5.0 \times 10^{14} \text{ s}^{-1}$$

Frequency is measured here as cycles per second, or s^{-1} (the word "cycle" being omitted). In SI the unit of frequency is a *hertz*, abbreviated Hz: 1 Hz = 1 s^{-1}. So in this example we could express the answer as 5.0×10^{14} Hz.

A. Spectroscopy

Spectroscopy is the study of how matter emits and absorbs electromagnetic radiation. Spectroscopic analysis requires a source of electromagnetic radiation and a device for analyzing the variation of the intensity of that radiation as a function of wavelength or frequency. Such a device is called a **spectrometer**.

The source of radiation in a simple spectrometer may be a hot gas, such as a bunsen burner flame. The light from the gas is *dispersed*—i.e., spread out as a function of wavelength—by a glass prism. A photovoltaic cell, which produces a voltage when visible light strikes its surface, is often used as a detector. The record of the variation of radiation intensity with wavelength or frequency is called a **spectrum** (plural is spectra).

B. Atomic spectra

When the light source in a spectrometer is a hot gas consisting of a single atomic species, the hot (excited) atoms emit energy at specific wavelengths, which are recorded in the **atomic spectrum**. These emission spectra are simple—consisting of series of lines that show regularities in their spacing—and characteristic for each element. The simplest atom hydrogen, for example, produces the simplest atomic spectrum. There are five series of lines observed in its atomic spectrum. Each of these series is named, and the series in the visible region is called the **Balmer series**. The lines in the Balmer series occur at specific wavelengths, which we can find by using an empirical formula to calculate the wave number:

BALMER SERIES (FOR H)
$$\bar{v} = R_{\mathrm{H}}\left(\frac{1}{2^2} - \frac{1}{n^2}\right) \tag{9-1}$$

where $\bar{v}$ is the wave number, defined as $1/\lambda$ or v/c; n is an integer larger than 2, and R_{H} is the experimentally determined **Rydberg constant**, which equals $1.097 \times 10^7 \mathrm{~m}^{-1}$.

EXAMPLE 9-2: Calculate the wavelength, in meters and nanometers, of the first line of the Balmer series (a bright red line), assuming that $n = 3$.

Solution: Using the wave number calculation of the Balmer series (Eq. 9-1)

$$\bar{v} = R_{\mathrm{H}}\left(\frac{1}{2^2} - \frac{1}{n^2}\right)$$

we can write

$$\frac{1}{\lambda} = (1.097 \times 10^7 \mathrm{~m}^{-1})\left(\frac{1}{2^2} - \frac{1}{3^2}\right) = 1.524 \times 10^6 \mathrm{~m}^{-1}$$

Thus

$$\lambda = \frac{1}{1.524 \times 10^6 \mathrm{~m}^{-1}} = 6.562 \times 10^{-7} \mathrm{~m}$$

Converting to nanometers, we get

$$\lambda = (6.562 \times 10^{-7} \mathrm{~m})(1 \times 10^9 \mathrm{~nm\,m}^{-1}) = 656.2 \mathrm{~nm}$$

note: The 2 and 3 in Eq. (9-1) are exact numbers and do not affect the number of significant figures.

9-2. Quantum Theory of Radiation

The **quantum theory** states that radiation of frequency v comes in discrete packets, or **quanta**, the form of which accounts for characteristic line spectra of atoms. Through the emission or absorption of discrete quanta, *transitions* of electrons between well-defined energy levels within the atoms produce well-defined lines. If an electron undergoes a transition from some higher energy level E_2 to a lower one E_1 the atom emits the *energy difference* as a quantum of light, a **photon**, of frequency v. Conversely, if an atom absorbs a photon of frequency v, an electron is excited from some lower energy level E_3 to a higher one E_4. The energy difference in both cases is given by **Planck's equation**:

PLANCK'S EQUATION $$E_f - E_i = \Delta E = hv \qquad \qquad (9\text{-}2)$$

where E_f is the final energy level, E_i is the initial level, ΔE is the energy difference, and h is **Planck's constant**, a fundamental universal constant equal to 6.626×10^{-34} J s in SI, or 6.626×10^{-27} erg s in cgs. The value of hv gives the *size* of the quantized energy of the radiation; the radiation can only be transferred in small integer multiples of hv ($1hv$, $2hv$, etc.).

EXAMPLE 9-3: Determine the energy difference for an electron transition that releases a quantum of light of wavelength 6.0×10^2 nm.

Solution: Recall that wavelength and frequency are related by $v = c/\lambda$. Now we can substitute into Planck's equation (9-2):

$$\Delta E = hv = hc/\lambda$$

Then make sure λ and c are expressed in compatible units:

$$\lambda = (6.0 \times 10^2 \text{ nm})(10^{-9} \text{ m/nm}) = 6.0 \times 10^{-7} \text{ m}$$

Solving for the energy difference ΔE, we get

$$\Delta E = \frac{(6.626 \times 10^{-34} \text{ J s})(3.00 \times 10^8 \text{ m s}^{-1})}{6.0 \times 10^{-7} \text{ m}} = 3.3 \times 10^{-19} \text{ J}$$

A. The photoelectric effect

When specific frequencies of light shine on the surfaces of certain metals (particularly the alkali metals), electrons, called **photoelectrons**, are given off. This phenomenon is called the **photoelectric effect**, about which experiments have established the following:

1. For a given metal there is a *threshold frequency* v_0 below which no electrons are given off.
2. When the incident radiation has a frequency higher than v_0, the photoelectrons produced have kinetic energy. The maximum kinetic energy of the electrons, $\frac{1}{2}mv^2$ (m = mass; v = velocity), is directly proportional to frequency v.
3. If the light intensity is increased, the number of photoelectrons increases but their maximum kinetic energy is unaffected.

Einstein explained these properties of the photoelectric effect by extending the quantum concept to electromagnetic radiation. One quantum ($1hv$), or photon, of a given frequency has the same energy as any other photon of the same frequency. High-intensity light simply has *more* photons, not photons of higher energy. Each photon that collides with a single surface electron can deliver a maximum of $1hv$ of energy. So if v is low—below the threshold frequency—the quantum $1hv$ will be too low to supply the energy required for an electron to escape. If v is above the threshold frequency, each photon will have enough energy to liberate an electron, thus imparting kinetic energy to the electron.

Fortunately, we've got equations to quantify these observations on the nature of the photoelectric effect:

PHOTOELECTRIC EFFECT $$\Delta E = hv = hv_0 + \tfrac{1}{2}mv^2 \qquad \qquad (9\text{-}3)$$

The energy (ΔE) of a quantum of radiation (hv) goes into ejecting a photoelectron out of the metal (the *work function* hv_0) and into giving it kinetic energy ($\frac{1}{2}mv^2$). Equation (9-3) is another example of the first law of thermodynamics.

EXAMPLE 9-4: The maximum kinetic energy of the photoelectrons emitted from a metal is 1.03×10^{-19} J when light that has a 656-nm wavelength shines on the surface. Determine the threshold frequency v_0 for this metal.

Solution: Solve $c = \nu\lambda$ for ν:

$$\nu = \frac{c}{\lambda} = \frac{3.00 \times 10^8 \text{ m s}^{-1}}{(656 \text{ nm})(10^{-9} \text{ m nm}^{-1})} = 4.57 \times 10^{14} \text{ s}^{-1}$$

Rearrange Eq. (9-3) and solve for ν_0:

$$\nu_0 = \frac{h\nu - \frac{1}{2}mv^2}{h}$$

$$= \frac{(6.626 \times 10^{-34} \text{ J s})(4.57 \times 10^{14} \text{ s}^{-1}) - (1.03 \times 10^{-19} \text{ J})}{6.626 \times 10^{-34} \text{ J s}}$$

$$= 3.02 \times 10^{14} \text{ s}^{-1}$$

So a frequency of 3.02×10^{14} Hz is required to evoke the photoelectric effect for this metal.

note: The wavelength λ_0 corresponding to ν_0 is given by $\lambda_0 = c/\nu_0$. For Example 9-4, $\lambda_0 = (3.00 \times 10^8 \text{ m s}^{-1})/(3.02 \times 10^{14} \text{ s}^{-1}) = 9.93 \times 10^{-7}$ m or 993 nm. Photoelectrons will not be emitted from the surface of this metal unless the wavelength of the light is shorter than 993 nm. Remember that higher energies are associated with higher frequencies and shorter wavelengths.

B. Energy levels in the hydrogen atom

Now that we have the conceptual clue of the quantization of energy, we can return to the hydrogen spectrum. To do this we look at the Bohr model, which accounts for the hydrogen spectrum by making three assumptions.

1. The electron in the hydrogen atom travels about the nucleus in any one of several stable circular orbits, but ONLY in these orbits.
2. The **momentum** (i.e., mass times velocity mv) of the electron (as you might expect by now) is quantized, which means that it has to be a whole number times the quantity $h/2\pi$, where h is Planck's constant.
3. Electron transitions from one stable orbit to another require the absorption or emission of one photon ($1h\nu$) of radiation. The lowest energy orbit—the one in which the electron normally resides—is the **ground state**; higher energy levels are **excited states**.

Using Bohr's assumptions, we can now rewrite Eq. (9-1) to provide the wave numbers of all the other lines in the hydrogen emission spectrum:

HYDROGEN SPECTRUM
$$\bar{\nu} = \frac{1}{\lambda} = R_H\left(\frac{1}{n_f^2} - \frac{1}{n_i^2}\right) \tag{9-4}$$

where integers n_i and n_f are the initial and final energy levels, respectively, of an electron and $n_f < n_i$. The lines in the atomic spectrum of hydrogen in the visible region (the Balmer series) arise from the transitions of electrons from higher energy levels to the level $n_f = 2$. The **Lyman series** of spectral lines for hydrogen occurs in the ultraviolet part of the spectrum and arises from the transition of electrons to the level $n_f = 1$; the **Paschen** and **Brackett** series occur in the infrared part of the spectrum as a result of transitions to the $n_f = 3$ and 4 levels, respectively.

EXAMPLE 9-5: Calculate the wavelength of the first line of the Lyman series ($n_f = 1$ and $n_i = 2$) in the hydrogen atom emission spectrum.

Solution: Solving Eq. (9-4),

$$\frac{1}{\lambda} = R_H\left(\frac{1}{n_f^2} - \frac{1}{n_i^2}\right) = (1.097 \times 10^7 \text{ m}^{-1})\left(\frac{1}{1^2} - \frac{1}{2^2}\right) = 8.228 \times 10^6 \text{ m}^{-1}$$

$$\lambda = \frac{1}{8.228 \times 10^6 \text{ m}^{-1}} = 1.215 \times 10^{-7} \text{ m}$$

The integer n, which characterizes the energy level of the electron in the hydrogen atom, is the **principal quantum number** of that level.

You can observe line spectra for atoms other than hydrogen, even though they are more complicated than the atomic spectrum of hydrogen. The fact that other elements have line spectra demonstrates that only discrete energy levels exist for electrons in all atoms.

C. Wave mechanics

Although the mathematics of wave mechanics are beyond the scope of this book, you should be aware of some of the more important consequences of the theory:

- A particle has a wavelength λ related to its momentum and given by the **de Broglie equation**:

DE BROGLIE EQUATION $$\lambda = \frac{h}{mv} \tag{9-5}$$

where h is, once again, Planck's constant.

- We can't know both the exact momentum and the exact position of an electron; we can discuss these parameters only in terms of probabilities. If we try to determine the position and momentum simultaneously, our precision will be no better than $h/4\pi$. This law of nature is known as the **Heisenberg uncertainty principle**. In quantitative terms, if you perform the best possible experiment, you can do no better than the following:

$$\Delta x \, \Delta(mv) = \frac{h}{4\pi} \tag{9-6}$$

where Δx is the uncertainty in your measurement of the particle's position, $\Delta(mv)$ is the uncertainty in your measurement of its momentum, and h is (you guessed it) Planck's constant 6.626×10^{-34} J s.

note: Mass is never zero in wave mechanics. Remember that an electron does have mass: In SI, $1\ e^-$ is 9.11×10^{-31} kg.

- It is possible, however, to determine the energy level and the probability of locating an electron in a particular area around the nucleus. Schrödinger used wave mechanics to replace the Bohr assumptions and extend the mathematics to atoms with multiple electrons. Instead of circular orbits we have **atomic orbitals**, the space in which an electron of a specific energy is most likely to be found. Orbitals correspond to specific energy levels and specific quantum numbers for a particular electron.

Finally, while wave mechanics will seem far too complicated for your first course in chemistry, there are some practical applications that you may have already read about, including electron microscopy, electron diffraction, and neutron diffraction.

EXAMPLE 9-6: Calculate the wavelength of a neutron that has a velocity of 2.8 m s^{-1} and a mass of 1.01 amu.

Solution: Apply the de Broglie equation (9-5):

$$\lambda = \frac{h}{mv}$$

The mass of the neutron is 1.01 amu, which must be changed to kilograms:

$$m = \frac{(1.01\ \text{amu})(10^{-3}\ \text{kg g}^{-1})}{6.02 \times 10^{23}\ \text{amu g}^{-1}} = 1.68 \times 10^{-27}\ \text{kg}$$

Substituting, we get

$$\lambda = \frac{6.626 \times 10^{-34}\ \text{J s}}{(1.68 \times 10^{-27}\ \text{kg})(2.8\ \text{m s}^{-1})} = 1.4 \times 10^{-7}\ \text{m}$$

D. Quantum numbers

For the Bohr circular orbits a single quantum number, n, is sufficient to indicate the principal energy level of the electron. For a complete description of orbitals, we use a set of four quantum numbers to describe the energy level of an electron and the high-probability regions of its spatial distribution. The first of these quantum numbers is n, the **principal quantum number**, which we've already discussed. Recall that n is a positive integer that can range from 1 to (in principle) infinity, although only values of n from 1 to 7 are known to exist in ground-state atoms. The value of n is the main factor that determines the energy level of the electron.

The second quantum number l is the **angular momentum quantum number**. The value of l, which can take all positive integral values from 0 to $n - 1$, controls the general spatial distribution of electron probability. The possible values of l depend on the value of n; there are n possible values of l associated with each value of n. For example, if $n = 4$, there are four l values: 0, 1, 2, and 3. Chemists often use the letters s, p, d, and f for the numerical values of l (0, 1, 2, and 3, respectively). A full explanation of this nomenclature for $n = 1$ to 5 is given in Table 9-1.

TABLE 9-1. Quantum Numbers

n	l	Letter description	m	s	Number of combinations
1	0	$1s$	0	$\pm\frac{1}{2}$	2
2	0	$2s$	0	$\pm\frac{1}{2}$	2 ⎫ 8
2	1	$2p$	$-1, 0, +1$	$\pm\frac{1}{2}$	6 ⎭
3	0	$3s$	0	$\pm\frac{1}{2}$	2 ⎫
3	1	$3p$	$-1, 0, +1$	$\pm\frac{1}{2}$	6 ⎬ 18
3	2	$3d$	$-2, -1, 0, +1, +2$	$\pm\frac{1}{2}$	10 ⎭
4	0	$4s$	0	$\pm\frac{1}{2}$	2 ⎫
4	1	$4p$	$-1, 0, +1$	$\pm\frac{1}{2}$	6 ⎬ 32
4	2	$4d$	$-2, -1, 0, +1, +2$	$\pm\frac{1}{2}$	10 ⎪
4	3	$4f$	$-3, \ldots, 0, \ldots, +3$	$\pm\frac{1}{2}$	14 ⎭
5	0	$5s$	0	$\pm\frac{1}{2}$	2 ⎫
5	1	$5p$	$-1, 0, +1$	$\pm\frac{1}{2}$	6 ⎬ 32
5	2	$5d$	$-2, -1, 0, +1, +2$	$\pm\frac{1}{2}$	10 ⎪
5	3	$5f$	$-3, \ldots, 0, \ldots, +3$	$\pm\frac{1}{2}$	14 ⎭
5	4*				

* Electrons with $l = 4$ (designated g) or larger (designated h, i, j, etc.) are not known to be of any significance in chemical processes.

The third quantum number m is the **magnetic quantum number**. It describes the particular orientation of the spatial distribution described by l. The possible values of m are all integral and are given by $-l, -(l-1), \ldots, 0, \ldots, l-1$, and l. There are $2l + 1$ values of m associated with each l; you'll see the symbol m_l used for the value of m that corresponds to a specific value of l. The array of possible combinations is given in Table 9-1.

The fourth quantum number, unrelated to energy or spatial distribution, is the **spin quantum number** s (sometimes m_s), which can take only two values, $+\frac{1}{2}$ or $-\frac{1}{2}$. It designates the two possible spin orientations of an electron such that the angular momentum of the spinning electron is $s(h/2\pi)$.

EXAMPLE 9-7: In an atom there is an energy level for which $n = 4$. Determine the permissible values of l. Then determine the permissible values of m for $l = 2$.

Solution: For $n = 4$ the quantum number l may be 0 to $n - 1$, that is, 0 to 3. The value of l must be positive or zero. If $l = 2$, then m_2 is $-l, -(l-1), 0, l-1$, and l, that is, $-2, -1, 0, 1$, and 2. Note that five different values of m are permitted for $l = 2$ ($2l + 1 = 2(2) + 1 = 5$).

9-3. Orbitals

Each **atomic orbital** (AO), i.e., the probability distribution of finding an electron with a specific energy, defined by three specific values of the quantum numbers n, l, and m, has a specific designation and often a distinctive "shape." The orbital is described by stating the value of n with the letter designation of l. For example, if $n = 1$, $l = 0$, the result is a spherically symmetrical $1s$ orbital (all s orbitals are spherical). The case for $n = 2$ is more complex: For $l = 0$, the result is a $2s$ orbital in which the electron, on average, will be further away from the nucleus than an electron in the $1s$ orbital; there are three m values for $l = 1$, so there are three $2p$ orbitals, each having two lobes, directed along the x, y, or z axes of a three-dimensional coordinate system (see Figure 9-1). The electrons in each p orbital are as far as possible from those in the other p orbitals. The general shapes of s, p, and d orbitals do not change greatly with changes in n; a $3p$ and a $4p$ orbital have many features in common. You can see the shapes of some s, p, and d orbitals in Figure 9-1.

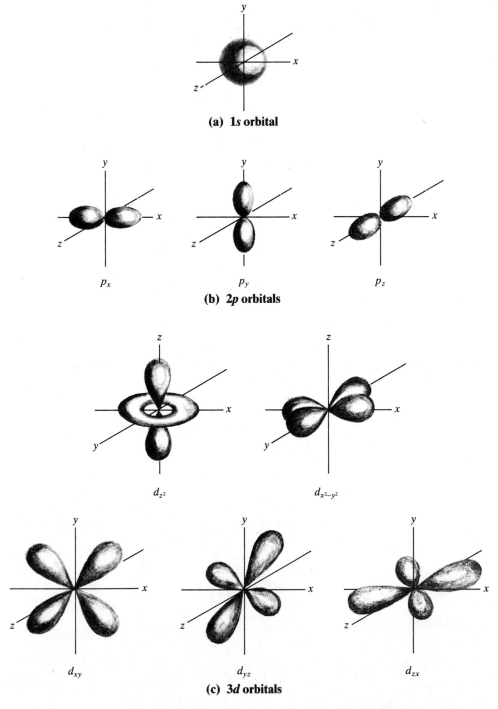

(a) $1s$ orbital

p_x p_y p_z

(b) $2p$ orbitals

d_{z^2} $d_{x^2-y^2}$

d_{xy} d_{yz} d_{zx}

(c) $3d$ orbitals

FIGURE 9-1. Shapes of s, p, and d orbitals.

'ing up atoms

need two new concepts to understand the way in which electrons are arranged to build up
complex atoms: the Pauli exclusion principle and the energy sequence of orbitals. The **Pauli
exclusion principle** states that no two electrons in any given atom can have the same four quantum
numbers (n, l, m_l, and s). Thus if you are given the first three out of the four quantum numbers, you
know that there can be only two electrons with these particular quantum numbers in a given atom
because there are only two possible values of s ($+\frac{1}{2}$ or $-\frac{1}{2}$) to distinguish these two electrons.

In a hydrogen atom the energy sequence of orbitals depends only on the value of n: Larger
values of n have higher-energy (i.e., less stable) orbitals. For multielectron atoms, the values of both
n and l affect the orbital energy level. The general order of the energy level sequence is given in
Figure 9-2. A useful way to remember this order is shown in Figure 9-3. To build up an atom, you

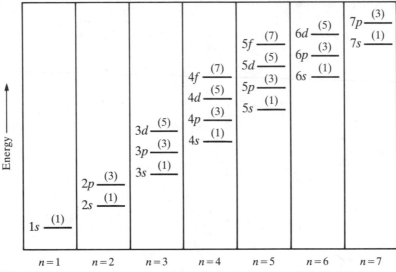

FIGURE 9-2. Energy-level sequence in multielectron atoms. Note that $4s$ is lower than
$3d$. The numbers in parentheses show the number of orbitals available at each energy
level.

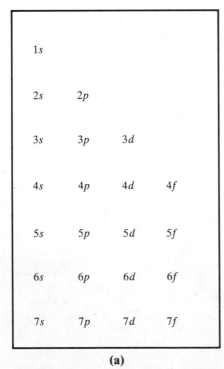

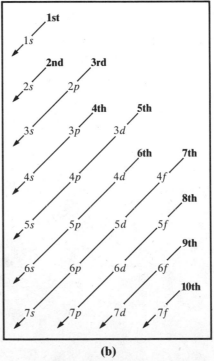

(a)   (b)

FIGURE 9-3. How to remember the orbital occupancy sequence: **(a)** Write out all the
available orbitals (s, p, d, f) for each n value (1–7); **(b)** draw diagonals labeled "1st,
2nd," etc., as shown.

simply put as many electrons as possible into the most stable (lowest-energy) orbitals available, remembering that no two of the electrons can have the same set of quantum numbers.

B. Electronic configurations of atoms

The **electronic configuration** of an atom shows the way in which the electrons in the atom occupy, in order of increasing energy, the available orbitals and spin states. The electronic configurations of ground-state atoms from $Z = 1$ (hydrogen) to $Z = 103$ (lawrencium) are given in Table 9-2. (Remember that the number of electrons is equal to Z in a neutral atom.)

The electronic configuration for any atom follows three principles:

1. **Aufbau principle:** In general, electrons occupy the lowest-energy orbitals available before entering the higher-energy orbitals.
2. **Hund principle:** Equal-energy orbitals are each occupied by a single electron before the second electron of opposite spin (s) enters the orbital. For example, each of the three $2p$ orbitals ($2p_x$, $2p_y$, and $2p_z$) will hold a single electron before any receives a second electron.
3. **Pauli exclusion principle:** No two electrons can have the same four quantum numbers.

Chemists use the notation nl^x to represent electrons in atomic orbitals, where n is the principal quantum number, l is the letter designation of the angular momentum quantum number, and x indicates the number of electrons in the particular orbitals.

EXAMPLE 9-8: Derive the electronic configuration of H, B, Ne, Na, and Mn.

Solution:

For H, $Z = 1$: We have only one electron to place. It will go into the most stable (lowest-energy) orbital, the $1s$ orbital. Thus the electronic configuration of hydrogen is $1s^1$.

For B, $Z = 5$: We have five electrons to place. Since the maximum occupancy of an s orbital is two electrons, $1s^2 2s^2 2p^1$ is the configuration.

For Ne, $Z = 10$. s Orbitals can hold a maximum of two electrons and p orbitals a maximum of six, so $1s^2 2s^2 2p^6$ is the configuration.

For Na, $Z = 11$. Following the reasoning for Ne, the configuration is $1s^2 2s^2 2p^6 3s^1$.

note: A convenient short notation uses the electronic configuration of the noble gases He, Ne, Ar, Kr, Xe, and Rn (2, 10, 18, 36, 54, and 86 total electrons, respectively) as part of the desired electronic configuration. Our configuration for Na could be written $[\text{Ne}]3s^1$.

For Mn, $Z = 25$, seven electrons more than argon, so we can express its configuration as $[\text{Ar}]4s^2 3d^5$. Recall that the $3d$ orbitals fill AFTER the $4s$ orbital (see Figure 9-2).

9-4. Ionization Energies

The energy needed to remove an electron from a gaseous atom, leaving behind a positive gaseous ion, is called the **ionization energy**. Atoms can be stripped of all their electrons, one by one, giving ions with one, two, or more positive charges. An ionization energy can be measured for each ionization step; the nth ionization energy corresponds to an ion that has an n^+ charge. Thus we can write

$$\text{Li}(g) \longrightarrow \text{Li}^+(g) + e^- \qquad \text{first ionization energy} = 519 \text{ kJ mol}^{-1}$$
$$\text{Li}^+(g) \longrightarrow \text{Li}^{2+}(g) + e^- \qquad \text{second ionization energy} = 7300 \text{ kJ mol}^{-1}$$
$$\text{Li}^{2+}(g) \longrightarrow \text{Li}^{3+}(g) + e^- \qquad \text{third ionization energy} = 11\,800 \text{ kJ mol}^{-1}$$

Because Li^{3+} is a bare nucleus with no more electrons, the third ionization energy is the last possible one for Li. Each successive ionization energy is larger than the preceding one; this reflects the growing electrostatic attraction of the increasingly positive ion for its negative electrons. Ionization energy values are given in Table 9-3.

A. Electronic configuration of atomic ions

1. *Positive ions*: In general, the electrons lost first in ionization are the ones added last in the building up of the neutral atom. When Na ($1s^2 2s^2 2p^6 3s^1$) is ionized to Na$^+$, the electronic

TABLE 9-2. Electronic Configurations of the Elements in Their Ground States

Atomic number	Element	n = 1	2	3	4	5	6	7
		1s	2s 2p	3s 3p 3d	4s 4p 4d 4f	5s 5p 5d 5f	6s 6p 6d	7s
1	Hydrogen	1						
2	**Helium**	**2**						
3	Lithium	2	1					
4	Beryllium	2	2					
5	Boron	2	2 1					
6	Carbon	2	2 2					
7	Nitrogen	2	2 3					
8	Oxygen	2	2 4					
9	Fluorine	2	2 5					
10	**Neon**	**2**	**2 6**					
11	Sodium	2	2 6	1				
12	Magnesium	2	2 6	2				
13	Aluminum	2	2 6	2 1				
14	Silicon	2	2 6	2 2				
15	Phosphorus	2	2 6	2 3				
16	Sulfur	2	2 6	2 4				
17	Chlorine	2	2 6	2 5				
18	**Argon**	**2**	**2 6**	**2 6**				
19	Potassium	2	2 6	2 6	1			
20	Calcium	2	2 6	2 6	2			
21	Scandium	2	2 6	2 6 1	2			
22	Titanium	2	2 6	2 6 2	2			
23	Vanadium	2	2 6	2 6 3	2			
24	Chromium	2	2 6	2 6 5	1			
25	Manganese	2	2 6	2 6 5	2			
26	Iron	2	2 6	2 6 6	2			
27	Cobalt	2	2 6	2 6 7	2			
28	Nickel	2	2 6	2 6 8	2			
29	Copper	2	2 6	2 6 10	1			
30	Zinc	2	2 6	2 6 10	2			
31	Gallium	2	2 6	2 6 10	2 1			
32	Germanium	2	2 6	2 6 10	2 2			
33	Arsenic	2	2 6	2 6 10	2 3			
34	Selenium	2	2 6	2 6 10	2 4			
35	Bromine	2	2 6	2 6 10	2 5			
36	**Krypton**	**2**	**2 6**	**2 6 10**	2 6			
			8	18				
37	Rubidium	2	8	18	2 6	1		
38	Strontium	2	8	18	2 6	2		
39	Yttrium	2	8	18	2 6 1	2		
40	Zirconium	2	8	18	2 6 2	2		
41	Niobium	2	8	18	2 6 4	1		
42	Molybdenum	2	8	18	2 6 5	1		
43	Technetium	2	8	18	2 6 6	1		
44	Ruthenium	2	8	18	2 6 7	1		
45	Rhodium	2	8	18	2 6 8	1		
46	Palladium	2	8	18	2 6 10			
47	Silver	2	8	18	2 6 10	1		
48	Cadmium	2	8	18	2 6 10	2		
49	Indium	2	8	18	2 6 10	2 1		
50	Tin	2	8	18	2 6 10	2 2		
51	Antimony	2	8	18	2 6 10	2 3		
52	Tellurium	2	8	18	2 6 10	2 4		
53	Iodine	2	8	18	2 6 10	2 5		
54	**Xenon**	**2**	**8**	**18**	**2 6 10**	**2 6**		

TABLE 9-2 (*Continued*)

Atomic number	Element	$n=1$	2	3	4	5	6	7
		$1s$	$2s\ 2p$	$3s\ 3p\ 3d$	$4s\ 4p\ 4d\ 4f$	$5s\ 5p\ 5d\ 5f$	$6s\ 6p\ 6d$	$7s$
55	Cesium	2	8	18	2 6 10	2 6	1	
56	Barium	2	8	18	2 6 10	2 6	2	
57	Lanthanum	2	8	18	2 6 10	2 6 1	2	
58	Cerium	2	8	18	2 6 10 2	2 6	2	
59	Praseodymium	2	8	18	2 6 10 3	2 6	2	
60	Neodymium	2	8	18	2 6 10 4	2 6	2	
61	Promethium	2	8	18	2 6 10 5	2 6	2	
62	Samarium	2	8	18	2 6 10 6	2 6	2	
63	Europium	2	8	18	2 6 10 7	2 6	2	
64	Gadolinium	2	8	18	2 6 10 7	2 6 1	2	
65	Terbium	2	8	18	2 6 10 9	2 6	2	
66	Dysprosium	2	8	18	2 6 10 10	2 6	2	
67	Holmium	2	8	18	2 6 10 11	2 6	2	
68	Erbium	2	8	18	2 6 10 12	2 6	2	
69	Thulium	2	8	18	2 6 10 13	2 6	2	
70	Ytterbium	2	8	18	2 6 10 14	2 6	2	
71	Lutetium	2	8	18	2 6 10 14	2 6 1	2	
72	Hafnium	2	8	18	2 6 10 14	2 6 2	2	
73	Tantalum	2	8	18	2 6 10 14	2 6 3	2	
74	Tungsten	2	8	18	2 6 10 14	2 6 4	2	
75	Rhenium	2	8	18	2 6 10 14	2 6 5	2	
76	Osmium	2	8	18	2 6 10 14	2 6 6	2	
77	Iridium	2	8	18	2 6 10 14	2 6 9		
78	Platinum	2	8	18	2 6 10 14	2 6 9	1	
79	Gold	2	8	18	2 6 10 14	2 6 10	1	
80	Mercury	2	8	18	2 6 10 14	2 6 10	1	
81	Thallium	2	8	18	2 6 10 14	2 6 10	2 1	
82	Lead	2	8	18	2 6 10 14	2 6 10	2 2	
83	Bismuth	2	8	18	2 6 10 14	2 6 10	2 3	
84	Polonium	2	8	18	2 6 10 14	2 6 10	2 4	
85	Astatine	2	8	18	2 6 10 14	2 6 10	2 5	
86	**Radon**	**2**	**8**	**18**	**2 6 10 14**	**2 6 10**	**2 6**	
87	Francium	2	8	18	32	2 6 10	2 6	1
88	Radium	2	8	18	32	2 6 10	2 6	2
89	Actinium	2	8	18	32	2 6 10	2 6 1	2
90	Thorium	2	8	18	32	2 6 10	2 6 2	2
91	Protoactinium	2	8	18	32	2 6 10 2	2 6 1	2
92	Uranium	2	8	18	32	2 6 10 3	2 6 1	2
93	Neptunium	2	8	18	32	2 6 10 4	2 6 1	2
94	Plutonium	2	8	18	32	2 6 10 6	2 6	2
95	Americium	2	8	18	32	2 6 10 7	2 6	2
96	Curium	2	8	18	32	2 6 10 7	2 6 1	2
97	Berkelium	2	8	18	32	2 6 10 8	2 6 1	2
98	Californium	2	8	18	32	2 6 10 10	2 6	2
99	Einsteinium	2	8	18	32	2 6 10 11	2 6	2
100	Fermium	2	8	18	32	2 6 10 12	2 6	2
101	Mendelevium	2	8	18	32	2 6 10 13	2 6	2
102	Nobelium	2	8	18	32	2 6 10 14	2 6	2
103	Lawrencium	2	8	18	32	2 6 10 14	2 6 1	2

TABLE 9-3. First Ionization Energies of the Elements (in kJ/mol)

								H	He								
								1310	2370								

Li	Be											B	C	N	O	F	Ne
519	900											799	1090	1400	1310	1680	2080

Na	Mg											Al	Si	P	S	Cl	Ar
494	736											577	786	1060	1000	1260	1520

K	Ca	Sc	Ti	V	Cr	Mn	Fe	Co	Ni	Cu	Zn	Ga	Ge	As	Se	Br	Kr
418	590	632	661	648	653	716	762	757	736	745	908	577	762	966	941	1140	1350

Rb	Sr	Y	Zr	Nb	Mo	Tc	Ru	Rh	Pd	Ag	Cd	In	Sn	Sb	Te	I	Xe
402	548	636	669	653	694	699	724	745	803	732	866	556	707	833	870	1010	1170

Cs	Ba	La	Hf	Ta	W	Re	Os	Ir	Pt	Au	Hg	Tl	Pb	Bi	Po	At	Rn
376	502	540	531	577	770	762	841	887	866	891	1010	590	716	774	812	—	1040

Fr	Ra	Ac
381	510	669

Ce	Pr	Nd	Pm	Sm	Eu	Gd	Tb	Dy	Ho	Er	Tm	Yb	Lu
665	556	607	556	540	548	594	648	657	—	—	—	598	481

Th	Pa	U	Np	Pu	Am	Cm	Bk	Cf	Es	Fm	Md	No	Lr
674	—	385	—	—	—	—	—	—	—	—	—	—	—

configuration of the ion is $1s^2 2s^2 2p^6$ (or [Ne]). In the gas phase Na^{2+} can be formed with a configuration of $1s^2 2s^2 2p^5$, and so on.

> *note:* For those atoms having incomplete d shells (Sc through Cu, Y through Ag, and Hf through Au), the electrons of the $(n+1)s$ orbital are lost on ionization BEFORE those of the nd orbital. For example, the configuration of Mn is $[Ar]4s^2 3d^5$; the configuration of Mn^{2+} is $[Ar]3d^5$, NOT $[Ar]4s^2 3d^3$.

2. *Negative ions*: Extra electrons are simply added to the lowest available orbitals, just as for neutral atoms. For Cl, $Z = 17$ and the configuration is $[Ne]3s^2 3p^5$; Cl^- has 18 electrons and may be written [Ar].

B. Electronic structure and magnetism

Monatomic species (atoms or ions) can be *diamagnetic*, which means, roughly, that they are not attracted toward a magnet; or they can be *paramagnetic*, which means, roughly, that they are attracted toward a magnet. These properties reflect the detailed electronic structure of the species.

Two electrons in an atom that have the same values of n, l, and m_l will necessarily have s values of $+\frac{1}{2}$ and $-\frac{1}{2}$. They are said to have *opposite spins* that are *paired*. If an electron has a particular set of n, l, and m_l values and there is no electron in that atom with the same quantum numbers, the spin of the electron is *unpaired*. Atoms, ions, or molecules having electrons with unpaired spins are paramagnetic; species having all paired spins are diamagnetic.

You'll need to recall one additional principle to understand these properties. Hund's principle states that the most stable state for a set of electrons in a group of equal-energy orbitals—sometimes called a *degenerate* group—is the state with the *maximum* number of unpaired spins. For example, consider the nitrogen atom: $Z = 7$ and the configuration is $1s^2 2s^2 2p^3$. The electrons in the $2p$ orbitals ($m_l = -1, 0$, and $+1$) each have the same energy. According to Hund's rule, in the most stable state of atomic nitrogen, the three $2p$ electrons will be unpaired, each having a different value of m_l. So the nitrogen atom has three unpaired electrons. You can designate this state in pictorial form by using boxes for the different m_l values and vertical half arrows (↑ and ↓) to signify an s of $+\frac{1}{2}$ or $-\frac{1}{2}$, respectively (see Figure 9-4).

5 = 2   p = 8   d = 12   f =

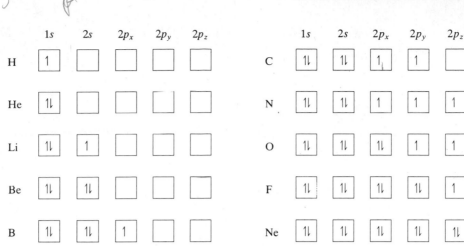

FIGURE 9-4. Box diagrams of electronic configurations. Half-arrows represent electrons; two half-arrows in opposite directions represent paired spins.

EXAMPLE 9-9: Determine the number of unpaired electrons in the sulfur atom.

Solution: For S, $Z = 16$ and the configuration is $[Ne]3s^23p^4$. The 3p orbitals each have the same energy. Three of the four electrons will be in separate 3p orbitals, but the fourth must pair with one of them. That will leave two electrons unpaired in sulfur. In terms of boxes:

$$[Ne] \quad 3s \quad 3p_x \quad 3p_y \quad 3p_z$$

$$\boxed{\uparrow\downarrow} \ \boxed{\uparrow\downarrow} \ \boxed{\uparrow} \ \boxed{\uparrow}$$

9-5. The Periodic Table

The **periodic law** states that the physical and chemical properties of the elements are periodic (repeating) functions of their atomic numbers. The **periodic table** of the elements is an array of the elements arranged in order of increasing atomic number (increasing number of electrons) with elements having similar chemical properties arranged in vertical columns called **groups**. A **period** is a horizontal row, where the period number (1 through 7) corresponds to the principal energy level (*n*) being filled.

You can understand the form of the periodic table when you appreciate that similar chemical properties are found in elements with similar electronic configuration. For example, if you look up the chemical properties of the elements of the group marked IA in the table (i.e., Li, Na, K, Rb, Cs, and Fr), you'll find that all of them are rather soft, silvery metals with low melting points. They are very reactive toward water and oxygen, and all of them easily form monopositive ions. They all have an electronic configuration of $[\text{noble gas}]ns^1$, where $n = 2$ through 7.

You can find other similar groupings within all of the main-group elements, groups IA through VIIIA, in which the *s* and *p* orbitals are being filled. Because there can be a maximum of two *s* electrons and six *p* electrons for a given principal quantum number, there are $2 + 6 = 8$ main groups.

The periodic table shows how the filling of *d* orbitals leads to other families of elements, called the *transition elements*; the filling of *f* orbitals leads to the *lanthanide* and *actinide* series of elements.

SUMMARY

1. All electromagnetic radiation can be described by its frequency (*v*) and its wavelength (*λ*), where $c = v\lambda$.
2. Atomic spectra contain regularly spaced lines that indicate the existence of specific energy levels in atoms.
3. Radiation of frequency *v* comes in discrete quanta, where the energy is given by $E = hv$ (*h* is Planck's constant).
4. The magnitude of the photoelectric effect is given by $\Delta E = hv_0 + \frac{1}{2}mv^2$.
5. The wavelengths of the lines in the atomic spectrum of hydrogen are given by $1/\lambda = R_H[(1/n_i^2) - (1/n_f^2)]$, where R_H is the Rydberg constant.

6. The relationship between wavelength and momentum (*mv*) for a particle is given by the de Broglie equation: $\lambda = h/mv$.

7. The Heisenberg uncertainty principle limits the accuracy with which the position and momentum of a particle can be determined.

8. The energy levels of electrons are specified by four quantum numbers: the principal (*n*), angular momentum (*l*), magnetic (*m*), and spin (*s* or m_s).

9. An atomic orbital is the probability distribution of finding an electron with a specific energy level as defined by its quantum numbers.

10. The ranges of the quantum numbers are

Quantum number	Range
n	1 to 7 (in ground-state atoms)
l	0 to $n - 1$ for each *n*
m	$-l$ to $+l$ for each *l*
s	$+\frac{1}{2}$ or $-\frac{1}{2}$

11. The Pauli exclusion principle states that no two electrons in any given atom can have the same four quantum numbers.

12. The electronic configuration for any atom is governed by the aufbau, Pauli exclusion, and Hund principles.

13. The energy required to remove an electron from an atom in the gaseous state, leaving a positive ion, is the ionization energy.

14. Atomic species having unpaired electrons are paramagnetic; atoms having all paired electrons are diamagnetic.

15. The periodic law states that the physical and chemical properties of the elements are periodic functions of their atomic numbers.

16. In the periodic table, a period is a horizontal row, a group is a vertical column. Elements in a group show physical and chemical similarities to the other group members.

RAISE YOUR GRADES

Can you . . . ?

☑ calculate wavelength from frequency of electromagnetic radiation, and vice versa
☑ use the Balmer equation
☑ calculate the energy of a quantum of given frequency
☑ calculate the magnitude of the photoelectric effect
☑ determine energy level differences from spectral data
☑ use the de Broglie equation
☑ describe the relationships between possible values of the quantum numbers *n*, *l*, *m*, and *s*
☑ deduce the energy level sequence for a multielectron atom
☑ deduce the ground-state electronic configuration for an atom or ion
☑ write out a box diagram for an atom or ion and determine its number of unpaired electrons

SOLVED PROBLEMS

PROBLEM 9-1 The normal human eye responds to visible light of wavelengths ranging from about 390 to 710 nm. Determine the frequency range of the human eye.

Solution: Wavelength λ and frequency v are related to the velocity of light c by

$$c = \lambda v$$

where c is a constant equal to 3.00×10^8 m s^{-1}. So change nanometers to meters:

$$390 \text{ nm} = 3.9 \times 10^{-7} \text{ m} \qquad 710 \text{ nm} = 7.1 \times 10^{-7} \text{ m}$$

then convert wavelength to frequency:

$$v = \frac{c}{\lambda} = \frac{3.00 \times 10^8 \text{ m s}^{-1}}{3.9 \times 10^{-7} \text{ m}} \qquad v = \frac{c}{\lambda} = \frac{3.00 \times 10^8 \text{ m s}^{-1}}{7.1 \times 10^{-7} \text{ m}}$$

$$= 7.7 \times 10^{14} \text{ s}^{-1} \qquad\qquad = 4.2 \times 10^{14} \text{ s}^{-1}$$

The frequency range of the normal human eye is 4.2×10^{14} to 7.7×10^{14} Hz.

PROBLEM 9-2   The wavelength of one of the Balmer series lines in the hydrogen spectrum is 410.2 nm. Determine the value of n for this line.

Solution: Use Eq. (9-1) for the Balmer series, where R_H is the Rydberg constant equal to 1.097×10^{-7} m^{-1}. You'll find it much easier simply to isolate the term containing n from Eq. (9-1) on one side of the equation, substitute the known values, and solve using your calculator. First, convert wavelength to meters:

$$410.2 \text{ nm} = 4.102 \times 10^{-7} \text{ m}$$

Next, isolate n and substitute for λ and R_H (remembering that the wave number $\bar{v}$ is the reciprocal of the wavelength λ):

$$\bar{v} = \frac{1}{\lambda} = R_H\left(\frac{1}{2^2} - \frac{1}{n^2}\right)$$

$$\frac{1}{R_H\lambda} = \frac{1}{4} - \frac{1}{n^2}$$

$$\frac{1}{n^2} = \frac{1}{4} - \frac{1}{R_H\lambda} = 0.2500 - \left(\frac{1}{(1.097 \times 10^7 \text{ m}^{-1})(4.102 \times 10^{-7} \text{ m})}\right) = 0.0278$$

$$n^2 = \frac{1}{0.0278} = 35.97$$

$$n = 6$$

PROBLEM 9-3   An energy change occurs in an atom, so that light is emitted. The energy change is 2.37×10^{-18} J. What is the wavelength (in nm) of the light emitted?

Solution: Because there is an energy change here, we need to find the energy difference using Planck's equation (9-2)

$$E_f - E_i = \Delta E = hv$$

where h is Planck's constant, equal to 6.626×10^{-34} J s. Then we can use the relationship $v = c/\lambda$ to write

$$\lambda = \frac{hc}{\Delta E} = \frac{(6.626 \times 10^{-34} \text{ J s})(3.00 \times 10^8 \text{ m s}^{-1})}{2.37 \times 10^{-18} \text{ J}}$$

$$= 8.39 \times 10^{-8} \text{ m}$$

which is $(8.39 \times 10^{-8}$ m$) \times (1 \times 10^9$ nm m$^{-1}) = 83.9$ nm. (Note that this is a relatively large energy difference that produces a short-wavelength photon).

PROBLEM 9-4   The threshold frequency for metallic potassium is 5.46×10^{14} s^{-1}. Calculate the maximum kinetic energy and velocity that the photoelectrons will have when the wavelength of light shining on the potassium surface is 350 nm. (The mass of an electron is 9.11×10^{-31} kg.)

Solution: Use the equation that quantifies the photoelectric effect (9-3):

$$\Delta E = hv = hv_0 + \tfrac{1}{2}mv^2$$

where hv_0 is the work function (Planck's constant times the threshold frequency) and $\tfrac{1}{2}mv^2$ is the maximum kinetic energy of the electrons. So, denoting the maximum kinetic energy as E_k, we can write

$$\tfrac{1}{2}mv^2 = hv - hv_0 = E_k$$

We're given the wavelength $\lambda = 350$ nm, so we convert λ to v:

$$v = \frac{c}{\lambda} = \frac{3.00 \times 10^8 \text{ m s}^{-1}}{(350 \text{ nm})(10^{-9} \text{ m nm}^{-1})} = 8.57 \times 10^{14} \text{ s}^{-1}$$

Now we can calculate E_k:

$$\begin{aligned}
E_k &= hv - hv_0 \\
&= (6.626 \times 10^{-34} \text{ J s})(8.57 \times 10^{14} \text{ s}^{-1}) - (6.626 \times 10^{-34} \text{ J s})(5.46 \times 10^{14} \text{ s}^{-1}) \\
&= 2.06 \times 10^{-19} \text{ J}
\end{aligned}$$

To find the velocity (v) of the electron, we rearrange the definition $E_k = \frac{1}{2}mv^2$ and substitute in the calculated E_k and the given mass:

$$v = \sqrt{\frac{2E_k}{m}} = \sqrt{\frac{(2)(2.06 \times 10^{-19} \text{ J})}{9.11 \times 10^{-31} \text{ kg}}}$$

$$= 6.73 \times 10^5 \text{ m s}^{-1}$$

PROBLEM 9-5   Calculate the ionization energy of hydrogen for the $n = 1$ state (ground state) for one atom and one mole of atoms. Express your answers in joules and kilojoules, respectively.

Solution: The ionization energy is, in effect, the energy required to move an electron from the ground state ($n = 1$) to the highest excited state ($n = \infty$), i.e., the energy required to remove the electron. We'll need to combine two equations: Planck's equation (9-2) and the wavelength calculation for the hydrogen spectrum (9-4):

$$\Delta E = hc\left(\frac{1}{\lambda}\right) \qquad \text{and} \qquad \frac{1}{\lambda} = R_H\left(\frac{1}{n_f^2} - \frac{1}{n_i^2}\right)$$

So we have

$$\Delta E = hcR_H\left(\frac{1}{n_f^2} - \frac{1}{n_i^2}\right)$$

Substitute and solve (note that $1/n_i^2 = 1/\infty = 0$)

$$\Delta E = (6.626 \times 10^{-34} \text{ J s})(3.00 \times 10^8 \text{ m s}^{-1})(1.907 \times 10^7 \text{ m}^{-1})\left(\frac{1}{1^2} - 0\right)$$

$$= 2.18 \times 10^{-18} \text{ J/atom}$$

For a mole of H atoms, multiply by Avogadro's number:

$$\begin{aligned}
\Delta E &= (2.18 \times 10^{-18} \text{ J/atom})(6.02 \times 10^{23} \text{ atoms/mol})(10^{-3} \text{ kJ/J}) \\
&= 1.31 \times 10^3 \text{ kJ/mol}
\end{aligned}$$

PROBLEM 9-6   A hydrogen atom in an excited state, $n = 6$, falls to a lower state where $n = 3$. How much energy has it lost? What is the wavelength of the light emitted?

Solution: Refer to Problem 9-5.

$$\frac{1}{\lambda} = R_H\left(\frac{1}{n_f^2} - \frac{1}{n_i^2}\right)$$

$$= (1.097 \times 10^7 \text{ m}^{-1})\left(\frac{1}{3^2} - \frac{1}{6^2}\right) = 9.14 \times 10^5 \text{ m}^{-1}$$

$$\lambda = \frac{1}{9.14 \times 10^5 \text{ m}^{-1}}$$

$$= 1.09 \times 10^{-6} \text{ m}$$

$$\Delta E = hc\left(\frac{1}{\lambda}\right)$$

$$= (6.626 \times 10^{-34} \text{ J s})(3.00 \times 10^8 \text{ m s}^{-1})(9.14 \times 10^5 \text{ m}^{-1})$$
$$= 1.82 \times 10^{-19} \text{ J}$$

PROBLEM 9-7   In the emission spectrum of mercury vapor a line is observed at 546 nm. (a) Calculate the energy difference between the two levels or states of the mercury atom that produces this light. (b) Explain why there is no light observed at 530 nm from the mercury vapor emission spectrum.

Solution:
(a) The wavelength = (546 nm) = 5.46×10^{-7} m.

$$\Delta E = \frac{hc}{\lambda} = \frac{(6.626 \times 10^{-34} \text{ J s})(3.00 \times 10^8 \text{ m s}^{-1})}{5.46 \times 10^{-7} \text{ m}}$$
$$= 3.64 \times 10^{-19} \text{ J}$$

(b) The mercury atom has many electrons and is much more complicated than the hydrogen atom. However, its energy is quantized; only certain discrete energies are permitted. The fact that you do not observe light at 530 nm means that there are not two energy levels allowed in the mercury atom that differ by exactly the right amount of energy to emit light at 530 nm.

PROBLEM 9-8   Electrons that have a wavelength of 12 pm are used for electron diffraction. What is the velocity of these electrons?

Solution: Use the de Broglie equation (9-5):

$$\lambda = \frac{h}{mv}$$

Changing 12 pm to 12×10^{-12} m and recalling that the mass of the electron is 9.11×10^{-31} kg, we get

$$v = \frac{h}{m\lambda} = \frac{6.626 \times 10^{-34} \text{ J s}}{(9.11 \times 10^{-31} \text{ kg})(12 \times 10^{-12} \text{ m})}$$
$$= 6.1 \times 10^7 \text{ m s}^{-1}$$

PROBLEM 9-9   What is the maximum number of electrons in an atom that can possess the following sets of quantum numbers: (a) $n = 4$, $l = 2$; (b) $n = 5$, $l = 3$, $m = -1$; and (c) $n = 3$, $l = 3$?

Solution:
(a) m_l can be $-2, -1, 0, 1$, or 2. For each m_l there can be two electrons, $m_s = +\frac{1}{2}$ or $-\frac{1}{2}$, so the maximum number of electrons is 10.
(b) Here $m_l = -1$ has been specified. The only other variable is m_s, which may be $+\frac{1}{2}$ or $-\frac{1}{2}$, so the maximum is 2.
(c) Trick question! If $n = 3$, l cannot equal 3. The l quantum number must be less than 3. The answer is zero.

PROBLEM 9-10   What are the values of n and l that correspond to the following orbitals: $5d$, $3s$, $2p$? What is the maximum number of electrons that can be in each of these orbitals?

Solution: The number is the value of n and the letter corresponds to the value of l (see Table 9-1). Once you know what l is, you have $2l + 1$ different values for m and two electrons for each m.

Orbital	n	l	Number of m values $(2l + 1)$	Total electrons
$5d$	5	2	5	10
$3s$	3	0	1	2
$2p$	2	1	3	6

PROBLEM 9-11   Write the electronic configurations of the Cl, Al, Fe, and Zn atoms.

Solution: You can look up the answers to this problem, but you must be able to do them without the textbook.

Atom	Z	Configuration		
Cl	17	$1s^2 2s^2 2p^6 3s^2 3p^5$	or	$[\text{Ne}]3s^2 3p^5$
Al	13	$1s^2 2s^2 2p^6 3s^2 3p^1$	or	$[\text{Ne}]3s^2 3p^1$
Fe	26	$1s^2 2s^2 2p^6 3s^2 3p^6 4s^2 3d^6$	or	$[\text{Ar}]4s^2 3d^6$
Zn	30	$1s^2 2s^2 2p^6 3s^2 3p^6 4s^2 3d^{10}$	or	$[\text{Ar}]4s^2 3d^{10}$

note: We are writing electronic configurations in their filling order, not by principal quantum number.

PROBLEM 9-12 Which of the following is the correct electronic configuration for the ground state of sulfur?

(a) $1s^2 2p^6 3s^2 3p^6$

(b) $1s^2 2s^2 2p^6 3s^2 3p^4$

(c) $1s^2 2s^2 2p^6 3s^2 3p^6$

(d) $1s^2 2s^2 2p^6 3s^2 3p^2 4s^2$

(e) $1s^2 2s^2 2p^6 3s^4 3p^2$

Solution: The correct answer is (b). In (a) the $2s$ orbital was omitted. In (c) there are 18 electrons total: For sulfur, $Z = 16$. In (d) the $3p$ orbital should fill completely before the $4s$ orbital; (d) would be an excited state of sulfur. In (e) the configuration violates the exclusion principle. There can be no more than two electrons in the $3s$ orbital.

PROBLEM 9-13 Derive the electronic configurations of the Os, Te, and Kr atoms.

Solution: You will find it convenient to use the shortened notation for these.

Atom	Z	Configuration	
Os	76	$[Xe]6s^2 4f^{14} 5d^6$	
Te	52	$[Kr]4d^{10} 5s^2 5p^4$	
Kr	36	$[Ar]3d^{10} 4s^2 4p^6$	([Kr] is hardly adequate!)

PROBLEM 9-14 Derive the electronic configuration for the Co^{3+}, Ru^{3+}, Al^+, and As^- ions.

Solution: Write the electronic configuration for the neutral atom. Then remove enough of the highest-energy electrons to give the proper positive charge, or add electrons to the lowest-energy orbital available to give the correct negative charge. For transition elements, the s electrons go first.

For Co^{3+}, $Z = 27$. The neutral atom is $[Ar]4s^2 3d^7$. The ion is $[Ar]3d^6$.

For Ru^{3+}, $Z = 44$. The neutral atom is $[Kr]5s^1 4d^7$. The ion is $[Kr]4d^5$. Note that the neutral Ru atom does not follow the rules. You'd expect the configuration to be $[Kr]5s^2 4d^6$ for the neutral atom. Transition elements are somewhat sneaky in this regard and bear watching.

For Al^+, $Z = 13$. The neutral atom is $1s^2 2s^2 2p^6 3s^2 3p^1$. The ion is $1s^2 2s^2 2p^6 3s^2$.

For As^-, $Z = 33$. The neutral atom is $[Ar]3d^{10} 4s^2 4p^3$. The ion is $[Ar]3d^{10} 4s^2 4p^4$.

PROBLEM 9-15 Use the ionization energies in Table 9-3 to determine the ionization energy (in joules) of one atom of (a) Ca, (b) N, and (c) Cl.

Solution: Convert kilojoules to joules and moles of atoms to atoms.

(a) The ionization energy of Ca is 590 kJ mol^{-1}, so

$$(590 \text{ kJ mol}^{-1})\left(\frac{10^3 \text{ J}}{1 \text{ kJ}}\right)\left(\frac{1 \text{ mol}}{6.02 \times 10^{23} \text{ atom}}\right) = 9.80 \times 10^{-19} \text{ J/atom}$$

(b) The ionization energy of N is 1400 kJ mol^{-1}, so

$$(1400 \text{ kJ mol}^{-1})\left(\frac{10^3 \text{ J}}{1 \text{ kJ}}\right)\left(\frac{1 \text{ mol}}{6.02 \times 10^{23} \text{ atom}}\right) = 2.33 \times 10^{-18} \text{ J/atom}$$

(c) The ionization energy of Cl is 1260 kJ mol^{-1}, so

$$(1260 \text{ kJ mol}^{-1})\frac{(10^3 \text{ J})(1 \text{ mol})}{(1 \text{ kJ})(6.02 \times 10^{23} \text{ atom})} = 2.09 \times 10^{-18} \text{ J/atom}$$

In an energy change such as this you should understand the difference between the ionization energy required for one mole and that for one atom.

PROBLEM 9-16 How many unpaired electrons are there in the carbon atom and the beryllium atom?

Solution: For C, $Z = 6$ and the configuration is $1s^2 2s^2 2p^2$. The two electrons in the set of $2p$ orbitals will have different values of m_l and be unpaired, so there are *two* unpaired electrons in the carbon atom.

For Be, $Z = 4$ and the configuration is $1s^2 2s^2$. There is only one possible value for m_l in the $2s$ orbital so the two $2s$ electrons must be paired: $m_s = +\frac{1}{2}$ and $-\frac{1}{2}$. There are *no* unpaired electrons in the beryllium atom.

$$2s$$

$$\text{Be} \quad \boxed{\uparrow\downarrow}$$

PROBLEM 9-17 Which of the following atoms has the greatest number of unpaired electrons: (a) Ne, (b) P, (c) O, or (d) B?

Solution: First determine the electronic configuration for each atom, then determine the number of unpaired electrons in the highest-energy orbital.

(a) For Ne, $Z = 10$; the configuration is $1s^2 2s^2 2p^6$. In the $2p$ orbital there are two electrons for each value of m_l. All of the electrons are paired, so there are *no* unpaired electrons in the Ne atom.

(b) For P, $Z = 15$; the configuration is $[\text{Ne}]3s^2 3p^3$. In the $3p$ orbital there is one electron for each m_l value, so there are *three* unpaired electrons in the P atom.

(c) For O, $Z = 8$; the configuration is $1s^2 2s^2 2p^4$. In the $2p$ orbital there are two electrons that must have the same m_l values and be paired. The other two $2p$ electrons have different m_l values and are unpaired, so there are *two* unpaired electrons in the O atom.

(d) For B, $Z = 5$; the configuration is $1s^2 2s^2 2p^1$. One electron in the $2p$ orbital is not paired, so there is *one* unpaired electron in the B atom.

The P atom has the greatest number of unpaired electrons.

PROBLEM 9-18 The number of unpaired electrons can be determined for ions as well as atoms. How many unpaired electrons are there in (a) F^- and (b) Fe^{2+}?

Solution:

(a) For F, $Z = 9$. The electronic configuration for the F^- ion, which has 10 electrons including the one for the negative charge, is $1s^2 2s^2 2p^6$. In the $2p$ orbital there are two electrons for each m_l value. All of them are paired, so there are *no* unpaired electrons in F^-.

(b) For Fe, $Z = 26$. The electronic configuration of the Fe^{2+} ion, which has 2 electrons less than the neutral atom, is $[\text{Ar}]3d^6$. In the $3d$ orbital there are five different values that m_l may take: $-2, -1, 0, +1, +2$. Two electrons must have the same m_l value and be paired; the other four electrons have different m values and are unpaired. There are *four* unpaired electrons in Fe^{2+}.

PROBLEM 9-19 Write the electronic configuration for the group IIIA elements. How are they similar?

Solution: The group IIIA elements are B, Al, Ga, In, and Tl.

Atom	Z	Configuration
B	5	$1s^2 2s^2 2p^1$
Al	13	$1s^2 2s^2 2p^6 3s^2 3p^1$
Ga	31	$[\text{Ar}]3d^{10}4s^2 4p^1$
In	49	$[\text{Kr}]4d^{10}5s^2 5p^1$
Tl	81	$[\text{Xe}]4f^{14}5d^{10}6s^2 6p^1$

The highest-energy orbitals occupied in all of these elements are $ns^2 np^1$.

PROBLEM 9-20 Which of the following, if any, are *not* transition elements: (a) Ti, (b) Ta, (c) Te, (d) Tc, (e) V?

Solution: Find the atomic numbers of these elements and locate them in the periodic table. It will help to become familiar with the periodic table. For Ti, $Z = 22$; Ta, $Z = 73$; Te, $Z = 52$; Tc, $Z = 43$; and for V, $Z = 23$. Only tellurium (Te), element number 52, is outside the groups classed as transition elements. The correct answer is (c).

Supplementary Exercises

PROBLEM 9-21   Excited sodium atoms emit light with a wavelength of 589 nm. Calculate (**a**) the frequency of this light and (**b**) the energy of one of these photons in joules.

Answer: (**a**) 5.09×10^{14} Hz   (**b**) 3.37×10^{-19} J

PROBLEM 9-22   A hydrogen atom is excited to the $n = 8$ energy level and then emits a photon of light as it falls to the $n = 2$ energy level. Calculate (**a**) the wavelength and (**b**) the frequency of the light emitted. (**c**) What series is this light a part of?

Answer: (**a**) 388.9 nm   (**b**) 7.71×10^{14} Hz   (**c**) Balmer

PROBLEM 9-23   The electron of a hydrogen atom is in the $n = 3$ level. What is its energy?

Answer: -2.423×10^{-19} J

PROBLEM 9-24   Calculate the wavelength of the light from a hydrogen atom that changes from the $n = 3$ to the $n = 1$ state.

Answer: 1.026×10^{-7} m

PROBLEM 9-25   The protons emitted by a cyclotron have a velocity 1.50×10^3 m s^{-1}. What is the wavelength of these protons?

Answer: 2.64×10^{-10} m

PROBLEM 9-26   Write the number and letter for the orbital that corresponds to the following pairs of n and l quantum numbers: (**a**) $n = 3$, $l = 1$; (**b**) $n = 4$, $l = 0$; (**c**) $n = 3$, $l = 2$; and (**d**) $n = 5$, $l = 3$.

Answer: (**a**) 3p (**b**) 4s (**c**) 3d (**d**) 5f

PROBLEM 9-27   What is the maximum number of electrons that can be in each of the following orbitals: (**a**) 2p, (**b**) 4p, (**c**) 6p, (**d**) 4s, (**e**) 4d, and (**f**) 4f.

Answer: (**a**) 6   (**b**) 6   (**c**) 6   (**d**) 2   (**e**) 10   (**f**) 14

PROBLEM 9-28   Write the electronic configurations for the following atoms and ions: (**a**) V, (**b**) Cu^{2+}, (**c**) As, and (**d**) Ni^{2+}.

Answer: (**a**) $[Ar]4s^2 3d^3$ (**b**) $[Ar]3d^9$ (**c**) $[Ar]3d^{10}4s^2 4p^3$ (**d**) $[Ar]3d^8$

PROBLEM 9-29   Derive the electronic configuration for the following atoms and ions: (**a**) Fe^{3+}, (**b**) Mg, (**c**) Al^-, and (**d**) Cr^{3+}.

Answer: (**a**) $[Ar]3d^5$ (**b**) $1s^2 2s^2 2p^6 3s^2$ (**c**) $1s^2 2s^2 2p^6 3s^2 3p^2$ (**d**) $[Ar]3d^3$

PROBLEM 9-30   Determine the number of unpaired electrons in the following atoms: (**a**) Mg, (**b**) Na, and (**c**) Si.

Answer: (**a**) 0   (**b**) 1   (**c**) 2

PROBLEM 9-31   Determine the number of unpaired electrons in the following ions: (**a**) F^+, (**b**) Mn^{2+}, and (**c**) Al^+.

Answer: (**a**) 2   (**b**) 5   (**c**) 0

PROBLEM 9-32   Pick out the transition elements from the following: Zr, Ne, Pd, Se, and Sb.

Answer: Zr and Pd

10 CHEMICAL BONDING

THIS CHAPTER IS ABOUT

☑ **Valence**
☑ **Bonding**
☑ **Molecular Structure**
☑ **Polarity and Electronegativity**
☑ **Bond Properties**
☑ **Molecular Orbitals**

You'll encounter few terms in chemistry as slippery as the "chemical bond." For our purposes, a **chemical bond** exists if any group of two or more species is held together for enough time to permit examination of the properties of the group. In prior chapters, we asked you to accept on faith the chemical bond for compounds such as NaCl and CCl_4. In this chapter we'll reveal all (well, nearly all) of the fundamental concepts of chemical bonding.

10-1. Valence

The **valence** of an atom is (roughly) the number of bonds that the atom actually forms. Many atoms can have more than one valence, depending on the conditions of bond formation and the number and location of their valence electrons. Electrons that can participate in any kind of bonding—i.e., electrons that can be lost, gained, or shared—are called **valence electrons**.

Valence electrons are located in the outermost occupied orbitals, or **valence shell**, of an atom—usually in the orbitals having the highest n value. For atoms of the main-group elements (IA through VIIIA), the number of valence electrons is equal to the group number and the number of the valence shell is that of the period. Thus, silicon, which is in group IVA and period 3, has four valence electrons located in the $n = 3$ shell ($3s^2$ and $3p^2$).

note: The transition elements are a special case. The electrons in the lower $(n - 1)d$ shell often take part in bonding and thus can be classified as valence electrons.

A. Lewis symbols

A **Lewis symbol** is a form of notation that shows valence electrons and the possibilities for bonding in an atom. The atomic symbol stands for the nucleus plus all the underlying, and normally filled, orbitals. Valence electrons are represented individually by dots placed in an imaginary square around the atomic symbol in a pairing pattern set by the aufbau and Hund principles. That is, the electron dots representing the lower-energy s orbital are paired first, while the dots representing the higher, equal-energy p orbitals are paired only after each side of the square has one. Table 10-1 shows some representative Lewis symbols.

note: We'll bend these rules quite often when we discuss Lewis formulas of *molecules*, but representations of compounds are easier to achieve if this system is used for individual atoms.

B. The octet rule

The **octet rule** for the formation of compounds states that atoms tend to gain, lose, or share electrons so that the outermost energy level holds or shares *four pairs* of electrons, i.e., an *octet*. This tendency leads to an ns^2np^6, or *noble gas configuration*, a particularly stable electronic state (demonstrated by the relative scarcity of noble gas compounds).

TABLE 10-1

Atom	Group number	Configuration	Lewis symbol*
H	I	$1s^1$	H·
Be	II	$[He]2s^2$	Be:
B	III	$[He]2s^22p^1$	Ḃ:
C	IV	$[He]2s^22p^2$	·Ċ:
N	V	$[He]2s^22p^3$	·Ṅ:
O	VI	$[He]2s^22p^4$	·Ö:
F	VII	$[He]2s^22p^5$	·F̈:
Ne	VIII	$[He]2s^22p^6$	:N̈e:

* Although no particular place is assigned for the one *s* and three *p* orbitals, we have chosen to put the *s* orbital electrons to the right of the symbol.

note: Hydrogen and helium have the 1*s* shell as their only low-energy orbital and are stable with two electrons, the maximum allowed in that orbital. Although you'll encounter numerous compounds that do not obey the octet rule (especially compounds involving those pesky transition elements), you should follow it as far as possible in deducing the bond arrangement of molecules composed of main-group elements (see Section 10-3D).

10-2. Bonding

There are three basic types of bonding—metallic, ionic, and covalent—each of which depends on types of electron behavior. Electrons in metallic and ionic bonds move away from their parent atoms, thus creating ions which are attracted to species of opposite charge. The forces of attraction in metallic and ionic bonds are *electrostatic*. In covalent bonds electrons do not leave their parent species, but shift around, or "migrate," to form a region of high electron density between atoms.

A. Metallic bonds

Electrons in metallic bonds are *freely mobile*. It takes little energy for metal atoms to surrender their outermost *s* and *p* electrons to a "sea" of free electrons. **Metallic bonds** thus consist of the attraction between a network of positively charged metal ions and the surrounding electron sea. This generalized electrostatic attraction to freely mobile electrons is characteristic of pure metals and most alloys.

B. Ionic bonds

An **ionic bond** results from the *transfer* of electrons from one species to another. When an electron is transferred from one atom to another, a positive ion and a negative ion are formed. The subsequent electrostatic attraction between the ions is the bond. For example, a sodium atom has a low ionization energy; i.e., only a small amount of energy is required to remove the $3s^1$ electron and leave a Na$^+$ ion with the noble gas configuration $2s^22p^6$ ([Ne]). So Na is a good *electron donor*. A chlorine atom has high electron affinity; i.e., much energy is released when an electron is added to the neutral atom $(3s^23p^5)$ to produce the noble gas configuration $3s^23p^6$. So Cl is a good *electron acceptor*. When Na and Cl atoms combine, an electron is transferred from Na to Cl, resulting in stable Na$^+$ and Cl$^-$ ions, which attract each other.

The energy required to free an electron for transfer from a donor to an acceptor atom must be relatively low for ionic bonding to occur. Obviously, it takes more energy to free two electrons than to free one, and still more energy to free three (or more). For this reason, ionic bonds involving one or two electron transfers are much more common than those requiring three or more electron transfers. The most common ionic bonds occurs in binary (two-element) compounds consisting of one metal atom from group IA or IIA (one- or two-electron donors) and one nonmetal atom from group VIA or VIIA (one- or two-electron acceptors).

The stoichiometry of ionic compounds is predictable because the compounds must be electrically neutral. All compounds that can be isolated as liquids or solids are electrically neutral.

EXAMPLE 10-1: Determine the formula of the ionic binary compound formed from magnesium and chlorine.

Solution: The stable magnesium ion is Mg^{2+}, which has the [Ne] configuration. The stable ion from chlorine, the chloride ion, is Cl^-, which has the [Ar] configuration. For neutrality, the $2+$ charge of magnesium must be balanced by *two* negative charges, which must be provided by *two* chloride ions. So the formula of the compound is $(Mg^{2+})(Cl^-)_2$ or, simply, $MgCl_2$.

C. Covalent bonds

Covalent bonds are the result of *electron-pair sharing*.

1. Single covalent bonds

A **single covalent bond** is a bond between two atoms that share a pair of electrons. Two electrons with paired spins form a region of high electron probability lying between the two nuclei. The positive nuclei are attracted to this region, so the atoms are held together.

Covalent bonds tend to occur between atoms that do not easily lose or gain electrons one from the other, i.e., between atoms having similar electron-attracting potential (electronegativity). Nonmetals, for example, tend to bond to each other—and with themselves, as in the simple case of the nonmetal hydrogen. The most common stable form of this element is the H_2 molecule, in which two H nuclei (protons) share the two available electrons in a single covalent bond. It takes more energy for hydrogen to exist as two H· atoms than as a molecule H:H.

The general "goal" of sharing is the attainment of the octet configuration on each atom. Usually, atoms can form as many single covalent bonds as it takes to make up a total of eight electrons. For example, oxygen has *six* valence electrons: It can attain an octet by forming *two* covalent bonds, sharing the electrons of two other atoms, as in H:Ö:H (H_2O). Nitrogen has *five* valence electrons, so it can form *three* covalent bonds, as in :N̈:H (NH_3, ammonia). Carbon has *four* valence electrons, so it can form *four* covalent bonds, as in H:C̈:H (CH_4, methane).

2. Multiple covalent bonds

More than one pair of electrons can be shared by two atoms to gain an octet. When two pairs of electrons are shared by two atoms, the result is a **double covalent bond**. For example, oxygen, which needs two electrons, can form a double bond with another oxygen, gaining its octet by sharing four electrons, as in :Ö::Ö: (O_2). Similarly, a **triple covalent bond** results when two atoms share three pairs of electrons, as in N̈:::N̈ (N_2).

The **bond order** of a covalent bond is the number of electron pairs shared between the two bonded atoms. It is possible for an atom to form molecules having different bond orders—depending on the number of bonds the molecule has. Versatile nitrogen, for example, has a bond order of 1 in H_2N—NH_2 (N_2H_4), 2 in HN=NH (N_2H_2), and 3 in N≡N (N_2).

3. Coordinate covalent bonds

There's one other possibility. One atom can "donate" two of its electrons to provide the shared pair between itself and an "acceptor" atom. When one atom provides both the electrons for a shared pair, the bond is called a **coordinate covalent bond**. Remember: All electrons are alike—there's no such thing as an "oxygen electron," for example. Consequently, the coordinate covalent bond is indistinguishable from a single covalent bond. But the idea of this type of bond will be handy when we try to account for electrons in Lewis formulas.

10-3. Molecular Structure

A. Lewis formulas

Lewis formulas are electron-dot pictures of molecules generated by putting the Lewis symbols of atoms together. For example, we can represent the H_2 molecule as H:H by combining two H· Lewis symbols. We can also represent H_2 as H—H, using a single line to indicate a *pair* of electrons. This short-form convention is called a **valence-bond formula**.

note: Valence-bond formulas are *shortcuts*. Learn to draw Lewis formulas, then use the shortcuts to save time. Valence-bond formulas do not give you all the information that Lewis formulas do. For example, unshared pairs are frequently left out of valence-bond formulas.

We can write the Lewis formula for a neutral covalent compound of known geometry by using the following rules. We'll use HCN, a linear molecule, to illustrate each step.

1. Place the Lewis symbols of the atoms in their proper geometric relationship to one another.

$$\text{H·} \quad \text{·}\overset{\displaystyle .}{\text{C}}\text{:} \quad \text{·}\overset{\displaystyle .}{\text{N}}\text{:}$$

2. Start with single bonds. Move the electrons around to form a single covalent bond between each pair of atoms.

$$\text{H:}\overset{\displaystyle .}{\text{C}}\text{:}\overset{\displaystyle .}{\text{N}}\text{:}$$
$$\uparrow \quad \uparrow$$
single bond

3. Using the octet rule, distribute the remaining electrons so that each main-group element has or shares eight electrons and each hydrogen atom shares two. If there aren't enough electrons to go around for octets, move unshared electrons and unshared terminal pairs to form double and triple bonds. (Note that an unshared pair of electrons is okay).

triple bond
$$\text{H:}\overset{\frown}{\text{C:::}}\text{N:} \quad \longleftarrow \text{ unshared pair}$$

4. Verify the formula by adding the group numbers of the atoms. This sum should equal the total number of electrons in the Lewis formula. For our example,

H	(I)	1
C	(IV)	4
N	(V)	+ 5
		10

and we have 10 electrons in our Lewis formula.

One more step: If the compound is charged, add one electron for each unit of negative charge or subtract one electron for each unit of positive charge. Place brackets round the entire structure, and indicate the charge outside the bracket. Consider the ammonium ion, NH_4^+. The H atoms surround the N atom:

Arrange the symbols	Remove one electron to account for the + charge	Form covalent bonds and satisfy the octet rule	Verify the formula
Ḧ ·H·N:H· Ḧ	$\left[\begin{array}{c} \dot{\text{H}} \\ \text{·H·}\overset{..}{\text{N}}\text{:H} \\ \dot{\text{H}} \end{array}\right]^+$	$\left[\begin{array}{c} \text{H} \\ \text{H:}\overset{..}{\text{N}}\text{:H} \\ \text{H} \end{array}\right]^+$	H (I) (1 × 3) N (V) +5 8

EXAMPLE 10-2: Draw Lewis and valence-bond structures for H_2, HBr, Br_2, O_2, and CO. Because each of these molecules contains only two atoms, the structure of each must be linear.

Solution:

Molecule	Total number of valence electrons	Lewis formula	Valence-bond formula
H_2	$1 + 1 = 2$	H:H	H—H
HBr	$1 + 7 = 8$	H:B̤r̤:	H—B̤r̤:
Br_2	$7 + 7 = 14$	:B̤r̤:B̤r̤:	:B̤r̤—B̤r̤:
O_2	$6 + 6 = 12$	:Ö::Ö:	:Ö=Ö:
CO	$4 + 6 = 10$	:C:::O:	:C≡O:

EXAMPLE 10-3: Draw Lewis and valence-bond structures for HOH, OCO, and OH^-.

Solution:

Molecules	Total number of valence electrons	Lewis formula	Valence-bond formula (all electrons shown)
HOH	$1 + 6 + 1 = 8$	H:Ö:H	H—Ö—H
OCO	$6 + 4 + 6 = 16$	:Ö::C::Ö:	Ö=C=Ö
OH^-	$6 + 1 + 1 = 8$ (Notice the extra electron)	[:Ö:H]⁻	[:Ö—H]⁻

B. Exceptions to the octet rule

You'll come across three groups of molecules that are exceptions to the octet rule.

1. *Molecules whose central atoms have fewer than eight electrons.* This group consists of molecules containing central atoms from groups IIA and IIIA. $BeCl_2(g)$, BF_3, and $AlCl_3(g)$, whose Lewis formulas are shown here, are typical examples:

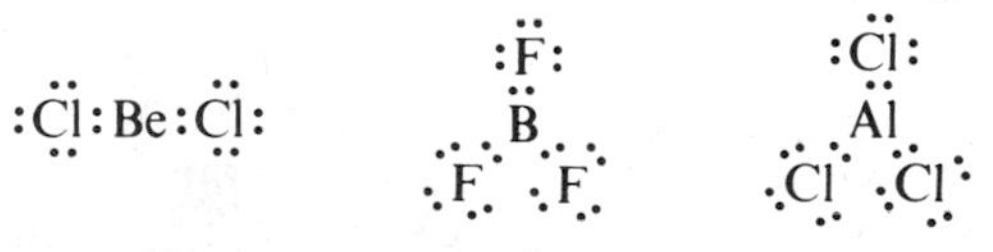

:C̤l̤:Be:C̤l̤:	:F̤: B̤ .F̤. .F̤.	:C̤l̤: A̤l̤ .C̤l̤. .C̤l̤.
4 electrons around Be	6 electrons around B	6 electrons around Al

2. *Molecules whose central atoms have more than eight electrons.* This group consists of molecules containing central atoms from periods 3, 4, 5, 6, or greater. PF_5, SF_6, and XeF_4 are typical:

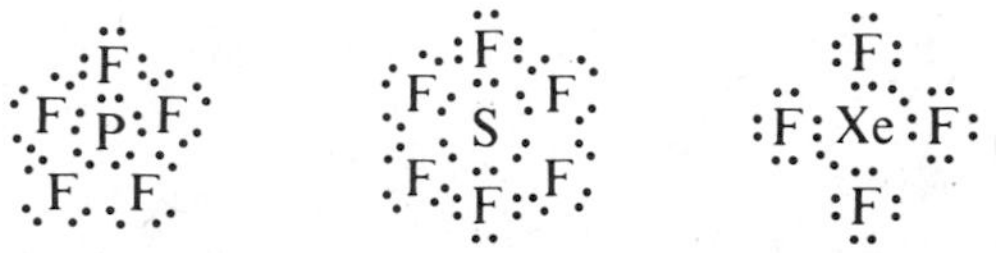

10 electrons around P	12 electrons around S	12 electrons around Xe

3. *Molecules containing an odd number of electrons.* While stable molecules of this kind are rare, they do exist. (They are paramagnetic.) Some examples are NO, which has a total of 11 valence electrons; NO_2, which has a total of 17 valence electrons; and ClO_2, which has a total of 19 valence electrons. If you have to suggest a Lewis-like structure for these molecules, the best you can do is this:

:N::Ö	:Ö:N::Ö	:Ö:C̤l̤:Ö:
NO	NO_2	ClO_2

C. Formal charge

Recall that in a coordinate covalent bond one atom provides both electrons. In such cases the molecule may be electrically neutral, but there are *local* charges, called **formal charges**, associated with particular atoms: A donor atom that has given up one electron has a formal charge of $+1$; the acceptor atom that has received one electron has a formal charge of -1.

Even if you can't identify the donor and acceptor atoms, you can calculate the formal charge on each atom in a molecule with the following formula:

FORMAL CHARGE

$$\text{formal charge} = \left(\begin{array}{c}\text{No. of}\\\text{valence } e^-\end{array}\right) - \frac{1}{2}\left(\begin{array}{c}\text{No. of}\\\text{shared } e^-\end{array}\right) - \left(\begin{array}{c}\text{No. of unshared}\\\text{valence } e^-\end{array}\right) \qquad (10\text{-}1)$$

Note that H atoms do not carry a formal charge in covalent molecules.

Let's work through the calculation of formal charge for HNO_3, whose geometric arrangement is given. The N atom must supply a pair of electrons to a covalent bond formed with the topmost O atom as follows:

Arrangement   Lewis formula

$$\cdot\ddot{O}\cdot$$

$$H\cdot \quad \cdot\ddot{O}\cdot \quad \cdot\ddot{N}\cdot \quad \cdot\ddot{O}\cdot \qquad\qquad H:\ddot{O}:\ddot{N}::\ddot{O}.$$

Now calculate the formal charge of each atom except hydrogen:

Structure	Atoms	$\left(\begin{array}{c}\text{No. of}\\\text{valence }e^-\end{array}\right)$	$-\frac{1}{2}\left(\begin{array}{c}\text{No. of}\\\text{shared }e^-\end{array}\right)$	$-\left(\begin{array}{c}\text{No. of unshared}\\\text{valence }e^-\end{array}\right)$	$=$ Formal charge
	N	5	$-$ 4	$-$ 0	$=$ $+1$
	O	6	$-$ 1	$-$ 6	$=$ -1
	O	6	$-$ 2	$-$ 4	$=$ 0
	O	6	$-$ 2	$-$ 4	$=$ 0

Formal charges on an atom are represented with pluses and minuses, as follows:

$$:\ddot{O}:^-$$
$$H:\ddot{O}:\underset{+}{\ddot{N}}::\ddot{O}.$$

The sum of all the formal charges in a neutral atom must equal zero. Notice that in HNO_3, $1 - 1 + 0 = 0$. The sum of all formal charges in a charged species must equal the total charge on the species.

Formal charge is useful when you have to choose the most reasonable Lewis formula for a number of possible alternative structures. ("Most reasonable" in this instance means closest to the actual distribution of electrons as shown by experiment.) The most reasonable structure is usually that having the lowest possible formal charges.

EXAMPLE 10-4: Which of the following is the most reasonable Lewis formula for FCN?

(a) $\ddot{F}::C::\ddot{N}$ (b) $:F:::C:\ddot{N}:$ (c) $:\ddot{F}:C:::N:$

Solution: Calculate the formal charges:

(a) F: $7 - 2 - 4 = +1$ (b) F: $7 - 3 - 2 = +2$ (c) F: $7 - 1 - 6 = 0$
       C: $4 - 4 - 0 = 0$ C: $4 - 4 - 0 = 0$ C: $4 - 4 - 0 = 0$
       N: $5 - 2 - 4 = -1$ N: $5 - 1 - 6 = -2$ N: $5 - 3 - 2 = 0$

All of these formulas obey the octet rule, but structure (c) has the lowest possible formal charges. So (c) is the most reasonable structure.

D. Resonance

For some molecules there is no best choice between a number of equally acceptable Lewis structures. Consider ozone O_3, in which the arrangement of nuclei is O O O. This molecule has $3 \times 6 = 18$ valence electrons, and it's possible to draw two precisely equivalent ozone structures:

$$\ddot{O}::\ddot{O}:\ddot{O}: \quad\text{and}\quad :\ddot{O}:\ddot{O}::\ddot{O} \qquad\text{or}\qquad O{=}O{-}O \quad\text{and}\quad O{-}O{=}O$$
$$\;1\;\;\;2\;\;\;3 \qquad\qquad\;\; 1\;\;\;2\;\;\;3 \qquad\qquad\qquad 1\quad 2\quad 3 \qquad\qquad 1\quad 2\quad 3$$

The octet rule is followed for both of these structures, as it must be for second-period elements.

The two structures for O_3 suggest that the bond between O-1 and O-2 differs from that between O-2 and O-3—one single and one double bond. But experimental evidence shows that the two bonds are exactly equivalent in length and strength. So we need a Lewis (or valence-bond) notation that can represent this equivalence.

When the experimentally observed properties of a molecule are not adequately represented by any one Lewis structure—and the actual properties can only be represented by a *superposition* of all the reasonable Lewis structures—we have a phenomenon called **resonance**. The Lewis structures you can draw do not represent any *real* molecule. The actual structure of the molecule is somewhere in between and is called a **resonance hybrid**. You represent the superposition of the Lewis structures by drawing all of them separated by a double-headed arrow ↔. So you write the Lewis structure of ozone as

$$\ddot{\text{O}}::\ddot{\text{O}}:\ddot{\text{O}}: \longleftrightarrow :\ddot{\text{O}}:\ddot{\text{O}}::\ddot{\text{O}} \qquad \text{or} \qquad \text{O}=\text{O}-\text{O} \longleftrightarrow \text{O}-\text{O}=\text{O}$$

note: Resonance is *not* equilibrium. There are not, for example, two different ozone molecules with a pair of electrons jumping back and forth between the two bonds. There is only one ozone structure, which we have to represent by two pictures.

10-4. Polarity and Electronegativity

A. Polarity and partial charge

When bonding electron pairs in a covalent bond are shared equally, the result is a **nonpolar covalent bond**. For example, in Cl_2 (Cl—Cl) the two Cl nuclei have the same magnitude of positive charge, so they attract the shared pair of electrons with equal force and the Cl—Cl bond is nonpolar. When bonding electron pairs in a covalent bond are not shared equally, the result is a **polar covalent bond**. For example, in HCl(*g*) (H—Cl) the highly positively charged Cl nucleus attracts the bonding electrons more strongly than does the small H nucleus. This unequal attraction causes the Cl atom to acquire a **partial negative charge**, symbolized by $\delta-$, while the H atom acquires a **partial positive charge**, symbolized by $\delta+$. Partial charges do not affect the electrical neutrality of a molecule.

note: Partial charges are *not* formal charges. Polarity is simply an imbalance in electron-attracting ability between covalently bonded atoms.

B. Electronegativity

Electronegativity values are a measure of the ability of an atom in a molecule to attract bonding electrons. Atoms can be ranked according to their experimentally determined electronegativities, as shown in Table 10-2. Although the electronegativity of an element changes somewhat, depending on the compound it's in, the listed value is a reasonable average. As the table shows, electronegativities

TABLE 10-2. Electronegativities of Some Elements

				H 2.1			
Li 1.0	Be 1.5	B 2.0		C 2.5	N 3.0	O 3.5	F 4.0
Na 0.9	Mg 1.2	Al 1.5		Si 1.9	P 2.1	S 2.5	Cl 3.0
K 0.8	Ca 1.0			Ge 2.0	As 2.0	Se 2.4	Br 2.8
Rb 0.8	Sr 1.0			Sn 1.7	Sb 1.9	Te 2.1	I 2.4
Cs 0.7	Ba 0.9						

generally increase across periods and decrease down groups: Nonmetals are more electronegative than metals, and lighter elements tend to be more electronegative than heavier ones.

EXAMPLE 10-5: Which atom will attract bonding electrons more strongly in C—H and C—F?

Solution: According to Table 10-2, the electronegativities are 2.1 for H, 2.5 for C, and 4.0 for F. In the C—H bond, carbon will attract the bonding electrons more strongly than hydrogen. In the C—F bond, fluorine will attract the bonding electrons more strongly than carbon. The polarity can be indicated by the $\delta\pm$ convention, where $\delta-$ represents higher electronegativity and $\delta+$ lower electronegativity, or by a crossed arrow:

$$^{\delta-}\text{C—H}^{\delta+} \qquad ^{\delta+}\text{C—F}^{\delta-} \qquad \text{or} \qquad \text{C—H} \qquad \text{C—F}$$

where the arrowhead points in the direction of the more electronegative element.

Electronegativity values also provide useful data for characterizing bonds. In general, when two adjacent atoms in a compound have similar electronegativities, the bond between them is covalent; and when adjacent atoms have electronegativities that differ by 2 or more, the bond is probably ionic.

C. Moments

1. Bond moments

The **bond moment** (also called a **dipole**) is a measure of the polarity of a *diatomic* covalent bond. Such measurements are made by subjecting the bond to an electric field and determining the amount of force required to align the bond in the field. Bond moments are expressed in units called debyes (D; $1D = 3.336 \times 10^{-30}$ coulomb-meters) and are relatively constant from compound to compound.

2. Dipole moments

The **dipole moment** μ (of a molecule) is the *vector sum* of the bond moments in a molecule. (Vector sums have *direction* as well as *magnitude*.) Dipole moments measure the polarity of molecules as a whole and, like bond moments, are expressed in debyes. Some representative bond and dipole moments are given in Table 10-3.

TABLE 10-3: Selected Bond Moments and Dipole Moments

Bond	Bond moment (D)	Compound	Dipole moment (D)
C—N	0.22	CO_2	0
H—C	0.4	CCl_4	0
C—O	0.74	CO	0.10
H—N	1.31	NO	0.16
C—F	1.51	NO_2	0.29
C—Cl	1.56	NH_3	1.46
H—O	1.51	H_2O	1.84
C=O	2.4	HF	1.9
C≡N	3.5	HCl	1.05

note: For a diatomic *compound*, the bond moment *is* the dipole moment.

EXAMPLE 10-6: Determine the dipole moment (magnitude and direction) for the following diatomic molecules: O_2, F_2, NO, CO, HF.

Solution: First look up the electronegativities of each atom (see Table 10-2): 2.1 for H, 2.5 for C, 3.0 for N, 3.5 for O, and 4.0 for F. Then draw the valence-bond structure for each molecule, showing where partial charges should be. The distribution of partial charges gives you the *direction*:

$$\text{O=O} \qquad \text{F—F} \qquad \text{N=O} \qquad \text{C=O} \qquad \text{H—F}$$

Molecular O_2 and F_2 are easy: These molecules are nonpolar, and so $\mu = 0$ for $O{=}O$ and $F{-}F$. (You didn't need electronegativities for that!) But NO, CO, and HF are polar diatomic compounds, so their dipole moments must be equal to their bond moments, which you look up (see Table 10-3): $\mu = 0.16$ for NO, $\mu = 0.10$ for CO, and $\mu = 1.9$ for HF.

The dipole moment of a polyatomic molecule (three or more atoms) depends on the geometry of the molecule. If the bond moments are equal in magnitude and opposite in direction, they will cancel and the resultant dipole moment will be zero.

EXAMPLE 10-7: Both CO_2 and BCl_3 have zero dipole moments, but the $C{=}O$ and $B{-}Cl$ bond moments are not zero. Explain.

Solution: The dipole moment of a molecule will be zero if the vector sum of the bond moments is zero. Since the bond moments in each molecule are equal in magnitude, they must be opposite in direction to cancel to zero. So the molecules must be arranged *symmetrically*:

$$\overset{\longleftarrow \ \longrightarrow}{O{=}C{=}O}, \quad \mu = 0 \qquad\qquad \underset{Cl}{\overset{Cl}{\diagdown}} B{\rightarrow}Cl, \quad \mu = 0$$

10-5. Bond Properties

A. Atomic radius, bond length, and bond angle

- The **atomic radius** may be considered to be the distance from the center of the nucleus to the outermost electrons. [*Note:* The notion of atomic radius is a rather vague concept—atoms don't have sharp boundaries. Consequently, it's usually defined as half the internuclear distance between like atoms that are directly bonded.] Atomic radii decrease across periods and increase down groups in the periodic table.
- In a covalent bond the **bond length** is the average internuclear distance. Bond lengths range from 75 pm to 300 pm (0.75 Å to 3 Å). For H_2, Cl_2, and similar diatomic molecules the bond length is twice the atomic radius. Bond length decreases as bond order increases. For example, $C{=}C$ bonds are shorter than $C{-}C$ bonds, and $C{\equiv}C$ are shorter than $C{=}C$ bonds.
- The bonds in a polyatomic molecule form angles called **bond angles**, which can vary from $<50°$ to $180°$. Molecules with more than three atoms are three-dimensional and are represented by a special notation. Consider carbon tetrachloride (CCl_4), which we draw as follows:

$$\underset{Cl \quad H}{\overset{Cl}{\diagup}} C{\text{---}}H$$

The solid line (—) represents a bond in the plane of the paper, the wedge (◄) represents a bond coming out of the paper toward you, and the broken line (---) represents a bond receding into the paper away from you.

B. VSEPR geometry

The number of valence electron pairs is the most important factor in determining the geometric arrangement of atoms in covalently bonded molecules. Because each pair stays as far away from other pairs as possible, the angle of separation depends on how many pairs there are. The **valence-shell electron-pair repulsion (VSEPR)** theory is an approach to predicting molecular geometry based on the principle of maximum electron-pair separation.

note: Instead of talking about "electron-pair" repulsion, we'll use the term "electron-*set*" repulsion, so double and triple bonds won't destroy our faith in our nice theory.

In VSEPR theory one *set* of electrons around the central atom in a Lewis structure is defined as ANY covalent bond (single, double, or triple) or any lone pair of electrons. For example, one double covalent bond plus two lone pairs of nonbonding valence electrons constitutes three sets.

The geometries of compounds having 2 to 6 sets of electrons around the central atom are given in Table 10-4. Notice how the bonding electron sets are maximally separated. The VSEPR theory works for molecules that obey the octet rule (and some few that don't), but you can't apply it easily to molecules having an odd number of valence electrons or to central transition elements.

TABLE 10-4: VSEPR Geometry
(A = central atom; X = terminal atom; E = lone pair of valence electrons)

Number of sets	Molecule type	Geometry	Sketch	Example(s)
2	AX_2	Linear	X—A—X	F—Be—F or O=C=O
3	AX_3	Trigonal planar (basic geometry for three sets)		
3	AX_2E	Bent or angular		
4	AX_4	Tetrahedral (basic geometry for four sets)		
4	AX_3E	Trigonal pyramidal		
4	AX_2E_2	Bent or angular		
5	AX_5	Trigonal bipyramidal (basic geometry for five sets)		
5	AX_4E	Distorted tetrahedral		
5	AX_3E_2	T-shaped		
5	AX_2E_3	Linear		
6	AX_6	Octahedral (basic geometry for six sets)		
6	AX_5E	Square pyramidal		
6	AX_4E_2	Square planar		

EXAMPLE 10-8: Explain why the H_2O molecule is polar and has a nonzero dipole moment, while CO_2 is nonpolar and has zero dipole moment.

Solution: We draw the Lewis structures of H_2O and CO_2 as

$$H:\ddot{O}:H \qquad \text{and} \qquad :\ddot{O}::C::\ddot{O}:$$

Counting the number of electron sets around the central atom, we see that the O in H_2O has four and the C in CO_2 has two. Then we count central atoms, terminal atoms, and lone pairs, determine the molecule type, and look up the appropriate shape (see Table 10-4). We see that H_2O is an AX_2E_2 molecule type and therefore angular, but that CO_2 is AX_2 and therefore linear.

Now that we've got the right shapes, we can think about polarity. Both the H—O and C—O bonds are polar. But the H—O bond moments are directed so that there is a net dipole moment μ for the molecule:

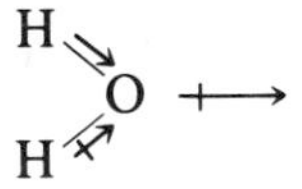

$$\begin{array}{c} H \\ \searrow \\ \quad O \longrightarrow \qquad \mu = 1.84 \text{ D} \\ \nearrow \\ H \end{array}$$

On the other hand, in CO_2 the C—O bonds are at a 180° angle, so that the bond moments exactly cancel, giving a zero dipole moment:

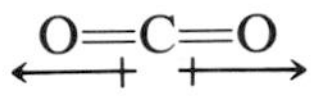

$$\underset{\longleftarrow \; + \; \longrightarrow}{O{=}C{=}O} \qquad \mu = 0$$

EXAMPLE 10-9: Which of the following molecules, if any, have nonzero dipole moments: methane CH_4, dichloromethane CH_2Cl_2, ethylene C_2H_4?

Solution: The valence-bond structures are

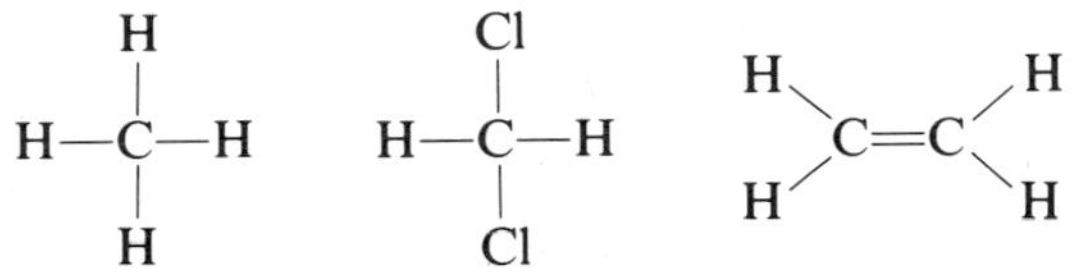

$$\begin{array}{ccc}
\begin{array}{c} H \\ | \\ H{-}C{-}H \\ | \\ H \end{array} &
\begin{array}{c} Cl \\ | \\ H{-}C{-}H \\ | \\ Cl \end{array} &
\begin{array}{c} H \qquad\quad H \\ \diagdown \quad\diagup \\ C{=}C \\ \diagup \quad\diagdown \\ H \qquad\quad H \end{array}
\end{array}$$

Both CH_4 and CH_2Cl_2 are tetrahedral (AX_4 type). In CH_4 the bonds are all identical and the vector sum of the bond moments equals zero, so the net dipole moment is zero. In CH_2Cl_2 the bond moments are not identical; the molecule has a dipole moment directed between the C—Cl bonds:

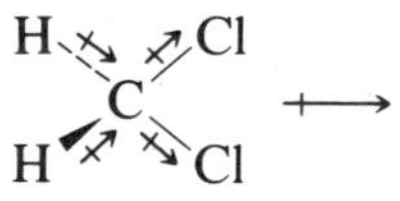

$$\begin{array}{c} H\searrow \quad \nearrow Cl \\ \quad C \\ H\nearrow \quad \searrow Cl \end{array} \longrightarrow \qquad \mu \neq 0$$

In C_2H_4 each carbon atom is trigonal planar (AX_3 type), and the molecule is planar. The C—H bonds are all equivalent

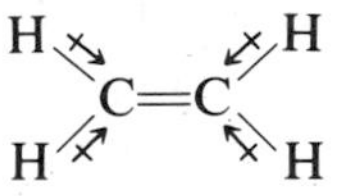

$$\begin{array}{c} H\searrow \qquad \nearrow H \\ \quad C{=}C \\ H\nearrow \qquad \searrow H \end{array} \qquad \mu = 0$$

and the vector sum of the bond moments equals zero, so C_2H_4 has a zero dipole moment.
Only CH_2Cl_2 will have a nonzero dipole moment.

C. Bond dissociation energy

In certain reactions it is possible to break a single chemical bond to give free atoms. Consider the following reaction:

$$H_2(g) \longrightarrow 2H(g), \qquad \Delta H_r^\circ = 436 \text{ kJ/mol}$$

The **bond dissociation energy** is the enthalpy per mole for breaking *one and only one* bond of the same type in a molecule. So the bond dissociation energy for H—H is equal to $\Delta H_r^\circ = 436$ kJ/mol because there is only one H—H bond to be broken in the H_2 molecule. Now consider the set of reactions

$$NH_3(g) \longrightarrow NH_2(g) + H(g), \qquad \Delta H_r^\circ = 435 \text{ kJ/mol}$$
$$NH_2(g) \longrightarrow NH(g) + H(g), \qquad \Delta H_r^\circ = 377 \text{ kJ/mol}$$
$$NH(g) \longrightarrow N(g) + H(g), \qquad \Delta H_r^\circ = 356 \text{ kJ/mol}$$

Here the bond dissociation energy is equal to the ΔH_r° of each step in which only *one* N—H bond is broken.

But suppose we have a reaction in which several bonds of the same type are broken:

$$NH_3(g) \longrightarrow N(g) + 3H(g), \qquad \Delta H_r^\circ = 1168 \text{ kJ/mol}$$

If this is all the information we have, we can find only the **bond energy**, which is the *average* amount of energy required to break one mole of bonds of the same type in a molecule (see Section 8-3E). Thus the bond energy is $\Delta H_r^\circ/3 = 389$ kJ/mol, which is equal to $(435 + 377 + 356)/3$.

note: The amount of energy it takes to break the same type of bond can vary—even in the same molecule. The exact bond dissociation energy, i.e., the strength of a particular bond, depends on the surroundings.

Now, remembering Hess' law (Section 8-3C), we can reformulate the calculation of the standard enthalpy of reaction:

STANDARD ENTHALPY
OF REACTION $\Delta H_r^\circ = -\left(\sum \text{product bond energies}\right) + \left(\sum \text{reactant bond energies}\right)$

Note that this formula must be only an estimate because bond energies are just averages.

Bond energy is related to bond length: The higher the bond energy, the shorter the bond. Note in the comparison below that as bond order increases, lengths go down and bond energies go up.

Molecule	Bond order	Bond length (pm)	Bond energy (kJ/mol)
H_3C—CH_3	1	154	347
H_2C=CH_2	2	134	610
HC≡CH	3	120	836

Thus, double and triple bonds (which have greater electron densities) are shorter and stronger than single bonds, and the triple bond is the shortest and strongest of all.

10-6. Molecular Orbitals

Molecular orbital theory is a method of accounting for covalent bonds that depends on quantum theory and mathematical principles. This theory is based on the fact that electrons are NOT the substantive little dots we portray in Lewis structures—and that we can speak only about energy levels/spatial probabilities when reckoning with them. We do this reckoning by thinking of quantized electron distribution as atomic orbitals (AOs) [*s, p, d, f*: see Section 9-3], which are capable of combining, or overlapping, to produce new electron distributions called **molecular orbitals** (MOs)—*one MO for every AO*. We assign electrons to MOs by energy level (representing the combined electronic configurations with a variation of the *spdf* nomenclature) in the same way that we do for AOs, using the rules that apply for all orbital filling: the aufbau, Hund, and Pauli exclusion principles. We can also account for the shapes of certain covalent bonds by describing the types of AO overlap in geometric terms.

A. σ Orbitals and σ bonds

• When two AOs overlap *end-to-end*, they form two **sigma molecular orbitals** (σ MOs).

Consider the H_2 molecule, which has two H atoms and therefore two $1s$ AOs. The two $1s$ AOs combine (see Figure 10-1) to produce two σ MOs, which differ in energy and location. One of the σ MOs is a **bonding orbital**, denoted σ_{1s}; the other is an **antibonding orbital**, denoted σ_{1s}^*.

The relative energy levels of these two MOs are different. The σ_{1s} MO has a lower energy level than the original $1s$ AOs, while the σ_{1s}^* MO has a higher energy. The best way to see this is by looking at an energy diagram.

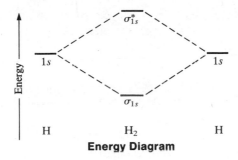

Energy Diagram

Applying two principles of orbital filling (which we can do because, AO or MO, orbitals are orbitals)—

* the *aufbau principle*: lower-energy orbitals fill first, and
* the *Pauli exclusion principle*: each orbital can hold a maximum of two electrons—

we can predict that both electrons in H_2 will go into the lower-energy σ_{1s} orbital. The filled σ_{1s} orbital is denoted by σ_{1s}^2. There are no more electrons in H_2, so the σ_{1s}^* orbital remains empty in the ground state.

note: The number of MOs formed *must equal* the number of AOs available for combination. Unfilled MOs are considered to be there, even when there are no electrons in them.

The redistribution of electrons by end-to-end overlap of two AOs produces a type of bond characterized distinctively by the shape of the σ MOs. σ Molecular orbitals are cylindrically symmetrical about the internuclear axis (a line joining the two nuclei), so that the whole area of electron probability completely surrounds that axis.

* Any bond in which the distribution of electron density is cylindrically symmetrical around the internuclear axis is a **σ bond**.

The H_2 model shows the shape of the σ MOs and the σ bond (see Figure 10-1b). As we can see from this model, the probability of finding electrons in the lower-energy σ_{1s} MO is greatest in the internuclear space—between the two positively charged nuclei—so that the nuclei are drawn together by a mutual attraction to the concentration of negatively charged electrons. The probability of finding electrons in the higher-energy σ_{1s}^* MO is greatest outside the internuclear space (see Figure 10-1c), so that the nuclei repel each other: hence the name, "antibonding" orbital. (Electrons in an antibonding orbital can weaken a covalent bond.)

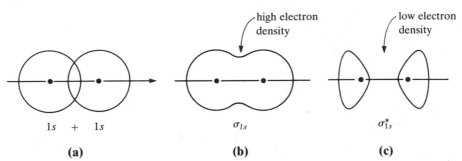

FIGURE 10-1. σ-Molecular orbitals: **(a)** Overlap of two $1s$ AOs; **(b)** bonding σ_{1s} MO; **(c)** antibonding σ_{1s}^* MO.

note: The overlap of two s AOs always results in σ MOs, but other combinations are also possible. Two p orbitals can overlap end-to-end if they approach along the same axis, or one s and one p AO in the same plane can overlap.

B. π Orbitals and π bonds

* When two AOs overlap from *parallel* positions, they form two **pi molecular orbitals (π MOs)**.

Any two parallel AOs (e.g., two p_x AOs) can overlap to produce two π MOs—one bonding π MO (e.g., π_{p_x} MO) and one antibonding π^* MO (e.g., $\pi_{p_x}^*$)—whose areas of electron density are two distinct regions separated by the internuclear axis. The bond formed as a result of parallel AO overlap has a shape characterized by a region of zero electron density at the internuclear axis. (Consult your textbook for the shapes of π MOs).

* A bond in which the distribution of electron density is separated by the internuclear axis is called a **π bond**.

Consider a hypothetical diatomic molecule composed of two second-period atoms whose electronic configuration is $1s^2 2s^2 2p^6$. There are therefore ten AOs (20 electrons) to account for. The electrons in the two $1s$ AOs are inner-shell (nonvalence, or *nonbonding*) electrons and are not available for bonding. So the 4 nonbonding electrons stay in the two $1s$ orbitals. That leaves 16 electrons in *eight* AOs that can overlap to form *eight* MOs, as follows:

1. The two $2s$ AOs overlap end-to-end, producing one σ_{2s} and one σ_{2s}^* MO (just as in H_2).
2. The six $2p$ AOs (two $2p_x$, two $2p_y$, and two $2p_z$) overlap by spatial orientation—either parallel or end-to-end:
   (a) four $2p$ AOs, say $2p_y$ and $2p_z$, overlap in parallel fashion to give two bonding (π_{2p_y} and π_{2p_z}) and two antibonding ($\pi_{2p_y}^*$ and $\pi_{2p_z}^*$) MOs;
   (b) two $2p$ AOs lying along the same axis overlap end-to-end, producing one σ_{2p_x} and one $\sigma_{2p_x}^*$ MO.

C. Building up electronic configurations of molecules

The order of filling MOs is dictated by the rules of all orbital filling—the aufbau and Pauli exclusion principles (which we saw in the H_2 model) and the *Hund principle* (equal-energy orbitals each receive one electron before any electrons are paired in those orbitals). This order is best seen in an energy diagram, such as that for the second-period elements (Figure 10-2). Notice that the σ_s (bonding) MOs are lower in energy than the s AOs and that the π_p (bonding) MOs are lower in energy than the p AOs.

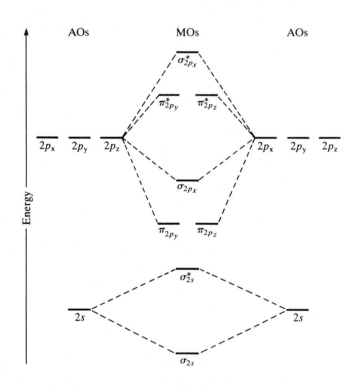

FIGURE 10-2. Energy diagram: Relative energy levels for second-period elements.

Molecular electronic configurations are represented with a σ,π notation that identifies the combined AOs with subscripts and the number of electrons with superscripts. Thus, our hypothetical diatomic second-period element would have an electronic configuration written, in order of filling, as

$$\text{Order:} \quad \underbrace{1s^2 1s^2}_{\substack{\text{nonbonding} \\ e^- \text{ (AOs)}}} \quad \overset{1}{\sigma_{2s}^2} \quad \overset{2}{\sigma_{2s}^{*2}} \quad \underbrace{\overset{3}{\pi_{2p_y}^2 \pi_{2p_z}^2}}_{\substack{\text{equal-energy} \\ \pi \text{ MOs}}} \quad \overset{4}{\sigma_{2p_x}^2} \quad \underbrace{\overset{5}{\pi_{2p_y}^{*2} \pi_{2p_z}^{*2}}}_{\substack{\text{equal-energy} \\ \pi^* \text{ MOs}}} \quad \overset{6}{\sigma_{2p_x}^{*2}}$$

EXAMPLE 10-10: Experimental evidence shows that molecular oxygen is paramagnetic. (**a**) Write the Lewis structure for O_2. (**b**) Write the electronic configuration for O_2. (**c**) Account for the paramagnetism of O_2.

Solution:

(a) $:\ddot{O}: + :\ddot{O}: \longrightarrow :\ddot{O}::\ddot{O}:$

(b) Start with the electronic configuration of atomic oxygen, $1s^2 2s^2 2p^4$, which shows a total of 8 electrons. So the electronic configuration of two atoms must account for 16 electrons. Of these 16 electrons, we know that 4 are nonbonding (and will stay in the two $1s$ AOs), 4 must come from the two $2s$ AOs, and 8 must come from the six $2p$ AOs. Remembering the energy diagram and the rules of orbital filling, we write

$$\underbrace{1s^2 1s^2}_{4e^-} \underbrace{\sigma_{2s}^2 \sigma_{2s}^{*2}}_{4e^-} \underbrace{\pi_{2p_y}^2 \pi_{2p_z}^2}_{4e^-} \underbrace{\sigma_{2p_x}^2}_{2e^-} \underbrace{\pi_{2p_y}^{*1} \pi_{2p_z}^{*1}}_{2e^-} \underbrace{(\sigma_{2p_x}^*)}_{0e^- = 16e^-}$$

(It is NOT customary to write the unfilled orbital, but we need to remember that it's there: We must have the same number of MOs as AOs.)

(c) Here we couldn't fill up the $\pi_{2p_y}^*$ orbital before the $\pi_{2p_z}^*$ orbital had one electron. So these two orbitals each have one electron, whose spins are unpaired, thereby accounting for the paramagnetism of O_2.

note: It is a fairly common convention to list σ MOs before the π MOs when writing electronic configurations of molecules. You can follow this σ,π order, but you should keep the relative energies in mind.

D. Bond order in MOs

Having built an electronic configuration for a molecule, we can now calculate bond order by using a simple equation:

BOND ORDER $\qquad$ Bond order $= \dfrac{1}{2}\left[\left(\begin{array}{c}\text{No. of } e^- \text{ in}\\ \text{bonding MOs}\end{array}\right) - \left(\begin{array}{c}\text{No. of } e^- \text{ in}\\ \text{antibonding MOs}\end{array}\right)\right]$ $\qquad$ **(10-2)**

For example, the bond order of O_2, whose electronic configuration is given in Example 10-10, is $\frac{1}{2}(8 - 4) = 2$.

SUMMARY

1. Valence electrons occupy the outermost orbitals—usually the orbital with the highest value of n—of a given atom.
2. Lewis symbols represent atoms: The nucleus and all the underlying (normally filled) orbitals are represented by the atomic symbol, and valence electrons are represented by dots placed around the atomic symbol.
3. The octet rule states that atoms tend to gain, lose, or share electrons so that the outermost energy level holds or shares four pairs of electrons.
4. In a metallic bond free electrons surround a network of positive metal ions.
5. When electrons are transferred from one atom to another, the subsequent attraction of the positive ion for the negative one is an ionic bond.
6. A covalent bond results from the sharing of electron pairs by two atoms.
7. The bond order of a covalent bond is the number of electron pairs shared between two atoms. As bond order increases, bond energy increases and bond length decreases.
8. Lewis and valence-bond formulas are conventional representations of molecular structure.
9. Formal charges are local charges that result from coordinate covalent bonds. Formal charge = (No. of valence e^- in the neutral atom) $- \frac{1}{2}$(No. of shared e^-) $-$ (No. of unshared valence e^-).
10. Resonance refers to the intermediate electronic state of a molecule for which several valid Lewis structures are possible.
11. Polarity is the unequal distribution of charge in an electrically neutral molecule.
12. Electronegativity is the ability of an atom in a molecule to attract bonding electrons.
13. The bond moment is a measure of the polarity of a diatomic covalent bond. The dipole moment is the vector sum of the bond moments in a molecule. Both are measured in debyes (D).
14. The valence-shell electron-pair repulsion (VSEPR) theory is an approach to predicting molecular geometry based on the principle of maximum electron-set separation.

15. Bond dissociation energy is the energy needed to break one mole of a particular kind of bond.
16. In molecular orbital theory, two $1s$ atomic orbitals overlap end-to-end to form two σ molecular orbitals, one bonding (σ_{1s}) and one antibonding (σ_{1s}^*); two $2p$ atomic orbitals overlap end-to-end to form two σ molecular orbitals (σ_{2p_x} and $\sigma_{2p_x}^*$); and four $2p$ atomic orbitals overlap in parallel to form two bonding (π_{2p_y} and π_{2p_z}) and two antibonding ($\pi_{2p_y}^*$ and $\pi_{2p_z}^*$) π molecular orbitals.

RAISE YOUR GRADES

Can you . . . ?

☑ deduce formulas for ionic compounds of main-group elements
☑ draw Lewis structures of atoms, molecules, and ions
☑ determine whether the octet rule is obeyed or not
☑ calculate formal charges for alternative Lewis structures
☑ draw resonance forms of Lewis structures and decide when they are needed
☑ use electronegativity data to predict polarity
☑ deduce and sketch the shapes of molecules
☑ give molecular orbital descriptions of diatomic molecules
☑ correlate bond length and bond energy with bond order
☑ use the VSEPR theory to predict molecular geometry
☑ determine the polarity of a bond and of a molecule
☑ calculate the bond order for compounds of second-period elements

SOLVED PROBLEMS

PROBLEM 10-1   What are the formulas of the ionic compounds formed from **(a)** Ca and O and **(b)** Na and O?

Solution:
(a) The ion O^{2-} has the [Ne] configuration and should be the stable ionic form of oxygen. Calcium is a group II element and the stable ion should be Ca^{2+} with the [Ar] configuration. The neutral compound should be CaO.
(b) Sodium is a group I element; the stable ion should be Na^+ with the [Ne] configuration. For a neutral compound the -2 charge of the oxygen must be balanced by two Na^+ ions. The formula should be Na_2O.

PROBLEM 10-2   Draw Lewis formulas for NH_3, CN^-, and OF_2. The relative positions of the nuclei are

$$H$$
$$H \quad N \quad H \qquad C \quad N \qquad F \quad O \quad F$$

Solution: Count the number of valence electrons (taking into account the charge on ions). Assign one pair of electrons between each pair of bonded atoms. Then distribute the remaining electrons in lone pairs, double bonds, and triple bonds.

	No. of valence e^-	Lewis formula
NH_3	$5 + 3(1) = 8$	H H:N̈:H
CN^-	$4 + 5 + 1 = 10$	$[:C:::N:]^-$
OF_2	$6 + 2(7) = 20$	:F̈:Ö:F̈:

PROBLEM 10-3 Draw valence-bond formulas for SF_4, SiF_4, and ClO_3^-, showing all unshared electrons. The arrangements of the nuclei are such that the central atoms are respectively S, Si, and Cl. Which of these do(es) not obey the octet rule?

Solution:

	No. of valence e^-	Valence-bond formula
SF_4	$6 + 4(7) = 34$	
SiF_4	$4 + 4(7) = 32$	
ClO_3^-	$7 + 3(6) + 1 = 26$	

SF_4 does not obey the octet rule. There must be 10 electrons on S.

PROBLEM 10-4 Of the two possible valence-bond formulas for formic acid (HCO_2H), which one is the most reasonable structure?

$$H-C=\ddot{O}-H \quad \text{or} \quad H-C-\ddot{O}-H$$
$$(A) \qquad\qquad (B)$$

Solution: Calculate the formal charges on O in each formula:

	Valence e^-	$\frac{1}{2}$(unshared e^-)	Unshared e^-	Formal charge
(A) terminal O	6	$-$ 1	$-$ 6	$=$ -1
interior O	6	$-$ 3	$-$ 2	$=$ $+1$
(B) terminal O	6	$-$ 2	$-$ 4	$=$ 0
interior O	6	$-$ 2	$-$ 4	$=$ 0

Structure (**B**) with the lower formal charges is more reasonable.

PROBLEM 10-5 Deduce the most reasonable valence-bond structure(s) for the thiocyanate ion (SCN^-), in which the nuclei are arranged around the central C atom.

Solution: Since the question asks for the most reasonable structure, we'll be alert to the possibility that more than one valence-bond structure that obeys the octet rule can be drawn. Count valence electrons: 6 for S, 4 for C, 5 for N, and 1 for the negative charge. That gives a total of 16. When we start arranging these in the usual way, we see that three structures that obey the octet rule are readily produced. Then we calculate the formal charges associated with each atom in each structure (given below). [Note that the sum of the formal charges for each structure equals -1, the real (net) charge on the ion.]

$$\textbf{(A)} \begin{bmatrix} \ddot{S}=C=\ddot{N} \end{bmatrix}^- \qquad \textbf{(B)} \begin{bmatrix} :S\equiv C-\ddot{N}: \end{bmatrix}^- \qquad \textbf{(C)} \begin{bmatrix} :\ddot{S}-C\equiv N: \end{bmatrix}^-$$
$$\quad 0 \quad 0 \quad -1 \qquad\qquad +1 \quad 0 \quad -2 \qquad\qquad -1 \quad 0 \quad 0$$

Since structures (**A**) and (**C**) have equally low values of formal charges, they are both most reasonable structures for the SCN^- ion.

PROBLEM 10-6 What is the correct valence-bond formula for hydrazine (N_2H_4)?

$$\textbf{(a)} \ H-\ddot{N}-\ddot{N}-H \qquad \textbf{(b)} \ H-N=N-H \qquad \textbf{(c)} \ H-N\equiv N-H \qquad \textbf{(d)} \ H-\ddot{N}-N=H$$

Solution: The number of valence electrons in N_2H_4 is $2(5) + 4(1) = 14$. Assigning one pair of electrons between each pair of bonded electrons requires 10 electrons. That leaves two pairs of electrons to fill the N atom octets. The correct answer is (**a**).

In a multiple-choice problem such as this it is often good strategy to look first for the obvious mistakes. Answer (b) shows only 12 valence electrons, and (c) has 10 electrons on each N atom: A second-period element can have no more than 8 electrons. Answer (d) has 4 electrons on one of the H atoms: H is limited to 2 electrons.

PROBLEM 10-7   What is wrong with each of the following valence-bond formulas? (The nuclei are arranged in the correct relative positions.)

(a) ethylene (C_2H_4)

$$H-\overset{\cdot\cdot}{\underset{|}{C}}-\overset{-}{\underset{|}{C}}-H$$
$$\quad\;\; H\;\;\; H$$

(b) sulfite ion (SO_3^{2-})

$$\left[:\overset{\cdot\cdot}{\underset{\cdot\cdot}{O}}-S\overset{\nearrow\overset{\cdot\cdot}{O}:}{\underset{\searrow\;\underset{\cdot\cdot}{O}:}{}}\right]^{2-}$$

Solution:

(a) One of the C atoms has only 6 electrons. If you move the lone pair on the left C atom to make a double bond, both C atoms will have a full octet:

$$H-C{=}C-H$$
$$\;\;|\;\;\;\;|$$
$$\;\;H\;\;\;H$$

(b) The total number of valence electrons in SO_3^{2-} should be $6 + 3(6) + 2 = 26$. This structure shows only 24 electrons. (Perhaps the -2 charge was not counted.) The correct formula is

$$\left[:\overset{\cdot\cdot}{\underset{\cdot\cdot}{O}}-S\overset{:\overset{\cdot\cdot}{O}:}{\underset{\searrow\;:\overset{\cdot\cdot}{O}:}{}}\right]^{2-}$$

PROBLEM 10-8   Draw Lewis structures of the resonance hybrids of NO_3^-. (N is the central atom.)

Solution: The number of valence electrons is $5 + 3(6) + 1 = 24$. The three O atoms are bonded to the N atom:

$$O:N\overset{\cdot O}{\underset{\cdot\cdot\; O}{}}$$

Filling in the remaining valence electrons requires a double bond between any one of the O atoms and the N, so there are three equivalent structures:

$$\left[\overset{\cdot\cdot}{\underset{\cdot\cdot}{O}}::N\overset{\cdot\cdot O\cdot}{\underset{\cdot\cdot O\cdot}{}}\right]^- \longleftrightarrow \left[:\overset{\cdot\cdot}{O}:N\overset{\cdot O\cdot}{\underset{\cdot\cdot O\cdot}{}}\right]^- \longleftrightarrow \left[:\overset{\cdot\cdot}{O}:N\overset{\cdot\cdot O\cdot}{\underset{\cdot O\cdot}{}}\right]$$

PROBLEM 10-9   Determine the VSEPR molecule type and geometry of

(a) O   C   S   (b) O   N   O   (c) F   N $\overset{F}{\underset{F}{}}$

(OCS)   (NO_2^-)   (NF_3)

The nuclei are arranged in the correct relative positions.

Solution: Draw the valence-bond formulas. Then determine the number of *sets* of electrons on the central atom.

	Valence-bond formula	No. of e^- sets	Molecule type
OCS	$:\overset{\cdot\cdot}{O}{=}C{=}\overset{\cdot\cdot}{S}:$	2	AX_2
NO_2^-	$\left[:\overset{\cdot\cdot}{O}{=}\overset{\cdot\cdot}{N}-\overset{\cdot\cdot}{O}:\right]^-$	3	AX_2E
NF_3	$:\overset{\cdot\cdot}{F}-\overset{\cdot\cdot}{\underset{|}{N}}-\overset{\cdot\cdot}{F}:$ $\quad\;:\overset{\cdot\cdot}{F}:$	4	AX_3E

Refer to Table 10-4 for the appropriate shape: (a) OCS, linear; (b) NO_2^-, bent; (c) NF_3, trigonal pyramidal. You should be able to sketch these molecules and ions to show their shapes:

(a) $O=C=S$ (b) $\left[\,O\stackrel{\ddots}{N}\,O\,\right]^-$ (c) $F\stackrel{\stackrel{\ddot{N}}{|}}{\diagup}\stackrel{}{\diagdown}F$

PROBLEM 10-10   Which of the following is the best description of the shape of a ClO_3^- ion (the Cl atom is bonded to the three O atoms): **(a)** linear, **(b)** trigonal planar, **(c)** trigonal pyramidal, **(d)** tetrahedral, **(e)** bent?

Solution: Determine the valence-bond formula: $\left[\,:\ddot{O}-\ddot{Cl}-\ddot{O}:\,\right]^-$ (26 valence electrons). The central atom has four sets of electrons and the ion is type AX_3E (see Table 10-4), so according to VSEPR theory the appropriate shape is trigonal pyramidal (**c**). It should be apparent that answers **(a)**, **(d)**, and **(e)** are not appropriate answers for an ion or molecule consisting of three atoms bonded to a central atom. Answer **(b)** is incorrect because the VSEPR theory requires that a tetraatomic species be nonplanar when the central atom has a lone pair of electrons.

PROBLEM 10-11   Draw the valence-bond formulas for the following molecules and ions. Describe the shape of each. **(a)** PCl_6^-, **(b)** XeF_4, **(c)** BF_4^-, **(d)** NO_2^+.

Solution: In each of these cases it should be obvious that the first atom is the central atom. Determine the number of valence electrons and distribute the valence electrons according to the octet rule where possible. Count the number of electron sets on the central atom and decide the molecule type. Then use VSEPR theory to choose the appropriate shape.

(a) PCl_6^- has $5 + 6(7) + 1 = 48$ valence electrons, which can only be accounted for by a valence-bond formula showing six Cl atoms singly bonded to the central P atom. The octet rule doesn't work for period-3 P here (see Section 10-3B): It has 12 electrons around it (including the extra electron from the -1 charge).

According to VSEPR theory, there are six electron sets around the central P, so the ion is type AX_6 and therefore *octahedral*, which allows for maximal electron-set separation.

PCl_6^- valence-bond structure

PCl_6^- VSEPR geometry

(b) XeF_4 has $8 + 4(7) = 36$ valence electrons, accounted for by a valence-bond formula showing four F atoms singly bonded to the central Xe atom, which has 4 electrons left over. (Again, the octet rule doesn't work for the period-5 Xe, which has 12 electrons around it.)

There are six electron sets around the central Xe atom; four sets are shared by F atoms, while two sets are lone pairs. So according to VSEPR theory, XeF_4 is type AX_4E_2 and therefore *square planar*.

XeF_4 valence-bond structure

XeF_4 VSEPR geometry

(c) BF_4^- has $3 + 4(7) + 1 = 32$ valence electrons. The valence bond formula shows four F atoms singly bonded to a central B atom, which has an extra electron because of the -1 charge. (Note that the octet rule works with the group-IIIA B here only because there is a -1 charge on the ion.)

There are four electron sets around the central atom, each of which is shared by a F atom, so BF_4^- is type AX_4 and therefore *tetrahedral*.

BF_4^- valence-bond structure

BF_4^- VSEPR geometry

(d) NO_2^+ has $5 + 2(6) - 1 = 16$ valence electrons. The valence-bond structure shows that the two O atoms are each doubly bonded to a central N atom. (The octet rule works!)

There are two electron sets (one for each *bond*, even though they are double bonds), so NO_2^+ is type AX_2 and therefore *linear*.

$[:\ddot{O}=N=\ddot{O}:]^+$

NO_2^+ valence-bond structure

$O=N=O$

NO_2^+ VSEPR geometry

note: Be sure you understand the definition of an electron *set*, and remember that the only unshared pairs that count toward determining the shape are those on the central atom.

PROBLEM 10-12 Determine the valence-bond formula and the shape of each of the following species (C is the central atom in each): (a) $CClFH_2$, (b) CH_2O, (c) CO_3^{2-}, (d) CCl_3^-.

Solution: Determine the number of valence electrons and distribute them in accordance with the octet rule. (Since carbon is the central atom, we can expect all of them to obey the octet rule.) We'll indicate the shape by determining the type of each and applying the appropriate geometric description.

(a) $CClFH_2$ has $4 + 7 + 7 + 2(1) = 20$ valence electrons. The valence-bond formula is $:\ddot{C}l-\overset{\displaystyle :\ddot{F}:}{\underset{\displaystyle H}{C}}-H$. This is an

AX$_4$-type structure, so the shape is tetrahedral. [*Note:* The geometry of this molecule is not that of a regular tetrahedron; however, for purposes of description, we'll call it "tetrahedral."]

(b) CH_2O has $4 + 2(1) + 6 = 12$ valence electrons. The valence-bond formula is $H-\overset{\displaystyle }{\underset{\displaystyle H}{C}}=\ddot{O}$. This is an AX$_3$ type, so the shape is trigonal planar.

(c) CO_3^{2-} has $4 + 3(6) + 2 = 24$ valence electrons. The valence-bond formula is $\left[:\ddot{O}=C\overset{\displaystyle :\ddot{O}:}{\underset{\displaystyle :O:}{\Big\langle}}\right]^{2-}$. This is an

AX$_3$-type ion, so the shape is trigonal planar. Note that this ion has three resonance forms

$$:\ddot{O}=C\overset{\displaystyle :\ddot{O}:}{\underset{\displaystyle :\ddot{O}:}{\Big\langle}} \longleftrightarrow :\ddot{O}-C\overset{\displaystyle \ddot{O}:}{\underset{\displaystyle :O:}{\Big\langle}} \longleftrightarrow :\ddot{O}-C\overset{\displaystyle \ddot{O}:}{\underset{\displaystyle O:}{\Big\langle}}$$

but resonance does not affect geometry.

(d) CCl_3^- has $4 + 3(7) + 1 = 26$ valence electrons. The valence-bond formula is $\left[:\ddot{C}l-\overset{\displaystyle }{\underset{\displaystyle :\ddot{C}l:}{\ddot{C}}}-\ddot{C}l:\right]^-$. This is an

AX$_3$E type, and the shape is trigonal pyramidal.

PROBLEM 10-13 The molecular orbital description of NO is similar to that of O_2 (see Section 10-6) except that the energies of the oxygen atomic orbitals are slightly lower than those of the nitrogen atomic orbitals and the NO molecule contains a different number of electrons than O_2. (a) Draw an energy diagram to show how the atomic orbitals of N and O combine to form molecular orbitals of NO. (b) How many unpaired electrons are there in the NO molecule? (c) What is the bond order of NO?

Solution:
(a) Draw the energy diagram showing the relative energy levels for second-period elements (see Figure 10-2). Now count the total number of electrons in NO: $7 + 8 = 15$. Of these 15 electrons, 4 are nonbonding—which leaves 11 to be distributed among the σ and π MOs. Write the number of electrons in each MO (as shown in Figure 10-3), being sure to follow the aufbau, Pauli exclusion, and Hund principles. All 8 electrons in the s AOs go into the σ_s and σ_s^* MOs, and the remaining 7 electrons in the p AOs are distributed in the π and π^* MOs.

(b) There's one electron left over after the lower-energy π_{2p_y} and π_{2p_z} and σ_{2p_x} MOs are filled with paired electrons. The single electron, which goes into the $\pi_{2p_y}^*$ antibonding MO, is unpaired.

(c) The bond-order equation (10-2) gives the bond order:

$$\frac{1}{2}\left[\left(\begin{array}{c}\text{No. of } e^- \text{ in}\\ \text{bonding MOs}\end{array}\right) - \left(\begin{array}{c}\text{No. of } e^- \text{ in}\\ \text{antibonding MOs}\end{array}\right)\right]$$

$$= \frac{8-3}{2} = 2.5$$

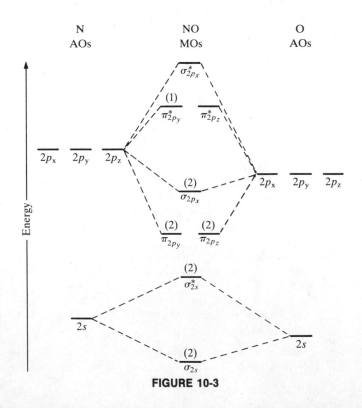

FIGURE 10-3

PROBLEM 10-14 The MO description of F_2 is

$$1s^2 1s^2 \sigma_{2s}^2 \sigma_{2s}^{*2} \pi_{2p_y}^2 \pi_{2p_z}^2 \sigma_{2p_x}^2 \pi_{2p_y}^{*2} \pi_{2p_z}^{*2} \sigma_{2p_x}^{*0}$$

What are the bond orders of F_2 and Ne_2?

Solution: The bond order of F_2 is $\frac{1}{2}(8 - 6) = 1$. In other words, the F—F bond is a single bond. The MO scheme for Ne_2 is the same as that of F_2 except that there are 2 more electrons in Ne_2, which go into the $\sigma_{2p_x}^*$ MO. The bond order therefore is $\frac{1}{2}(8 - 8) = 0$. The molecular orbital theory correctly shows that the Ne_2 molecule has zero bond order and therefore should not be stable.

PROBLEM 10-15 The following are the valence-bond formulas of methanol CH_3OH, formaldehyde CH_2O, and carbon monoxide CO:

$$CH_3OH \qquad CH_2O \qquad CO$$

How would you expect the C—O bonds in these molecules to compare in bond length and bond energy?

Solution: From the structures you would expect the C—O bond to be single in methanol, double in formaldehyde, and triple in carbon monoxide. Therefore the order of *decreasing* C—O bond length would be $CH_3OH > CH_2O > CO$. The order of *increasing* bond energy would be $CH_3OH < CH_2O < CO$.

PROBLEM 10-16 List the following bonds in order of increasing bond polarity: N—F, F—F, H—F, Cs—F.

Solution: Refer to a table of electronegativities (Table 10-2). Determine the difference in electronegativity between the two bonded atoms in each species.

Bond	Electronegativity difference
N—F	$4.0 - 3.0 = 1.0$
F—F	$4.0 - 4.0 = 0$
H—F	$4.0 - 2.1 = 1.9$
Cs—F	$4.0 - 0.7 = 3.3$

F—F has the least difference in electronegativity; the F—F bond will be nonpolar. The bond polarities will increase with increasing electronegativity difference: F—F (0), N—F (1.0), H—F (1.9), Cs—F (3.3). Notice how the electronegativities of the elements are related to positions in the periodic table: O, F, and Cl have the highest electronegativities, while the metals (groups 1 and 2) have the lowest electronegativities.

PROBLEM 10-17 Give VSEPR structures for the following. Indicate the shape and show the direction of the dipole moment for the polar molecules. **(a)** PF_3, **(b)** HCN, **(c)** BF_3.

Solution:
(a) The number of valence electrons in PF_3 is $5 + 3(7) = 26$. The valence-bond formula is :F̈—P̈—F̈: . This is

an AX_3E type, so its shape is trigonal pyramidal. The P—F bond moments are directed from the P to F; and since the molecule is nonplanar, the bond moments do not cancel and the dipole moment is directed between the three F atoms:

(b) The number of valence electrons in HCN is $1 + 4 + 5 = 10$. The valence-bond formula is H—C≡N:. This is an AX_2 type, so it is linear. Since N is the most electronegative atom, the dipole moment will be directed toward the N atom:

(c) The number of valence electrons in BF_3 is $3 + 3(7) = 24$. The valence-bond formula is :F̈—B . This is an

AX$_3$ type, and so the shape is trigonal planar. The B—F bond moments are directed at 120° toward the corners of an equilateral triangle to give a resultant zero dipole moment:

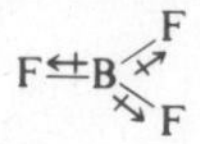

$$F \rightleftharpoons B \overset{F}{\underset{F}{\diagdown}} \qquad \mu = 0$$

PROBLEM 10-18   Answer the following questions. [*Hint*: Draw the valence-bond formulas for each species first. The central atom is italicized in each case.]

(a) O$_2$*N*Cl: Is the shape planar or pyramidal?
(b) *Cl*O$_2^-$: Is the shape linear or bent?
(c) *Si*H$_2$Cl$_2$: Is the dipole moment zero or nonzero?

Solution:

(a) The number of valence electrons in O$_2$NCl is $2(6) + 5 + 7 = 24$. The valence-bond formula is

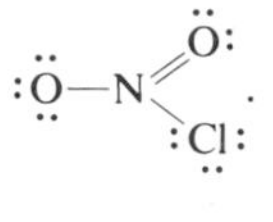

You should be able to give both resonance structures:

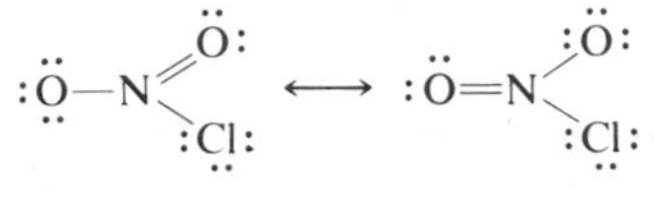

This is an AX$_3$ type and so the shape is *trigonal planar*.

(b) The number of valence electrons in ClO$_2^-$ is $7 + 2(6) + 1 = 20$. The valence-bond formula is

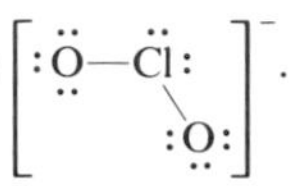

This is an AX$_2$E$_2$ type and so the shape is *bent*.

(c) The number of valence electrons in SiH$_2$Cl$_2$ is $4 + 2(1) + 2(7) = 20$. The valence-bond formula is

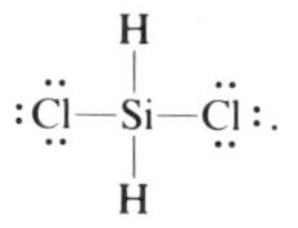

This is an AX$_4$ type, so the shape is *tetrahedral*.

Look at a sketch or model of this molecule. The Si—H and Si—Cl bonds are not equivalent, so their bond moments are not equivalent and do not cancel out. The dipole moment is *not zero*. This is a polar molecule.

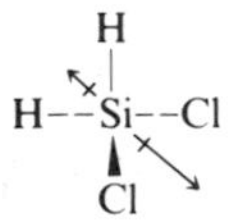

Supplementary Exercises

PROBLEM 10-19   What are the ionic compounds formed from Li and Cl and from Ca and Cl?

Answer: LiCl; CaCl$_2$

PROBLEM 10-20   Draw valence-bond formulas for ClO$_2^-$, SF$_5^+$, and NCl$_3$. Show all of the unshared electrons. (In each case the first atom is the central atom.)

Answer:

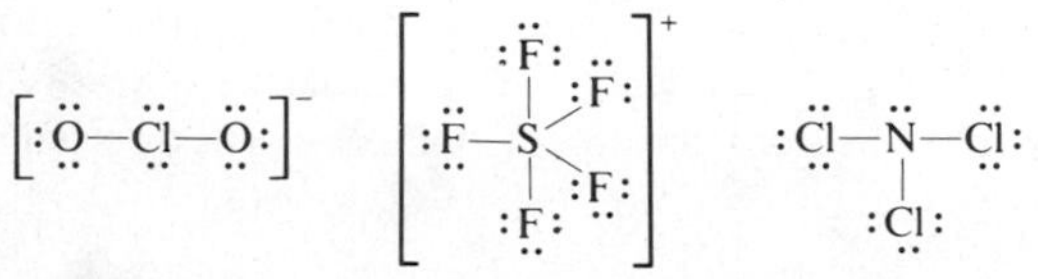

PROBLEM 10-21   Draw valence-bond formulas for BCl$_3$, SeCl$_4$, and SnCl$_4$. (The first atom is central in each case.) Which do not obey the octet rule?

Answer: :Cl—B (with :Cl: groups)   :Se (with :Cl: groups)   :Cl—Sn—Cl: BCl$_3$ and SeCl$_4$ do not obey the octet rule.

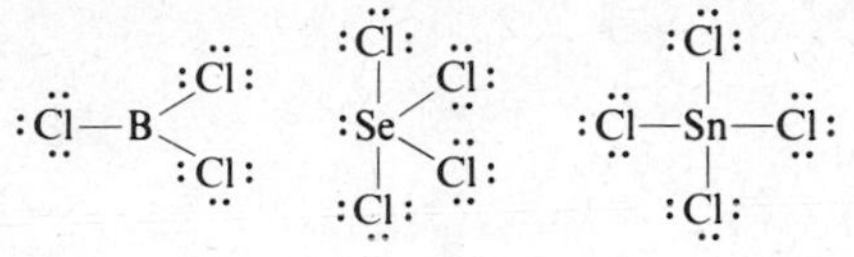

PROBLEM 10-22   Calculate the formal charges on the atoms in the valence-bond formulas given for nitrosyl chloride (NOCl). Which one is the more reasonable structure?

(a) :Ö—N̈=C̈l: **(b)** :Ö=N̈—C̈l:

Answer: **(a)** O −1   **(b)** O 0
  N 0   N 0
  Cl +1   Cl 0

Structure **(b)** is the more reasonable formula

PROBLEM 10-23   Draw valence-bond formulas for the following molecules, showing all the unshared electrons. The nuclei are shown in their correct relative positions. **(a)** acetylene (C$_2$H$_2$), HCCH; **(b)** hydrogen peroxide (H$_2$O$_2$), HOOH; **(c)** tetrafluorohydrazine (N$_2$F$_4$), FNNF.

F F

Answer: **(a)** H—C≡C—H   **(b)** H—Ö—Ö—H   **(c)** 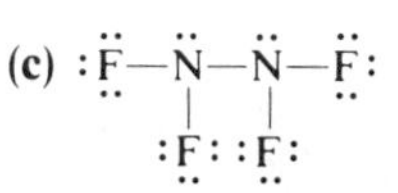 :F̈—N̈—N̈—F̈:

:F: :F:

PROBLEM 10-24   Draw the resonance hybrids of SO$_2$. (Sulfur is bonded to the two oxygen atoms.)

Answer: S̈ ⟷ S̈
  :O: O: :O :O:

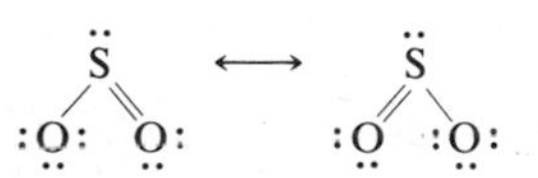

PROBLEM 10-25   Indicate the shape of the following ions and molecules with the appropriate term or a sketch (the central atom is italicized in each case): **(a)** *S*Cl$_2$, **(b)** *S*F$_3^-$, **(c)** *Xe*O$_3$.

Answer: **(a)** bent   **(b)** T-shaped   **(c)** trigonal pyramidal

PROBLEM 10-26   Which, if any, of the following molecules and ions are linear? **(a)** OF$_2$ (FOF), **(b)** C$_2$H$_2$ (HCCH), **(c)** H$_2$O$_2$ (HOOH), **(d)** N$_2$O (NNO).

Answer: C$_2$H$_2$ and N$_2$O are linear

PROBLEM 10-27   Draw the resonance hybrids of formate ion HCO$_2^-$. Describe the shape by the appropriate term. (Carbon is the central atom.)

Answer: [H—C with :Ö: and O: groups ⟷ H—C with Ö: and :O: groups]$^-$ trigonal planar

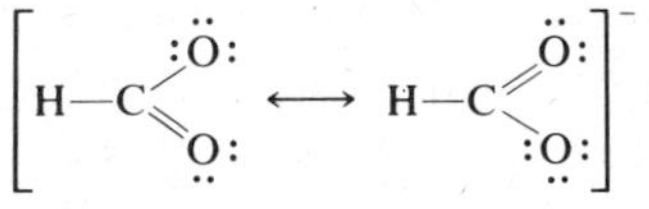

PROBLEM 10-28   It is possible to form the ions O$_2^+$ and O$_2^-$. Use the molecular orbital theory to determine the bond order and the number of unpaired electrons in these two ions.

Answer: O$_2^+$, bond order $= 2\frac{1}{2}$; one unpaired electron
  O$_2^-$, bond order $= 1\frac{1}{2}$; one unpaired electron

PROBLEM 10-29   List the following in order of increasing O—O bond length: O$_2$, O$_2^+$, O$_2^-$.

Answer: O$_2^+ <$ O$_2 <$ O$_2^-$

PROBLEM 10-30   Show the direction of the bond moment in each of the following bonds: N—F, N—H, N—Si, N—O.

Answer: N—F   N—H   N—Si   N—O

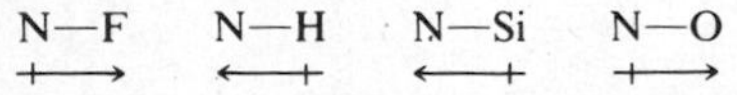

PROBLEM 10-31   Answer the following questions, giving the valence-bond formulas for each species. (The central atom is shown in italics.)

(a) OF_2: What is the direction of the dipole moment?
(b) SiF_4: Is the shape square planar or tetrahedral?
(c) XeF_4: Is the shape square planar or tetrahedral?

Answer: (a) 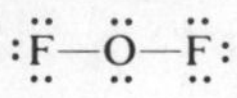 :F̈—Ö—F̈:　(b) 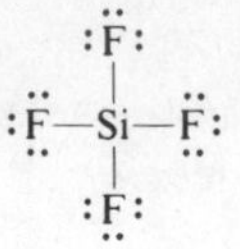 :F̈—Si—F̈:　(c) 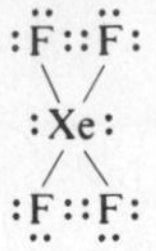 ：Xe：

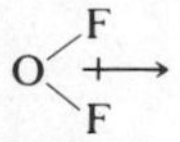

tetrahedral　　square planar

PROBLEM 10-32　Answer the following questions, giving the valence-bond formulas of each species. (The central atom is shown in italics.)

(a) PF_3: Is this a polar or nonpolar molecule?
(b) SO_3: Is this a polar or nonpolar molecule?
(c) SiH_3^-: Is this ion planar or nonplanar?

Answer: (a) 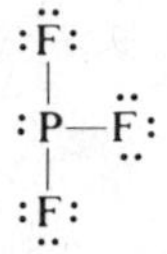 　(b) 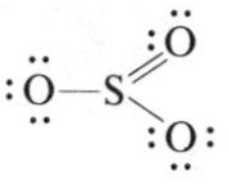 　(c)

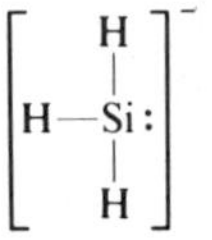

polar ($\mu \neq 0$)　　nonpolar ($\mu = 0$)　　nonplanar; trigonal pyramidal

11 SOLUTIONS

THIS CHAPTER IS ABOUT

☑ **Definitions**
☑ **Concentration Expressions**
☑ **Changes in Solubility**
☑ **Colligative Properties**
☑ **Electrolytes**

11-1. Definitions

A **solution** is a homogeneous mixture of two or more components. The components in a solution can be separated by purely physical processes such as distillation, evaporation, freezing, or diffusion. Although there can be solutions of any phase in any other phase, solutions involving liquids are of the most interest.

The components of solutions are differentiated by which is dissolved in what. For a solution of a gas or a solid in a liquid, the liquid is the **solvent**, and the gas or solid is the **solute**. For a solution of one liquid in another liquid, the choice of the term solvent or solute is arbitrary, but generally the liquid taken in larger amount is considered to be the solvent.

EXAMPLE 11-1: Give an example of a solution (solvent/solute) consisting of the following phases: (a) gas/gas, (b) liquid/gas, (c) liquid/liquid, (d) liquid/solid, (e) solid/solid.

Solution:
(a) Gas/gas: Air (O_2, CO_2, etc., in N_2)
(b) Liquid/gas: Soda ($CO_2(g)$ in $H_2O(l)$)
(c) Liquid/liquid: Rubbing alcohol ($H_2O(l)$ in i-$C_3H_7OH(l)$)
(d) Liquid/solid: Seawater (NaCl, etc., in $H_2O(l)$)
(e) Solid/solid: 14-Karat "gold" (Cu in Au)

When solids or gases are dissolved in liquid solvents (and often when liquids are dissolved in other liquids), there is a limit to the amount of solute that can dissolve in a fixed amount of solvent at a fixed temperature. A **saturated solution** is a solution that has reached its maximum capacity for dissolving a solute, such that further addition of solute results in no further dissolving. The amount of solute that can be dissolved in a given amount of saturated solution at a fixed temperature is the **solubility** of the solute in the solvent.

Solutes dissolve in solvents because, in very rough and approximate terms, the energy of interaction between solute and solvent particles makes the solution more stable than the pure solute phase (with its energy of interaction between solute particles) plus the pure solvent phase (with its energy of interaction between solvent particles). So the choice of an appropriate solvent for a particular solute depends upon the kinds of forces that exist between the various particles. Consider, for instance, ionic solids. The forces that exist between the ions in ionic solids are strong electrostatic forces, which can be broken only by strong forces between the ions and the solvent molecules. Thus a polar solvent (like water or ammonia) is chosen to dissolve an ionic solid.

note: There's an old rule-of-thumb that is useful when you're trying to choose a solvent: *Like dissolves like.* Polar solutes dissolve in polar solvents; nonpolar solutes dissolve in nonpolar solvents.

11-2. Concentration Expressions

A. Mole fraction

Mole fraction is defined for solutions just as it is for gases: The **mole fraction** of a component A (X_A) of a solution is the ratio of the number of moles of A in the solution (n_A) to the sum of the numbers of moles of all components of the solution.

MOLE FRACTION
$$X_A = \frac{n_A}{n_A + n_B + \cdots}$$

EXAMPLE 11-2: Calculate the mole fractions of benzene (X_B) and of toluene (X_T) in a mixture of 0.884 mol of benzene and 1.657 mol of toluene.

Solution: Using the definition of a mole fraction,

$$X_B = \frac{n_B}{n_B + n_T} = \frac{0.884 \text{ mol}}{0.884 \text{ mol} + 1.657 \text{ mol}} = 0.348$$

$$X_T = \frac{n_T}{n_T + n_B} = \frac{1.657 \text{ mol}}{1.657 \text{ mol} + 0.884 \text{ mol}} = 0.652$$

Notice that $X_B + X_T = 0.348 + 0.652 = 1.000$. The sum of the mole fractions of all the components of a solution always equals exactly 1. So you could have determined X_T very simply as $X_T = 1.000 - X_B = 1.000 - 0.348 = 0.652$.

B. Percentage by mass and volume

The **percentage by mass** of a solute in a solution is calculated as follows:

PERCENTAGE BY MASS
$$\text{mass \%} = \frac{\text{mass of solute}}{\text{mass of soln}} \times 100\%$$

EXAMPLE 11-3: Calculate the percentage by mass of (a) 1.4 g of cane sugar in 50.0 g of sugar syrup (sugar + water) and (b) 9.2 g of sugar dissolved in 105 g of water.

Solution:

(a) $\text{mass \% (sugar)} = \dfrac{1.4 \text{ g sugar}}{50.0 \text{ g syrup}} \times 100\% = 2.8\%$

(b) $\text{mass \% (sugar)} = \dfrac{9.2 \text{ g sugar}}{114 \text{ g soln (sugar + water)}} \times 100\% = 8.1\%$

The **percentage by volume** of a solute is calculated as follows:

PERCENTAGE BY VOLUME
$$\text{vol \%} = \frac{\text{vol of solute}}{\text{vol of soln}} \times 100\%$$

EXAMPLE 11-4: When 50.00 mL of H_2O is mixed with 50.00 mL of ethanol C_2H_5OH, the final volume is only 96.54 mL. Compute the percentage by volume of each liquid.

Solution:

$$\text{vol \% } (H_2O) = \frac{50 \text{ mL } H_2O}{96.54 \text{ mL soln}} \times 100\% = 51.79\%$$

$$\text{vol \% } (C_2H_5OH) = \frac{50 \text{ mL } C_2H_5OH}{96.54 \text{ mL soln}} \times 100\% = 51.79\%$$

note: When you add volume x of liquid A to volume y of liquid B you will generally NOT end up with $x + y$ volume of solution. Percentages by volume do not necessarily total 100%. (This is why chemists don't use volume percentage very often.)

C. Molarity

The **molarity** (M) of a solution is the number of moles of solute per liter of solution, that is, $M = n/(\text{L of solution})$. For example, if 1.84 mol of glucose is dissolved in enough water to make 5.45 L of solution, $M = (1.84 \text{ mol})/(5.45 \text{ L}) = 0.338 \text{ mol L}^{-1}$. The number of moles of solute available in a solution of given molarity can be found from $n = M \times (\text{L of solution})$.

EXAMPLE 11-5: What is the molarity of a solution made up by dissolving 9.52 g of NaCl in enough H_2O to form 575 mL of solution?

Solution: Use the definition of molarity, $M = n/(\text{L of solution})$, setting up so that the volume is expressed in liters and the given mass of NaCl is changed to number of moles:

$$M = n/\text{L}$$

$$= \left(\frac{9.52 \text{ g}}{58.4 \text{ g mol}^{-1}}\right)\bigg/\left(575 \text{ mL} \times \frac{1 \text{ L}}{10^3 \text{ mL}}\right)$$

$$= \frac{9.52 \text{ g}}{58.4 \text{ g mol}^{-1}} \times \frac{1}{575 \text{ mL}} \times \frac{10^3 \text{ mL}}{1 \text{ L}}$$

$$= 0.284 \text{ mol L}^{-1}$$

EXAMPLE 11-6: How many moles of H_2SO_4 are present in a 37.0-mL sample of a 0.104 M solution?

Solution:

$$n = M \times \text{L}$$

$$= (0.104 \text{ mol L}^{-1})\left(37.0 \text{ mL} \times \frac{1 \text{ L}}{10^3 \text{ mL}}\right)$$

$$= 3.85 \times 10^{-3} \text{ mol}$$

D. Molality

The **molality** (m) of a solution is the number of moles of solute per kilogram of *solvent* (NOT solution), or $m = n_{\text{sol}}/(\text{kg solv})$.

EXAMPLE 11-7: Determine the molality of a solution made by adding 0.345 mol of C_2H_5OH to 168 g of H_2O.

Solution:

$$m = \frac{n_{\text{sol}}}{\text{kg solv}}$$

$$= \frac{0.345 \text{ mol } C_2H_5OH}{168 \text{ g } H_2O} \times \frac{10^3 \text{ g}}{\text{kg}}$$

$$= 2.05 \text{ mol kg}^{-1}$$

EXAMPLE 11-8: Commercial concentrated sulfuric acid contains 95.0% by mass of H_2SO_4 (5.0% water) and the density of the solution is 1.83 g/mL. Calculate the molarity and molality of concentrated sulfuric acid.

Solution: The density tells you that 1 L of the acid solution has a mass of 1830 g. So in 1 L,

$$n_{H_2SO_4} = 0.950 \times 1830 \text{ g} \times \frac{1 \text{ mol}}{98.1 \text{ g}} = 17.7 \text{ mol}$$

Therefore

$$M = 17.7 \text{ mol L}^{-1}$$

This liter of acid contains 1740 g H_2SO_4 (17.7 mol) and $(5.0/100) \times 1830 = 92$ g of water; therefore

$$m = \frac{17.7 \text{ mol}}{92 \text{ g H}_2\text{O}} \times \frac{10^3 \text{ g H}_2\text{O}}{1 \text{ kg H}_2\text{O}}$$

$$= 1.9 \times 10^2 \text{ mol kg}^{-1}$$

note: For very dilute solutions molarity and molality are numerically almost the same, but for concentrated solutions such as this they are widely divergent.

___

E. Normality

In reactions involving acids and bases we're concerned about the number of moles of H^+ or OH^- ions that a compound contains, so we express acid and base concentrations in terms of **equivalents**. For our purposes, one equivalent of an acid will generate one mole of H^+ ions; one equivalent of a base will generate one mole of OH^- ions (see Chapter 12). The **normality** (N) of a solution is the concentration of the solution expressed in equivalents per liter.

___

EXAMPLE 11-9: A solution of H_2SO_4 is made by dissolving 196 g of acid in enough water to produce 500.0 mL of solution. Determine the normality of the solution.

Solution: We know that 1 mol of H_2SO_4 will produce 2 mol of H^+, so

$$\text{No. of equiv} = \frac{2 \text{ mol H}^+}{1 \text{ mol H}_2\text{SO}_4} \times \frac{1 \text{ equiv}}{1 \text{ mol H}^+} \times \frac{196 \text{ g H}_2\text{SO}_4}{98.1 \text{ g/mol H}_2\text{SO}_4}$$

$$= 4.00$$

Now using the definition of normality

$$N = \frac{\text{equiv}}{\text{L}} = \frac{4.00 \text{ equiv}}{(500.0 \text{ mL})(10^{-3} \text{ L/mL})}$$

$$= 8.00$$

___

11-3. Changes in Solubility

A. Solubility and temperature: Solid/liquid solutions

Solubility varies with temperature. The variation is linked with the **enthalpy of solution** ΔH_{sol} of the substance, which is the heat released or absorbed when one mole of solute is dissolved in a large excess of solvent (infinite dilution) at one atmosphere pressure and a fixed temperature. If ΔH_{sol} is positive, the solution process is endothermic and solubility increases with increasing temperature. If ΔH_{sol} is negative, the solution process is exothermic and solubility decreases with increasing temperature.

B. Solubility and pressure: Gas/liquid solutions

The solubility of a gas in a liquid depends on both temperature and pressure. At a fixed temperature, **Henry's law** states that the solubility (C) of a gas A is directly proportional to the partial pressure of the gas (p_A) in contact with the liquid:

HENRY'S LAW: $\qquad\qquad\qquad C_A = kp_A \qquad\qquad\qquad$ (11-1)

where k is a constant specific to the solvent and gas at a given temperature. Real gases don't follow Henry's law exactly, but their behavior may be reasonably approximated by using this law.

EXAMPLE 11-10: The solubility of pure nitrogen at 1.00 atm in water at 298 K is 6.8×10^{-3} *M*. What is the value of k for N_2 at 298 K?

Solution: By Henry's law

$$C_{N_2} = kp_{N_2} = 6.8 \times 10^{-3}\ M = k(1.00\ \text{atm})$$

So

$$k = \frac{6.8 \times 10^{-3}\ M}{1.00\ \text{atm}} = 6.8 \times 10^{-3}\ M\ \text{atm}^{-1}$$

EXAMPLE 11-11: Given the solubility of N_2 (6.8×10^{-3} *M*) at 298 K and 1 atm, calculate its solubility at 298 K and 25 atm.

Solution: Using Henry's law and the value of k calculated in Example 11-10,

$$C_{N_2} = (6.8 \times 10^{-3}\ M\ \text{atm}^{-1})(25\ \text{atm}) = 0.17\ M$$

But we don't really need that value of k. Instead, we can set up a proportion, using the given C_{N_2} as the initial solubility, $C_1 = 6.8 \times 10^{-3}$ *M*:

$$\frac{C_1}{p_1} = \frac{C_2}{p_2}$$

So

$$C_2 = \frac{C_1 p_2}{p_1} = \frac{(6.8 \times 10^{-3}\ M)(25\ \text{atm})}{1\ \text{atm}} = 0.17\ M$$

11-4. Colligative Properties

A **colligative property** is any property of a solution that varies in proportion to the concentration of the solute, i.e., the number of solute particles (atoms, molecules, or ions) present in a given volume of solution. For dilute solutions, the change in any colligative property with changing concentration is *directly proportional* to the amount of solute dissolved in a specific amount of solvent.

A. Solution vapor pressure

When nonvolatile solutes (i.e., solutes with negligible vapor pressure) are dissolved in volatile solvents, the vapor pressure of the solution is lower than the vapor pressure of the pure solvent. The extent of vapor pressure lowering is given by **Raoult's law**: At a constant temperature the lowering of the vapor pressure in a dilute solution is directly proportional to the amount of solute dissolved in a a specific amount of solvent:

RAOULT'S LAW $$p_A = X_A p_A^{\circ} \qquad\qquad (11\text{-}2)$$

where p_A is the vapor pressure over the solution, X_A is the mole fraction of solvent, and p_A° is the vapor pressure of pure solvent at the stated temperature.

EXAMPLE 11-12: Calculate the vapor pressure of a solution of 0.250 mol of sucrose in 1.100 mol of water at 50°C. The vapor pressure of H_2O at 50°C is 92.5 torr. (Sucrose is not volatile).

Solution: In this case, A is the solvent water, so

$$X_A = \frac{1.100\ \text{mol}\ H_2O}{1.100\ \text{mol}\ H_2O + 0.250\ \text{mol sucrose}} = 0.815$$

Thus by Raoult's law

$$P_A = X_A p_A^{\circ}$$
$$= (0.815)(92.5\ \text{torr}) = 75.4\ \text{torr}$$

B. Boiling point elevation

Solutions containing nonvolatile solutes have higher boiling points than pure solvents. The **boiling point elevation** (ΔT_b) at a constant pressure is directly proportional to the *molal* (NOT molar) concentration of the solute:

BOILING POINT ELEVATION $$\Delta T_b = K_b m \tag{11-3}$$

where ΔT_b is the amount of change in the boiling point, K_b is a characteristic constant of the solvent, and m is the molal concentration of the solute (moles of solute per kilogram of solvent).

EXAMPLE 11-13: Calculate the boiling point of a solution of 0.0150 mol of anthracene (which is nonvolatile) in 45.0 g of toluene (K_b for toluene is $3.33°$ kg mol^{-1}); the normal boiling point of toluene is 110.63°C.

Solution: Find m:

$$m = \frac{\text{mol sol}}{\text{kg solv}} = \frac{0.0150 \text{ mol anthracene}}{45.0 \text{ g toluene}} \times \frac{10^3 \text{ g}}{1 \text{ kg}} = \frac{0.333 \text{ mol}}{\text{kg}}$$

From the boiling point elevation equation (11-3),

$$\Delta T_b = \left(\frac{3.33° \text{ kg}}{\text{mol}}\right)\left(\frac{0.333 \text{ mol}}{\text{kg}}\right) = 1.11°$$

Now that we know the boiling point elevation, we can calculate the boiling point of this anthracene solution:

$$110.63°C + 1.11° = 111.74°C$$

C. Freezing point depression

Solutions have lower freezing points than pure solvents. The freezing point depression is directly proportional to molal concentration of the solute:

FREEZING POINT DEPRESSION $$\Delta T_f = K_f m \tag{11-4}$$

where the symbols are analogous to those for boiling point elevation.

EXAMPLE 11-14: Calculate the freezing point of a solution of 0.047 mol of lactose (a sugar) in 25.0 g of H_2O (K_f for $H_2O = 1.86°$ kg mol^{-1}).

Solution: Find m:

$$m = \frac{0.047 \text{ mol lactose}}{25.0 \text{ g } H_2O} \times \frac{10^3 \text{ g}}{\text{kg}} = \frac{1.88 \text{ mol}}{\text{kg}}$$

From the freezing point depression equation (11-4),

$$\Delta T_f = \left(\frac{1.88 \text{ mol}}{\text{kg}}\right)\left(\frac{1.86° \text{ kg}}{\text{mol}}\right) = 3.50°$$

Since pure water freezes at 0.00°, the freezing point of this lactose solution is $-3.50°C$.

D. Osmotic pressure

A *semipermeable membrane* is a membrane that allows solvent molecules to pass through but prohibits the passage of solute molecules. *Osmosis* is the tendency of solvent molecules to pass through a semipermeable membrane from a dilute solution to a more concentrated one at a greater rate than in the reverse direction. The result of osmosis is the tendency for solutions on both sides of a semipermeable membrane to achieve the same concentration. **Osmotic pressure** Π is the exact external pressure needed to just stop osmosis. For a dilute solution Π is calculated by the following law:

OSMOTIC PRESSURE $$\Pi = \frac{n}{V} RT = MRT$$ **(11-5)**

where n is the number of moles of solute, V is the solution volume, R is the gas constant, and T is the Kelvin temperature. Because n/V is the molar concentration of solute, it can be replaced with M. [Notice the similarity of Eq. (11-5) to the ideal gas law ($PV = nRT$).]

EXAMPLE 11-15: Calculate the osmotic pressure of a 0.100 M solution of sucrose in water at 298 K.

Solution: Given $M = 0.100$ mol L^{-1} and $T = 298$ K, we choose an R whose units are consistent, 0.0821 L atm mol^{-1}K^{-1}, and use the osmotic pressure law (11-5):

$$\Pi = MRT$$
$$= (0.100 \text{ mol L}^{-1})(0.0821 \text{ L atm mol}^{-1}\text{K}^{-1})(298 \text{ K}) = 2.45 \text{ atm}$$

> *note:* Even for fairly dilute solutions, the osmotic pressure can be large. This makes osmotic pressure one of the most useful colligative properties for the study of large molecules, which have large molar masses.

11-5. Electrolytes

An **electrolyte** is any substance that conducts electricity in solution. When a solute is an electrolyte, some percentage of the solute will ionize and each ion will act independently to produce its own colligative effect. (Recall that colligative properties depend on the number of solute particles—atoms, molecules, or ions—present in solution.) We need ways to deal with these "extra" effects.

A. The van't Hoff factor: Complete dissociation

The **van't Hoff factor** i is the number of moles of particles formed per mole of solute electrolyte formula units in solution. For very dilute solutions of strong electrolytes, the van't Hoff factor approaches the number of moles of ions predicted on the basis of *complete dissociation* and can be introduced into every colligative property equation (Eqs. 11-2 through 11-5) as the number of moles of solute involved.

note: For nonionic solutions, $i = 1$, always.

EXAMPLE 11-16: Determine the i factors for (a) LiBr, (b) $Ca(NO_3)_2$, and (c) Na_3PO_4. Assume complete dissociation.

Solution:
(a) LiBr gives 1 Li$^+$ particle and 1 Br$^-$ particle in solution, i.e., 2 mol of ions per 1 mol of LiBr formula units, so $i = 2$.
(b) $Ca(NO_3)_2$ gives 1 Ca$^+$ and 2 NO$_3^-$ ions in solution, so $i = 3$.
(c) Na_3PO_4 gives 3 Na$^+$ and 1 PO$_4^{3-}$ in solution, so $i = 4$. (But watch out for this one!)

B. The van't Hoff factor as fudge factor

Unfortunately, the world does not consist of strong electrolytes in very dilute solutions. Weak electrolytes do not dissociate completely; neither do strong electrolytes in relatively concentrated solution. For example, NaCl is a strong electrolyte. In solution you might expect 1 mol of NaCl to produce 1 mol of Na$^+$ ions and 1 mol of Cl$^-$ ions. But experimental determination of the degree of ionization for various concentrations of NaCl produces the following results:

Concentration of NaCl solution (m)	Moles of ions (i)
0.1	1.87
0.01	1.94
0.001	1.97

The van't Hoff factor i, which is the *actual* number of moles of particles formed per mole of formula units of solute, must be introduced into every colligative property equation in order to bring the theoretical formula into agreement with experimental reality. Thus, for example, we insert the i factor into the boiling point elevation (11-3) and freezing point depression (11-4) equations before the solvent constants, so we have

BOILING POINT ELEVATION (CORRECTED)
$$\Delta T_b = i K_b m \tag{11-3a}$$

FREEZING POINT DEPRESSION (CORRECTED)
$$\Delta T_f = i K_f m \tag{11-4a}$$

Now we can use the results of colligative property experiments to determine the actual **degree of ionization** (or **dissociation**) for an electrolyte, α.

DEGREE OF IONIZATION
$$\alpha = \frac{i - 1}{v - 1} \tag{11-6}$$

where α is the degree of ionization, v is the (ideal) number of ions per formula unit, and i is the van't Hoff (fudge) factor found from a colligative property equation.

EXAMPLE 11-17: Determine the degree of ionization of Na_3PO_4 in a 0.40 m aqueous solution if the boiling point elevation is $0.78°C$ (K_b for $H_2O = 0.51°$ kg mol^{-1}).

Solution: Rearrange $\Delta T_b = i K_b m$ to isolate i:

$$i = \frac{\Delta T_b}{K_b m} = \frac{0.78°}{(0.51° \text{ kg mol}^{-1})(0.40 \text{ mol kg}^{-1})} = 3.82$$

From the degree of ionization equation (11-6)

$$\alpha = \frac{i - 1}{v - 1} = \frac{3.82 - 1}{4 - 1} = 0.94$$

Thus Na_3PO_4 is 94% ionized when the concentration is 0.40 m.

SUMMARY

1. A solution is a homogeneous mixture of two or more components that can be separated by physical processes.
2. A saturated solution is one in which so much solute is dissolved that the addition of more solute results in no further dissolving.
3. Solubility is the amount of solute that can be dissolved in a given amount of a saturated solution at a fixed temperature.
4. Like dissolves like: We select solvents to dissolve solutes having similar properties.
5. The mole fraction of a component of a solution is the ratio of moles of component to total moles:

$$X_A = \frac{n_A}{n_A + n_B + \cdots}$$

6. Percentage by mass is the mass of solute divided by the mass of solution times 100%; percentage by volume is the volume of solute divided by the volume of solution times 100%.
7. Molarity is the moles of solute per liter of solution.
8. Molality is the moles of solute per kilogram of solvent.
9. Normality is the number of equivalents per liter of solution.
10. In an endothermic solution process, solubility increases with increasing temperature and ΔH_{sol} is positive; in an exothermic solution process, solubility decreases with increasing temperature and ΔH_{sol} is negative.

11. Henry's law states that the solubility of a gas is directly proportional to its partial pressure:

$$C_A = kp_A$$

12. Colligative properties of solution are directly proportional to the number of solute particles and are quantified by the colligative property equations:

vapor pressure lowering	$p_A = X_A p_A^\circ$	(Raoult's law)
boiling point elevation	$\Delta T_b = K_b m$	
freezing point depression	$\Delta T_f = K_f m$	
osmotic pressure	$\Pi = MRT$	

13. The van't Hoff factor i is the actual number of moles of particles formed per mole of solute formula units in solution. This factor must be inserted into the solute terms of the colligative property equations.

14. The degree of ionization, or dissociation, of an electrolytic compound is given by

$$\alpha = \frac{i - 1}{v - 1}$$

RAISE YOUR GRADES

Can you ...?

☑ calculate and interconvert mass %, vol %, M, m, and X
☑ decide which solvent is best for a given solute
☑ use Henry's law to calculate gas solubility
☑ use Raoult's law to calculate vapor pressure
☑ calculate boiling point elevation, freezing point depression, and osmotic pressure
☑ use ΔT_b, ΔT_f, and Π data to determine molar masses
☑ determine and use i factors for electrolytes
☑ determine degree of dissociation for electrolytes

SOLVED PROBLEMS

PROBLEM 11-1   The solubility of potassium chloride (KCl) in water is 45.5 g/100.0 g H_2O at 60°C and 31.0 g/100.0 g H_2O at 10°. What happens when a saturated solution of 45.5 g KCl in 100.0 g H_2O at 60° is cooled to 10°?

Solution: At 10°, 100.0 g of H_2O can dissolve only 31.0 g of KCl. Consequently, 45.5 g − 31.0 g = 14.5 g of KCl will crystallize from a saturated solution at 60° cooled to 10°.
note: **Recrystallization** is an important method of purifying crystalline solids. It consists of dissolving a solid in a hot solvent and then cooling the solution so that the solute crystallizes in pure form from the cool solution.

PROBLEM 11-2   What mass of sucrose is present in 27.0 g of a 12.5% by mass solution?

Solution: Percentage by mass is calculated by

$$\text{mass \%} = \frac{\text{mass of solute}}{\text{mass of soln}} \times 100\%$$

So, rearranging,

$$\text{mass of solute} = \left(\frac{\text{mass \%}}{100\%}\right)(\text{mass of soln})$$

$$\text{mass of sucrose} = \left(\frac{12.5\%}{100\%}\right)(27.0 \text{ g}) = 3.38 \text{ g}$$

PROBLEM 11-3   Determine the molarity of a solution of 63.2 g of formic acid (HCO_2H) in 500.0 mL of solution.

Solution: Molarity is defined as

$$M = \frac{\text{mol solute}}{\text{L soln}} \quad \text{or} \quad M = \frac{n_{sol}}{L_{soln}}$$

Thus we have to calculate the number of moles of formic acid, and for that we need the molar mass of HCO_2H, which is 46.03 g mol^{-1}. We also need to convert to liters. So

$$M = \left(\frac{63.2 \text{ g formic acid}}{46.03 \text{ g mol}^{-1}}\right)\left(\frac{10^3 \text{ mL L}^{-1}}{500.0 \text{ mL}}\right) = 2.75 \text{ mol L}^{-1}$$

PROBLEM 11-4   (a) How many moles of sodium hydroxide (NaOH) are present in 11.8 mL of a 0.0864 M solution? (b) How many millimoles?

Solution:

(a) $$n = (M)(L) = (0.0864 \text{ } M)\left(\frac{11.8 \text{ mL}}{10^3 \text{ mL/L}}\right) = 1.02 \times 10^{-3} \text{ mol}$$

(b) Since 1 mol = 10^3 mmol, the number of millimoles is

$$(1.02 \times 10^{-3} \text{ mol})(10^3 \text{ mmol/mol}) = 1.02 \text{ mmol}$$

An alternative solution to the question of how many millimoles makes use of the fact that

$$M = \frac{\text{mol}}{\text{L}} = \frac{\text{mmol}}{\text{mL}}$$

So

$$\text{mmol} = (M)(\text{mL}) = (0.0864 \text{ } M)(11.8 \text{ mL}) = 1.02 \text{ mmol}$$

This second approach is often very useful in handling real laboratory-scale problems, where you're more likely to be measuring in milliliters than in liters.

PROBLEM 11-5   What is the molality of a solution of 1.295 g of naphthalene ($C_{10}H_8$) in 21.4 g of toluene (C_7H_7)?

Solution: Molality is calculated as

$$m = \frac{n_{sol}}{\text{kg solv}}$$

The molar mass of $C_{10}H_8$ is 128.2 g/mol, so (converting to kilograms) we get

$$m = \frac{1.295 \text{ g } C_{10}H_8}{128.2 \text{ g/mol}} \times \frac{10^3 \text{ g/kg}}{21.4 \text{ g toluene}} = 0.472 \text{ mol/kg}$$

[Note that neither the formula nor the molar mass of the solvent toluene is involved.]

PROBLEM 11-6   What are the mole fractions of H_2O and methanol (CH_3OH) in a solution of 50.0 g H_2O in 50.0 g CH_3OH?

Solution: Using the definition of mole fraction, we know that

$$X_{H_2O} = \frac{n_{H_2O}}{n_{H_2O} + n_{CH_3OH}}$$

So we need to calculate the number of moles of H_2O and CH_3OH:

$$n_{H_2O} = \frac{50.0 \text{ g}}{18.0 \text{ g/mol}} = 2.78 \text{ mol} \quad \text{and} \quad n_{CH_3OH} = \frac{50.0 \text{ g}}{32.0 \text{ g/mol}} = 1.56 \text{ mol}$$

Substituting into the definition,

$$X_{H_2O} = \frac{2.78 \text{ mol}}{(2.78 \text{ mol} + 1.56 \text{ mol})} = 0.641$$

Since this is a two-component mixture and the sum of all mole fractions must equal 1 exactly,

$$X_{CH_3OH} = 1 - X_{H_2O} = 1 - 0.641 = 0.359$$

Check:

$$X_{CH_3OH} = \frac{n_{CH_3OH}}{n_{H_2O} + n_{CH_3OH}} = \frac{1.56\ mol}{2.78\ mol + 1.56\ mol} = 0.359$$

PROBLEM 11-7   You have three solvents available: water, ethanol, and hexane (C_6H_{14}). Choose the best solvent for (a) KBr and (b) naphthalene.

Solution:
(a) Water is the best solvent for KBr because it is the most polar of the available solvents. Potassium bromide is an ionic solid and consequently dissolves best in a very polar solvent. The strong ion–ion forces in KBr are replaced by strong ion–dipole forces in the solution.
(b) Hexane is the best solvent for naphthalene. Like dissolves like: A hydrocarbon like naphthalene $C_{10}H_8$ dissolves best in a nonpolar hydrocarbon solvent like hexane because the weak van der Waals attractions between solvent molecules are easily broken and replaced by weak van der Waals attractive forces between solute and solvent molecules.

PROBLEM 11-8   The solubility of KCl in water increases with increasing temperature. Is ΔH_{sol} for the following process positive or negative? Is the solution process exothermic or endothermic?

$$KCl(s) \xrightarrow{H_2O(l)} K^+(aq) + Cl^-(aq)$$

Solution: Since the solubility of the solute increases with increasing temperature, the enthalpy change for the process given, ΔH_{sol}, will be positive. The solution process is endothermic.

PROBLEM 11-9   At 20°C, 1.00 L of H_2O dissolves 0.028 L of O_2 at a pressure of 1.00 atm. Calculate the molar concentration of O_2 in H_2O in contact with air at a total pressure of 1.00 atm, if the partial pressure of O_2 in the air is 0.19 atm. (Neglect any water vapor pressure corrections).

Solution: According to Henry's law, the solubility of O_2 (C_{O_2}) is directly proportional to p_{O_2} at a fixed temperature (20°C + 273 = 293 K). So we know we need Eq. (11-1):

$$C_{O_2} = kp_{O_2}$$

We're given the value of p_{O_2} at 0.19 atm, but we need to know the number of moles of O_2 in aqueous solution at 1.00 atm, which we find by using the ideal gas law:

$$n_{O_2} = \frac{PV}{RT} = \frac{(1.00\ atm)(0.028\ L)}{(0.0821\ L\ atm\ mol^{-1}\ K^{-1})(293\ K)}$$

$$= 1.2 \times 10^{-3}\ mol$$

Given 1 L of solution, we now have the solubility at 1.00 atm, so we rearrange Henry's law to find k:

$$k = \frac{C_{O_2}}{p_{O_2}} = \frac{1.2 \times 10^{-3}\ mol\ L^{-1}}{1.00\ atm}$$

Then at a partial pressure of 0.19 atm

$$C_{O_2} = kp_{O_2} = \left(\frac{1.2 \times 10^{-3}\ mol\ L^{-1}}{1.00\ atm}\right)0.19\ atm$$

$$= 2.3 \times 10^{-4}\ mol\ L^{-1}$$

which can be read as 2.3×10^{-4} *M*.

PROBLEM 11-10   The vapor pressure of ethyl acetate (molar mass 88.1 g mol^{-1}) is 4.00×10^2 torr at 59°C. What mass of biphenyl (molar mass 154.2 g mol^{-1}), a nonvolatile compound, would reduce the vapor pressure of a 50.0-g portion of ethyl acetate at 59°C to 3.50×10^2 torr?

Solution: Raoult's law (Eq. 11-2) is the key equation here, since it connects vapor pressures of solutions and of pure solvents:

$$p_A = X_A p_A^\circ$$

In this case, the solution vapor pressure $p_A = 3.50 \times 10^2$ torr, and the solvent vapor pressure $p_A^\circ = 4.00 \times 10^2$ torr. Consequently, the mole fraction of *solvent* in the solution must be equal to

$$X_A = \frac{p_A}{p_A^\circ} = \frac{3.50 \times 10^2 \text{ torr}}{4.00 \times 10^2 \text{ torr}} = 0.875$$

and the mole fraction of solvent is defined as

$$X_A = X_{\text{solv}} = \frac{n_{\text{solv}}}{n_{\text{solv}} + n_{\text{sol}}}$$

Given the molar mass of solvent ethyl acetate, we know that

$$n_{\text{solv}} = \frac{50.0 \text{ g}}{88.1 \text{ g mol}^{-1}} = 0.568 \text{ mol}$$

So we can find the required number of moles of solute n_{sol}:

$$0.875 = \frac{0.568 \text{ mol}}{0.568 \text{ mol} + n_{\text{sol}}}$$

$$0.497 + 0.875 n_{\text{sol}} = 0.568$$

$$n_{\text{sol}} = 0.081 \text{ mol}$$

Since the solute biphenyl has molar mass 154.2 g mol^{-1}, the mass needed is $(0.081 \text{ mol})(154.2 \text{ g mol}^{-1}) = 12$ g.

PROBLEM 11-11   You are handed an unknown nonvolatile organic solid and asked to determine its molar mass. Describe a technique for doing this.

Solution: You know you have an organic unknown, so you try dissolving a weighed amount of the stuff in a known amount of an organic solvent, toluene. You find that 23.2 g of toluene dissolves 1.54 g of the solid. Then you heat the solution and find that it has a boiling point of 112.04°C at 1.00 atm. But your reference tables tell you that pure toluene has a normal boiling point of 110.63° and a K_b of 3.33° kg mol^{-1}. Obviously, you can use boiling point elevation to find the molar mass required.
You know from Eq. (11-3) that

$$\Delta T_b = m K_b$$
$$= 112.04° - 110.63° = 1.41°$$

So the molality is

$$m = \frac{\Delta T_b}{K_b} = \frac{1.41°}{3.33° \text{ kg mol}^{-1}} = 0.423 \text{ mol kg}^{-1}$$

You also know that molar mass is expressed in grams per mole. So call the molar mass of the unknown "X." Then a solution of 1.54 g of the unknown in 23.2 g toluene must have a molality that equals the observed m from the bp elevation of 0.423 mol kg^{-1}:

$$\frac{1.54 \text{ g}}{X} \times \frac{10^3 \text{ g kg}^{-1}}{23.2 \text{ g}} = 0.423 \text{ mol kg}^{-1}.$$

and

$$X = \frac{(1.54 \text{ g})(10^3 \text{ g kg}^{-1})}{(23.2 \text{ g})(0.423 \text{ mol kg}^{-1})} = 157 \text{ g mol}^{-1}$$

which is the molar mass.

PROBLEM 11-12   How many liters of ethylene glycol (molar mass 62 g/mol; density 1.1 g/mL), an antifreeze, must you add to 10.0 L of radiator water to lower its freezing point to -5.0°C? (K_f for H_2O is 1.86° kg mol^{-1}.)

Solution: Since you want a freezing point (fp) depression here, you'll need Eq. (11-4):

$$\Delta T_f = K_f m$$

You want a ΔT_f of 0.00°C (fp of pure water) $- (-5.0°C)$ (fp of solution required) $= 5.0°$. Then

$$m = \frac{\Delta T_f}{K_f} = \frac{5.0°}{1.86° \text{ kg mol}^{-1}} = 2.7 \text{ mol kg}^{-1}$$

You have 10.0 L of radiator water, whose total mass is $(10.0 \text{ L})(1.00 \text{ kg L}^{-1}) = 10.0$ kg; consequently, you need $(10.0 \text{ kg})(2.7 \text{ mol kg}^{-1}) = 27$ mol glycol. Converting to liters, you need

$$(27 \text{ mol glycol})\left(\frac{62 \text{ g mol}^{-1}}{1.1 \text{ g mL}^{-1}}\right)\left(\frac{1 \text{ L}}{10^3 \text{ mL}}\right) = 1.5 \text{ L of glycol}$$

PROBLEM 11-13 A solution of 15.3 mg of a protein in 1.50 mL of water is found to have an osmotic pressure of 9.7 torr at 293 K. Calculate the molar mass of the protein.

Solution: From the osmotic pressure law

$$\Pi = MRT$$

so

$$M = \frac{\Pi}{RT} = \frac{(9.7 \text{ torr})/(760 \text{ torr atm}^{-1})}{(0.0821 \text{ L atm mol}^{-1}\text{K}^{-1})(293 \text{ K})}$$

$$= 5.3 \times 10^{-4} \text{ mol L}^{-1}$$

If the molar mass of the protein is Y, then its molarity is (watch out for the units!)

$$M = = \frac{(15.3 \text{ mg})(10^3 \text{ mL L}^{-1})}{(10^3 \text{ mg g}^{-1})(Y)(1.50 \text{ mL})} = 5.3 \times 10^{-4} \text{ mol L}^{-1}$$

For this very dilute solution, you can assume that the volume of the solution is the same as the volume of the solvent, so

$$Y = \frac{(15.3 \text{ mg})(10^3 \text{ mL L}^{-1})}{(10^3 \text{ mg g}^{-1})(1.50 \text{ mL})(5.3 \times 10^{-4} \text{ mol L}^{-1})} = 1.9 \times 10^4 \text{ g mol}^{-1}$$

note: The osmotic pressure of this dilute protein solution is quite large and measurable. The measurement of osmotic pressure is a valuable way of determining the molar masses of large molecules, including those of biological importance.

PROBLEM 11-14 What is the expected freezing point of a solution of 1.50 g of KBr in 7.6 g of water? K_f for $H_2O = 1.86° \text{ kg mol}^{-1}$. Assume complete dissociation.

Solution: KBr is an ionic solid that will have an i factor of 2 if it is completely ionized in solution to K^+ and Br^-. Using the fp depression equation with the i factor (Eq. 11-4a),

$$\Delta T_f = iK_f m$$

$$= 2(1.86° \text{ kg mol}^{-1})\left(\frac{1.50 \text{ g KBr}}{119 \text{ g mol}^{-1} \text{ KBr}}\right)\left(\frac{10^3 \text{ g kg}^{-1}}{7.6 \text{ g}}\right) = 6.2°$$

Since pure water freezes at 0.0°C, the expected freezing point of the solution is $-6.2°$C.

PROBLEM 11-15 A 0.10 m solution of H_3PO_4 in water is observed to freeze at $-0.24°$C (K_f for water is 1.86° kg mol^{-1}). Determine the degree of dissociation in the 0.10 m H_3PO_4 solution. (Assume the complete dissociation would yield three H^+ ions and one PO_4^{3-} ion.)

Solution: In general, $\Delta T_f = iK_f m$; and if you know T_f, K_f, and m you can calculate i:

$$i = \frac{\Delta T_f}{K_f m} = \frac{0.00° - (-0.24°)}{(1.86° \text{ kg mol}^{-1})(0.10 \text{ } m)} = 1.3$$

Now use the degree of ionization equation (11-6) to calculate the degree of dissociation:

$$\alpha = \frac{i-1}{v-1}$$

$$= \frac{1.3-1}{4-1} = 0.10$$

So the H_3PO_4 is $(0.10 \times 100\%) = 10\%$ ionized.

PROBLEM 11-16 The solubility of ammonia (NH_3) in water at 20°C and 1.00 atm is 0.531 g per 1.00 g of water. The density of the solution obtained is 0.882 g/mL. Calculate the **(a)** molarity, **(b)** molality, and **(c)** mole fraction of ammonia in this solution.

Solution:

(a) The number of moles of NH_3 per gram of solution is

$$\frac{n_{sol}}{g\ soln} = \frac{(0.531\ g\ NH_3)/(17.03\ g\,mol^{-1}\ NH_3)}{(1.00\ g\ H_2O + 0.531\ g\ NH_3)}$$

Convert this mass into volume by using the given density, thus finding the molarity in moles per liter:

$$M = \left(\frac{0.531\ g\ NH_3}{17.03\ g/mol}\right)\left(\frac{0.882\ g\ soln/mL}{1.531\ g\ soln}\right)\left(\frac{10^3\ mL}{1\ L}\right) = 18.0\ mol/L$$

(b) The molality m is

$$m = \frac{n_{sol}}{kg\ solv} = \left(\frac{0.531\ g\ NH_3}{17.03\ g/mol}\right)\left(\frac{10^3\ g/kg}{1.00\ g\ H_2O}\right) = 31.2\ mol/kg$$

(c) Since the mole fraction of NH_3 is

$$X_{NH_3} = \frac{n_{NH_3}}{n_{NH_3} + n_{H_2O}}$$

and

$$n_{NH_3} = \frac{0.531\ g\ NH_3}{17.03\ g/mol} = 0.0312\ mol \qquad n_{H_2O} = \frac{1.00\ g\ H_2O}{18.02\ g/mol} = 0.0555\ mol$$

then

$$X_{NH_3} = \frac{0.0312\ mol}{(0.0312 + 0.0555)mol} = 0.360$$

PROBLEM 11-17   Poly(ethylene oxide) is a synthetic polymer whose formula is $HO(CH_2CH_2O)_nH$, where n is a large integer. You prepare a sample of this polymer for which $n = 180$. Calculate the freezing point depression and the osmotic pressure at 293 K of a solution of 150 mg of the polymer in 5.0 mL of water—and comment on the utility of the two results in characterizing the polymer. (K_f for $H_2O = 1.86°\ kg\,mol^{-1}$; the polymer solution has a volume of 5.1 mL.)

Solution: The freezing point depression is $\Delta T_f = K_f m$. Now the molar mass of the polymer $HO(CH_2CH_2O)_{180}H$ is the molar mass of the compound $C_{360}H_{722}O_{181}$, which is quite large: $7.95 \times 10^3\ g\,mol^{-1}$. Remembering that the density of water is close to $1.00\ g\,mL^{-1}$, calculate m first:

$$m = \frac{(150\ mg)(10^3\ g\,kg^{-1})}{(10^3\ mg\,g^{-1})(7.95 \times 10^3\ g\,mol^{-1})(5.0\ mL)(1.00\ g\,mL^{-1})} = 3.77 \times 10^{-3}\ mol\,kg^{-1}$$

Then

$$\Delta T_f = (1.86°\ kg\,mol^{-1})(3.77 \times 10^{-3}\ mol\,kg^{-1}) = 7.0° \times 10^{-3} \quad \text{(or 7.0 millidegrees)}$$

$$\Pi = MRT = \frac{(150\ mg)(10^3\ mL\,L^{-1})}{(10^3\ mg\,g^{-1})(7.95 \times 10^3\ g\,mol^{-1})(5.1\ mL)}(0.0821\ L\,atm\,mol^{-1}K^{-1})(293\ K)$$

$$= 0.089\ atm$$

Comment: 7.0×10^{-3} degrees would be hard to measure. But 0.089 atm, which equals 68 torr, is quite easy to measure. So osmotic pressure is more useful in characterizing polymer solutions than is freezing point depression.

PROBLEM 11-18   Magnesium chloride ($MgCl_2$) has an i factor of 3. Calculate the molarity of (a) Mg^{2+} and (b) Cl^- in a solution containing 10.86 g $MgCl_2$/L.

Solution: The i factor of 3 indicates that $MgCl_2$ is completely ionized in solution. Each mole of $MgCl_2$ gives rise to 1 mol of Mg^{2+} and 2 mol of Cl^- in solution. The molar mass of $MgCl_2$ is 95.21 g/mol.

(a) For Mg^{2+}

$$M = \frac{10.86\ g\ MgCl_2}{(95.21\ g/mol\ MgCl_2)L} \times \frac{1\ mol\ Mg^{2+}}{1\ mol\ MgCl_2} = 0.1141\ mol\ Mg^{2+}/L$$

(b) For Cl^-

$$M = \frac{10.86\ g\ MgCl_2}{(95.21\ g/mol\ MgCl_2)L} \times \frac{2\ mol\ Cl^-}{1\ mol\ MgCl_2} = 0.2281\ mol\ Cl^-/L$$

(Of course, the formula $MgCl_2$ tells you at once that if this compound ionizes completely, the Cl^- molarity must be exactly twice the Mg^{2+} molarity.)

PROBLEM 11-19   What volume of each of the following solutions would you have to take to obtain 10.0 mmol of Li^+ if each of the salts is completely ionized in solution?

$$\text{(a) } 0.250 \, M \text{ LiCl} \qquad \text{(b) } 0.382 \, M \text{ Li}_2\text{SO}_4 \qquad \text{(c) } 1.05 \times 10^{-4} \, M \text{ LiHCO}_3$$

Solution: Since $M = \text{mol/L} = \text{mmol/mL}$, then $\text{mL} = \text{mmol}/M$—but be careful about M.

(a) $\text{Volume} = \dfrac{10.0 \text{ mmol Li}^+}{0.250 \, M \text{ LiCl}} \times \dfrac{1 \text{ mol LiCl}}{1 \text{ mol Li}^+} = 40.0 \text{ mL}$

(b) $\text{Volume} = \dfrac{10.0 \text{ mmol Li}^+}{0.382 \, M \text{ Li}_2\text{SO}_4} \times \dfrac{1 \text{ mol Li}_2\text{SO}_4}{2 \text{ mol Li}^+} = 13.1 \text{ mL}$

(c) $\text{Volume} = \dfrac{10.0 \text{ mmol Li}^+}{1.05 \times 10^{-4} \, M \text{ LiHCO}_3} \times \dfrac{1 \text{ mol LiHCO}_3}{1 \text{ mol Li}^+} = 9.52 \times 10^4 \text{ mL}$

or more conveniently

$$\frac{9.52 \times 10^4 \text{ mL}}{10^3 \text{ mL/L}} = 95.2 \text{ L}$$

note: If M for a salt is given and you need to consider only one of its ions, you must always include an explicit statement about the stoichiometric relationship between the salt and the ion in question.

PROBLEM 11-20   What is the vapor pressure of a saturated solution of magnesium nitrate $Mg(NO_3)_2$ in water at 20°C? The solubility of $Mg(NO_3)_2$ is 70.0 g/100.0 g H_2O. The vapor pressure of water at 20°C is 2.34 kN m^{-2}; $Mg(NO_3)_2$ is a strong electrolyte and has an i factor of 3.

Solution: Solution vapor pressure is given by Raoult's law (Eq. 11-2):

$$p_A = X_A p_A^\circ$$

You're given the value of p_A°, the vapor pressure of the solvent. Consequently, you have to calculate the mole fraction of water in the solution, X_A. The molar mass of $Mg(NO_3)_2$ is 148.3 g/mol. If you consider the saturated solution of 70.0 g $Mg(NO_3)_2$/100.0 g H_2O,

$$n_{\text{sol ions}} = \frac{70.0 \text{ g Mg(NO}_3)_2}{148.3 \text{ g/mol}} \times 3 = 1.416 \text{ mol} \qquad \text{(the } i \text{ factor is 3)}$$

$$n_{\text{H}_2\text{O}} = \frac{100.0 \text{ g H}_2\text{O}}{18.02 \text{ g/mol H}_2\text{O}} = 5.549 \text{ mol}$$

$$X_{\text{H}_2\text{O}} = \frac{n_{\text{H}_2\text{O}}}{n_{\text{H}_2\text{O}} + n_{\text{ions}}} = \frac{5.549}{6.966} = 0.797$$

So

$$p_{\text{soln}} = X_{\text{H}_2\text{O}} p_{\text{H}_2\text{O}}^\circ = (0.797)(2.34 \text{ kN m}^{-2}) = 1.86 \text{ kN m}^{-2}$$

Supplementary Exercises

PROBLEM 11-21   The solubility of lithium carbonate (Li_2CO_3) is 1.33 g/100.0 g H_2O at 20°C and 0.85 g/100.0 g H_2O at 80°C. (a) Is the solution process for Li_2CO_3 exo- or endothermic? (b) How many millimoles of Li_2CO_3 will precipitate when 15.0 g of a solution of Li_2CO_3 saturated at 20°C is heated to 80°C?

Answer: (a) exothermic   (b) 0.96 mmol

PROBLEM 11-22   Calculate the mass of ethanol present in 125 g of a 37.5% by mass solution.

Answer: 46.9 g

PROBLEM 11-23   A sample of 8.655 g of sulfamic acid (H_2NSO_3H) is dissolved in water and the solution is carefully diluted to 100.00 mL. What is the molarity of the solution?

Answer: 0.8914 M

PROBLEM 11-24 How many millimoles of H_2SO_4 are there in 27.6 mL of a 0.239 M solution?

Answer: 6.60 mmol

PROBLEM 11-25 What is the molality of a solution of 1.97 g of glucose ($C_6H_{12}O_6$) in 12.55 g of H_2O?

Answer: 0.871 m

PROBLEM 11-26 Calculate the mole fractions of sucrose ($C_{12}H_{22}O_{11}$), acetic acid (CH_3CO_2H), and H_2O in a solution of 5.5 g sucrose, 4.7 g acetic acid, and 10.2 g of H_2O.

Answer: $n_{sucrose} = 0.024$; $n_{acetic} = 0.12$; $n_{H_2O} = 0.86$

PROBLEM 11-27 At 0°C the solubility of nitrogen gas in water is 0.024 L of N_2 per liter of water under a pressure of 1.00 atm of N_2. Calculate the molar solubility of N_2 in water at a pressure of 25.0 atm of N_2. (Neglect any effects of water vapor pressure.)

Answer: 0.027 M

PROBLEM 11-28 The vapor pressure of phosphorus trichloride (PCl_3) is 98 torr at 20°C. What is the vapor pressure of a solution of 2.50 g of $PSBr_3$ (which is nonvolatile at 20°C) in 10.0 g of PCl_3 at 20°C?

Answer: 88 torr

PROBLEM 11-29 The boiling point elevation for a solution of 1.5 g of the dye alizarin in 7.7 g of toluene is 2.70°C. What is the molar mass of alizarin? (K_b for toluene $= 3.33°$ kg mol^{-1}.)

Answer: 2.4×10^2 g mol^{-1}

PROBLEM 11-30 What will be the freezing point of a solution of 5.8 mg of the amino acid cystine (molar mass $= 240.3$ g/mol) in 0.20 mL of water? (K_f for $H_2O = 1.86°$ kg/mol; freezing point of $H_2O = 0.00°C$; density of $H_2O = 1.00$ g/mL.)

Answer: $-0.22°C$

PROBLEM 11-31 A solution of 26.4 mg of paraldehyde, a polymer of formaldehyde having a formula $(CH_2O)_n$, in 3.16 mL of water is found to have an osmotic pressure of 135 torr at 301 K. What is the value of n?

Answer: $n = 38.7$

PROBLEM 11-32 Calcium chloride ($CaCl_2$) is used to melt snow. What is the freezing point of a solution of 250 g of $CaCl_2$ (which is a strong electrolyte; $i = 3$) in 1.0 kg of water? (K_f for $H_2O = 1.86°$ kg/mol.)

Answer: $-13°C$

PROBLEM 11-33 The observed freezing point of a saturated solution of $HgCl_2$, which contains 3.50 g of $HgCl_2$ per 100.0 g of H_2O, is $-0.24°C$. Calculate the i factor for $HgCl_2$. Is $HgCl_2$ ionized in aqueous solution?

Answer: $i = 1.0$; not ionized

PROBLEM 11-34 The density of a 15.0% by mass solution of nitric acid (HNO_3) in water is 1.084 g/mL at 20°C. Calculate the molarity, the molality, and the mole fraction of HNO_3 in this solution.

Answer: $M = 2.58$ mol/L; $m = 2.80$ mol/kg; $X_{HNO_3} = 4.80 \times 10^{-2}$

PROBLEM 11-35 What volume of each of the following would you have to take to obtain 5.00 mmol of Cl^- if each of the salts is completely ionized in solution? (a) 0.150 M NaCl, (b) 0.0502 M AlCl$_3$, (c) 4.38×10^{-3} M MgCl$_2$.

Answer: (a) 33.3 mL   (b) 33.2 mL   (c) 571 mL

SEMESTER EXAM
(Chapters 1–11)

1. The diameter of an atom is 1.8×10^{-8} cm. Determine its diameter in nanometers.

2. Express a pressure of 3.11×10^{-3} atm in torr.

3. Determine the mass of one C_2H_4 molecule in grams.

4. Give the number of neutrons, protons, and electrons in Xe^{2+}. The atomic number is 54 and the mass number is 131.

5. Determine the percentage by mass of carbon present in formaldehyde (CH_2O).

6. Determine the empirical formula of a compound that contains 90.4% Si and 9.6% H by mass.

7. Calculate the number of Cl atoms present in 7.35×10^{-5} g of phosphorus trichloride (PCl_3).

8. Determine the empirical formula of a compound that contains 29.1% Na, 40.5% S, and 30.4% O by mass.

9. Balance the following equations:

   (a) $P(OH)_3 \rightarrow PH_3 + H_3PO_4$

   (b) $Fe_2O_3 + HCl \rightarrow FeCl_2 + Cl_2 + H_2O$

   (c) $H_2SO_4 + KOH \rightarrow K_2SO_4 + H_2O$

10. Aluminum metal and molecular bromine react to form $AlBr_3$. Calculate the maximum mass of $AlBr_3$ (in grams) that can be obtained from 10.0 g of Al and 30.0 g of Br_2.

11. Consider the following reaction:

$$CS_2 + Cl_2 \longrightarrow CCl_4 + SCl_2$$

Calculate the maximum amount of CCl_4 (in moles and grams) that can be obtained from 311 g of CS_2. How many moles of Cl_2 will be consumed in the reaction?

12. Calculate the percentage yield of $CrCl_3$ made from 27.3 g of Cr and excess Cl_2 if 45.0 g of $CrCl_3$ is obtained.

13. Consider the following reaction:

$$I_2O_5 + SF_4 \longrightarrow IF_5 + SO_2$$

Calculate the amount of SF_4 (in grams) required to react completely with 10.00 g of I_2O_5.

14. For a sample containing 4.77×10^{22} molecules of HCF_3 gas, calculate (a) the number of moles in the sample, (b) the mass of one HCF_3 molecule in grams, (c) the number of F atoms in the sample, and (d) the pressure (in atm) exhibited by the sample in a 1.00-L container at 25°C.

15. A pure sample of $COCl_2(g)$ decomposes to $CO(g)$ and $Cl_2(g)$ such that the total pressure is 0.900 atm and the CO pressure is 0.375 atm. Determine the partial pressure of the $Cl_2(g)$ and that of the remaining $COCl_2(g)$.

16. A sample of a gas that behaves ideally is sealed in a container (constant volume) at 27°C and 135 torr and heated to 566°C. Determine the pressure exerted by the gas.

17. Determine the volume (in liters) occupied by 30.5 g of $SO_2(g)$ at 308 K and 675 torr.

18. A 483-cm^3 container holds 2.16 g of a gas at 20°C and 623 torr. Calculate the density of the gas (in grams per liter) and the molar mass of the gas.

19. The average speed of a Br_2 gas molecule at 30°C is 2.00×10^2 m s^{-1}. Calculate the average speed of a Br_2 molecule at 120°C; determine the molar mass of a gas X that effuses 2.5 times as fast as Br_2 under identical conditions.

20. A gaseous compound contains 6.42% H, 25.5% C, and 68.1% S. A 0.1912-g sample of the compound exhibits a pressure of 187 torr at 77°C in a 237-mL container. Determine the molecular formula of the compound.

21. A rigid container holding a mixture of Ne(g) and SF_6(g), each of which has a partial pressure of 1.00×10^{-2} atm, is punctured with a small pin and placed in a vacuum. Calculate the ratio of the rates at which the molecules will escape.

22. Sketch the hydrogen bonding that takes place between acetic acid (CH_3—C—OH) and water (H—O—H).

$$\overset{\text{O}}{\underset{\|}{}}$$

23. The vapor pressure of water is 24 torr at 25°C and 72 torr at 45°C. Determine the vapor pressure of water at 35°C.

24. At high temperatures the structure of solid lanthanum ($Z = 57$) is face-centered cubic (fcc). The edge length of the unit cell is 5.303 Å. Calculate the density of solid La (in g/cm^3) and the radius of the lanthanum atom (in angstroms).

25. Lithium ($Z = 3$) crystallizes in a body-centered cubic (bcc) structure. The unit cell edge length is 3.53 Å. Calculate the density of solid Li (in g/cm^3) and the radius of the lithium atom (in angstroms).

26. For the reaction

$$2H_2S(g) + 3O_2(g) \longrightarrow 2H_2O(l) + 2SO_2(g)$$

the enthalpy change is -1125.1 kJ/mol. The standard enthalpies of formation for $H_2S(g)$ and $H_2O(l)$ are -20.2 kJ/mol and -285.8 kJ/mol, respectively. Calculate the standard enthalpy of formation of $SO_2(g)$.

27. For the reaction

$$3N_2H_4(l) \longrightarrow 4NH_3(g) + N_2(g)$$

the enthalpy change is -336.0 kJ/mol. The standard enthalpy of formation of $NH_3(g)$ is -46.2 kJ/mol. Calculate the standard enthalpy of formation of $N_2H_4(l)$.

28. Calculate the wavelength of light emitted when a hydrogen atom changes from the $n = 5$ to the $n = 3$ state. (The Rydberg constant is 1.0968×10^7 m^{-1}.)

29. Calculate the frequency and wavelength of a photon that possesses 8.23×10^{-19} J of energy. Use 6.63×10^{-34} J s for Planck's constant. Can this photon ionize a neon atom (ionization potential equals 2081 kJ/mol)?

30. Give the ground-state electronic configurations for the following atoms and ions: I, Ti, Fe^{4+}, Na^{2-}, Se, Co, Zn^-, Si^{2+}.

31. Which species has the greatest number of unpaired electrons in the ground state, S^+, S, or S^-? Explain your answer.

32. Give valence-bond formulas (showing all the electrons) for the following: (a) SO_3^{2-}, (b) OF_2, (c) $COCl_2$, (d) PH_3, (e) HCOOH, (f) PF_5, (g) NH_2CH_3.

33. For the following compounds (central atoms italicized), give the valence-bond formula, the formal charge on each nonhydrogen atom present, and the VSEPR geometry; indicate the polarity of each compound: (a) CF_2H_2, (b) NF_2^-, (c) SF_2, (d) SiF_4, (e) $NClO_2$.

34. Give valence-bond formulas for the resonance hybrids of the following (central atoms italicized): (a) HCO_2^-, (b) H_3CNO_2, (c) SO_3.

35. Determine the molecular orbital electronic configuration, the number of electrons in the π_{2p} MOs, the number of unpaired electrons, and the bond order for O_2.

36. For CN, CN^-, and CN^+, place the molecules in order of increasing bond length. The molecular orbital electronic configuration of CN is $1s^2 1s^2 \sigma_{2s}^2 \sigma_{2s}^{*2} \pi_{2p_y}^2 \pi_{2p_z}^2 \sigma_{2p_x}^1$.

37. The solubility of NaCl at $0°C$ is 34.2 g per 100.0 g of H_2O. Calculate the mole fraction X of NaCl and its molality in a saturated solution at $0°C$.

38. Water was added to a 30.0-mL sample of 0.177 M HCl solution until the molarity was 2.12×10^{-2} M. How much water was added?

39. When 0.356 g of $(C_5H_3NO_2)_x$ is dissolved in 14.6 g of CCl_4, the boiling point of the solution is $0.560°C$ higher than the boiling point of pure CCl_4 ($K_b = 5.03°$ kg mol^{-1}). Find x.

40. A 375-mL sample of a sugar solution contains 2.33 g of sugar and has an osmotic pressure of 35.6 torr at $37°C$. Calculate the molar mass of the sugar.

Periodic Table

s Orbitals being filled — d Orbitals being filled — p Orbitals being filled

Transition elements · f Orbitals being filled · Noble gases

Period number = n, the highest occupied electron level.

Group →	IA ns^1	IIA ns^2	IIIB $(n-1)d^1ns^2$	IVB $(n-1)d^2ns^2$	VB $(n-1)d^3ns^2$	VIB $(n-1)d^4ns^1$	VIIB $(n-1)d^5ns^2$	VIIIB $(n-1)d^6ns^2$	VIIIB $(n-1)d^7ns^2$	VIIIB $(n-1)d^8ns^2$	IB $(n-1)d^{10}ns^1$	IIB $(n-1)d^{10}ns^2$	IIIA ns^2np^1	IVA ns^2np^2	VA ns^2np^3	VIA ns^2np^4	VIIA ns^2np^5	VIIIA ns^2np^6
1	1 **H** $1s^1$ 1.0079																	2 **He** $1s^2$ 4.0026
2	3 **Li** $2s^1$ 6.941	4 **Be** $2s^2$ 9.01218											5 **B** $2s^22p^1$ 10.81	6 **C** $2s^22p^2$ 12.011	7 **N** $2s^22p^3$ 14.0067	8 **O** $2s^22p^4$ 15.9994	9 **F** $2s^22p^5$ 18.9984	10 **Ne** $2s^22p^6$ 20.179
3	11 **Na** $3s^1$ 22.9898	12 **Mg** $3s^2$ 24.305											13 **Al** $3s^23p^1$ 26.9815	14 **Si** $3s^23p^2$ 28.086	15 **P** $3s^23p^3$ 30.9738	16 **S** $3s^23p^4$ 32.06	17 **Cl** $3s^23p^5$ 35.453	18 **Ar** $3s^23p^6$ 39.948
4	19 **K** $4s^1$ 39.098	20 **Ca** $4s^2$ 40.08	21 **Sc** $3d^14s^2$ 44.959	22 **Ti** $3d^24s^2$ 47.90	23 **V** $3d^34s^2$ 50.9414	24 **Cr** $3d^54s^1$ 51.996	25 **Mn** $3d^54s^2$ 54.938	26 **Fe** $3d^64s^2$ 55.847	27 **Co** $3d^74s^2$ 58.9332	28 **Ni** $3d^84s^2$ 58.70	29 **Cu** $3d^{10}4s^1$ 63.546	30 **Zn** $3d^{10}4s^2$ 65.38	31 **Ga** $4s^24p^1$ 69.72	32 **Ge** $4s^24p^2$ 72.59	33 **As** $4s^24p^3$ 74.9216	34 **Se** $4s^24p^4$ 78.96	35 **Br** $4s^24p^5$ 79.904	36 **Kr** $4s^24p^6$ 83.80
5	37 **Rb** $5s^1$ 85.4678	38 **Sr** $5s^2$ 87.62	39 **Y** $4d^15s^2$ 88.9059	40 **Zr** $4d^25s^2$ 91.22	41 **Nb** $4d^45s^1$ 92.9064	42 **Mo** $4d^55s^1$ 95.94	43 **Tc** $4d^55s^2$ (97)	44 **Ru** $4d^75s^1$ 101.07	45 **Rh** $4d^85s^1$ 102.905	46 **Pd** $4d^{10}$ 106.4	47 **Ag** $4d^{10}5s^1$ 107.868	48 **Cd** $4d^{10}5s^2$ 112.40	49 **In** $5s^25p^1$ 114.82	50 **Sn** $5s^25p^2$ 118.69	51 **Sb** $5s^25p^3$ 121.75	52 **Te** $5s^25p^4$ 127.60	53 **I** $5s^25p^5$ 126.904	54 **Xe** $5s^25p^6$ 131.30
6	55 **Cs** $6s^1$ 132.905	56 **Ba** $6s^2$ 137.33	57 **La*** $5d^16s^2$ 138.905	72 **Hf** $4f^{14}5d^26s^2$ 178.49	73 **Ta** $5d^36s^2$ 180.948	74 **W** $5d^46s^2$ 183.85	75 **Re** $5d^56s^2$ 186.207	76 **Os** $5d^66s^2$ 190.2	77 **Ir** $5d^76s^2$ 192.22	78 **Pt** $5d^96s^1$ 195.09	79 **Au** $5d^{10}6s^1$ 196.967	80 **Hg** $5d^{10}6s^2$ 200.59	81 **Tl** $6s^26p^1$ 204.37	82 **Pb** $6s^26p^2$ 207.19	83 **Bi** $6s^26p^3$ 208.980	84 **Po** $6s^26p^4$ (209)	85 **At** $6s^26p^5$ (210)	86 **Rn** $6s^26p^6$ (222)
7	87 **Fr** $7s^1$ (223)	88 **Ra** $7s^2$ (226)	89 **Ac†** $6d^17s^2$ (227)	104 **Rf** (260)	105 **Ha** (260)													

* Lanthanides ~ $4f^{\,n}5d^{0-1}6s^2$

58 **Ce** $4f^15d^16s^2$ 140.12	59 **Pr** $4f^35d^06s^2$ 140.907	60 **Nd** $4f^45d^06s^2$ 144.24	61 **Pm** $4f^55d^06s^2$ (145)	62 **Sm** $4f^65d^06s^2$ 150.35	63 **Eu** $4f^75d^06s^2$ 151.96	64 **Gd** $4f^75d^16s^2$ 157.25	65 **Tb** $4f^95d^06s^2$ 158.925	66 **Dy** $4f^{10}5d^06s^2$ 162.50	67 **Ho** $4f^{11}5d^06s^2$ 164.930	68 **Er** $4f^{12}5d^06s^2$ 167.26	69 **Tm** $4f^{13}5d^06s^2$ 168.934	70 **Yb** $4f^{14}5d^06s^2$ 173.04	71 **Lu** $4f^{14}5d^16s^2$ 174.97

† Actinides ~ $5f^{\,n}6d^{0-1}7s^2$

90 **Th** $5f^06d^27s^2$ 232.038	91 **Pa** $5f^26d^17s^2$ (231)	92 **U** $5f^36d^17s^2$ 238.03	93 **Np** $5f^46d^17s^2$ (237)	94 **Pu** $5f^66d^07s^2$ (244)	95 **Am** $5f^76d^07s^2$ (243)	96 **Cm** $5f^76d^17s^2$ (247)	97 **Bk** $5f^96d^07s^2$ (247)	98 **Cf** $5f^{10}6d^07s^2$ (251)	99 **Es** $5f^{11}6d^07s^2$ (254)	100 **Fm** $5f^{12}6d^07s^2$ (257)	101 **Md** $5f^{13}6d^07s^2$ (258)	102 **No** $5f^{14}6d^07s^2$ (255)	103 **Lr** $5f^{14}6d^07s^2$ (260)

Solutions to Semester Exam

1.
$$(1.8 \times 10^{-8} \text{ cm})\left(\frac{1 \text{ m}}{10^2 \text{ cm}}\right)\left(\frac{10^9 \text{ nm}}{1 \text{ m}}\right) = 0.18 \text{ nm}$$

2.
$$(3.11 \times 10^{-3} \text{ atm})\left(\frac{760 \text{ torr}}{1 \text{ atm}}\right) = 2.36 \text{ torr}$$

3. The molar mass of C_2H_4 is $2(12.0) + 4(1.0) = 28.0$ g/mol. A mole of C_2H_4 molecules contains Avogadro's number of molecules:

$$\text{mass of 1 molecule} = \frac{28.0 \text{ g/mol}}{6.02 \times 10^{23} \text{ molecules/mol}}$$
$$= 4.65 \times 10^{-23} \text{ g/molecule}$$

4. The number of neutrons is given by $A - Z$, the mass number minus the atomic number, so $131 - 54 = 77$ neutrons. The number of protons is given by Z, so there are 54 protons. In the neutral Xe atom, the number of electrons equals the number of protons, so neutral Xe has 54 electrons; Xe^{2+} has lost two electrons, leaving 52.

5. The molar mass of CH_2O is $12.0 + 2(1.0) + 16.0 = 30.0$ g/mol.

$$\%C = \frac{12.0 \text{ g/mol}}{30.0 \text{ g/mol}} \times 100\% = 40.0\%$$

6. Calculate the number of moles of Si and H atoms in 100.0 g of compound:

$$n_{Si} = \frac{90.4 \text{ g Si}}{28.04 \text{ g Si/mol Si atoms}} = 3.22 \text{ mol Si atoms}$$

$$n_H = \frac{9.6 \text{ g H}}{1.008 \text{ g H/mol H atoms}} = 9.5 \text{ mol H atoms}$$

Determine the relative number of moles of each atom:

$$\text{relative number of moles of Si} = \frac{3.22}{3.22} = 1.00$$

$$\text{relative number of moles of H} = \frac{9.5}{3.22} = 2.95 \cong 3$$

The integer ratio is $1\text{Si} : 3\text{H}$, so the empirical formula is SiH_3.

7. The molar mass of PCl_3 is $30.97 + 3(35.45) = 137.32$ g/mol.

$$\text{No. Cl atoms} = \left(\frac{7.35 \times 10^{-5} \text{ g PCl}_3}{137.32 \text{ g/mol PCl}_3}\right)\left(\frac{6.02 \times 10^{23} \text{ molecules PCl}_3}{1 \text{ mol PCl}_3}\right)\left(\frac{3 \text{ Cl atoms}}{1 \text{ molecule PCl}_3}\right)$$
$$= 9.67 \times 10^{17}$$

Only 3 significant figures are justified because the mass of PCl_3 was given to 3 sig. figs.

8. Determine the number of moles of each element in 100 g of compound:

$$n_{Na} = \frac{29.1 \text{ g Na}}{22.99 \text{ g Na/mol Na atoms}} = 1.27 \text{ mol Na atoms}$$

$$n_S = \frac{40.5 \text{ g S}}{32.06 \text{ g S/mol S atoms}} = 1.26 \text{ mol S atoms}$$

$$n_O = \frac{30.4 \text{ g O}}{16.00 \text{ g O/mol O atoms}} = 1.90 \text{ mol O atoms}$$

Determine the relative number of moles of each atom:

$$\text{relative number of moles of Na} = \frac{1.27}{1.26} = 1.01$$

$$\text{relative number of moles of S} = \frac{1.26}{1.26} = 1.00$$

$$\text{relative number of moles of O} = \frac{1.90}{1.26} = 1.51$$

Multiplying by 2 will give integer multiples: 2 for Na and S, 3 for O. The empirical formula is $Na_2S_2O_3$.

9. (a) $4P(OH)_3 \rightarrow PH_3 + 3H_3PO_4$

   (b) $Fe_2O_3 + 6HCl \rightarrow 2FeCl_2 + Cl_2 + 3H_2O$

   (c) $H_2SO_4 + 2KOH \rightarrow K_2SO_4 + 2H_2O$

10. Write the balanced equation for the reaction and determine the molar relationships of the reactants and product:

Equation:	$2Al$	+	$3Br_2$	$\longrightarrow$	$2AlBr_3$
g/mol:	27.0		$2(79.9) = 159.8$		$27.0 + 3(79.9) = 266.7$
Moles of reactant available:	$\frac{10.0}{27.0} = 0.370$		$\frac{30.0}{159.8} = 0.188$		

Determine how much Br_2 would be required to react completely with the available Al:

$$(0.370 \text{ mol Al})\left(\frac{3 \text{ mol Br}_2}{2 \text{ mol Al}}\right) = 0.555 \text{ mol Br}_2$$

Only 0.188 mol of Br_2 is available, so Br_2 is the limiting reagent. Now determine how much $AlBr_3$ can be produced stoichiometrically from the limited Br_2 available:

$$(0.188 \text{ mol Br}_2)\left(\frac{2 \text{ mol AlBr}_3}{3 \text{ mol Br}_2}\right)\left(\frac{266.7 \text{ g AlBr}_3}{1 \text{ mol AlBr}_3}\right) = 33.4 \text{ g AlBr}_3$$

11. Proceed as in Problem 10:

Equation:	CS_2	+	$4Cl_2$	$\longrightarrow$	CCl_4	+	$2SCl_2$
g/mol:	76.13				153.81		

Each mole of CS_2 consumed will produce 1 mol of CCl_4, so

$$\text{yield of } CCl_4 = \left(\frac{311 \text{ g CS}_2}{76.13 \text{ g/mol CS}_2}\right)\left(\frac{1 \text{ mol CCl}_4}{1 \text{ mol CS}_2}\right) = 4.09 \text{ mol}$$

$$= (4.09 \text{ mol})(153.81 \text{ g mol}^{-1}) = 629 \text{ g}$$

Four moles of Cl_2 will be consumed for each mole of CCl_4 produced, so

$$\text{consumption of } Cl_2 = (4.09 \text{ mol})\left(\frac{4 \text{ mol Cl}_2}{1 \text{ mol CCl}_4}\right) = 16.4 \text{ mol}$$

12. Balance the equation and determine the molar relationships of the reactants and product:

Equation:	$2Cr$	+	$3Cl_2$	$\longrightarrow$	$2CrCl_3$
g/mol:	52.00				158.35

Calculate the theoretical yield (the maximum amount of $CrCl_3$ that can be made):

$$\text{theoretical yield} = \left(\frac{27.3 \text{ g Cr}}{52.00 \text{ g/mol Cr}}\right)\left(\frac{2 \text{ mol CrCl}_3}{2 \text{ mol Cr}}\right)\left(\frac{158.35 \text{ g CrCl}_3}{1 \text{ mol CrCl}_3}\right) = 83.1 \text{ g}$$

$$\% \text{ yield} = \frac{\text{actual yield}}{\text{theoretical yield}} \times 100\% = \frac{45.0 \text{ g}}{83.1 \text{ g}} \times 100\% = 54.1\%$$

13. Balance the equation and determine the molar relationships:

$$\text{Equation:} \quad 2I_2O_5 \quad + \quad 5SF_4 \quad \longrightarrow \quad 4IF_5 \quad + \quad 5SO_2$$

$$\text{g/mol:} \quad \ \ 333.8 \qquad \quad 108.1$$

$$\text{required SF}_4 = \left(\frac{10.00 \text{ g } I_2O_5}{333.8 \text{ g/mol } I_2O_5}\right)\left(\frac{5 \text{ mol SF}_4}{2 \text{ mol } I_2O_5}\right)\left(\frac{108.1 \text{ g SF}_4}{1 \text{ mol SF}_4}\right)$$

$$= 8.096 \text{ g}$$

14. (a)

$$n = \frac{4.77 \times 10^{22} \text{ molecules}}{6.02 \times 10^{23} \text{ molecules/mol}} = 7.92 \times 10^{-2} \text{ mol}$$

(b) The molar mass of HCF_3 is $1.0 + 12.0 + 3(19.0) = 70.0$ g/mol, so

$$\text{mass of 1 HCF}_3 \text{ molecule} = \frac{70.0 \text{ g mol}^{-1}}{6.02 \times 10^{23} \text{ mol}^{-1}} = 1.16 \times 10^{-22} \text{ g}$$

(c)

$$(4.77 \times 10^{22} \text{ molecules HCF}_3)\left(\frac{3 \text{ F atoms}}{1 \text{ molecule HCF}_3}\right) = 1.43 \times 10^{23} \text{ F atoms}$$

(d) Use the ideal gas equation:

$$P = \frac{nRT}{V} = \frac{(7.92 \times 10^{-2} \text{ mol})(0.0821 \text{ L atm mol}^{-1}\text{K}^{-1})(298 \text{ K})}{1.00 \text{ L}}$$

$$= 1.94 \text{ atm}$$

15. Write the balanced equation:

$$COCl_2(g) \longrightarrow CO(g) + Cl_2(g)$$

From the stoichiometry, the partial pressures of CO and Cl_2 must be equal, so

$$p_{Cl_2} = p_{CO} = 0.375 \text{ atm}$$

Use Dalton's law to determine the partial pressure of the remaining $COCl_2$:

$$P_{tot} = p_{COCl_2} + p_{CO} + p_{Cl_2}$$
$$0.900 = p_{COCl_2} + 2(0.375)$$
$$p_{COCl_2} = 0.900 - 2(0.375) = 0.150 \text{ atm}$$

16. At constant volume, the combined gas law reduces to $P_1/T_1 = P_2/T_2$:

$$P_2 = P_1\left(\frac{T_2}{T_1}\right) = (135 \text{ torr})\left(\frac{273 + 566 \text{ K}}{273 + 27 \text{ K}}\right) = 378 \text{ torr}$$

17. Use the ideal gas law:

$$V = \frac{nRT}{P} = \left(\frac{30.5 \text{ g SO}_2}{64.06 \text{ g mol}^{-1} \text{ SO}_2}\right)(0.0821 \text{ L atm mol}^{-1}\text{K}^{-1})\left(\frac{308 \text{ K}}{675 \text{ torr}}\right)\left(\frac{760 \text{ torr}}{1 \text{ atm}}\right)$$

$$= 13.6 \text{ L}$$

18.

$$d = \frac{\text{mass}}{\text{volume}} = \left(\frac{2.16 \text{ g}}{483 \text{ cm}^3}\right)\left(\frac{1000 \text{ cm}^3}{1 \text{ L}}\right) = 4.47 \text{ g/L}$$

Use the ideal gas law, substituting density for mass/volume:

$$PV = nRT = \frac{m}{M}RT$$

$$M = \left(\frac{m}{V}\right)\left(\frac{RT}{P}\right) = d\frac{RT}{P}$$

$$= (4.47 \text{ g L}^{-1})\left(\frac{62.4 \text{ L torr}}{\text{mol K}}\right)\left(\frac{(20 + 273) \text{ K}}{623 \text{ torr}}\right)$$

$$= 131 \text{ g/mol}$$

19. From the kinetic theory, the average speed ($\bar{v}$) of a molecule is directly proportional to the square root of the absolute temperature:

$$\frac{\bar{v}_2}{\bar{v}_1} = \sqrt{\frac{T_2}{T_1}}$$

$$\bar{v}_2 = \bar{v}_1 \sqrt{\frac{T_2}{T_1}} = (2.00 \times 10^2 \text{ m s}^{-1}) \sqrt{\frac{(273 + 120) \text{ K}}{(273 + 30) \text{ K}}}$$

$$= 2.28 \times 10^2 \text{ m s}^{-1}$$

From Graham's law, the rates of effusion are inversely proportional to the square roots of molar masses:

$$\frac{M_{Br_2}}{M_X} = \left(\frac{r_X}{r_{Br_2}}\right)^2 \quad \text{and} \quad M_X = M_{Br_2}\left(\frac{r_{Br_2}}{r_X}\right)^2$$

You are given that $r_X = 2.5 r_{Br_2}$, so $r_{Br_2}/r_X = 1/2.5$, and

$$M_X = (159.8 \text{ g/mol})\left(\frac{1}{2.5}\right)^2 = 2.6 \text{ g/mol}$$

20. You need to determine the empirical formula and molar mass of the compound to derive the molecular formula:

Element	g/100 g	mol/100 g	Relative composition
H	6.42	$\frac{6.42}{1.008} = 6.37$	$\frac{6.37}{2.12} = 3.00$
C	25.5	$\frac{25.5}{12.01} = 2.12$	$\frac{2.12}{2.12} = 1.00$
S	68.1	$\frac{68.1}{32.06} = 2.12$	$\frac{2.12}{2.12} = 1.00$

The empirical formula is CH_3S, which has a mass of 47.1 g/mol.

Use the ideal gas law to find the molar mass:

$$M = \frac{mRT}{PV} = \frac{(0.1912 \text{ g})(62.4 \text{ L torr mol}^{-1}\text{K}^{-1})(273 + 77 \text{ K})(10^3 \text{ mL L}^{-1})}{(187 \text{ torr})(237 \text{ mL})}$$

$$= 94.2 \text{ g/mol}$$

There are $94.2/47.1 = 2.00$ empirical formula units per molecular formula unit, so the molecular formula is $(CH_3S)_2$ or $C_2H_6S_2$.

21. Use Graham's law:

$$\frac{r_{Ne}}{r_{SF_6}} = \frac{\sqrt{M_{SF_6}}}{\sqrt{M_{Ne}}} = \sqrt{\frac{146}{20.2}} = \sqrt{7.23} = 2.69$$

22. Hydrogen atoms that are attached to a strongly electronegative atom in one molecule can form hydrogen bonds with electronegative atoms in another molecule:

23. The equation for the variation of vapor pressure with temperature is

$$\log P_{vap} = \frac{-a}{T} + b$$

You are given P_{vap} at two temperatures, so you can solve two equations simultaneously to determine a and b:

At 25°C $\log 24 = 1.38 = \dfrac{-a}{298} + b$ At 45°C $\log 72 = 1.86 = \dfrac{-a}{318} + b$

So

$$298(1.38) = -a + 298b$$
$$\underline{-318(1.86) = a - 318b}$$
$$411.2 - 591.5 = -20b$$
$$b = 9.015$$
$$a = -318(1.86) + 318(9.015) = 2275$$

Now solve for P_{vap} at 35°C:

$$\log P_{vap} = \frac{-2275}{308} + 9.015$$

$$P_{vap} = \text{antilog}(1.629)$$
$$= 43 \text{ torr}$$

24. The atomic mass of La is 138.9; solids with *fcc* structures contain 4 atoms per unit cell. First, convert edge length a to centimeters:

$$a = (5.303 \text{ Å})(10^{-8} \text{ cm/Å}) = 5.303 \times 10^{-8} \text{ cm}$$

Now determine the density:

$$d = \frac{m}{V} = \frac{4(138.9 \text{ g mol}^{-1})}{(6.02 \times 10^{23} \text{ mol}^{-1})(5.303 \times 10^{-8} \text{ cm})^3}$$
$$= 6.19 \text{ g/cm}^3$$

For an *fcc* structure, $4r = \sqrt{2}\,a$, and

$$r = \frac{\sqrt{2}\,a}{4} = \frac{\sqrt{2}(5.303 \text{ Å})}{4} = 1.87 \text{ Å}$$

25. The atomic mass of Li is 6.94; solids with *bcc* structures contain 2 atoms per unit cell. Proceed as in Problem 24.

$$d = \frac{m}{V} = \frac{2(6.94 \text{ g mol}^{-1})}{(6.02 \times 10^{23} \text{ mol}^{-1})(3.53 \text{ Å})^3}\left(\frac{10^8 \text{ Å}}{1 \text{ cm}}\right)^3$$
$$= 0.524 \text{ g/cm}^3$$

For *bcc* structures, $d_b = \sqrt{3}\,a$; the body diagonal is 4 times the radius, so $d_b = 4r$, and

$$4r = (\sqrt{3})(3.53 \text{ Å})$$
$$r = 1.53 \text{ Å}$$

26. First note that $\Delta H_f^\circ(O_2)$ is zero. Now use Hess' law:

$$\Delta H_r^\circ = \sum(\Delta H_f^\circ \text{ products}) - \sum(\Delta H_f^\circ \text{ reactants})$$

Therefore,

$$\Delta H_r^\circ = 2\,\Delta H_f^\circ(H_2O) + 2\,\Delta H_f^\circ(SO_2) - [3\,\Delta H_f^\circ(O_2) + 2\,\Delta H_f^\circ(H_2S)]$$
$$-1125.1 = 2(-285.8) + 2\,\Delta H_f^\circ(SO_2) - [3(0) + 2(-20.2)]$$
$$\Delta H_f^\circ(SO_2) = \frac{-1125.1 + 571.6 - 40.4}{2}$$
$$= -297.0 \text{ kJ/mol}$$

27. Proceed as in Problem 26:

$$\Delta H_r^\circ = 4\,\Delta H_f^\circ(NH_3) + \Delta H_f^\circ(N_2) - 3\,\Delta H_f^\circ(N_2H_4)$$

$$-336.0 = 4(-46.2) + 0 - 3\,\Delta H_f^\circ(N_2H_4)$$

$$\Delta H_f^\circ(N_2H_4) = \frac{4(-46.2) + 336.0}{3} = 50.4\ \text{kJ/mol}$$

28. Use the equation for the hydrogen spectrum:

$$\frac{1}{\lambda} = R_H\left(\frac{1}{n_f^2} - \frac{1}{n_i^2}\right)$$

$$= (1.0968 \times 10^7\ \text{m}^{-1})\left(\frac{1}{3^2} - \frac{1}{5^2}\right)$$

$$= 7.7995 \times 10^5\ \text{m}^{-1}$$

$$\lambda = (7.7995 \times 10^5\ \text{m}^{-1})^{-1} = 1.2821 \times 10^{-6}\ \text{m}$$

29. Use Planck's equation, $\Delta E = h\nu$, with ΔE equal to the total energy of the photon:

$$\nu = \frac{\Delta E}{h} = \frac{8.23 \times 10^{-19}\ \text{J}}{6.63 \times 10^{-34}\ \text{J s}} = 1.24 \times 10^{15}\ \text{s}^{-1}$$

$$\lambda = \frac{c}{\nu} = \frac{3.00 \times 10^8\ \text{m s}^{-1}}{1.24 \times 10^{15}\ \text{s}^{-1}} = 2.42 \times 10^{-7}\ \text{m}$$

The energy needed to ionize a Ne atom is

$$\frac{2081 \times 10^3\ \text{J mol}^{-1}}{6.02 \times 10^{23}\ \text{atoms mol}^{-1}} = 3.46 \times 10^{-18}\ \text{J}$$

The photon has less energy than required, so it will not ionize a Ne atom.

30.

Species	Z	Number of e^-	Configuration
I	53	53	$[Kr]4d^{10}5s^25p^5$
Ti	22	22	$[Ar]4s^23d^2$
Fe^{4+}	26	$26 - 4 = 22$	$[Ar]3d^4$
Na^{2-}	11	$11 + 2 = 13$	$[Ne]3s^23p^1$
Se	34	34	$[Ar]4s^23d^{10}4p^4$
Co	27	27	$[Ar]4s^23d^7$
Zn^-	30	$30 + 1 = 31$	$[Ar]4s^23d^{10}4p^1$
Si^{2+}	14	$14 - 2 = 12$	$[Ne]3s^2$

31. Write the ground-state electronic configurations. Hund's rule gives the number of unpaired electrons:

	Number of e^-	Configuration	Unpaired e^-
S^+	$16 - 1 = 15$	$[Ne]3s^23p^3$	3
S	16	$[Ne]3s^23p^4$	2
S^-	$16 + 1 = 17$	$[Ne]3s^23p^5$	1

So S^+ has the greatest number of unpaired electrons.

32. (a) (b) (c)

(d) (e) (f) (g)

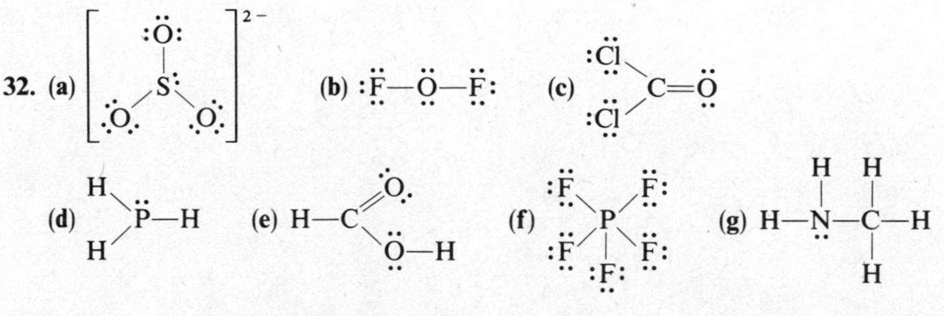

33.

	Valence-bond formula	Formal charge	VSEPR geometry	Polarity
(a)	H $:\ddot{F}-C-H$ $:\ddot{F}:$	C: $4 - \frac{1}{2}(8) - 0 = 0$ F: $7 - \frac{1}{2}(2) - 6 = 0$	tetrahedral AX_4	polar
(b)	$\left[:\ddot{F}-\ddot{N}-\ddot{F}:\right]^-$	N: $5 - \frac{1}{2}(4) - 4 = -1$ F: $7 - \frac{1}{2}(2) - 6 = 0$	angular AX_2E_2	polar
(c)	$:\ddot{F}-\ddot{O}-\ddot{F}:$	S: $6 - \frac{1}{2}(4) - 4 = 0$ F: $7 - \frac{1}{2}(2) - 6 = 0$	angular AX_2E_2	polar
(d)	$:\ddot{F}$ $\ddot{F}:$ S $:\ddot{F}$ $\ddot{F}:$	Si: $4 - \frac{1}{2}(8) - 0 = 0$ F: $7 - \frac{1}{2}(2) - 6 = 0$	tetrahedral AX_4	nonpolar
(e)	$:\ddot{O}$ $N-\ddot{C}l:$ $\ddot{O}$	N: $5 - \frac{1}{2}(8) - 0 = +1$ Cl: $7 - \frac{1}{2}(2) - 6 = 0$ —O: $6 - \frac{1}{2}(2) - 6 = -1$ =O: $6 - \frac{1}{2}(4) - 4 = 0$	planar AX_3	polar

34. (a) $\left[H-C\genfrac{}{}{0pt}{}{\diagup\ddot{O}}{\diagdown\ddot{O}:}\right]^- \longleftrightarrow \left[H-C\genfrac{}{}{0pt}{}{\diagup\ddot{O}:}{\diagdown\ddot{O}}\right]^-$ **(b)** $H-C\genfrac{}{}{0pt}{}{}{}-N\genfrac{}{}{0pt}{}{\diagup\ddot{O}:}{\diagdown\ddot{O}} \longleftrightarrow H-C\genfrac{}{}{0pt}{}{}{}-N\genfrac{}{}{0pt}{}{\diagup\ddot{O}}{\diagdown\ddot{O}:}$

(c) $\genfrac{}{}{0pt}{}{:O:}{\ddot{O}\diagup S \diagdown \ddot{O}:} \longleftrightarrow \genfrac{}{}{0pt}{}{:\ddot{O}:}{:\ddot{O}\diagup S \diagdown O} \longleftrightarrow \genfrac{}{}{0pt}{}{:\ddot{O}:}{\ddot{O}\diagup S \diagdown \ddot{O}:}$

35. Diatomic oxygen has 16 electrons, distributed as follows:

$$\underbrace{1s^2 1s^2}_{\text{nonbonding}} \sigma_{2s}^2 \sigma_{2s}^{*2} \pi_{2p_y}^2 \pi_{2p_z}^2 \sigma_{2p_x}^2 \pi_{2p_y}^{*1} \pi_{2p_z}^{*1}$$

There are four electrons in the π_{2p} orbitals and two unpaired electrons in the π_{2p}^* orbitals. The bond order is

$$\text{bond order} = \frac{1}{2}\left[\left(\begin{array}{c}\text{No. of } e^- \text{ in}\\\text{bonding MOs}\end{array}\right) - \left(\begin{array}{c}\text{No. of } e^- \text{ in}\\\text{antibonding MOs}\end{array}\right)\right]$$

$$= \frac{1}{2}(8-4) = 2$$

36. The bond order of CN is $\frac{1}{2}(7-2) = \frac{5}{2}$. For CN^-, the additional e^- goes into the σ_{2p_x} MO, a bonding MO, and the bond order is $\frac{1}{2}(8-2) = 3$. For CN^+, the σ_{2p_x} MO is empty and the bond order is $\frac{1}{2}(6-2) = 2$. The larger the bond order, the shorter the bond length; in order of increasing bond length, $CN^+ < CN < CN^-$.

37. $$X_{NaCl} = \frac{n_{NaCl}}{n_{H_2O} + n_{NaCl}} = \frac{34.2/58.44}{(100.0/18.02) + (34.2/58.44)} = 9.54 \times 10^{-2}$$

$$m = \frac{n_{sol}}{\text{kg solv}} = \frac{(34.2 \text{ g})(10^3 \text{ g kg}^{-1})}{(58.44 \text{ g mol}^{-1})(100.0 \text{ g})} = 5.85 \text{ mol kg}^{-1}$$

38. This is a dilution problem, so $M_1 V_1 = M_2 V_2$ and

$$V_2 = \frac{(0.177 \text{ M})(30.0 \text{ mL})}{(2.12 \times 10^{-2} \text{ M})} = 2.50 \times 10^2 \text{ mL}$$

The amount of water added is $V_2 - V_1 = (2.50 - 0.30) \times 10^2 \text{ mL} = 2.20 \times 10^2 \text{ mL}$.

39. The molar mass of $C_5H_3NO_2$ is

$$5(12.0) + 3(1.0) + 14.0 + 2(16.0) = 109.0 \text{ g mol}^{-1}$$

Use the formula for boiling point elevation to find the molality (m):

$$\Delta T_b = K_b m$$

$$m = \frac{0.560°}{5.03°\ \text{kg mol}^{-1}} = 0.111\ \text{mol kg}^{-1}$$

The molar mass of the unknown is found as follows:

$$\left(\frac{0.356\ \text{g}}{M}\right)\left(\frac{10^3\ \text{g kg}^{-1}}{14.6\ \text{g}}\right) = 0.111\ \text{mol kg}^{-1}$$

$$M = \frac{(0.356)(10^3)}{(14.6)(0.111)} = 2.20 \times 10^2\ \text{g mol}^{-1}$$

The ratio is $220/109 \cong 2$, so $x = 2$ and the compound is $(C_5H_3NO_2)_2$.

40. Use the formula for osmotic pressure to find n:

$$n = \frac{\Pi V}{RT}$$

$$= \frac{(35.6\ \text{torr})(375\ \text{mL})(10^{-3}\ \text{L mL}^{-1})}{(62.4\ \text{L torr mol}^{-1}\text{K}^{-1})(310\ \text{K})}$$

$$= 6.90 \times 10^{-4}\ \text{mol}$$

Now use n to find M in grams per mole:

$$M = \frac{2.33\ \text{g}}{6.90 \times 10^{-4}\ \text{mol}} = 3.38 \times 10^3\ \text{g mol}^{-1}$$

12 ACIDS AND BASES

THIS CHAPTER IS ABOUT

☑ **Acid–Base Systems**
☑ **Acid–Base Strength**
☑ **Titration**
☑ **Stoichiometry**

12-1. Acid–Base Systems

Acid–base reactions constitute one of the most important types of reactions in chemistry. There are three systems for classifying acids and bases in current use. Each has its own peculiar advantages, so you'll need to be familiar with all of them.

A. Arrhenius system

An **Arrhenius acid** is a proton (H^+) donor: When added to water, an Arrhenius acid gives up hydrogen ions. An **Arrhenius base** is a hydroxide ion (OH^-) donor: When added to water, an Arrhenius base gives up hydroxide ions. Arrhenius acids and bases neutralize each other to produce water and a **salt**, an ionic compound containing a positive and a negative ion other than H^+ or OH^-.

EXAMPLE 12-1: Hydrogen chloride (HCl) is an Arrhenius acid; in water, it is completely ionized into protons and chloride ions:

$$HCl(aq) \longrightarrow H^+(aq) + Cl^-(aq)$$

Sodium hydroxide (NaOH) is an Arrhenius base; in water, it is completely ionized into hydroxide ions and sodium ions:

$$NaOH(s) \longrightarrow OH^-(aq) + Na^+(aq)$$

What happens when HCl and NaOH are mixed in the proper proportions?

Solution: The neutralization reaction occurs:

$$NaOH(aq) + HCl(aq) \longrightarrow \underset{\text{(salt)}}{NaCl(aq)} + H_2O$$

B. Brønsted–Lowry system

A **Brønsted–Lowry acid** is any molecule or ion that can donate a proton. A **Brønsted–Lowry base** is any molecule or ion that can accept a proton. In a Brønsted–Lowry acid–base reaction, two bases compete for a proton: The acid loses a proton to form the **conjugate base** of that acid; the base that gains the proton is converted into the **conjugate acid** of that base. For example,

Brønsted–Lowry acid		Conjugate base	Brønsted–Lowry base			Conjugate acid
HCl	$\longrightarrow H^+ +$	Cl^-	OH^-	$+ H^+$	$\longrightarrow$	H_2O
H_2O	$\longrightarrow H^+ +$	OH^-	H_2O	$+ H^+$	$\longrightarrow$	H_3O^+
H_2SO_4	$\longrightarrow H^+ +$	HSO_4^-	NH_3	$+ H^+$	$\longrightarrow$	NH_4^+

note: Water can lose or gain a proton and thus can be either an acid or a base. **Amphoterism** is the ability to act as either an acid or a base. **Amphiprotism** is the ability of a species either to gain or to lose a proton. *Monoprotic* Brønsted–Lowry acids can donate 1 mol, *diprotic* acids donate 2 mol, and *triprotic* acids donate 3 mol of protons per mole of acid.

EXAMPLE 12-2: Identify all the components in the following Brønsted–Lowry reaction:

$$HCl(aq) + H_2O(l) \longrightarrow H_3O^+ + Cl^-$$

Solution: First note that there are indeed two acids, HCl and H_3O^+ (the hydronium ion), and two bases, H_2O and Cl^-. In all Brønsted–Lowry acid–base reactions, there are two conjugate acid–base pairs in equilibrium, so you can write

$$\underset{HCl}{\overset{acid_1}{}} + \underset{H_2O}{\overset{base_2}{}} \rightleftharpoons \underset{H_3O^+}{\overset{acid_2}{}} + \underset{Cl^-}{\overset{base_1}{}}$$

where the subscripts denote conjugate acid–base pairs, and the bidirectional arrow denotes the equilibrium condition, indicating that the reaction is reversible and can proceed in either direction.

C. Lewis system

A **Lewis acid** is any species that can accept an electron pair. A **Lewis base** is any species that can donate an electron pair. A Lewis acid–base reaction creates a *coordinate covalent bond* in the reaction product, which is often called a *"complex"* rather than a *"salt"* if the Lewis acid is a metal ion. The Lewis system is useful for classifying reactions that occur in solvents other than water—or in the complete absence of a solvent. In the Lewis system, all cations (positive charge) are acids and all anions (negative charge) are bases.

EXAMPLE 12-3: Boron trifluoride (BF_3) is a Lewis acid. It acts as an electron-pair acceptor because the boron atom shares only six electrons with the three fluorine atoms, and has a vacant $2p$ orbital that can accept a pair of electrons from a Lewis base. Ammonia ($:NH_3$) is a Lewis base. The lone pair of electrons on the nitrogen atom can be donated to a suitable Lewis acid, such as BF_3.

Write the Lewis structures of BF_3 and NH_3 and of the product formed as a result of a Lewis acid–base reaction between them.

Solution: The reaction between BF_3 and NH_3 is a Lewis acid–base reaction in which a complex is formed:

$$
\begin{array}{ccccc}
& \overset{\cdot\cdot}{:}\overset{\cdot\cdot}{F}\overset{\cdot\cdot}{:} & & H & \\
\cdot\cdot & & & \cdot\cdot & \\
:F:B & + & :N:H & \longrightarrow & :F:B : N:H \\
\cdot\cdot & & & \cdot\cdot & \\
& :F: & & H & \\
\end{array}
$$

:F:	H	:F: H
:F:B	:N:H	:F:B : N:H
:F:	H	:F: H
Lewis acid	Lewis base	complex

[Note that the octets of both B and N are complete in the complex.]

12-2. Acid–Base Strength

A. Ionization of water

In any sample of pure water we always find a relatively small number of H^+ and OH^- ions, which are produced by the ionization of H_2O. We can write the reaction as

$$H_2O \rightleftharpoons H^+ + OH^-$$

where the long arrow indicates the preferred direction of the reaction. In order to express the concentration of ions, we use the **ion product constant** K: For water K_w is the concentration of hydrogen ions multiplied by the concentration of hydroxide ions:

ION PRODUCT CONSTANT FOR H₂O
$$K_w = [H^+][OH^-] \tag{12-1}$$

where the brackets are used to indicate concentration. At 25°C, $K_w = 1.008 \times 10^{-14}$ when the concentration is in moles per liter (M). K_w increases with temperature, but in most cases you can use $K_w = 1.00 \times 10^{-14}$.

Pure water is neither acidic nor basic because the concentrations of H^+ and OH^- are equal. Knowing this, you can easily compute the value of each from the value and definition of K_w:

$$[H^+] = [OH^-]$$
$$K_w = [H^+][OH^-] = [H^+]^2$$
$$[H^+]^2 = 1.00 \times 10^{-14}$$
$$[H^+] = 1.00 \times 10^{-7}\,\text{mol/L} = [OH^-]$$

B. pH and pK

Because concentrations are often very small numbers, chemists simplify concentration expressions by using a base-10 logarithmic scale. Thus, for example, **pH** and **pOH** are defined as the negative (reciprocal) logarithm of the H^+ and OH^- concentrations, respectively:

pH
$$pH = -\log[H^+] = \log \frac{1}{[II^+]} \qquad \text{(12-2)}$$

pOH
$$pOH = -\log[OH^-] = \log \frac{1}{[OH^-]} \qquad \text{(12-3)}$$

The values of pH and pOH usually fall between 1 and 14. For pure water $[H^+] = [OH^-] = 1.00 \times 10^{-7}\,\text{mol/L}$, so $pH = pOH = -\log(1.00 \times 10^{-7}) = 7.00$. Any aqueous solution for which $pH = pOH$ is a *neutral* solution. In an *acidic* solution $pH < 7.00$; in a basic (alkaline) solution $pH > 7.00$.

EXAMPLE 12-4: Calculate the pH of solutions in which the molar concentration of H^+ is (a) $1.0 \times 10^{-2}\,M$, (b) $1.0 \times 10^{-7}\,M$, (c) $3.0 \times 10^{-10}\,M$.

Solution:
(a) $[H^+] = 1.0 \times 10^{-2}$; $pH = -\log(1.0 \times 10^{-2}) = -(-2.00) = 2.00$
(b) $[H^+] = 1.0 \times 10^{-7}$; $pH = -\log(1.0 \times 10^{-7}) = -(-7.00) = 7.00$
(c) $[H^+] = 3.0 \times 10^{-10}$; $pH = -\log(3.0 \times 10^{-10}) = -(\log 3) - (-10.00) = 10.00 - 0.477 = 9.52$

You'll encounter the same logarithmic convention for other expressions, especially those involving K, the concentration constant. The ion product constant for water K_w, for instance, can be expressed as pK_w:

$$K_w = [H^+][OH^-] = 1.00 \times 10^{-14}$$
$$\log K_w = \log[H^+] + \log[OH^-] = \log(1.00 \times 10^{-14}) = -14.00$$
$$-\log K_w = -\log[H^+] - \log[OH^-] = -(-14.00)$$
$$pK_w = pH + pOH = 14.00 \qquad \text{(12-4)}$$

C. Strong and weak acids

The strength of an acid in aqueous solution is measured by the extent to which it ionizes. For the reaction

$$HA(aq) \rightleftharpoons H^+ + A^-$$

the **acid ionization constant** K_a is given by

ACID IONIZATION CONSTANT
$$K_a = \frac{[H^+][A^-]}{[HA]} \qquad \text{(12-5)}$$

Strong acids have K_a values greater than 1 and are, for all practical purposes, completely ionized in dilute aqueous solution. **Weak acids** have K_a values considerably less than 1 and are usually ionized to only a small extent. Values of K_a are also often expressed as logarithmic expressions:

$$pK_a = -\log K_a = \log \frac{1}{K_a} \qquad (12\text{-}6)$$

- The larger the K_a, the stronger the acid; the larger the pK_a, the weaker the acid.

EXAMPLE 12-5: Arrange the following acids in order of increasing strength: HCl (completely ionized in water), HCN ($pK_a = 9.40$), HNO_2 ($K_a = 4.57 \times 10^{-4}$).

Solution: The larger the K_a, the stronger the acid. Convert the pK_a for HCN to K_a:

$$pK_a = 9.40 = -\log K_a$$
$$\log K_a = -9.40 = -10.00 + 0.60$$
$$K_a = \text{antilog}(0.60 - 10.00) = 3.98 \times 10^{-10}$$

Since HCl is completely ionized, it is a strong acid. Both HCN and HNO_2 are weak acids. The order of increasing strength is HCN < HNO_2 < HCl.

D. Strong and weak bases

The strength of a base in aqueous solution is measured by the extent to which it dissociates. For the reaction

$$MOH \rightleftharpoons M^+ + OH^-$$

the **base ionization constant** K_b is given by

BASE IONIZATION CONSTANT
$$K_b = \frac{[M^+][OH^-]}{[MOH]} \qquad (12\text{-}7)$$

Strong bases have K_b values greater than 1 and are completely dissociated in water. **Weak bases** have K_b values considerably less than 1 and are only partially dissociated. Like K_a values, K_b values are also expressed logarithmically:

$$pK_b = -\log K_b = \log \frac{1}{K_b} \qquad (12\text{-}8)$$

- The larger the K_b value, the stronger the base; the larger the pK_b, the weaker the base.

EXAMPLE 12-6: Determine K_b for the following reaction:

$$CN^- + H_2O(l) \rightleftharpoons HCN(aq) + OH^-$$

The equilibrium concentrations are $[CN^-] = 1.85\,M$, $[HCN] = 6.5 \times 10^{-4}\,M$, and $[OH^-] = 4.6 \times 10^{-2}\,M$.

Solution: Using the definition of the base ionization constant (12-7),

$$K_b = \frac{[HCN][OH^-]}{[CN^-]}$$
$$= \frac{(6.5 \times 10^{-4})(4.6 \times 10^{-2})}{1.85}$$
$$= 1.6 \times 10^{-5}$$

12-3. Titration

Titration is the measurement of the solution volume of one reactant that is required to react completely with a specified amount of another reactant. In an acid–base reaction the **equivalence point** (or **endpoint**) is the point at which chemically equivalent (stoichiometric) amounts of base and acid have been mixed.

The general titration relationship can be expressed by

TITRATION RELATIONSHIP
$$N_a V_a = N_b V_b \quad \text{or} \quad N_1 V_1 = N_2 V_2 \tag{12-9}$$

where N is normality (in equivalents/liter) and V is solution volume (in any convenient unit). [If V is in liters, Eq. (12-9) tells you that the number of equivalents of acid equals the number of equivalents of base.]

note: For monoprotic acids, normality equals molarity, so relationship (12-9) becomes $M_a V_a = M_b V_b$ or $M_1 V_1 = M_2 V_2$.

EXAMPLE 12-7: A 50.0-mL amount of NaOH solution was completely neutralized by 25.0 mL of a 0.200 N HCl solution. Determine the normality of the NaOH solution.

Solution: From the titration relation (12-9)

$$N_a V_a = N_b V_b$$

So

$$N_{\text{NaOH}} = \frac{N_{\text{HCl}} V_{\text{HCl}}}{V_{\text{NaOH}}} = \frac{(0.200\ N)(25.0\ \text{mL})}{50.0\ \text{mL}} = 0.100\ N$$

EXAMPLE 12-8: You have a 0.563 M solution of NaOH, and you want to prepare 1.500 L of 0.0500 M NaOH from it. What volume of the initial 0.563 M solution should you take to dilute to 1.500 L?

Solution: Let the unknown initial volume be V_1. Since the number of moles of NaOH in V_1 will be unchanged during the dilution (and since the number of moles = L × mol/L), $M_1 V_1 = M_2 V_2$, where $M_1 = 0.563\ M$, $M_2 = 0.0500\ M$, and $V_2 = 1.500$ L. So

$$(0.563\ M)(V_1) = (0.0500\ M)(1.500\ \text{L})$$

$$V_1 = \left(\frac{0.0500\ M}{0.563\ M}\right)(1.500\ \text{L}) = 0.133\ \text{L} \quad \text{or} \quad 133\ \text{mL}$$

12-4. Stoichiometry

Most Brønsted–Lowry acid–base reactions in aqueous solution can be balanced by inspection. If the base is a hydroxide (as it often will be), each mole of hydroxide requires a mole of protons from the acid for complete reaction.

EXAMPLE 12-9: Balance the acid–base reaction between phosphoric acid H_3PO_4 and calcium hydroxide $Ca(OH)_2$.

Solution: Each mole of hydroxide in $Ca(OH)_2$ requires a mole of protons from the H_3PO_4. Since there are 2 mol of hydroxide in 1 mol of $Ca(OH)_2$, you'll need 2 mol of protons, which can be donated from $\frac{2}{3}$ mol of H_3PO_4, per mole of $Ca(OH)_2$. Expressed in integer terms,

$$3Ca(OH)_2 + 2H_3PO_4 = 6H_2O + Ca_3(PO_4)_2$$

note: In calculations based upon such equations, concentration is almost always expressed in terms of molarity.

A. Spectator ions

The reaction between an Arrhenius acid and base always yields a salt plus water:

$$HCl(aq) + NaOH(aq) \longrightarrow NaCl(aq) + H_2O$$
$$H_2SO_4(aq) + 2KOH(aq) \longrightarrow K_2SO_4(aq) + 2H_2O$$

But notice that the actual reaction is the same for both equations; i.e,

$$H^+(aq) + OH^-(aq) \longrightarrow H_2O$$

Consider the nature of Arrhenius acids and bases in aqueous solution. A strong acid like HCl is completely ionized in aqueous solution. There are no HCl *molecules* in HCl(aq); all of the acid is ionized to $H^+(aq)$ and $Cl^-(aq)$. (Only HCl(g) can be considered molecular.) The situation is similar for a strong base like NaOH; un-ionized NaOH(aq) doesn't exist; all the NaOH exists as $Na^+(aq)$ and $OH^-(aq)$.

If you now write the reaction between HCl(aq) and NaOH(aq) in terms of the actual species present in solution, you get

$$H^+(aq) + Cl^-(aq) + Na^+(aq) + OH^-(aq) \longrightarrow Na^+(aq) + Cl^-(aq) + H_2O$$

because the product NaCl is a salt that exists only as Na^+ and Cl^- in the solid state and in solution. Looking at this equation, you'll see that $Na^+(aq)$ and $Cl^-(aq)$ appear both as reactants and as products; but these ions have simply "gone along for the ride" during the reaction. They are **spectator ions**, which play no role in the reaction.

EXAMPLE 12-10: Which ions are the spectator ions in the following reaction?

$$H_2SO_4(aq) + 2KOH(aq) \longrightarrow K_2SO_4(aq) + 2H_2O$$

Solution: Write out all the reagents and products in their ionic forms:

$$2H^+(aq) + SO_4^{2-}(aq) + 2K^+(aq) + 2OH^-(aq) \longrightarrow 2K^+(aq) + SO_4^{2-}(aq) + 2H_2O$$

The ions remaining unaffected by the reaction are the spectator ions K^+ and SO_4^{2-}.

B. Net ionic equations

Reactants that are written without the accompanying spectator ions are called **net ionic equations** (or, sometimes, *essential* ionic equations). These equations are useful because they focus attention on the actual chemical changes occurring in a system. The complete equations with spectator ions are important because they represent the true stoichiometry of the reaction. For example, you can weigh out a sample of NaOH; you can't weigh out a sample of free OH^- ions. Consequently, you must understand and be able to use both stoichiometric equations and net ionic equations.

SUMMARY

1. An Arrhenius acid is a proton donor; an Arrhenius base is a hydroxide ion donor.
2. A Brønsted–Lowry acid is a proton donor; a Brønsted–Lowry base is a proton acceptor.
3. A Lewis acid is an electron-pair acceptor; a Lewis base is an electron-pair donor.
4. The ion product constant for water is given by $K_w = [H^+][OH^-]$, and $pK_w = -\log K_w = 14.00$.
5. pH is defined as $-\log[H^+]$; pOH is defined as $-\log[OH^-]$. For a neutral solution pH = 7.00; for an acidic solution pH < 7.00; for a basic (alkaline) solution pH > 7.00.
6. The acid ionization constant for the reaction $HA(aq) \rightleftharpoons H^+ + A^-$ is given by

$$K_a = \frac{[H^+][A^-]}{[HA]} \quad \text{and} \quad pK_a = -\log K_a$$

7. The base ionization constant for the reaction $MOH \rightleftharpoons M^+ + OH^-$ is given by

$$K_b = \frac{[M^+][OH^-]}{[MOH]} \quad \text{and} \quad pK_b = -\log K_b$$

8. Titration is the measurement of the volume of a solution of one reactant that is required to react completely with a specified amount of another reactant.
9. Ions that appear unchanged as both reactants and products in a chemical reaction are spectator ions.
10. Reactions written without spectator ions are called net ionic equations.

RAISE YOUR GRADES

Can you define ... ?

☑ salt
☑ conjugate acid
☑ conjugate base
☑ amphoterism
☑ amphiprotism

☑ mono-, di-, and triprotic acids
☑ complex
☑ strong and weak acids
☑ strong and weak bases
☑ equivalence point (endpoint)

Can you ... ?

☑ identify Arrhenius, Brønsted–Lowry, and Lewis acids and bases
☑ determine pH, pK_a, pK_b, or $[H^+]$ for a stoichiometric reaction
☑ use the results of titration experiments to determine molarity
☑ write net ionic equations

SOLVED PROBLEMS

PROBLEM 12-1 Identify the Arrhenius acid, the Arrhenius base, and the salt in the following reaction

$$Ca(OH)_2 + 2HNO_3 \longrightarrow Ca(NO_3)_2 + 2H_2O$$

Solution: An Arrhenius acid is a proton donor; consequently, HNO_3 (nitric acid) must be the Arrhenius acid. An Arrhenius base is a hydroxide donor; thus $Ca(OH)_2$ (calcium hydroxide) must be the Arrhenius base. The products of an Arrhenius acid–base reaction are a salt and water. So $Ca(NO_3)_2$ (calcium nitrate) must be the salt.

PROBLEM 12-2 Identify the Brønsted–Lowry acid and its conjugate base as well as the Brønsted–Lowry base and its conjugate acid in the following reaction:

$$CH_3CO_2H + OH^- \rightleftharpoons CH_3CO_2^- + H_2O$$

Solution: A Brønsted–Lowry acid is a proton donor, so CH_3CO_2H (acetic acid) is the Brønsted–Lowry acid. The species produced when a Brønsted–Lowry acid loses its proton is its conjugate base. Consequently, $CH_3CO_2^-$ (acetate ion) is the conjugate base of CH_3CO_2H (acetic acid).

A Brønsted–Lowry base is a proton acceptor, so OH^- is the Brønsted–Lowry base. The species produced when a Brønsted–Lowry base accepts a proton is its conjugate acid. Thus H_2O is the conjugate acid of the Brønsted–Lowry base OH^- (hydroxide ion).

PROBLEM 12-3 Identify the Lewis acid and Lewis base in each of the following reactions:

(a) $BCl_3 + PH_3 \longrightarrow Cl_3BPH_3$
(b) $H^+ + H_2O \rightleftharpoons H_3O^+$
(c) $PtCl_2 + 2NH_3 \longrightarrow (H_3N)_2PtCl_2$

Solution: A Lewis base is an electron-pair donor; a Lewis acid is an electron-pair acceptor. Since the electron pair is usually a lone pair, you'd better rewrite the reactions using valence-bond structures:

(a)

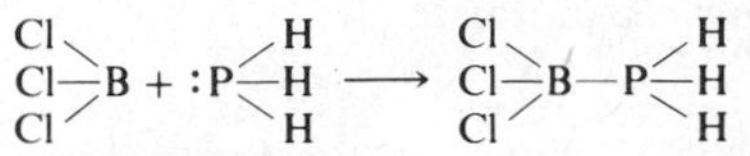

:PH_3 is the electron-pair donor, and therefore it is the Lewis base. BCl_3 is the electron-pair acceptor, and therefore the Lewis acid.

(b)

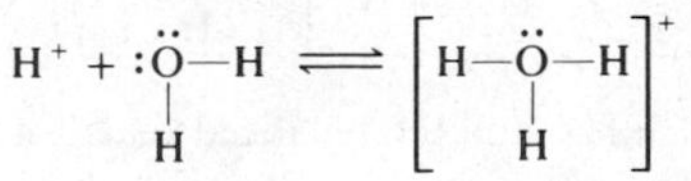

H_2O is the electron-pair donor, or Lewis base. H^+ is the electron-pair acceptor, or Lewis acid.

(c) This one may puzzle you for a moment, since you haven't seen a Lewis structure for a transition-metal compound like $PtCl_2$. But never mind: Just write what you *do* know and see where that leads.

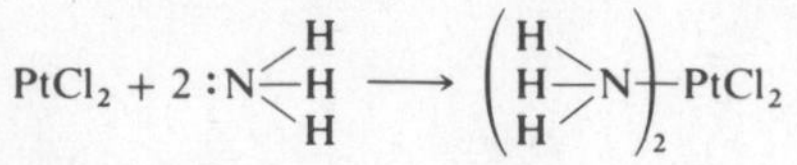

$$PtCl_2 + 2 :N{\overset{H}{\underset{H}{\diagdown}}}H \longrightarrow \left(H{\overset{H}{\diagdown}}N{\diagup}PtCl_2 \right)_2$$

Somehow, since the product is written as $(H_3N)_2PtCl_2$, the nitrogen atoms in each $:NH_3$ have become bonded to Pt, and this can be only through donation of each nitrogen's lone pair of electrons. So $:NH_3$ is the electron-pair donor, or Lewis base; and the whole $PtCl_2$ is the electron-pair acceptor, or Lewis acid.

PROBLEM 12-4 Classify the Brønsted–Lowry acids in the following reactions as mono-, di-, or triprotic:

(a) $HCO_2H + OH^- \rightleftharpoons HCO_2^- + H_2O$
(b) $H_3AsO_4 + 3OH^- \rightleftharpoons AsO_4^{3-} + 3H_2O$
(c) $H_2CO_3 + 2OH^- \rightleftharpoons CO_3^{2-} + 2H_2O$
(d) $HF + OH^- \rightleftharpoons F^- + H_2O$

Solution:
(a) One mole of HCO_2H (formic acid) donates 1 mol of protons to the base OH^-; formic acid is monoprotic.
(b) One mole of H_3AsO_4 (arsenic acid) donates 3 mol of protons to the base OH^-; arsenic acid is triprotic.
(c) One mole of H_2CO_3 (carbonic acid) donates 2 mol of protons to the base OH^-; carbonic acid is diprotic.
(d) One mole of HF (hydrogen fluoride) donates 1 mol of protons to the base OH^-; hydrogen fluoride is monoprotic.

PROBLEM 12-5 Complete and balance the equations for the following acid–base reactions:

(a) $HF + LiOH$
(b) $HCl + Ca(OH)_2$
(c) $H_2SO_4 + KOH$
(d) $H_2CO_3 + Ba(OH)_2$
(e) $H_2SO_4 + NH_3$
(f) $H_2SO_4 + Al(OH)_3$

Solution:
(a) HF is monoprotic; LiOH has 1 mol of OH per mole. Thus the stoichiometry will be 1:1 HF:LiOH, and the balanced reaction is

$$HF + LiOH \longrightarrow LiF + H_2O$$

(b) HCl is monoprotic, but $Ca(OH)_2$ has 2 mol of OH per mole. Thus each mole of $Ca(OH)_2$ will react with 2 mol of HCl, and the balanced reaction is

$$2HCl + Ca(OH)_2 \longrightarrow CaCl_2 + 2H_2O$$

(c) H_2SO_4 is diprotic, while KOH has 1 mol of OH per mole. Thus 2 mol of KOH is required to react with each mole of H_2SO_4, and the balanced reaction is

$$H_2SO_4 + 2KOH \longrightarrow K_2SO_4 + 2H_2O$$

(d) H_2CO_3 is diprotic, and $Ba(OH)_2$ has 2 mol of OH per mole. So you expect a 1:1 stoichiometry, the balanced reaction being

$$H_2CO_3 + Ba(OH)_2 \qquad BaCO_3 + 2H_2O$$

(e) This is a tougher one. H_2SO_4 is obviously diprotic, but NH_3 has no OH groups. However, NH_3 is a Brønsted–Lowry base and can accept a proton. Consideration of its Lewis structure shows that $:NH_3$ has *one* lone pair, which means that it can accept only *one* proton to give NH_4^+:

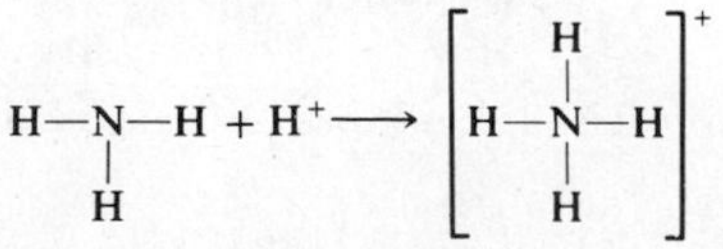

$$H{-}\underset{\underset{H}{|}}{N}{-}H + H^+ \longrightarrow \left[H{-}\underset{\underset{H}{|}}{\overset{\overset{H}{|}}{N}}{-}H \right]^+$$

So the stoichiometry is 1:2 $H_2SO_4:NH_3$, and the balanced reaction is

$$H_2SO_4 + 2NH_3 \longrightarrow (NH_4)_2SO_4$$

(f) H_2SO_4 is diprotic, whereas $Al(OH)_3$ has 3 mol of OH per mole and needs 3 mol of protons for reaction. Three moles of protons can be provided by $\frac{3}{2}$ mol of H_2SO_4, so you get

$$\tfrac{3}{2}H_2SO_4 + Al(OH)_3 \longrightarrow Al(SO_4)_{3/2} + 3H_2O$$

This is clearly awkward, so you double it to get rid of the fractions, ending up with

$$3H_2SO_4 + 2Al(OH)_3 \longrightarrow Al_2(SO_4)_3 + 6H_2O$$

PROBLEM 12-6   What volume of 0.500 M standard sodium hydroxide solution would you need to prepare 1.000 L of 0.0250 M sodium hydroxide solution?

Solution: This is a dilution problem in which the number of moles of NaOH remains constant, and so the titration formula $M_1V_1 = M_2V_2$ (12-9) applies.

$$M_1 = 0.500\ M \qquad V_1 \text{ is the unknown}$$
$$M_2 = 0.0250\ M \qquad V_2 = 1.000\ L$$

So

$$(0.500\ M)V_1 = (0.0250\ M)(1.000\ L)$$

$$V_1 = \frac{(0.0250\ M)(1.000\ L)}{0.500\ M} = 0.0500\ L = 50.0\ mL$$

PROBLEM 12-7   You have to make exactly 250 mL of 0.100 M sulfamic acid. You are given pure crystalline sulfamic acid (H_2NSO_3H) and a 250.00-mL graduated flask. Describe your procedure.

Solution: In 250 mL of 0.100 M sulfamic acid there is

$$\frac{0.100\ mol}{L} \times \frac{250\ mL}{10^3\ mL\,L^{-1}} = 0.0250\ mol \text{ sulfamic acid}$$

The molar mass of H_2NSO_3H is 97.0 g/mol, so the mass of H_2NSO_3H you need is

$$0.0250\ mol \times 97.0\ g/mol = 2.43\ g$$

You carefully weigh out this amount of the acid, transfer it quantitatively to the graduated flask, and then carefully add just enough water to bring the solution volume to 250.00 mL at the graduation line.

PROBLEM 12-8   A commonly used standard acid in analytical chemistry is "KHP"—potassium hydrogen phthalate ($KC_8H_5O_4$)—a monoprotic acid of molar mass 204.2 g/mol, which is a crystalline solid available in high purity. In a titration experiment you find that a sample of 0.100 g of potassium hydrogen phthalate requires 8.85 mL of sodium hydroxide to neutralize it. What is the molarity of the sodium hydroxide solution?

Solution: Since the KHP is monoprotic, its reaction with NaOH must be in a 1:1 molar ratio:

$$KC_8H_5O_4 + NaOH \longrightarrow KNaC_8H_4O_4 + H_2O$$

The number of moles of KHP used is

$$\frac{0.100\ g}{204.2\ g/mol} = 4.90 \times 10^{-4}\ mol$$

Therefore the number of moles of NaOH in 8.85 mL of solution is

$$4.90 \times 10^{-4}\ mol\ KHP \times \frac{1\ mol\ NaOH}{1\ mol\ KHP} = 4.90 \times 10^{-4}\ mol\ NaOH$$

And so

$$M_{NaOH} = \frac{4.90 \times 10^{-4}\ mol\ NaOH}{8.85\ mL\ soln} \times \frac{10^3\ mL}{L} = 0.0553\ mol/L$$

note: This is the most usual method of determining the exact molarity of NaOH or KOH solutions. You can't weigh these bases out accurately because they rapidly absorb water from the air.

PROBLEM 12-9   You're now going to use the NaOH solution you standardized in Problem 12-8 (0.0553 M) to determine the molarity of some HCl solution. You find that it takes 36.4 mL of the NaOH solution to neutralize 25.0 mL of the HCl solution. What is the molarity of the HCl solution?

Solution: Your first step in questions like this is to write a balanced reaction equation to establish stoichiometry:

$$NaOH + HCl \longrightarrow NaCl + H_2O$$

Then find the number of moles of NaOH used:

$$36.4 \text{ mL} \times 0.0553 \text{ mmol/mL} = 2.01 \text{ mmol}$$

(Remember that $M = \text{mol/L} = \text{mmol/mL}$.)

Now find the number of moles needed to establish the unknown molarity:

$$2.01 \text{ mmol NaOH} \times \frac{1 \text{ mol HCl}}{1 \text{ mol NaOH}} = 2.01 \text{ mmol HCl}$$

(The result in this case is obvious—but you should get into the habit of including stoichiometric ratios in your calculations; they aren't always 1:1.)
   Finally, determine the molarity:

$$M_{\text{HCl}} = \frac{2.01 \text{ mmol}}{25.0 \text{ mL}} = 0.0805 \text{ mol/L}$$

PROBLEM 12-10   You use your standard NaOH solution (0.0553 M) in a titration with some H_2SO_4 solution. A 25.0-mL sample of the H_2SO_4 solution requires 20.7 mL of the NaOH to neutralize it. What's the molarity of the H_2SO_4 solution?

Solution: H_2SO_4 is diprotic, so the balanced equation is

$$H_2SO_4 + 2NaOH \longrightarrow Na_2SO_4 + 2H_2O$$

The number of moles of NaOH used is

$$20.7 \text{ mL} \times 0.0553 \text{ } M = 1.14 \text{ mmol}$$

The number of moles of H_2SO_4 in the 25.0-mL sample is

$$1.14 \text{ mmol NaOH} \times \frac{1 \text{ mmol H}_2\text{SO}_4}{2 \text{ mmol NaOH}} = 0.572 \text{ mmol}$$

So

$$M_{\text{H}_2\text{SO}_4} = \frac{0.572 \text{ mmol}}{25.0 \text{ mL}} = 0.0229 \text{ mol/L}$$

PROBLEM 12-11   Determine the pH of solutions in which the H^+ concentration is (a) 1.0 M, (b) 4.5×10^{-3} M, (c) 1.0×10^{-12} M.

Solution: Use the definition of pH (Eq. 12-2):

$$pH = -\log[H^+]$$

(a) If $[H^+]$ is 1.0 M,

$$pH = -\log 1.0 = -0 = 0 \qquad \text{(Remember that } 10^0 = 1\text{)}$$

(b) If $[H^+] = 4.5 \times 10^{-3}$ M,

$$pH = -\log(4.5 \times 10^{-3})$$

Using your scientific calculator, take the log of the reciprocal of 4.5×10^{-3}, which gives you a pH of 2.35. (Be sure to check the instruction book that came with your calculator for the right procedure.)
   Alternatively: Remember that $\log ab = \log a + \log b$, so that $-\log ab = -\log a - \log b$,

$$pH = -\log 4.5 - \log 10^{-3}$$

Using your simple calculator or log tables, you get

$$pH = -0.653 - (-3) = 2.35$$

note: Two significant figures are justified. In pH calculations, the significant figures are those that *follow* the decimal. The number to the left of the decimal is an exponent and is not considered part of the significant figures.

(c) If $[H^+] = 1.0 \times 10^{-12}$ *M*,

$$pH = -\log(1.0 \times 10^{-12}) = -(-12.00) = 12.00$$

PROBLEM 12-12 What is the hydrogen ion concentration in solutions whose pH values are **(a)** 5.00, **(b)** 8.43, **(c)** 12.01?

Solution: Getting $[H^+]$ from pH is just the inverse of getting pH from $[H^+]$. The whole numbers you can do in your head, and your scientific calculator will give you the required antilog for the others. But if you look for negative fractions between 0 and 1 in a logarithm table, you won't find any. You'll need to transform your fractions into positive ones, as shown in examples **(b)** and **(c)**.

(a) If $pH = -\log[H^+] = 5.00$,

$$\log[H^+] = -5.00$$

and, taking the antilog,

$$[H^+] = \text{antilog}(-5.00) = 1.0 \times 10^{-5}$$

(b) If $pH = -\log[H^+] = 8.43$,

$$\log[H^+] = -8.43 = -9.00 + 0.57 \qquad \text{(Here's where your positive fractions come in.)}$$

Now you can take antilogs:

$$[H^+] = \text{antilog}(-9.00) \times \text{antilog}(0.57) = 10^{-9} \times 3.7 = 3.7 \times 10^{-9}$$

(c) If $pH = -\log[H^+] = 12.01$,

$$\log[H^+] = -12.01 = -13.00 + 0.99$$

Taking antilogs,

$$[H^+] = \text{antilog}(-13.00) \times \text{antilog}(0.99) = 10^{-13} \times 9.8 = 9.8 \times 10^{-13}$$

PROBLEM 12-13 Classify the following solutions as acidic or basic: **(a)** pH = 0.00, **(b)** $[H^+] = 3.8 \times 10^{-9}$, **(c)** $[H^+] = 1.00 \times 10^{-7}$, **(d)** pH = 7.33, **(e)** $[H^+] = 8.6 \times 10^{-4}$.

Solution: The general conditions for acidic solutions are pH < 7.00 or $[H^+] > 1.00 \times 10^{-7}$; for basic solutions, pH > 7.00 or $[H^+] < 1.00 \times 10^{-7}$. Consequently, solutions **(a)** and **(e)** are acidic, and solutions **(b)** and **(d)** are basic. Solution **(c)** is neither acidic nor basic—it is neutral.

PROBLEM 12-14 Complete and balance the stoichiometric equations for the following reactions in aqueous solution. Write the net ionic equation and identify the spectator ions in each reaction.

(a) $Ba(OH)_2 + HClO_4$ ($HClO_4$ is a monoprotic acid)
(b) $H_2SO_3 + LiOH$ (H_2SO_3 is a diprotic acid)

Solution:
(a) Balance the stoichiometric equation following the methods used in Problem 12-5:

$$Ba(OH)_2 + 2HClO_4 \longrightarrow Ba(ClO_4)_2 + 2H_2O$$

Now write all the species in their ionic forms:

$$Ba^{2+} + 2OH^- + 2H^+ + 2ClO_4^- \longrightarrow Ba^{2+} + 2ClO_4^- + 2H_2O$$

The unchanged species are the spectator ions, Ba^{2+} and ClO_4^-. Strike out the spectator ions to obtain the net ionic reaction:

$$2OH^- + 2H^+ \longrightarrow 2H_2O$$

(b)

$$H_2SO_3 + 2LiOH \longrightarrow Li_2SO_3 + 2H_2O \qquad \text{(stoichiometric)}$$
$$2H^+ + SO_3^{2-} + 2Li^+ + 2OH^- \longrightarrow 2Li^+ + SO_3^{2-} + 2H_2O \qquad \text{(fully ionized)}$$
$$2H^+ + 2OH^- \longrightarrow 2H_2O \qquad \text{(net ionic)} \quad Li^+ \text{ and } SO_3^{2-} \text{ are spectator ions}$$

PROBLEM 12-15   Complete and balance the stoichiometic equation for the reaction

$$AgNO_3 + HCl \longrightarrow AgCl(s) \qquad (AgCl \text{ is insoluble})$$

Write the net ionic equation and identify the spectator ions.

Solution: Notice that this isn't an acid–base reaction. However, the ideas of stoichiometric and net ionic reactions aren't limited to acids and bases, but include all reactions that involve ions. Carry on as usual, and see what happens.

$$AgNO_3 + HCl \longrightarrow AgCl(s) + HNO_3 \qquad (\text{stoichiometric})$$

Now, remember that AgCl is insoluble, so it is *not* ionized:

$$Ag^+ + NO_3^- + H^+ + Cl^- \longrightarrow AgCl(s) + H^+ + NO_3^- \qquad (\text{fully ionized})$$

$$Ag^+ + Cl^- \longrightarrow AgCl(s) \qquad (\text{net ionic}) \quad H^+ \text{ and } NO_3^- \text{ are spectator ions}$$

PROBLEM 12-16   Solutions of HCl and NaOH whose molar concentrations are unknown are available in your laboratory. You also have pure solid calcium carbonate ($CaCO_3$). You find that 25.0 mL of the HCl solution is neutralized by 36.2 mL of the NaOH solution. You then add 0.200 g of $CaCO_3$ to a fresh 25.0-mL portion of the HCl solution so that some of the HCl is used up in a reaction:

$$2HCl + CaCO_3 \longrightarrow CaCl_2 + H_2O + CO_2(g)$$

The remaining HCl in the 25.0-mL portion requires 24.9 mL of NaOH for neutralization. Calculate the molarities of the HCl and NaOH solutions.

Solution: This problem looks formidable at first, but it yields to careful analysis. The one direct stoichiometric determination you can make involves the $CaCO_3$: You can calculate the number of moles in 0.200 g of pure $CaCO_3$:

$$\frac{0.200 \text{ g}}{100.1 \text{ g/mol}} = 1.99 \times 10^{-3} \text{ mol}$$

By the stoichiometry of the given reaction of $CaCO_3$ with HCl, this has reacted with and neutralized $2(1.99 \times 10^{-3} \text{ mol}) = 3.99 \times 10^{-3}$ mol of HCl, or 3.99 mmol.

Now that you've got some (numerical) facts, you can start dealing with the molarities. Let the molarity of the HCl be x mmol/mL and the molarity of the NaOH be y mmol/mL. From the first titration result you know that $M_1V_1 = M_2V_2$, or

$$(25.0 \text{ mL})\left(\frac{x \text{ mmol HCl}}{\text{mL}}\right) = (36.2 \text{ mL})\left(\frac{y \text{ mmol NaOH}}{\text{mL}}\right)\left(\frac{1 \text{ mmol HCl}}{1 \text{ mmol NaOH}}\right)$$

So

$$25.0x = 36.2y \qquad\qquad (1)$$

You have calculated that 3.99 mmol HCl has reacted with $CaCO_3$ before the second titration, so the HCl remaining in the solution is $(25.0x - 3.99 \text{ mmol})$, which is neutralized by 24.9 mL NaOH. Thus

$$(25.0x - 3.99) = (24.9 \text{ mL})(y)$$

$$(25.0x - 3.99) \text{ mmol} = 24.9y \text{ mL} \qquad\qquad (2)$$

Subtracting eq. (2) from eq. (1), you get

$$3.99 \text{ mmol} = (11.3y) \text{ mL}$$

and so

$$y = \frac{3.99 \text{ mmol}}{11.3 \text{ mL}} = 0.353 \text{ } M$$

And from eq. (1)

$$x = \frac{36.2y}{25.0} = 0.511 \text{ } M$$

So HCl is 0.511 M, and NaOH is 0.353 M.

PROBLEM 12-17   A 15.0-mL sample of a sulfuric acid solution of unknown molarity was neutralized by 10.8 mL of 0.113 M sodium hydroxide. Which of the following gives the correct numerical answer for the molarity of the H_2SO_4 solution?

(a) $0.113\left(\dfrac{10.8}{15.0}\right)$ **(b)** $0.113\left(\dfrac{15.0}{10.8}\right)$ **(c)** $\dfrac{15.0 + 10.8}{2}$ **(d)** $\dfrac{0.113 \times 10.8}{2 \times 15.0}$

(e) $2\left(\dfrac{0.113 \times 10.8}{15.0}\right)$

Solution: This is a test of your understanding of acid–base stoichiometry. Balance the equation first:

$$H_2SO_4 + 2NaOH \longrightarrow Na_2SO_4 + 2H_2O$$

There's 1 mmol of H_2SO_4 for every 2 mmol of NaOH used. The number of millimoles of NaOH required is $0.113\ M \times 10.8$ mL. So the number of millimoles of H_2SO_4 required is

$$(0.113\ M \times 10.8\ \text{mL NaOH})\left(\frac{1\ \text{mmol } H_2SO_4}{2\ \text{mmol NaOH}}\right)$$

and so

$$M_{H_2SO_4} = \frac{0.113\ M \times 10.8\ \text{mL}}{2(15.0\ \text{mL})}$$

which shows that **(d)** is the correct answer. [Answers **(a)** and **(b)** don't involve the 2:1 $NaOH:H_2SO_4$ ratio; answer **(c)** is simply nonsensical; answer **(e)** has the 2:1 ratio inverted.]

PROBLEM 12-18   What is the pH of 100.0 mL of aqueous solution containing 0.215 g of trifluoromethane sulfonic acid (CF_3SO_3H), a completely ionized monoprotic acid?

Solution: The molar mass of CF_3SO_3H is 150.1 g/mol, so the number of moles of CF_3SO_3H in 100.0 mL of solution is

$$\frac{0.215\ \text{g}}{150.1\ \text{g/mol}} = 1.43 \times 10^{-3}\ \text{mol}$$

The number of moles of H^+ in 100.0 mL of solution is

$$(1.43 \times 10^{-3}\ \text{mol } CF_3SO_3H)\left(\frac{1\ \text{mol } H^+}{1\ \text{mol } CF_3SO_3H}\right) = 1.43 \times 10^{-3}\ \text{mol}$$

Hence

$$[H^+] = \frac{\text{mol } H^+}{\text{L soln}} = \frac{(1.43 \times 10^{-3}\ \text{mol})(10^3\ \text{mL/L})}{100.0\ \text{mL}} = 1.43 \times 10^{-2}\ M$$

Thus

$$pH = -\log[H^+] = -\log(1.43 \times 10^{-2}) = 1.84$$

PROBLEM 12-19   Calculate the pH of a mixture of 15.0 mL of 0.126 M NaOH and 21.0 mL of 0.051 M H_2SO_4. (Assume that the volume of the mixture is simply the sum of the volumes mixed).

Solution: First examine the stoichiometry and see if there's a limiting reagent:

$$2NaOH + H_2SO_4 \longrightarrow Na_2SO_4 + 2H_2O$$

The number of millimoles of NaOH taken is $(15.0\ \text{mL})(0.126\ M) = 1.89$. This would require $(1.89\ \text{mmol NaOH}) \times (1\ \text{mmol } H_2SO_4)/(2\ \text{mmol NaOH}) = 0.945$ mmol of H_2SO_4 for complete neutralization. The amount of H_2SO_4 taken is $(21.0\ \text{mL})(0.051\ M) = 1.071$ mmol. So NaOH is limiting.

The amount of H_2SO_4 neutralized by the 1.89 mmol NaOH in the mixture is $(1.89\ \text{mmol NaOH}) \times (1\ \text{mmol } H_2SO_4)/(2\ \text{mmol NaOH}) = 0.945$ mmol. The amount of unreacted H_2SO_4 is $1.071\ \text{mmol} - 0.945\ \text{mmol} = 0.126$ mmol; therefore, the amount of H^+ is $(0.126\ \text{mmol } H_2SO_4) \times (2\ \text{mmol } H^+)/(1\ \text{mmol } H_2SO_4) = 0.252$ mmol. Thus

$$[H^+] = \frac{0.252\ \text{mmol}}{(15.0 + 21.0)\ \text{mL}} = 7.00 \times 10^{-3}$$

and

$$pH = -\log[H^+] = -\log(7.00 \times 10^{-3}) = 2.15$$

Supplementary Exercises

PROBLEM 12-20   What are the conjugate bases of the following Brønsted–Lowry acids: (a) HI, (b) HSO_3^-, (c) NH_4^+, (d) OH^-, (e) $N_2H_6^{2+}$?

Answer: (a) I^- (b) SO_3^{2-},   (c) NH_3 (d) O^{2-} (e) $N_2H_5^+$

PROBLEM 12-21   Identify the Lewis bases in each of the following Lewis acid–base complexes: (a) H_2OAlCl_3, (b) NH_4^+, (c) H_3PAgI, (d) BF_4^-.

Answer: (a) H_2O (b) NH_3 (c) H_3P (d) F^-

PROBLEM 12-22   Complete and balance the following acid–base reaction equations:

(a) $Al(OH)_3 + H_2SO_4 \longrightarrow Al_2(SO_4)_3$
(b) $CH_3CO_2H + NH_3$
(c) $NH_4^+ + OH^- \longrightarrow$

Answer: (a) $2Al(OH)_3 + 3H_2SO_4 \longrightarrow Al_2(SO_4)_3 + 6H_2O$
(b) $CH_3CO_2H + NH_3 \longrightarrow CH_3CO_2^- \, NH_4^+$
(c) $NH_4^+ + OH^- \longrightarrow NH_3 + H_2O$

PROBLEM 12-23   How much water would you add to 100.0 mL of 0.300 M NaOH to make the resulting solution 0.100 M NaOH (assuming that the volumes are additive)?

Answer: 2.00×10^2 mL

PROBLEM 12-24   What volume of 0.0641 M $HClO_4$ solution would you take to make 250.0 mL of 0.0250 M solution by dilution with water?

Answer: 97.5 mL

PROBLEM 12-25   How much solid oxalic acid $(C_2O_4H_2)$ would you weigh out to make 500.0 mL of a 0.100 M solution?

Answer: 4.50 g

PROBLEM 12-26   A sample of 0.242 g of $KC_8H_5O_4$, a monoprotic acid, requires 31.4 mL of KOH solution to neutralize it. What is the molarity of the KOH solution?

Answer: 0.0377 M

PROBLEM 12-27   A 15.0-mL sample of 0.0377 M KOH is neutralized by 7.84 mL of a hydrochloric acid solution. What is the molarity of the HCl solution?

Answer: 0.0721 M

PROBLEM 12-28   A 25.0-mL sample of 0.0377 M KOH is neutralized by 18.1 mL of a sulfuric acid solution. What is the molarity of the H_2SO_4 solution?

Answer: 0.0260 M

PROBLEM 12-29   What is the pH of the following solutions: (a) $[H^+] = 1.0 \times 10^{-2}$ M, (b) 0.30 M HCl solution, (c) 1.0×10^{-3} M H_2SO_4 solution, (d) $[H^+] = 7.9 \times 10^{-10}$ M?

Answer: (a) 2.00   (b) 0.52   (c) 2.70   (d) 9.10

PROBLEM 12-30   Calculate the hydrogen ion concentration in solutions of the following pH values: (a) 4.64, (b) 7.07, (c) 9.90.

Answer: (a) 2.3×10^{-5} M (b) 8.5×10^{-8} M (c) 1.3×10^{-10} M

PROBLEM 12-31 Write the balanced ionic equations for the following reactions and identify the spectator ions:

(a) $KOH + H_3PO_4 \longrightarrow K_3PO_4$

(b) $(NH_4)_2SO_4 + NaOH \longrightarrow NH_3 + Na_2SO_4$

(c) $Mg(OH)_2 + HI \longrightarrow MgI_2$

Answer: (a) $3OH^- + 3H^+ \longrightarrow 3H_2O$; spectator ions K^+, PO_4^{3-}

(b) $2NH_4^+ + 2OH^- \longrightarrow 2NH_3 + 2H_2O$; spectator ions SO_4^{2-}, Na^+

(c) $2OH^- + 2H^+ \longrightarrow 2H_2O$; spectator ions Mg^{2+}, I^-

13 REDOX REACTIONS

THIS CHAPTER IS ABOUT

☑ **Redox Reactions: Definitions**
☑ **Oxidation Numbers**
☑ **Balancing Redox Equations by Half-Reactions**
☑ **Balancing Redox Equations by Oxidation Numbers**
☑ **Analytical Uses of Redox Reactions**

13-1. Redox Reactions: Definitions

Reactions in which electrons are transferred from one atom to another are called **oxidation–reduction** or **redox reactions**. **Oxidation** is a loss of electrons, and **reduction** is a gain of electrons. Because electrons are always conserved in ordinary chemical reactions, the processes of oxidation and reduction always occur simultaneously. Any species that can supply electrons in a redox reaction is a **reducing agent**; any species that can accept electrons in a redox reaction is an **oxidizing agent**.

- The oxidizing agent oxidizes the reducing agent, and the reducing agent reduces the oxidizing agent; the reducing agent is oxidized by the oxidizing agent, and the oxidizing agent is reduced by the reducing agent.
- The reducing agent donates electrons to the oxidizing agent while the oxidizing agent accepts electrons from the reducing agent.

If you try to put this information together with what you know about chemical bonding, you may conclude that redox reactions can only produce ionic compounds, i.e., compounds formed by the transfer of electrons. Don't fall for this. An atom is oxidized during ANY change that makes it more positive or less negative. Recall that in covalent bonds electrons are most likely to be found near the more electronegative atom, resulting in an unequal sharing of electrons. This unequal sharing affects the electrical character of each atom and can be accounted for in redox reactions.

EXAMPLE 13-1: The reaction between metallic lithium and chlorine gas proceeds as follows:

$$2Li(s) + Cl_2(g) \longrightarrow 2Li^+Cl^-(s)$$

Which element is (**a**) oxidized; (**b**) reduced? Which element is the (**c**) oxidizing agent; (**d**) reducing agent?

Solution:
(**a**) Oxidation is the loss of electrons. In going from metallic Li to Li^+, each lithium atom has lost an electron, so lithium is oxidized.
(**b**) Reduction is the gain of electrons. In going from gaseous Cl_2 to $2Cl^-$, each chlorine atom has gained an electron, so chlorine is reduced.
(**c**) The oxidizing agent accepts electrons. In going from Cl_2 to $2Cl^-$, each chlorine atom has accepted an electron from a lithium atom, so gaseous chlorine is the oxidizing agent.
(**d**) The reducing agent supplies electrons. In going from Li to Li^+, each lithium atom has supplied an electron to a chlorine atom, so metallic lithium is the reducing agent.

13-2. Oxidation Numbers

A. Definition of the oxidation number

The **oxidation number**, or **oxidation state**, of an atom in a compound is an assigned numerical representation of the positive or negative character of the atom; in other words, it is the number of electrons that an atom *appears* to have gained over or lost from its normal complement when it is combined with other atoms. There are two general criteria for assigning electrons to atoms in combination:

- Electrons shared between two different atoms are counted with the more electronegative atom.
- Electrons shared between two identical atoms are divided equally between the two atoms.

EXAMPLE 13-2: Determine the oxidation numbers of each atom in (a) H_2 and (b) H_2O.

Solution:
(a) Because the electron pair in H_2 is shared by two identical atoms, each H atom is assigned 1 electron. A hydrogen atom having 1 electron is a neutral free hydrogen atom, so the oxidation number for each H is 0.
(b) The Lewis structure of H_2O is

$$H : \overset{\cdot\cdot}{\underset{\cdot\cdot}{O}} : H$$

Since oxygen is more electronegative than hydrogen, you count all the electrons with O. Each H atom then appears to have lost 1 electron, giving it a $+1$ charge; and the O appears to have gained 2 electrons, giving it a -2 charge. So the oxidation numbers are $+1$ for H and -2 for O.

> *note:* DON'T equate oxidation number and formal charge. You use the concept of formal charge to map out the *real* charge distribution of a molecule or ion based on a detailed understanding of the structure and chemical bonds involved. For example, the formal charge on each atom in H_2O is 0, while the oxidation numbers of H and O in H_2O are $+1$ and -2, respectively. Oxidation numbers are simply devices, derived from a set of arbitrary rules, that make the treatment of redox reactions quantitative: They are used to determine which atoms are being oxidized or reduced in a redox reaction and to keep track of the electrons being transferred in the redox process.

B. Rules for determining oxidation numbers

You could determine oxidation numbers by drawing the complete Lewis structure of any compound and counting the electrons lost or gained by each atom, but you'll find it easier to assign oxidation numbers in accordance with a set of rules derived from the two criteria for assigning electrons to atoms in combination:

1. *The oxidation number of atoms in neutral elements is always zero.*
2. *The oxidation number for atoms in simple ions consisting of only one atom is the same as the charge on the ion:* H^+ $(+1)$, Ca^{2+} $(+2)$, Al^{3+} $(+3)$, Cd^{2+} $(+2)$, F^- (-1), Cl^- (-1), and O^{2-} (-2). *Elements of groups IA and IIA always have oxidation numbers of $+1$ and $+2$, respectively, in compounds.*
3. *The oxidation number of the oxygen atom is usually -2.* Exceptions occur in hydrogen peroxide (H_2O_2), where the oxidation number of oxygen is -1, and in oxygen difluoride (OF_2), where the oxidation number of oxygen is $+2$.
4. *The oxidation number of hydrogen is usually $+1$.* The one exception occurs in metal hydrides like Na^+H^- or $Ca^{2+}(H^-)_2$, where the oxidation number of hydrogen is -1, consistent with general criterion 1.
5. *The sum of the oxidation numbers of all of the atoms in a molecule or ion equals the charge on the molecule or ion.* In a neutral molecule the sum of the oxidation numbers must equal zero. For example, in H_2O

$$\begin{array}{ccc} H & H & O \\ (+1) & + (+1) & + (-2) = 0 \end{array}$$

You can use this rule to determine the oxidation number of other atoms as long as they are combined only with elements covered by Rules 1 through 4. [You'll find that 99% of the compounds you deal with will be covered by these five rules.]

6. *Some compounds contain atoms that are not covered by the rules. For these atoms you assign electrons according to the general criteria.* Consider the compound chlorine trifluoride, ClF_3, whose central atom does not follow the octet rule. Recall that the most electronegative atom in a group of the periodic table is the one highest in the group. Consequently, F is more electronegative than Cl. In ClF_3, then, you assign both electrons of each Cl—F bonding pair to F:

$$\begin{matrix}
\ddot{\underset{\cdot\cdot}{F}}: & & & \ddot{\underset{\cdot\cdot}{F}}:^- \\
| & & & \\
:\underset{\cdot\cdot}{Cl}—\ddot{\underset{\cdot\cdot}{F}}: \longrightarrow & :\underset{\cdot\cdot}{Cl}^{3+} \quad \text{and} & & :\ddot{\underset{\cdot\cdot}{F}}:^- \\
| & & & \\
:\ddot{\underset{\cdot\cdot}{F}}: & & & :\ddot{\underset{\cdot\cdot}{F}}:^-
\end{matrix}$$

Since each F atom started with 7 valence electrons and now has 8, each F appears to have a -1 charge; therefore, F has an oxidation number of -1. Since Cl started with 7 valence electrons and now has 4, it appears to have a $+3$ charge, and therefore an oxidation number of $+3$.

note: Remember that this is a hypothetical process devised to assign oxidation numbers. In ClF_3 there are no Cl^{3+} or F^- ions; it is a covalent compound. Rule 6 is certainly the most complicated of the oxidation number rules, and the one that you'll find the hardest initially. Don't worry: It's the one you'll need the least, since most compounds of interest in redox reactions do have hydrogen or oxygen and Rules 1 through 5 work fine for them.

EXAMPLE 13-3: Assign oxidation numbers to each of the atoms in (a) N_2, (b) HNO_2, (c) N_2H_4, (d) NO_3^-, and (e) NH_3.

Solution:

(a) N_2: Nitrogen is not combined with any other element. According to Rule 1, its oxidation number is 0.

(b) HNO_2: Following Rules 3 and 4, the oxidation numbers of O and H are -2 and $+1$, respectively. The oxidation number of N is found from Rule 5:

$$\begin{matrix} N & H & O & HNO_2 \\ x + (+1) + 2(-2) = & & & 0 \end{matrix}$$

So the oxidation number (x) of N in HNO_2 is $+3$.

(c) N_2H_4: The oxidation number of H is $+1$ (Rule 4). Apply Rule 5: $2x + 4(+1) = 0$; $x = -4(+1)/2 = -2$. So the oxidation number of N in N_2H_4 is -2.

(d) NO_3^-: The oxidation number of O is -2 (Rule 3). The charge on the NO_3^- ion is -1. Apply Rule 5: $x + 3(-2) = -1$; $x = -1 - 3(-2) = +5$. So the oxidation number of N in NO_3^- is $+5$.

(e) NH_3: The oxidation number of H is $+1$ (Rule 4). From Rule 5: $x + 3(+1) = 0$; $x = -3$. So the oxidation number of N in NH_3 is -3.

---

note: As you can see from Example 13-3, a single element can display quite a range of oxidation numbers.

13-3. Balancing Redox Equations by Half-Reactions

Every complete redox reaction has two **half-reactions**—one reaction representing oxidation alone and the other representing reduction alone. Individual half-reactions, most often written as net ionic equations, can be quickly balanced in terms of atoms and charge, but free electrons appear as reactants (reduction half-reactions) or products (oxidation half-reactions). The two balanced half-reactions are then combined, the free electrons eliminated, and a complete redox reaction produced. The pH of the solution in which a reaction takes place is a major factor in balancing half-reactions.

A. Half-reactions in acidic solution

Many common oxidizing agents are used in acidic solution, so that there's a good supply of H^+. Many of these agents contain **oxyanions**—anions having oxygen atoms bonded to another atom with a large positive oxidation number; permanganate (MnO_4^-) and dichromate ($Cr_2O_7^{2-}$) are typical oxidizing agents that contain oxyanions. The reduced products of such oxyanion-containing agents rarely contain oxygen (e.g., Mn^{2+}; Cr^{3+}); the oxygen atoms are accounted for by combination with H^+ to produce H_2O. [The production of water is NOT a redox process: The oxidation numbers of oxygen and hydrogen remain unchanged in this process (-2 and $+1$, respectively).]

The steps for balancing half-reactions in acidic solution are

Step 1: Calculate oxidation numbers if necessary to identify the oxidation and reduction half-reactions. Write half-reactions from the unbalanced equation.

Step 2: Balance each half-reaction for all atoms except H and O.

Step 3a: Add H_2O to the side deficient in O and add H^+ to the side deficient in H to account for O and H.

Step 4: Balance each half-reaction for total electrical charge: Add electrons to the reactants of the reduction half-reaction and to the products of the oxidation half-reaction.

Step 5: Balance the electrons in the two half-reactions: Multiply the half-reactions by appropriate integers so that the numbers of electrons in each equation are equal.

Step 6: Add the two half-reactions, canceling where appropriate. All free electrons must cancel.

EXAMPLE 13-4: Balance the following redox reaction in acidic solution:

$$Fe^{2+} + Cl_2 \longrightarrow Fe^{3+} + Cl^-$$

Solution: Follow the steps for balancing half-reactions in acid solution.

Step 1. Calculate oxidation numbers:

Species	Oxidation number
Fe^{2+}	$+2$
Fe^{3+}	$+3$
Cl_2	0
Cl^-	-1

Chlorine has gained electrons, so it is being reduced; Fe has lost electrons, so it is being oxidized. Now write the half-reactions:

Oxidation half-reaction	**Reduction half-reaction**
$Fe^{2+} \longrightarrow Fe^{3+}$	$Cl_2 \longrightarrow Cl^-$

Step 2. Balance each half-reaction:

$$Fe^{2+} \longrightarrow Fe^{3+} \qquad | \qquad Cl_2 \longrightarrow 2Cl^-$$

Step 3a. There are no O atoms to account for.

Step 4. Add enough e^- to balance the charge (add e^- to the products in the oxidation reaction and to the reactants in the reduction reaction):

$$Fe^{2+} \longrightarrow Fe^{3+} + e^- \qquad | \qquad Cl_2 + 2e^- \longrightarrow 2Cl^-$$
$$(+2) = (+3) + (-1) \qquad | \qquad (0) + 2(-1) = 2(-1)$$

Step 5. Balance electrons in the two half-reactions. Here, there are $2e^-$ in the reduction half-reaction, so we multiply the oxidation half-reaction by 2:

$$2Fe^{2+} \longrightarrow 2Fe^{3+} + 2e^- \qquad | \qquad Cl_2 + 2e^- \longrightarrow 2Cl^-$$

Step 6.   Add the two half-reactions and cancel where appropriate:

$$2Fe^{2+} \longrightarrow 2Fe^{3+} + 2e^-$$
$$Cl_2 + 2e^- \longrightarrow 2Cl^-$$
$$\overline{2Fe^{2+} + Cl_2 + \cancel{2e^-} \longrightarrow 2Fe^{3+} + 2Cl^- + \cancel{2e^-}}$$
$$2Fe^{2+} + Cl_2 \longrightarrow 2Fe^{3+} + 2Cl^-$$

A quick check shows that all atoms and charges are balanced.

EXAMPLE 13-5: Balance the following redox reaction in acidic solution:

$$MnO_4^- + S^{2-} \longrightarrow Mn^{2+} + SO_4^{2-}$$

Solution: This reaction is rather complex, but don't panic: Just follow the steps.

Step 1.   Calculate oxidation numbers (remember, O is -2):

Species			Oxidation number
MnO_4^-	$x + 4(-2) = -1$;	$x = +7$	$+7$ for Mn
Mn^{2+}			$+2$ for Mn
S^{2-}			-2 for S
SO_4^{2-}	$x + 4(-2) = -2$;	$x = +6$	$+6$ for S

Mn has gained electrons, so it is being reduced; S has lost electrons, so it is being oxidized (MnO_4^- is the oxidizing agent, and S^{2-} is the reducing agent):

Oxidation half-reaction	**Reduction half-reaction**
$S^{2-} \longrightarrow SO_4^{2-}$	$MnO_4^- \longrightarrow Mn^{2+}$

Step 2.   The half-reactions are balanced for Mn and S.

Step 3a. Add H_2O to balance O; then add H^+ to balance H:

$$S^{2-} + 4H_2O \longrightarrow SO_4^{2-} + 8H^+ \qquad \vert \qquad MnO_4^- + 8H^+ \longrightarrow Mn^{2+} + 4H_2O$$

Step 4.* Add e^- to balance the charge:

$$S^{2-} + 4H_2O \longrightarrow SO_4^{2-} + 8H^+ + 8e^- \qquad \vert \qquad MnO_4^- + 8H^+ + 5e^- \longrightarrow Mn^{2+} + 4H_2O$$
$$(-2) + (0) = (-2) + (8) + (-8) \qquad \qquad (-1) + (8) + (-5) = (+2) + (0)$$

Step 5.   Balance the electrons in the two half-reactions:

$$5S^{2-} + 20H_2O \longrightarrow 5SO_4^{2-} + 40H^+ + 40e^- \quad \vert \quad 8MnO_4^- + 64H^+ + 40e^- \longrightarrow 8Mn^{2+} + 32H_2O$$

Step 6.   Add the two half-reactions and cancel where appropriate:

$$5S^{2-} + 20H_2O + 8MnO_4^- + 64H^+ + 40e^- \longrightarrow 8Mn^{2+} + 32H_2O + 5SO_4^{2-} + 40H^+ + 40e^-$$
$$-20H_2O \qquad\qquad -40H^+ - 40e^- \longrightarrow \qquad -20H_2O \qquad -40H^+ - 40e^-$$
$$\overline{5S^{2-} \qquad\qquad + 8MnO_4^- + 24H^+ \qquad \longrightarrow 8Mn^{2+} + 12H_2O + 5SO_4^{2-}}$$

* *Important*: Step 4 is crucial. The numbers of electrons you use to balance the charges *must* reflect the changes in oxidation numbers determined in Step 1. Thus each Mn, in changing from $+7$ to $+2$, gains 5 electrons; and each S, in changing from -2 to $+6$, gives up 8 electrons. If the numbers of electrons needed in Step 4 do not show this correlation at this stage, you have made a mistake somewhere. Check your work before going on to Step 5.

B. Half-reactions in basic solution

The steps for balancing half-reactions in basic solution are the same as those for acidic solutions— with one exception: You can't use H^+ because there aren't any protons available in basic solution. So we replace Step 3a with

Step 3b: Add sufficient OH^- and H_2O to balance the half-reactions for H and O.

EXAMPLE 13-6: Balance the following redox reaction in basic solution:

$$HS^- + NO_3^- \longrightarrow S + NO_2^-$$

Solution: Follow the steps for balancing half-reactions in basic solution.

Step 1.   Calculate the oxidation numbers:

Species			Oxidation number
HS^-	$+1 + x = -1;$	$x = -2$	-2 for S
S			0 for S
NO_3^-	$x + 3(-2) = -1;$	$x = +5$	$+5$ for N
NO_2^-	$x + 2(-2) = -1;$	$x = +3$	$+3$ for N

S has lost electrons so it is being oxidized; N has gained electrons so it is being reduced. (NO_3^- is the oxidizing agent and HS^- is the reducing agent.)

Oxidation half-reaction	**Reduction half-reaction**
$HS^- \longrightarrow S$	$NO_3^- \longrightarrow NO_2^-$

Step 2.   The half-reactions are already balanced for S and N.

Step 3b. Balance the half-reaction for H and O with H_2O and OH^-:

$$HS^- + OH^- \longrightarrow S + H_2O \qquad \Big| \qquad NO_3^- + H_2O \longrightarrow NO_2^- + 2OH^-$$

> *note:* The Step 3b procedure for the reduction half-reaction may not be obvious. Proceed this way. You see that NO_3^- has one more O than NO_2^- does; so add OH^- to the right side: $NO_3^- \to NO_2^- + OH^-$. Now the left side is deficient in H, so you add H_2O to the reactants: $NO_3^- + H_2O \to NO_2^- + OH^-$. Now balance the sides by changing the coefficient of OH^- until you come out even in H's and O's: $NO_3^- + H_2O \to NO_2^- + 2OH^-$.

Step 4.   Add e^- to balance the charge:

$$HS^- + OH^- \longrightarrow S + H_2O + 2e^- \qquad \Big| \qquad NO_3^- + H_2O + 2e^- \longrightarrow NO_2^- + 2OH^-$$
$$(-1) + (-1) = (-2) \qquad\qquad\qquad (-1) \quad + (-2) = (-1) + (-2)$$

(*Check:* S loses $2e^-$ in going from -2 to 0, and N gains $2e^-$ in going from $+5$ to $+3$.)

Step 5.   Both half-reactions hold the same number of e^-.

Step 6.   Add and cancel where appropriate:

$$HS^- + \cancel{OH^-} + NO_3^- + \cancel{H_2O} + \cancel{2e^-} \longrightarrow S + \cancel{H_2O} + \cancel{2e^-} + NO_2^- + \cancel{2}OH^-$$
$$HS^- + NO_3^- \longrightarrow S + NO_2^- + OH^-$$

13-4. Balancing Redox Equations by Oxidation Numbers

You can save some time and effort by using oxidation numbers directly to balance redox equations. The direct oxidation-number method is useful because the reactions need not occur in solution. Another advantage is that you don't have to know whether the reactants or products are ionic or covalent. There are only three steps:

Step 1: Find the oxidation numbers for all oxidized and reduced substances and thus determine the oxidation-number change for each.

Step 2: Add integer coefficients to reactants and products so that the total increase in oxidation number equals the total decrease in oxidation number.

Step 3: Balance all remaining atoms by inspection; save H^+, OH^-, and H_2O until last and add as needed according to the pH of the solution.

EXAMPLE 13-7: Balance the following redox reaction in acidic solution using the oxidation-number method:

$$H_2S + NO_3^- \longrightarrow S + NO_2$$

Solution:

Step 1. Determine the oxidation numbers and oxidation-number changes for all oxidized and reduced substances:

$$\overset{-2}{H_2}\overset{+5}{S} + \overset{+5}{NO_3^-} \longrightarrow \overset{0}{S} + \overset{+4}{NO_2}$$

For $H_2S \rightarrow S$, in which S goes from -2 to 0, there is an increase of 2 in oxidation number.
For $NO_3^- \rightarrow NO_2$, in which N goes from $+5$ to $+4$, there is a decrease of 1 in oxidation number.

Step 2. Balance (equate) increase and decrease in oxidation-number change: You need two N atoms for each S atom to balance the oxidation-number change, so

$$H_2S + 2NO_3^- \longrightarrow S + 2NO_2$$

Step 3. Balance the remaining atoms. There are 6 O atoms on the left and 4 on the right, so add $2H_2O$ to the right:

$$H_2S + 2NO_3^- \longrightarrow S + 2NO_2 + 2H_2O$$

There are 2 H atoms on the left and 4 on the right. Because the reaction takes place in acidic solution, add $2H^+$ to the left side:

$$H_2S + 2NO_3^- + 2H^+ \longrightarrow S + 2NO_2 + 2H_2O$$

All atoms balance and so does the charge ($-2 + 2 = 0$). [If your charges don't balance at this stage, you did something wrong. Go back and check.]

EXAMPLE 13-8: Balance the following redox reaction in basic solution using the oxidation-number method:

$$Ag_2O + Co(OH)_2 \longrightarrow Ag + Co(OH)_3$$

Solution:

Step 1. Determine the oxidation numbers and oxidation-number changes for all oxidized and reduced substances:

$$\overset{+2}{Ag_2}\overset{+2}{O} + \overset{+2}{Co}(OH)_2 \longrightarrow \overset{0}{Ag} + \overset{+3}{Co}(OH)_3$$

$$\overset{+2}{Ag_2}O \longrightarrow \overset{0}{Ag}: \quad \text{decrease of 2 in oxidation number}$$

$$\overset{+2}{Co}(OH)_2 \longrightarrow \overset{+3}{Co}(OH)_3: \quad \text{increase of 1 in oxidation number}$$

Step 2. Balance increase and decrease in oxidation-number change: You need 2 Co atoms for each Ag atom to balance the oxidation-number change, so

$$Ag_2O + 2Co(OH)_2 \longrightarrow Ag + 2Co(OH)_3$$

Step 3. Balance the remaining atoms. There are 2 Ag atoms on the left and only 1 Ag on the right, so

$$Ag_2O + 2Co(OH)_2 + H_2O \longrightarrow 2Ag + 2Co(OH)_3$$

There are 5 O atoms on the left and 6 on the right, so add H_2O to the left:

$$Ag_2O + 2Co(OH)_2 + H_2O \longrightarrow 2Ag + 2Co(OH)_3$$

There are 6 H atoms on the left and on the right, and both sides are neutral. It's done.

---

note: Both the half-reaction and oxidation-number methods give the same answer, so use whichever you feel most comfortable with.

13-5. Analytical Uses of Redox Reactions

You can use a balanced redox equation for stoichiometric calculations just as you can use any balanced chemical equation. In fact, redox reactions are very useful. They are often both fast and quantitative, so they're widely used in the chemical analysis of elements or compounds that can be quantitatively oxidized or reduced, including most of the transition metals and a wide range of main-group inorganic and organic compounds. For these determinations a relatively narrow range of standard redox reagents is used. A listing of some of the more popular reagents is given in Table 13-1.

TABLE 13-1: Common Redox Reagents

Reagent	Product(s)	Conditions
Oxidizing agents		
Ce^{4+}	Ce^{3+}	acidic
I_2	I^-	acidic
$Cr_2O_7^{2-}$ (dichromate)	Cr^{3+}	acidic
MnO_4^- (permanganate)	Mn^{2+}	acidic
MnO_4^- (permanganate)	MnO_2	basic
Reducing agents		
As_2O_3	AsO_4^{3-}	acidic
$S_2O_3^{2-}$ (thiosulfate)	$S_4O_6^{2-}$	neutral
Cr^{2+}	Cr^{3+}	acidic
Zn	Zn^{2+}	acidic
HI	I_2	acidic

SUMMARY

1. Oxidation is a loss of electrons and reduction is a gain of electrons.
2. The oxidation number of an atom in a compound is a numerical representation of the positive or negative character of the atom as defined by a set of arbitrary rules.
3. Half-reactions in redox reactions represent oxidation alone or reduction alone.
4. The pH of the solution in which a redox reaction takes place affects the way in which individual half-reactions are balanced.
5. The total increase in oxidation number in a redox reaction must equal the total decrease in oxidation number.

RAISE YOUR GRADES

Can you define...?

☑ oxidizing agent
☑ reducing agent
☑ oxyanion

Can you...?

☑ recognize a redox reaction
☑ identify the atom oxidized and atom reduced
☑ determine oxidation numbers
☑ write and balance half-reactions
☑ use oxidation numbers to balance redox equations

SOLVED PROBLEMS

PROBLEM 13-1 In the following reaction between metallic Mg and S, which element is oxidized? Which is reduced? Identify the oxidizing and reducing agents.

$$Mg(s) + S(s) \longrightarrow Mg^{2+}S^{2-}(s)$$

Solution: Oxidation is defined as the loss of electrons. Consequently, you know that an element undergoing oxidation must lose negative charge, thereby gaining positive charge. Here it is Mg that increases in positive charge. Therefore, Mg is oxidized.

Reduction is defined as the gain of electrons. An element being reduced will increase in negative charge. Consequently, S, which goes from a neutral atom to an ion carrying a double negative charge, S^{2-}, is reduced.

A species that supplies electrons to reduce another is a reducing agent. It is the Mg that supplies the electrons to S in this reaction (and thus loses electrons itself), and so Mg is the reducing agent. A species that accepts electrons from a reducing agent is an oxidizing agent. In this reaction, S is the oxidizing agent.

PROBLEM 13-2   Assign oxidation numbers to each atom in (a) Br_2, (b) $Ca^{2+}(Br^-)_2$, (c) K^+OBr^-, (d) $Mg^{2+}(BrO_4^-)_2$, (e) HBr, (f) Br_2O, (g) CBr_4.

Solution: Inspect the compound under consideration, decide which rules are relevant (see Section 13-2B), and then start counting.

(a) Br_2 is a neutral element. By Rule 1, the oxidation number of a neutral atom is 0, so the oxidation number of each bromine atom is 0.

(b) $Ca^{2+}(Br^-)_2$ (calcium bromide) is clearly written as an ionic compound. (Even if it were written $CaBr_2$, you would probably realize that it is likely to be ionic.) Rule 2 states that elements in group IIA of the periodic table always have an oxidation number of +2. Consequently, the oxidation number of calcium in $Ca^{2+}(Br^-)_2$ is +2. Rule 2 also states that an atom in a simple ion consisting of only one atom has an oxidation number equal to the charge of the ion, so the oxidation number of bromine is −1. The total of the oxidation numbers is $+2 + 2(-1) = 0$, which is what you would expect, since the compound as a whole is neutral (Rule 5).

(c) By Rule 2, you see at once that the oxidation number of K in K^+OBr^- is +1. Then by Rule 5, the sum of the oxidation numbers of O and Br in OBr^- must be −1. Now apply Rule 3 to OBr^-: If O has an oxidation number of −2, then Br must have an oxidation number of +1 so that the total $(-2 + 1 = -1)$ is −1.

(d) As in example (c), Mg (group IIA) in $Mg^{2+}(BrO_4^-)_2$ must have oxidation number +2. The sum of the oxidation numbers of Br and the four O atoms in BrO_4^- must then equal −1, the charge on the ion. By Rule 3, each O is assigned an oxidation number of −2. Consequently, Br must have an oxidation number of +7 in this ion since $+7 + 4(-2) = -1$.

(e) By Rule 4 you assign +1 as the oxidation number of H in HBr. Since the compound is neutral (net charge = 0), the oxidation number of Br must be −1 so that the total $(+1 - 1)$ is 0.

(f) By Rule 3 you assign −2 as the oxidation number of O in Br_2O. The compound is neutral, so the oxidation number of each Br must be +1.

(g) CBr_4 is the only example that needs Rule 6. Because Br is in group VIIA at the right-hand side of the periodic table, and C is in group IVA in the middle, you expect Br to be more electronegative than C. Consequently, you assign both electrons in the C—Br bonds to the Br, making the oxidation number of each Br −1. Since the compound is neutral, the oxidation number of C must be +4 because $+4 + 4(-1) = 0$. In Lewis structural terms you could write

$$\text{:}\overset{\cdot\cdot}{\underset{\cdot\cdot}{Br}}\text{:}\underset{\text{:}\overset{\cdot\cdot}{\underset{\cdot\cdot}{Br}}\text{:}}{\overset{\text{:}\overset{\cdot\cdot}{\underset{\cdot\cdot}{Br}}\text{:}}{\text{:}Br\text{:}C\text{:}Br\text{:}}} \longrightarrow {}^-\text{:}Br\text{:} \quad C^{4+} \quad \text{:}Br\text{:}^-$$

Remember that this is only an arbitrary assignment; there are no free ions in CBr_4, which is obviously a covalent compound. The oxidation number assignment is made simply because it's useful in balancing redox reactions. Note that bromine displays a range of oxidation numbers, including +7, +1, 0, and −1 (the last being the most common).

PROBLEM 13-3   In which of the following molecules and ions is the oxidation number of sulfur +4? (a) SO_3^{2-}, (b) H_2SO_4, (c) SO_3, (d) SO_2, (e) SCl_2.

Solution: The only certain way to do this problem is to determine the oxidation number of S in all of the choices because it is possible that more than one answer is correct. For each choice then, let x be the oxidation number of S.

(a) The oxidation number of O in SO_3^{2-} is −2 (Rule 3), and the sum of the oxidation numbers is −2 (Rule 5). Therefore, $x + 3(-2) = -2$; $x = +4$. SO_3^{2-} is a correct answer.

(b) The oxidation number of H in H_2SO_4 is +1 (Rule 4) and O is −2 (Rule 3). The sum of the oxidation numbers must be 0: $2(+1) + x + 4(-2) = 0$, so $x = +6$. H_2SO_4 is not a correct answer.

(c) The oxidation number of O in SO_3 is −2. The sum of the oxidation numbers must equal 0: $x + 3(-2) = 0$, so the oxidation number of S is +6, and SO_3 is not a correct answer.

(d) For SO_2, $x + 2(-2) = 0$, so the oxidation number of S is +4. SO_2 is also a correct answer.

(e) For SCl_2, you must use Rule 6. The Lewis structure for the compound is $\text{:}\overset{\cdot\cdot}{\underset{\cdot\cdot}{Cl}}\text{:}\overset{\cdot\cdot}{\underset{\cdot\cdot}{S}}\text{:}\overset{\cdot\cdot}{\underset{\cdot\cdot}{Cl}}\text{:}$ Judging from their positions in the periodic table, you know that Cl is more electronegative than S; therefore, the bonding electron pair is assigned to Cl, and you have

$$:\ddot{\underset{..}{Cl}}:^- \quad \ddot{S}^{2+} \quad :\ddot{\underset{..}{Cl}}:^-$$

The oxidation number of S is $+2$ here.
The correct answers are (a) and (d).

PROBLEM 13-4 Balance the following oxidation half-reactions in acidic solution (where H^+ may be used as a reactant) and in basic solution (where OH^- may be used as a reactant):

(a) $H_2 \longrightarrow H^+$ (d) $I^- \longrightarrow I_2$

(b) $Fe^{2+} \longrightarrow Fe^{3+}$ (e) $H_2O \longrightarrow O_2$

(c) $Cd \longrightarrow Cd^{2+}$

Solution: Although it's fairly easy to balance most of these by inspection, it's important to remember that in oxidation half-reactions electrons must appear in the equation as products. The number of electrons appearing must be equal to the change in oxidation number(s) of the atom(s) undergoing oxidation.

(a) $\quad\quad H_2 \longrightarrow H^+$ (acidic solution) $H_2 \longrightarrow H^+$ (basic solution)

The oxidation-number change for each H atom is from 0 (in the element H_2) to $+1$ in H^+; thus each H atom undergoing oxidation produces one electron. Since two H atoms are present in H_2, the balanced half-reaction is

The H^+ produced would react with OH^- to produce H_2O, so a realistic half-reaction would be

$$H_2 + 2OH^- \longrightarrow 2H_2O + 2e^-$$

$$H_2 \longrightarrow 2H^+ + 2e^-$$

(b) $\quad\quad\quad Fe^{2+} \longrightarrow Fe^{3+}$

By inspection it's clear that

$$Fe^{2+} \longrightarrow Fe^{3+} + e^-$$

Neither H^+ nor OH^- is involved, so the reaction must be the same in either acidic or basic solution.

(c) $\quad\quad\quad Cd \longrightarrow Cd^{2+}$

By inspection

$$Cd \longrightarrow Cd^{2+} + 2e^-$$

Same in acidic or basic solution.

(d) $\quad\quad\quad\quad I^- \longrightarrow I_2$

By inspection

$$2I^- \longrightarrow I_2 + 2e^-$$

Same in acidic or basic solution.

(e) $\quad\quad H_2O \longrightarrow O_2$ (acidic solution) $H_2O \longrightarrow O_2$ (basic solution)

The oxidation-number change for O is from -2 in H_2O to 0 in O_2, producing $2e^-$. The H is not involved in redox change and must stay $+1$ throughout. Thus in acidic solution, since you must use two H_2O molecules to produce one O_2 molecule, you have

H^+ is not a realistic product, so you must add four OH^- ions to each side of the equation:

$$4OH^- + 2H_2O \longrightarrow O_2 + 4e^- + 4H^+ + 4OH^-$$
$$4OH^- + 2H_2O \longrightarrow O_2 + 4e^- + 4H_2O$$

And now you can subtract $2H_2O$ from each side, getting

$$2H_2O \longrightarrow O_2 + 4e^- + 4H^+$$

$$4OH^- \longrightarrow O_2 + 2H_2O + 4e^-$$

which is in fact an accurate description of the net oxidation of water to give O_2 in basic solution.

PROBLEM 13-5 Which elements are oxidized and which elements are reduced in the following equations?

(a) $2Cu_2O + O_2 \longrightarrow 4CuO$

(b) $Sn + 2H^+ \longrightarrow H_2 + Sn^{2+}$

Solution: To do this problem, you must be able to recognize where there is an increase in oxidation number, i.e., a loss of electrons (oxidation), and where there is a decrease in oxidation number, i.e., a gain of electrons (reduction).

(a) The oxidation number of Cu in Cu_2O is $+1$ and in CuO is $+2$. Copper has increased in oxidation number (lost electrons) and therefore been oxidized. The oxidation number of O in O_2 is zero. In Cu_2O and CuO the oxidation number of O is -2. The oxidation number of O in O_2 has decreased from 0 to -2. (The oxidation number of the O in Cu_2O has not changed.) Oxygen has therefore gained electrons and been reduced.

(b) The oxidation number of tin in Sn is zero. It increases in oxidation number to $+2$ in Sn^{2+}, so it loses electrons. Tin is oxidized. The oxidation number of H is $+1$ in H^+, and it decreases in oxidation number to 0 in H_2. Hydrogen gains electrons and is reduced.

PROBLEM 13-6 In the following equation which element is reduced? Which substance is the oxidizing agent? Which element is oxidized? Which substance is the reducing agent?

$$2CrO_3 + 3SO_2 \longrightarrow 2Cr^{3+} + 3SO_4^{2-}$$

Solution: You must understand the meanings of the terms oxidation, reduction, oxidizing agent, and reducing agent. You may assume that the oxidation number of oxygen is -2 and does not change.

Here Cr changes oxidation number from $+6$ in CrO_3 to $+3$ in Cr^{3+}. The Cr gains electrons, so it is the element that is reduced. The oxidizing *agent*, which contains the element that is reduced, is CrO_3. [Note that the oxidizing agent is the compound CrO_3, not Cr or Cr^{6+}.]

Sulfur changes from an oxidation number of $+4$ in SO_2 to $+6$ in SO_4^{2-}. So S is the element that is oxidized, and SO_2 is the reducing agent.

PROBLEM 13-7 Give the balanced oxidation half-reaction, the balanced reduction half-reaction, and the balanced overall equation for

(a) $Cu + NO_3^- \longrightarrow NO + Cu^{2+}$ (acidic solution)
(b) $PbO_2 + Cl^- \longrightarrow ClO^- + Pb(OH)_2$ (basic solution)

Solution: Use the steps for balancing redox reactions. Note that (a) is in acidic and (b) is in basic solution.

(a) $Cu + NO_3^- \longrightarrow NO + Cu^{2+}$ (acidic soln)

Step 1: Identify the oxidation and reduction half-reactions by finding oxidation numbers. (Assume O is -2.)

Species	Oxidation number
Cu	0
Cu^{2+}	$+2$
NO_3^-	$+5$ for N
NO	$+2$ for N

Copper has lost electrons, so it is being oxidized; N has gained electrons, so it is being reduced. Write the unbalanced half-reactions:

Oxidation half-reaction	**Reduction half-reaction**
$Cu \longrightarrow Cu^{2+}$	$NO_3^- \longrightarrow NO$

Step 2: Balance each half-reaction for all atoms except H and O. Here both half-reactions are already balanced:

$Cu \longrightarrow Cu^{2+}$ $NO_3^- \longrightarrow NO$

Step 3a: For acidic solution, add H_2O and H^+ to account for H and O:

$Cu \longrightarrow Cu^{2+}$ $NO_3^- + 4H^+ \longrightarrow NO + 2H_2O$
(no H or O involved)

Step 4: Balance each equation for electrical charge:

$Cu \longrightarrow Cu^{2+} + 2e^-$ $NO_3^- + 4H^+ + 3e^- \longrightarrow NO + 2H_2O$
$(0) = (+2) + (-2)$ $(-1) + (+4) + (-3) = (0) + (0)$

Now the two half-reactions are balanced.

Step 5: Balance the electrons in the two half-reactions. Here, multiply the oxidation half-reaction by 3, and the reduction half-reaction by 2:

$3Cu \longrightarrow 3Cu^{2+} + 6e^-$ $2NO_3^- + 8H^+ + 6e^- \longrightarrow 2NO + 4H_2O$

Step 6: Add the two half-reactions, canceling where appropriate:

$$3Cu \longrightarrow 3Cu^{2+} \qquad +\cancel{6e^-}$$
$$+ \quad 2NO_3^- + 8H^+ + \cancel{6e^-} \longrightarrow 2NO + 4H_2O$$
$$\overline{3Cu + 2NO_3^- + 8H^+ \longrightarrow 3Cu^{2+} + 2NO + 4H_2O}$$

And the overall equation is balanced.

(b)
$$PbO_2 + Cl^- \longrightarrow ClO^- + Pb(OH)_2 \qquad \text{(basic soln)}$$

Step 1:

Species	Oxidation number
PbO_2	+4 for Pb
$Pb(OH)_2$	+2 for Pb
Cl^-	−1 for Cl
ClO^-	+1 for Cl

Oxidation half-reaction	**Reduction half-reaction**
$Cl^- \longrightarrow ClO^-$	$PbO_2 \longrightarrow Pb(OH)_2$

Step 2: The non-H and non-O atoms are balanced;

$$Cl^- \longrightarrow ClO^- \qquad \Big| \qquad PbO_2 \longrightarrow Pb(OH)_2$$

Step 3b: Since the reaction is carried out in basic solution, add H_2O and OH^- to account for O and H:

$$Cl^- + 2OH^- \longrightarrow ClO^- + H_2O \qquad \Big| \qquad PbO_2 + 2H_2O \longrightarrow Pb(OH)_2 + 2OH^-$$

Step 4:

$$Cl^- + 2OH^- \longrightarrow ClO^- + H_2O + 2e^-$$
$$(-1) + (-2) \ = \ (-1) + (0) + (-2)$$

$$PbO_2 + 2H_2O + 2e^- \longrightarrow Pb(OH)_2 + 2OH^-$$
$$(0) + (0) + (-2) = (0) + (-2)$$

The two half-reactions are now balanced.

Step 5: There are already an equal number of electrons in both half-reactions: $2e^-$ in the oxidation and $2e^-$ in the reduction half-reaction.

Step 6:

$$Cl^- \qquad + \cancel{2OH^-} \longrightarrow ClO^- + \cancel{H_2O} \qquad + \cancel{2e^-}$$
$$PbO_2 \quad + \cancel{2}H_2O \quad + \cancel{2e^-} \longrightarrow Pb(OH)_2 \qquad \qquad + \cancel{2OH^-}$$
$$\overline{PbO_2 + Cl^- + \ H_2O \longrightarrow Pb(OH)_2 + ClO^-}$$

The overall reaction is now balanced.

PROBLEM 13-8 **(a)** Which element is reduced in the following redox reaction in acidic solution? **(b)** Write the balanced equation for the oxidation half-reaction.

$$CH_2O + Cr_2O_7^{2-} \longrightarrow Cr^{3+} + CO_2$$

Solution: Note that this problem does not ask for the balanced overall equation. However, you must be able to tell which part is oxidation and which part is reduction.

(a) Chromium is reduced from the +6 oxidation state (number) in $Cr_2O_7^{2-}$ to the +3 oxidation state in Cr^{3+}. [Note that it is the element Cr, not $Cr_2O_7^{2-}$, that is reduced.]

(b) Use the relevant steps for balancing redox half-reactions to write the balanced oxidation reaction.

Step 1:

Species		Oxidation number
CH_2O	$x + 2 + (-2) = 0$	0* for C
	$x = 0$	
CO_2	$x + (-4) = 0$	+4 for C
	$x = +4$	

---

* There is nothing wrong with an oxidation number of zero in a compound.

Oxidation half-reaction
$$CH_2O \longrightarrow CO_2 \qquad \text{(acidic soln)}$$

Step 2: The non-H and non-O atoms are already balanced.

Step 3a: The reactant side is oxygen-deficient, so you add H_2O to it, balancing the H_2O hydrogens with H^+ on the product side:

$$CH_2O + H_2O \longrightarrow CO_2 + 2H^+$$

But you're still short two hydrogens (from CH_2O), so you add two more H^+ ions, thus balancing all the atoms:

$$CH_2O + H_2O \longrightarrow CO_2 + 4H^+$$

Step 4: Balance the charges by adding electrons to the products:

$$CH_2O + H_2O \longrightarrow CO_2 + 4H^+ + 4e^-$$
$$(0) \;+\; (0) \;\;=\;\; (0) + (+4) + (-4)$$

The oxidation half-reaction is now balanced.

Check: The net change in oxidation number found in Step 1 is 4 (going from 0 to $+4$), which means that four electrons are produced. Since the charges balance with $4e^-$, all is well.

PROBLEM 13-9 Write the balanced oxidation and reduction half-reactions and the balanced overall equation for the following redox reaction in basic solution:

$$Cl_2 \longrightarrow Cl^- + ClO^-$$

Solution:

Step 1:

Species	Oxidation number
Cl_2	0
Cl^-	-1 for Cl
ClO^-	$+1$ for Cl

Chlorine, which is the only substance on the reactant side, gives two products here: Cl^-, which has a lower oxidation number, and ClO^-, which has a higher oxidation number. Therefore the element Cl is *both* reduced *and* oxidized—a phenomenon called **disproportionation**.

Oxidation half-reaction	**Reduction half-reaction**
$Cl_2 \longrightarrow ClO^-$ (base)	$Cl_2 \longrightarrow Cl^-$ (base)

Step 2:

$Cl_2 \longrightarrow 2ClO^-$ (base)	$Cl_2 \longrightarrow 2Cl^-$ (base)

Step 3b:

$$Cl_2 + 2OH^- \longrightarrow 2ClO^-$$
$$Cl_2 + 2OH^- \longrightarrow 2ClO^- + H_2O$$
$$Cl_2 + 4OH^- \longrightarrow 2ClO^- + 2H_2O$$

$$Cl_2 \longrightarrow 2Cl^-$$

Step 4:

$$Cl_2 + 4OH^- \longrightarrow 2ClO^- + 2H_2O + 2e^-$$
$$(0) + (-4) \;=\; (-2) + (0) + (-2)$$

$$Cl_2 + 2e^- \longrightarrow 2Cl^-$$
$$(0) + (-2) \;=\; (-2)$$

Check:

Each Cl, in going from 0 to $+1$, gives up one electron, producing a net total of two electrons.

Each Cl, in going from 0 to -1, gains one electron, giving a net total of two electrons.

Step 5: The electrons are already balanced.

Step 6:

$$Cl_2 + 4OH^- \longrightarrow 2ClO^- + 2H_2O + 2e^-$$
$$Cl_2 + 2e^- \longrightarrow 2Cl^-$$
$$\overline{2Cl_2 + 4OH^- \longrightarrow 2ClO^- + 2Cl^- + 2H_2O}$$

which can be simplified to

$$Cl_2 + 2OH^- \longrightarrow ClO^- + Cl^- + H_2O$$

PROBLEM 13-10   Which of these equations represents a redox reaction?

(a) $NaPO_3 + H_2O + H^+ \longrightarrow H_3PO_4 + Na^+$
(b) $HNO_2 + H_2O_2 \longrightarrow NO_3^- + H^+ + H_2O$
(c) $HF + OH^- \longrightarrow H_2O + F^-$

Solution: Look for changes in oxidation number (x).

(a) The only atom that might change oxidation number is P:

In $NaPO_3$: $\begin{aligned} x_P + x_{Na} + 3x_O &= 0 \\ x_P + 1 + 3(-2) &= 0 \\ x_P &= 5 \end{aligned}$ In H_3PO_4: $\begin{aligned} x_P + 3x_H + 4x_O &= 0 \\ x_P + 3 + (-8) &= 0 \\ x_P &= 5 \end{aligned}$

There is no change in oxidation number in choice (a), so it is not a redox reaction.
(b) Check the oxidation number of N:

In HNO_2: $\begin{aligned} x_N + x_H + 2x_O &= 0 \\ x_N + 1 + 2(-2) &= 0 \\ x_N &= +3 \end{aligned}$ In NO_3^-: $\begin{aligned} x_N + 3x_O &= -1 \\ x_N &= +5 \end{aligned}$

Since the oxidation number of N has increased from $+3$ to $+5$, choice (b) must be a redox reaction.
note: If there is an element that has been oxidized, there must be an element that is reduced. In this case, the oxidizing agent is H_2O_2, which contains oxygen in the -1 oxidation state—and is the exception to the rule the oxygen is *usually* in the -2 state. So oxygen is the element reduced because it goes from -1 in H_2O_2 to -2 in H_2O.

(c) All of the atoms in choice (c) have the same oxidation number in the reactants and products. This is an acid–base reaction.

PROBLEM 13-11   Balance the following redox reactions. Write the balanced oxidation and reduction half-reactions.

(a) $F_2 + H_2O \longrightarrow HF + O_2$ (acidic soln)
(b) $Ca + H_2O \longrightarrow Ca(OH)_2 + H_2$ (basic soln)

Solution:
(a) *Step 1:*

Species	Oxidation number
F_2	0
HF	-1 (for F)
H_2O	-2 (for O)
O_2	0

Oxidation half-reaction	**Reduction half-reaction**
$H_2O \longrightarrow O_2$ (acid)	$F_2 \longrightarrow HF$ (acid)

Step 2:

$2H_2O \longrightarrow O_2$ (acid)	$F_2 \longrightarrow 2HF$ (acid)

Step 3:

$2H_2O \longrightarrow O_2 + 4H^+$	$F_2 + 2H^+ \longrightarrow 2HF$

Step 4:

$2H_2O \longrightarrow O_2 + 4H^+ + 4e^-$	$F_2 + 2H^+ + 2e^- \longrightarrow 2HF$
$(0) = (0) + (4) + (-4)$	$(0) + (2) + (-2) = (0)$

Check:

Each O gives up two e^- in going from -2 to 0.	Each F gains one e^- in going from 0 to -1.

Step 5:

$2H_2O \longrightarrow O_2 + 4H^+ + 4e^-$	$2F_2 + 4H^+ + 4e^- \longrightarrow 4HF$

Step 6:

$$2F_2 + 2H_2O + 4\cancel{H^+} + \cancel{4e^-} \longrightarrow 4HF + \cancel{4H^+} + \cancel{4e^-} + O_2$$

$$2F_2 + 2H_2O \longrightarrow 4HF + O_2$$

(b) *Step 1:*

Species	Oxidation number
Ca	0
$Ca(OH)_2$	+2 (for Ca)
H_2O	+1 (for H)
H_2	0

Oxidation half-reaction	**Reduction half-reaction**
$Ca \longrightarrow Ca(OH)_2$ (base)	$H_2O \longrightarrow H_2$ (base)

Step 2: All non-O and -H atoms are balanced.

Step 3:

$Ca + 2OH^- \longrightarrow Ca(OH)_2$	$2H_2O \longrightarrow H_2 + 2OH^-$

Step 4:

$$Ca + 2OH^- \longrightarrow Ca(OH)_2 + 2e^-$$
$$(0) + (-2) \ = \ (0) + (-2)$$

$$2H_2O + 2e^- \longrightarrow H_2 + 2OH^-$$
$$(0) + (-2) \ = \ (0) + (-2)$$

Check:

Each Ca gives up two e^- in going from 0 to +2.   Each H gains one e^- in going from +1 to 0.

Step 5: The number of electrons is already equal.

Step 6:

$$Ca + \cancel{2OH^-} + 2H_2O + \cancel{2e^-} \longrightarrow H_2 + \cancel{2OH^-} + Ca(OH)_2 + \cancel{2e^-}$$
$$Ca + 2H_2O \longrightarrow H_2 + Ca(OH)_2$$

PROBLEM 13-12 $KMnO_4$ solution reacts with H_2O_2 in acidic solution to form O_2 and Mn^{2+}. Write the balanced net ionic equation for this reaction.

Solution: First we have to translate the information given into a skeletal ionic equation. Checking for oxidation-number change, we see that manganese changes from +7 to +2; so Mn is reduced. Now we have to find a species that is oxidized. Since the oxygen in H_2O_2 is in the unusual −1 state, we can safely assume that it is the element oxidized to the 0 state in O_2. Finally, since potassium does not change oxidation number, we can leave it out of the reaction because it plays no role in the redox process. Now we can write an unbalanced ionic equation

$$MnO_4^- + H_2O_2 \longrightarrow Mn^{2+} + O_2   \text{(acid)}$$

which we can balance by writing half-reactions and following the usual steps:

Oxidation half-reaction	**Reduction half-reaction**
$H_2O_2 \longrightarrow O_2$	$MnO_4^- \longrightarrow Mn^{2+}$
$H_2O_2 \longrightarrow O_2 + 2H^+$	$MnO_4^- + 8H^+ \longrightarrow Mn^{2+} + 4H_2O$
$H_2O_2 \longrightarrow O_2 + 2H^+ + 2e^-$	$MnO_4^- + 8H^+ + 5e^- \longrightarrow Mn^{2+} + 4H_2O$
$(0) \ = \ (0) + (2) + (-2)$	$(-1) + (8) + (-5) \ = \ (2) + (0)$
$5H_2O_2 \longrightarrow 5O_2 + 10H^+ + 10e^-$	$2MnO_4^- + 16H^+ + 10e^- \longrightarrow 2Mn^{2+} + 8H_2O$

Adding the balanced half-reactions, canceling $10e^-$ and $10H^+$, gives the balanced net ionic equation:

$$2MnO_4^- + 5H_2O_2 + 6H^+ \longrightarrow 2Mn^{+2} + 5O_2 + 8H_2O$$

PROBLEM 13-13   An acidic solution of Cr^{2+} is used to remove O_2 from a stream of inert argon gas:

$$Cr^{2+} + O_2 \longrightarrow Cr^{3+} + H_2O$$

Which answer best represents the number of moles of O_2 that may be removed by a solution containing 0.200 mol of Cr^{2+}?

(a) 0.200 mol O_2
(b) 0.0500 mol O_2
(c) 0.100 mol O_2
(d) 0.800 mol O_2
(e) This problem cannot be answered without knowing the molarity of the Cr^{2+} solution.

Solution: You can't do anything until the equation is balanced. Write the half-reactions

Oxidation half-reaction	**Reduction half-reaction**
$Cr^{2+} \longrightarrow Cr^{3+}$	$O_2 \longrightarrow H_2O$
$Cr^{2+} \longrightarrow Cr^{3+}$	$O_2 + 4H^+ \longrightarrow 2H_2O$
$Cr^{2+} \longrightarrow Cr^{3+} + e^-$	$O_2 + 4H^+ + 4e^- \longrightarrow 2H_2O$
$4Cr^{2+} \longrightarrow 4Cr^{3+} + 4e^-$	$O_2 + 4H^+ + 4e^- \longrightarrow 2H_2O$

and find the balanced equation:

$$4Cr^{2+} + O_2 + 4H^+ \longrightarrow 4Cr^{3+} + 2H_2O$$

Then from the balanced equation, you see that 0.200 mol of Cr^{2+} reacts with

$$(0.200 \text{ mol } Cr^{2+})\left(\frac{1 \text{ mol } O_2}{4 \text{ mol } Cr^{2+}}\right) = 0.0500 \text{ mol } O_2$$

So the correct answer is **(b)**.

note: You would obtain answers **(a)** or **(c)** if you did not balance the equation correctly. You would obtain answer **(d)** if you multiplied 0.200 mol Cr^{2+} by 4. Answer **(e)** would pertain only if you were given the *volume* of solution instead of the number of moles of Cr^{2+} present.

PROBLEM 13-14 A 0.3356-g sample of iron ore was dissolved in acid, and all of the Fe was converted to Fe^{2+}. The resulting Fe^{2+} solution was titrated with 0.01873 M $KMnO_4$ solution. The volume of $KMnO_4$ solution required to react with all of the iron was 21.24 mL. Calculate the percentage of iron in the iron ore. The unbalanced equation for the titration in acidic solution is

$$Fe^{2+} + MnO_4^- \longrightarrow Fe^{3+} + Mn^{2+}$$

Solution: First balance the equation by half-reactions:

Oxidation half-reaction	**Reduction half-reaction**
$Fe^{2+} \longrightarrow Fe^{3+}$	$MnO_4^- \longrightarrow Mn^{2+}$
$Fe^{2+} \longrightarrow Fe^{3+}$	$MnO_4^- + 8H^+ \longrightarrow Mn^{2+} + 4H_2O$
$Fe^{2+} \longrightarrow Fe^{3+} + e^-$	$MnO_4^- + 8H^+ + 5e^- \longrightarrow Mn^{2+} + 4H_2O$
$5Fe^{2+} \longrightarrow 5Fe^{3+} + 5e^-$	$MnO_4^- + 8H^+ + 5e^- \longrightarrow Mn^{2+} + 4H_2O$

$$5Fe^{2+} + MnO_4^- + 8H^+ \longrightarrow 5Fe^{3+} + Mn^{2+} + 4H_2O$$

Now calculate the number of moles of MnO_4^- used in the titration:

$$(0.01873 \text{ mol } MnO_4^-/L)(21.24 \text{ mL})(10^{-3} \text{ L/mL}) = 3.978 \times 10^{-4} \text{ mol } MnO_4^-$$

The balanced equation tells you that 5 mol of Fe^{2+} react with 1 mol of MnO_4^-:

$$(3.978 \times 10^{-4} \text{ mol } MnO_4^-)\left(\frac{5 \text{ mol } Fe^{2+}}{1 \text{ mol } MnO_4^-}\right) = 1.989 \times 10^{-3} \text{ mol } Fe^{2+}$$

This is also the number of moles of Fe in the sample of iron ore. The mass of Fe in grams is $(1.989 \times 10^{-3} \text{ mol})(55.85 \text{ g/mol}) = 0.1111 \text{ g Fe}$. So the % Fe in the sample is

$$\frac{0.1111 \text{ g}}{0.3356 \text{ g}} \times 100\% = 33.10\%$$

Supplementary Exercises

PROBLEM 13-15 Assign oxidation numbers to each element in (a) MgO, (b) Na_2SO_4, (c) P_4, (d) $C_2O_4^{2-}$.

Answer: (a) Mg +2; O −2   (b) Na +1, S +6, O −2   (c) P 0   (d) C +3; O −2

PROBLEM 13-16   Give **(a)** the balanced oxidation half-reaction, **(b)** the balanced reduction half-reaction, and **(c)** the balanced overall net ionic reaction for the following equation in acidic solution:

$$Zn + VO_2^+ \longrightarrow V^{2+} + Zn^{2+}$$

Answer: **(a)** $Zn \longrightarrow Zn^{2+} + 2e^-$
(b) $VO_2^+ + 4H^+ + 3e^- \longrightarrow V^{2+} + 2H_2O$
(c) $2VO_2^+ + 3Zn + 8H^+ \longrightarrow 2V^{2+} + 3Zn^{2+} + 4H_2O$

PROBLEM 13-17   Give **(a)** the balanced oxidation half-reaction, **(b)** the balanced reduction half-reaction, and **(c)** the balanced overall net ionic reaction for the following equation in basic solution:

$$SO_3^{2-} + O_2 \longrightarrow SO_4^{2-}$$

Answer: **(a)** $SO_3^{2-} + 2OH^- \longrightarrow SO_4^{2-} + H_2O + 2e^-$
(b) $O_2 + 2H_2O + 4e^- \longrightarrow 4OH^-$
(c) $2SO_3^{2-} + O_2 \longrightarrow 2SO_4^{2-}$

PROBLEM 13-18   **(a)** Which substance is the reducing agent, **(b)** which element is reduced, **(c)** which substance is the oxidizing agent, and **(d)** which element is oxidized in the following equation?

$$2MnO_4^- + 5H_2C_2O_4 + 6H^+ \longrightarrow 2Mn^{2+} + 10CO_2 + 8H_2O$$

Answer: **(a)** $H_2C_2O_4$ **(b)** Mn   **(c)** MnO_4^- **(d)** C

PROBLEM 13-19   Give **(a)** the balanced oxidation half-reaction, **(b)** the balanced reduction half-reaction, and **(c)** the balanced overall net ionic equation for the following reaction in acidic solution:

$$CuCl + HClO \longrightarrow Cl^- + Cu^{2+}$$

Answer: **(a)** $CuCl \longrightarrow Cu^{2+} + Cl^- + e^-$
(b) $HClO + 2e^- + H^+ \longrightarrow Cl^- + H_2O$
(c) $2CuCl + HClO + H^+ \longrightarrow 2Cu^{2+} + 3Cl^- + H_2O$

PROBLEM 13-20   Give the oxidation number of chromium in **(a)** $(Cr^{3+})_2(SO_4^{2-})_3$, **(b)** $K_2Cr_2O_7$, **(c)** CrO_2^-, **(d)** $(Cr^{2+})(Cl^-)_2$, **(e)** CrO_3, **(f)** Cr_2O_3.

Answer: **(a)** +3   **(b)** +6   **(c)** +3   **(d)** +2   **(e)** +6   **(f)** +3

PROBLEM 13-21   Give **(a)** the oxidation half-reaction, **(b)** the reduction half-reaction, and **(c)** the balanced overall net ionic equation for the following reaction in acidic solution:

$$NO_2 \longrightarrow NO_3^- + NO$$

Answer: **(a)** $NO_2 + H_2O \longrightarrow NO_3^- + 2H^+ + e^-$
(b) $NO_2 + 2H^+ + 2e^- \longrightarrow NO + H_2O$
(c) $3NO_2 + H_2O \longrightarrow 2NO_3^- + NO + 2H^+$

PROBLEM 13-22   Write **(a)** the balanced equation for the oxidation half-reaction and **(b)** the balanced net ionic equation for the overall reaction

$$MnO_4^- + CH_3CHO \longrightarrow MnO_2 + CH_3CO_2^- \quad \text{(basic solution)}$$

Answer: **(a)** $CH_3CHO + 3OH^- \longrightarrow CH_3CO_2^- + 2H_2O + 2e^-$
(b) $3CH_3CHO + 2MnO_4^- + OH^- \longrightarrow 3CH_3CO_2^- + 2MnO_2 + 2H_2O$

PROBLEM 13-23   Write **(a)** the balanced equation for the oxidation half-reaction and **(b)** the balanced equation for the reduction half-reaction for the reaction

$$Co^{3+} + H_2O \longrightarrow Co^{2+} + O_2 \quad \text{(acid solution)}$$

Answer: **(a)** $2H_2O \longrightarrow O_2 + 4H^+ + 4e^-$
(b) $Co^{3+} + e^- \longrightarrow Co^{2+}$

PROBLEM 13-24 Balance the reduction half-reaction $XeO_3 \longrightarrow Xe$ in acidic solution.

Answer: $XeO_3 + 6H^+ + 6e^- \longrightarrow Xe + 3H_2O$

PROBLEM 13-25 Given a redox reaction in acidic solution,

$$Br^- + HSO_4^- \longrightarrow SO_2 + Br_2$$

(a) Which substance is the reducing agent? (b) Which element is reduced? (c) Write the balanced equation for the reduction half-reaction. (d) Write the balanced overall net ionic equation. (e) How many grams of SO_2 can be made from 10.00 mL of 1.50 M NaBr solution and excess H_2SO_4?

Answer: (a) Br^- (b) S   (c) $HSO_4^- + 3H^+ + 2e^- \longrightarrow SO_2 + 2H_2O$
(d) $2Br^- + HSO_4^- + 3H^+ \longrightarrow Br_2 + SO_2 + 2H_2O$ (e) 0.481 g

PROBLEM 13-26 How many grams of Cl_2 can be made from 25.0 g of MnO_2 and excess HCl using the reaction $MnO_2 + Cl^- \rightarrow Mn^{2+} + Cl_2$?

Answer: 20.4 g

PROBLEM 13-27 An acidic solution containing $S_2O_3^{2-}$ (as $Na_2S_2O_3$) was titrated with 0.01753 M I_2 solution:

$$I_2 + S_2O_3^{2-} \longrightarrow I^- + S_4O_6^{2-}$$

Calculate the molarity of the $Na_2S_2O_3$ solution if 25.00 mL of it required 13.73 mL of I_2 solution.

Answer: 0.01925 M

14 CHEMICAL KINETICS

THIS CHAPTER IS ABOUT

☑ **Definition and Expression of Reaction Rates**
☑ **Reaction Rates and Concentration Effects**
☑ **Reaction Rates and Temperature Effects: The Arrhenius Equation**
☑ **How Reactions Take Place: Reaction Mechanisms**

Chemical kinetics is the study of reaction rates and reaction mechanisms. A reaction rate is the *velocity* with which reactants are used up and products produced. A reaction mechanism is the *pathway* by which reactants are converted to products.

14-1. Definition and Expression of Reaction Rates

A **reaction rate** is a velocity—the "distance" a reaction goes in an interval of time—expressed as the change in concentration of reactant or product per unit time. For example, the reaction rate of

$$H_2(g) + Br_2(g) \qquad 2HBr(g)$$

can be expressed as the change in HBr concentration over some time (t) interval, or as the change in H_2 or Br_2 concentration over some time interval. Using brackets to indicate concentration in moles per liter ($mol\ L^{-1}$) and Δ to indicate change, you can write different expressions for the rate of this reaction:

$$Rate_1 = \frac{\Delta[HBr]}{\Delta t} \qquad or \qquad Rate_2 = \frac{\Delta[H_2]}{\Delta t} = \frac{\Delta[Br_2]}{\Delta t}$$

There is, of course, a relationship between these expressions. $Rate_1$ and $Rate_2$ have opposite signs because [HBr] increases over time while $[H_2]$ and $[Br_2]$ decrease over time. And from the stoichiometry of the reaction, you can establish their equivalence by writing either $Rate_1 = -2(Rate_2)$ or $-(Rate_1) = 2(Rate_2)$.

Now consider a general reaction, in which $a, b,\ldots,$ are stoichiometric coefficients and A, B,$\ldots,$ are chemical species:

$$aA + bB + \cdots + \longrightarrow cC + dD + \cdots +$$

The reaction rate of the general reaction is given by

REACTION RATE
$$Rate = -\left(\frac{1}{a}\right)\left(\frac{\Delta[A]}{\Delta t}\right) = -\left(\frac{1}{b}\right)\left(\frac{\Delta[B]}{\Delta t}\right)\cdots = \left(\frac{1}{c}\right)\left(\frac{\Delta[C]}{\Delta t}\right) = \left(\frac{1}{d}\right)\left(\frac{\Delta[D]}{\Delta t}\right)\cdots \quad \textbf{(14-1)}$$

EXAMPLE 14-1: Write three *equivalent* expressions for the reaction rate of $H_2 + Br_2 \rightarrow 2HBr$.

Solution: From the general expression (14-1) of the reaction rate:

$$Rate = -\frac{\Delta[H_2]}{\Delta t} = -\frac{\Delta[Br_2]}{\Delta t} = \frac{\frac{1}{2}\Delta[HBr]}{\Delta t}$$

note: Some beginning chemistry texts use the notation of differential calculus for the rate of a reaction. The idea here is to shrink the time interval Δt to one that approaches zero. Then it is possible to speak of the *instantaneous rate* of the reaction rather than its average rate over a finite time interval. In these differential equations each Δ is replaced by d. Thus in Example 14-1, $-\Delta[H_2]/\Delta t$ becomes $-d[H_2]/dt$. Although you won't encounter any direct use of the calculus in this chapter, you will be given formulas, presented without proof, that were derived by use of the calculus.

The rate for any chemical reaction depends on the nature of the reactants and on external factors such as concentration, temperature, and the presence or absence of **catalysts**—substances that increase the reaction rate but can be recovered in their original form when the reaction has ended. The reaction rate for a *homogeneous* (single liquid or gas phase) reaction usually depends on the concentration of the reactants. The reaction rate for a *heterogeneous* (more than one phase) reaction may depend on the area of contact between the phases.

14-2. Reaction Rates and Concentration Effects

A. Rate equations and reaction order

Concentration effects on reaction rate *can only be established by experiment*. The experimental results can be reported as a **rate equation** (sometimes called a **rate law**), in which the reaction rate is expressed as a function of the concentrations of the reactants. Consider, for example, the homogeneous decomposition of gaseous dinitrogen pentoxide:

$$2N_2O_5(g) \longrightarrow 4NO_2(g) + O_2(g)$$

Chemists have determined by experiment that the rate of decomposition of N_2O_5 at any given time is proportional only to the concentration of N_2O_5—the concentrations of NO_2 and O_2 aren't significant. From the general rate expression (14-1),

$$\text{Rate} = -\frac{\frac{1}{2}\Delta[N_2O_5]}{\Delta t} = \frac{\frac{1}{4}\Delta[NO_2]}{\Delta t} = \frac{\Delta[O_2]}{\Delta t}$$

Since all three terms are proportional to $[N_2O_5]$, the reaction rate can be expressed as a rate equation:

$$\text{Rate} = -\frac{\frac{1}{2}\Delta[N_2O_5]}{\Delta t} = k[N_2O_5]$$

where k is the specific **rate constant** for the reaction. The rate constant k is characteristic of the specific reaction, but varies with temperature. For this particular reaction k has the units of $1/\text{time}$, so that the rate is properly expressed in moles per liter per unit time.

In general, rate equations take the form

GENERAL RATE EQUATION
$$\text{Rate} = k[A]^x[B]^y \cdots \tag{14-2}$$

where k is the rate constant and each exponent $(x, y, \ldots)$ represents the individual **reaction order** for its reactant. The value of any individual reaction order may be an integer or a fraction, positive or negative. The value of the sum of the individual reaction orders $(x + y + \cdots)$ is the **overall reaction order**. Although chemical reactions having overall orders of three, zero, and fractional values are known, the great majority are either first-order (exponent sum = 1) or second-order (exponent sum = 2) reactions.

note: The units of k vary, depending on the reaction order. For first-order reactions, k has units of reciprocal time (e.g., $1/\text{s}$ or s^{-1}).

EXAMPLE 14-2: You know that the balanced equation for the homogeneous decomposition of N_2O_5 is

$$2N_2O_5(g) \longrightarrow 4NO_2(g) + O_2(g)$$

You want to know the individual reaction order of N_2O_5 and the overall reaction order. Do you have enough information? How would you go about finding these values?

Solution: No, the balanced equation does not give you enough information to find either individual reaction orders or overall reaction orders. In order to find these values, you need some experimental data, which are given in the rate equation. In this case, the text above gives you the rate equation:

$$\text{Rate} = -\frac{\frac{1}{2}\Delta[N_2O_5]}{\Delta t} = k[N_2O_5]$$

Obviously, the exponent of $[N_2O_5]$ is 1 here, so the reaction is first-order for N_2O_5. And because there are no other significant species or exponents the reaction is first-order overall.

EXAMPLE 14-3: The reaction between ethyl iodide (C_2H_5I) and OH^- in aqueous ethanol gives $C_2H_5OH + I^-$ as products. The rate law is

$$\text{Rate} = -\frac{\Delta[C_2H_5I]}{\Delta t} = k[C_2H_5I][OH^-]$$

What is the reaction order in each component, and what is the overall reaction order?

Solution: The reaction is first-order in $[C_2H_5I]$ and also first-order in $[OH^-]$. It is therefore second-order ($1 + 1 = 2$) overall.

> *note:* Don't confuse the coefficients $a, b, \ldots$, in the general expression of the reaction rate (eq. 14-1) with the exponential values $x, y, \ldots$, in the general rate equation (eq. 14-2). The coefficients are derived from reaction stoichiometry, while the exponents are derived from the rate equation, which is experimentally determined.

B. Calculations for first-order reactions

The form of first-order reactions is A → products: The only significant concentration is that of the single reactant. (Many decomposition and rearrangement reactions are first-order.) Now we need to know how to use experimental data to find the rate equation and k for a first-order reaction. Figure 14-1a (see Example 14-4) is a graph of the concentration (expressed as partial pressure) of N_2O_5 as a function of time at 320 K. But this graph doesn't tell us much: It is not at all obvious from the graph that the decomposition of N_2O_5 is a first-order reaction or that $-\frac{1}{2}\Delta[N_2O_5]/\Delta t = k[N_2O_5]$. So we use the **integrated form** of the first-order rate equation (obtained by the methods of the calculus), which is much more useful. If the first-order rate equation is

$$\text{Rate} = -\frac{\Delta[A]}{\Delta t} = k[A]$$

then its integrated form is

FIRST-ORDER RATE EQUATION (integrated form)
$$\log\frac{[A]_0}{[A]} = \frac{kt}{2.303} \quad \text{or} \quad \log[A] = \log[A]_0 - \frac{kt}{2.303} \tag{14-3}$$

where $[A]_0$ is the initial concentration of reactant A, $[A]$ is the concentration of reactant A at time t, and 2.303 is the log conversion factor (which you should memorize). For any given initial conditions, $\log[A]_0$ is constant, so a graph of $\log[A]$ versus time will give a straight line whose SLOPE (ratio of change of concentration to change in time) is $-k/2.303$ (see Figure 14-1b). In other words, a reaction is first-order in A if a plot of $\log[A]$ against time is a straight line. The value of k can be determined from the slope of the line.

EXAMPLE 14-4: The experimental data for the decomposition of N_2O_5 plotted in Figure 14-1 are

t (min)	0.00	10.0	20.0	40.0	60.0
$p_{N_2O_5}$ (kPa)	46.4	32.8	24.6	14.0	7.71
$\log p_{N_2O_5}$	1.67	1.52	1.39	1.15	0.89

Determine the value of k for this decomposition at 320 K.

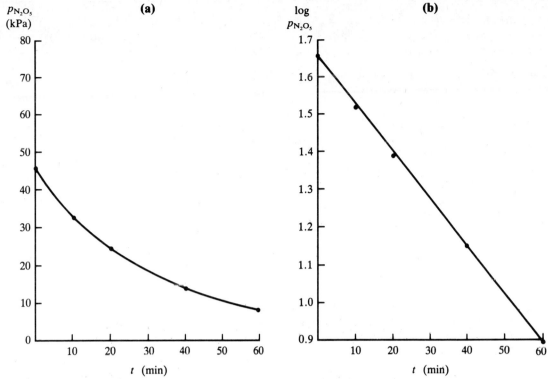

FIGURE 14-1. Concentration of N_2O_5 (expressed as partial pressure in kPa) versus time at 320 K. (a) $p_{N_2O_5}$ vs. time; (b) log $p_{N_2O_5}$ vs. time.

Solution: Find the slope of the line in Figure 14-1b:

$$\text{slope} = \frac{\Delta[N_2O_5]}{\Delta t} = \frac{\log[N_2O_5] - \log[N_2O_5]_0}{t - t_0}$$

$$= \frac{0.89 - 1.67}{60 - 0 \text{ min}} = -0.013 \text{ min}^{-1}$$

Now we know the slope is

$$-\frac{k}{2.303} = -0.013 \text{ min}^{-1}$$

So

$$k = 2.303(0.013 \text{ min}^{-1})$$
$$= 0.030 \text{ min}^{-1}$$

note: In this example the experimental data for [A] were calculated in terms of partial pressure. You can use *any* measure for [A] that is proportional to the concentration in moles per liter.

A useful way of thinking about first-order reactions is to concentrate on the time when the initial concentration of the reactant has fallen to half its starting value, i.e., when $[A] = \frac{1}{2}[A]_0$. Substituting this into the integrated form of the first-order rate equation, we get the **half-life** $t_{1/2}$ of the reactant:

$$\log \frac{[A]_0}{0.5[A]_0} = \log 2 = \frac{kt_{1/2}}{2.303}$$

HALF-LIFE $$t_{1/2} = \frac{0.693}{k}$$ (14-4)

EXAMPLE 14-5: The rate constant for the first-order decomposition of N_2O_5 at 320 K is 3.00×10^{-2} min^{-1}. If the initial concentration of N_2O_5 is 2.00×10^{-3} mol L^{-1}, how long will it take for the concentration to drop to 5.00×10^{-4} mol L^{-1}?

Solution: The half-life of this reaction is

$$t_{1/2} = \frac{0.693}{k}$$

$$= \frac{0.693}{3.00 \times 10^{-2} \text{ min}^{-1}} = 23.1 \text{ min}$$

So after 23.1 min the initial concentration of 2.00×10^{-3} mol L^{-1} falls to half that value, i.e., to 1.00×10^{-3} mol L^{-1}. After another 23.1 min, i.e., 46.2 min after the start, the concentration halves again, to 5.00×10^{-4} mol L^{-1}.

> *note:* In general terms, after a time period of $b \times t_{1/2}$, where b is an integer, the initial concentration of reactant in a first-order reaction falls to $1/(2)^b$ of its original value.

C. Calculations for second-order reactions

There are many types of overall second-order reactions. The two simplest have second-order rate equations

$$\text{Rate} = k[A]^2$$
$$\text{Rate} = k[A][B]$$

The units of k in second-order rate equations are 1/(concentration $\times$ time). For example, the units of k might be 1/[(mol/L)(s)], often expressed as L/mol s or L mol^{-1}s^{-1}.

For Rate $= k[A]^2$, the integrated form of the second-order rate equation is

SECOND-ORDER RATE EQUATION (integrated form)
$$\frac{1}{[A]} = \frac{1}{[A]_0} + kt \qquad (14\text{-}5)$$

For any given initial conditions, $[A]_0$ is constant. This means that a graph of $1/[A]$ versus time will give a straight line whose slope is equal to k.

EXAMPLE 14-6: The reaction for the decomposition of pure nitrogen dioxide is

$$2NO_2(g) \longrightarrow 2NO(g) + O_2(g)$$

The following concentration data were collected in a study of the decomposition of pure NO_2 at 592 K:

t (s):	0.0	100.0	200.0	400.0	800.0	1200.0	1600.0
$[NO_2]$ (mol L^{-1}):	1.5×10^{-3}	1.39×10^{-3}	1.30×10^{-3}	1.14×10^{-3}	9.22×10^{-4}	7.73×10^{-4}	6.66×10^{-4}

Determine the reaction order and the value of k.

Solution: Determine reciprocal values for the concentrations found and plot $1/[NO_2]$ versus time, as in Figure 14-2.

t (s)	$\dfrac{1}{[NO_2]}$ (L mol^{-1})
0.0	6.67×10^2
100.0	7.19×10^2
200.0	7.69×10^2
400.0	8.77×10^2
800.0	1.08×10^3
1200.0	1.29×10^3
1600.0	1.50×10^3

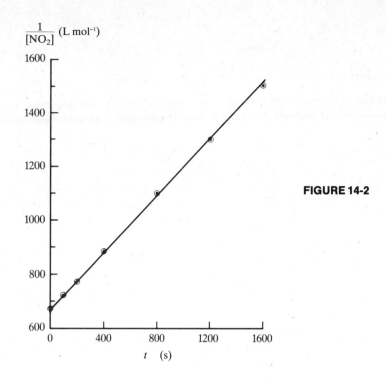

$\dfrac{1}{[NO_2]}$ (L mol⁻¹)

t (s)

FIGURE 14-2

Because the graph is a straight line, the reaction is second-order. Then find the slope from the data to determine k:

$$\text{slope} = k = \frac{1500 - 675 \ \text{L mol}^{-1}}{1600 - 0 \ \text{s}}$$

$$= 0.516 \ \text{L mol}^{-1}\text{s}^{-1}$$

14-3. Reaction Rates and Temperature Effects: The Arrhenius Equation

Most reactions go faster when the temperature is raised. In fact, the rate constant k changes with temperature.

note: Changing the concentration affects the *rate* of a reaction; changing the temperature affects the *rate constant* as well as the rate.

The effects of temperature changes on reaction rates can be quantitatively expressed (over moderate temperature ranges) by the **Arrhenius equation**:

ARRHENIUS EQUATION
$$\log\left(\frac{k_2}{k_1}\right) = -\left(\frac{E_a}{2.303R}\right)\left(\frac{1}{T_2} - \frac{1}{T_1}\right) \tag{14-6}$$

where k_1 and k_2 are the reaction rate constants at temperatures T_1 and T_2 (in degrees Kelvin), respectively; R is the universal gas constant; and E_a is the **activation energy**, a constant characteristic of the reaction. Since E_a is an energy parameter, the gas constant R is expressed in energy units ($8.314 \ \text{J K}^{-1}\text{mol}^{-1}$) in the Arrhenius equation.

The Arrhenius equation can be used in two ways. Given the values of the rate constant at two different temperatures, you can calculate the value of E_a. Or, given the value of E_a and that of one rate constant at one temperature, you can calculate the value of the rate constant at any other temperature.

EXAMPLE 14-7: The rate constant for the decomposition of NO_2 is $0.516 \ \text{L mol}^{-1}\text{s}^{-1}$ at 592 K (as found in Example 14-6) and is $1.70 \ \text{L mol}^{-1}\text{s}^{-1}$ at 627 K. Calculate the activation energy for the reaction.

Solution: Given that

$$k_1 = 0.516 \text{ L mol}^{-1}\text{s}^{-1}; \ T_1 = 592 \text{ K} \quad \text{and} \quad k_2 = 1.70 \text{ L mol}^{-1}\text{s}^{-1}; \ T_2 = 627 \text{ K}$$

use the Arrhenius equation (14-6):

$$\log\left(\frac{k_2}{k_1}\right) = -\left(\frac{E_a}{2.303R}\right)\left(\frac{1}{T_2} - \frac{1}{T_1}\right)$$

$$\log\left(\frac{1.70}{0.516}\right) = \frac{-E_a}{2.303(8.314 \text{ J mol}^{-1}\text{K}^{-1})}\left(\frac{1}{627 \text{ K}} - \frac{1}{592 \text{ K}}\right)$$

$$0.518 = \frac{-E_a}{2.303(8.314 \text{ J mol}^{-1}\text{K}^{-1})}(1.5949 \times 10^{-3} - 1.6892 \times 10^{-3}) \text{ K}^{-1}$$

$$= \frac{-E_a(-9.43 \times 10^{-5}) \text{ K}^{-1}}{2.303(8.314 \text{ J mol}^{-1}\text{K}^{-1})}$$

Thus

$$E_a = \frac{0.518(2.303)(8.314)}{9.43 \times 10^{-5}} \text{ J mol}^{-1} = 1.05 \times 10^5 \text{ J mol}^{-1} \quad \text{or} \quad 105 \text{ kJ mol}^{-1}$$

14-4. How Reactions Take Place: Reaction Mechanisms

A. Theories

It seems obvious that reactant species must collide with each other in order to react. While this is true, it isn't the whole story: Only a fraction of the collisions between reactant species are effective. How do we account for this theoretically?

1. Collision theory

The **collision theory** postulates that when particles of A collide with particles of B, sufficient kinetic energy must be available to allow for the rearrangement of bonds (electrons) to produce the products. A successful collision occurs only when colliding molecules possess enough kinetic energy to overcome the repulsive and bonding forces that maintain the identity of the reactants. As a result, the rate for any single step in a reaction is directly proportional (1) to the number of collisions between reactant particles per second AND (2) to the fraction of the collisions that possess sufficient energy for reaction. Collision theory readily accounts for the dependence of reaction rates on the concentration of the reactants (number of collisions per second) and on the temperature (kinetic energy available for successful collisions). The fraction of molecules with higher kinetic energies increases with an increase in temperature, as Figure 14-3 shows.

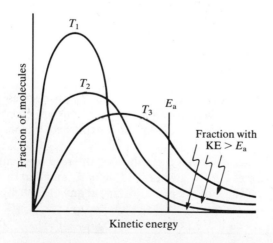

FIGURE 14-3.
Average kinetic energy for gas molecules at three temperatures, $T_1 < T_2 < T_3$.

The extra kinetic energy required for a successful reaction to occur is the **activation energy** of the reaction (the constant E_a in eq. 14-6). It is the energy necessary to alter the valence-electron configuration of the reactants to that of the *activated complex*.

2. Reaction-rate theory

Reaction-rate theory employs wave mechanics and the collision theory to explain why chemical reactions occur at observed rates. A collision between particle A and particle B is postulated to occur in steps:

- *The gradual approach of particle A to particle B*: The bond character of each particle is weakened and their valence-electron configurations are disrupted.
- *Formation of an activated complex*: If the sum of the energies that A and B possess is equal to the activation energy of the reaction, A and B will form a short-lived transient species, the **activated complex** or **transition state**, intermediate in form between reactants and products.
- *Breakdown of the activated complex*: Depending on reaction conditions, the transition state forms either products or reactants.

The activation energy of a reaction is effectively an energy hill that the reactants must climb in order to form products. The reactants can get over the hill only if they possess sufficient energy.

The energy profile for a typical reaction is shown in Figure 14-4. The vertical axis represents potential energy and the horizontal axis is the *reaction coordinate*, a measure of how far the reaction has moved from reactants to products. Note that if the sum of the energy of the products is lower than the energy of the reactants, the reaction is exothermic; if the products have higher energy than the reactants, the reaction is endothermic.

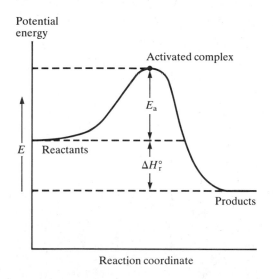

FIGURE 14-4.
Energy profile for an exothermic reaction. (ΔH_r is the heat of reaction.)

B. Reaction mechanisms

A **reaction mechanism** is a postulated sequence of single steps—or pathway—by which particular reactants are converted into particular products. The validity of a reaction mechanism depends upon its compatibility with experiment—and thus on its ability to account for the experimentally determined rate law.

Reaction mechanisms are written as a series of steps called **elementary reactions**, for which a rate law can be determined by inspection. If only one reactant molecule is involved, the elementary reaction is first-order and is called a *unimolecular* reaction. If two reactant molecules are involved, the elementary reaction is second-order and is called *bimolecular*. Occasionally, it is necessary to propose a step in a reaction mechanism that involves three molecules—a *termolecular* reaction that is third-order.

note: Elementary reactions are hypothetical steps in a reaction mechanism. They don't usually correspond to reactions that are directly observable.

EXAMPLE 14-8: (a) The decomposition of an activated ozone molecule (O_3^*), which has acquired excess kinetic energy by vigorous collision, is hypothesized to occur by the elementary reaction

$$O_3^* \longrightarrow O_2 + O$$

(b) The elementary reaction between a deuterium atom and hydrogen is hypothesized to be

$$H_2 + D \longrightarrow HD + H$$

Characterize and write rate equations for these elementary reactions.

Solution: Because these are elementary reactions, we know that we can determine their rate laws by inspection.

(a) Only one reactant molecule is involved in the decomposition of O_3^*, so this reaction is *unimolecular*, and the rate equation for a first-order reaction is

$$Rate = k[O_3^*]$$

(b) The reaction between H_2 and D is *bimolecular*, so the rate equation is (second-order)

$$Rate = k[D][H_2]$$

A complete reaction mechanism is a sequence of elementary reactions that begins with actual reactants, ends with actual products, and leads to a rate equation that accords with experimental observation. But—a reaction can only be as fast as the slowest step in the set of the elementary reactions. The slowest step in a reaction mechanism is called the *rate-limiting step* because it is the step that controls the rate of formation of products.

- If the rate-limiting step is much slower than the other steps in a reaction mechanism, the rate equation for the whole reaction will reflect the rate-limiting step directly.

EXAMPLE 14-9: At room temperature the rate law for the reaction $NO_2 + CO \rightarrow NO + CO_2$ is Rate $= k[NO_2]^2$. Is this rate equation compatible with the following mechanism?

$$NO_2 + NO_2 \longrightarrow NO_3 + NO \quad \text{slow}$$
$$NO_3 + CO \longrightarrow NO_2 + CO_2 \quad \text{fast}$$

Solution: Remember that a reaction mechanism is a sequence of elementary processes for which the rate law can be written simply by looking at the equation. In the mechanism proposed, the first (slow) step will be rate-limiting, and so will control the rate of formation of the products. The rate of that step, which is a bimolecular reaction between two molecules of NO_2, is $k[NO_2]^2$, which is exactly the observed rate equation. Consequently, the mechanism is one that leads to the observed rate equation, and the mechanism and rate equation are mutually compatible.

Catalysis

A **catalyst** is a substance that increases the rate of a reaction, but does not appear in the stoichiometric equation for a reaction—and can be fully recovered after a reaction.

EXAMPLE 14-10: Sucrose breaks down very slowly in hot water, forming a mixture of two other sugars, glucose and fructose. When a little hydrochloric acid is added to a sucrose/water mixture, the rate of decomposition increases substantially, but the products are the same and all of the hydrochloric acid is recovered at the end of the reaction. Account qualitatively for the effect of hydrochloric acid in this reaction.

Solution: Since the hydrochloric acid can be fully recovered, it is almost certainly a catalyst. Somehow, the hydrochloric acid must react with the sucrose and water to form an activated complex whose energy of activation is lower than that formed by sucrose and water alone, thereby increasing the rate of the reaction. This activated complex then breaks down to produce fructose and glucose, releasing hydrochloric acid quantitatively so the acid can be recovered without appearing to have been affected.

SUMMARY

1. The rate of a chemical reaction is measured by the change of reactant or product concentration in a certain period of time.
2. A rate equation (or law) gives the relationship between the rate of a reaction, the concentration of the reagents involved, and the rate constant k.
3. The exponent (power) of the concentration of each component in the general rate equation is the order of the reaction in that component.
4. The sum of the exponents in the general rate equation is the overall order of the reaction.
5. In a first-order reaction the sum of the exponents of the concentrations is one; in a second-order reaction the sum of the exponents of the concentrations is two.
6. The rate constant of a first-order reaction can be obtained from the integrated form of the rate law by plotting the logarithm of the concentration versus time.
7. The half-life of a reaction is the time it takes for one-half of the starting material to be consumed.
8. The rate constant of a second-order reaction can be obtained from the integrated rate equation by plotting reciprocal concentration versus time.
9. The rate of a chemical reaction increases when the temperature increases; the change in rate constant is given by the Arrhenius equation.
10. Reaction rate theory explains reaction rate in terms of an activated complex between reacting species. The reaction will proceed if the reactants possess the required activation energy.
11. Kinetic studies are used to determine the mechanism of a chemical reaction, which is a series of steps called elementary reactions. The proposed mechanism must be compatible with the experimentally observed rate law.
12. The rate of a reaction may be increased by adding a catalyst that interacts with the reagents to lower the activation energy. The catalyst does not appear in the overall stoichiometric equation of the reaction.

RAISE YOUR GRADES

Can you ...?

☑ name the factors that affect the rates of chemical reactions
☑ tell the difference between the rate of a reaction, the rate constant, and the rate equation
☑ determine the rate equation and order of a reaction, given the effects of changing the concentration of reagents on the rate of the reaction
☑ use first-order and second-order integrated rate equations to determine rate constants
☑ determine the half-life of a first-order reaction from the rate constant, or determine the first-order rate constant from the half-life
☑ use the Arrhenius equation to calculate the activation energy, or determine the rate constant at different temperatures given the activation energy
☑ explain why the fraction of molecules in a reaction having energies greater than the activation energy rises rapidly with increasing temperature.
☑ tell whether or not a reaction mechanism is compatible with an observed overall rate equation
☑ explain what a catalyst is

SOLVED PROBLEMS

PROBLEM 14-1 In the reaction of nitrogen(II) oxide (NO) with hydrogen at 1000 K,

$$2NO(g) + 2H_2(g) \longrightarrow N_2(g) + 2H_2O(g)$$

the rate of disappearance of NO is

$$\text{Rate} = -\frac{\Delta[\text{NO}]}{\Delta t} = 5.0 \times 10^{-5} \text{ mol L}^{-1}\text{s}^{-1}$$

What is the rate of formation of N_2?

Solution: Each mole of NO that reacts will form $\frac{1}{2}$ mol of N_2. Therefore, N_2 will form at half the rate of NO disappearance. Using the general reaction rate equation (14-1),

$$\frac{\Delta[\text{N}_2]}{\Delta t} = -\frac{\frac{1}{2}\Delta\text{NO}}{\Delta t} = -\frac{(-5.0 \times 10^{-5} \text{ mol L}^{-1}\text{s}^{-1})}{2}$$

$$= 2.5 \times 10^{-5} \text{ mol L}^{-1}\text{s}^{-1}$$

Note that the rate of formation of N_2 is opposite in sign to the rate of NO disappearance.

PROBLEM 14-2   The reaction $F_2(g) + 2ClO_2(g) \rightarrow 2FClO_2(g)$ is first-order in F_2 and first-order in ClO_2. Write the rate equation for this reaction.

Solution: The exponents on the concentrations of each reactant will be 1 since the reaction is first-order in F_2 and in ClO_2. Using the general expression for reaction rate (14-1) and the general rate equation (14-2),

$$\text{Rate} = -\frac{\Delta[\text{F}_2]}{\Delta t} = -\frac{\frac{1}{2}\Delta[\text{ClO}_2]}{\Delta t} = k[\text{F}_2][\text{ClO}_2]$$

Notice that the stoichiometric coefficient for ClO_2 is 2, but the exponent in the rate law is 1.

PROBLEM 14-3   Acetaldehyde (CH_3CHO) decomposes at 800 K according to the reaction

$$CH_3CHO(g) \longrightarrow CH_4(g) + CO(g)$$

The following experimental data were obtained for the reaction:

Concentration [CH$_3$CHO] (mol L^{-1})	Rate $-\dfrac{\Delta[\text{CH}_3\text{CHO}]}{\Delta t}$ (mol L^{-1} s^{-1})
0.010	9.0×10^{-7}
0.020	3.6×10^{-6}
0.040	1.43×10^{-5}

(a) Write the rate equation for the reaction. What is the overall reaction order?
(b) Calculate the rate constant k. Give the correct units.
(c) If $[CH_3CHO]$ is 0.030 mol L^{-1}, what will the rate of CH_3CHO decomposition be?

Solution:
(a) When the concentration of CH_3CHO is doubled, the rate is four times larger (don't be fooled by the change of exponents in the powers of 10):

Relative concentration	Relative rate
$\dfrac{0.020}{0.010} = 2$	$\dfrac{3.6 \times 10^{-6}}{9.0 \times 10^{-7}} = 4$
$\dfrac{0.040}{0.020} = 2$	$\dfrac{1.43 \times 10^{-5}}{3.6 \times 10^{-6}} = 4$

The rate is proportional to $[CH_3CHO]^2$, so the rate equation is

$$\text{Rate} = -\frac{\Delta[\text{CH}_3\text{CHO}]}{\Delta t} = k[\text{CH}_3\text{CHO}]^2$$

This is a second-order reaction.

(b) Solve the rate equation for k:

$$k = \frac{-\dfrac{\Delta[CH_3CHO]}{\Delta t}}{[CH_3CHO]^2} = \frac{\text{rate}}{[CH_3CHO]^2}$$

Then find k at each concentration:

$$k = \frac{9.0 \times 10^{-7}}{(0.010)^2} = 9.0 \times 10^{-3}$$

$$k = \frac{3.6 \times 10^{-6}}{(0.020)^2} = 9.0 \times 10^{-3}$$

$$k = \frac{1.43 \times 10^{-5}}{(0.040)^2} = 8.9 \times 10^{-3}$$

The average value of k (within experimental error) is 9.0×10^{-3}. The units of k are

$$\frac{mol\,L^{-1}s^{-1}}{(mol\,L^{-1})^2} = \frac{mol\,L^{-1}s^{-1}}{mol^2L^{-2}} = mol^{-1}L\,s^{-1}$$

(c)

$$\begin{aligned}
\text{Rate} &= k[CH_3CHO]^2 \\
&= (9.0 \times 10^{-3}\,mol^{-1}L\,s^{-1})(0.030\,mol\,L^{-1})^2 \\
&= 8.1 \times 10^{-6}\,mol\,L^{-1}s^{-1}
\end{aligned}$$

PROBLEM 14-4 For a hypothetical reaction $A + B \rightarrow C$, the following data apply:

Concentration [A] (mol L^{-1})	Concentration [B] (mol L^{-1})	Rate (mol $L^{-1}s^{-1}$)
0.100	1.0×10^{-2}	5.0×10^{-3}
0.300	1.0×10^{-2}	15×10^{-3}
0.200	2.0×10^{-2}	10×10^{-3}

Determine the rate equation for the reaction.

Solution: When [A] is tripled (from 0.100 to 0.300) and [B] is kept constant (1.0×10^{-2}), the rate of reaction triples (from 5.0×10^{-3} to 15×10^{-3}). Therefore the rate must be first-order in A. When the [A] and [B] are both doubled (from 0.100 to 0.200 for [A] and 1.0×10^{-2} to 2.0×10^{-2} for [B]), the rate is doubled. We already know that the rate is first-order in A, which would double the rate, so we can conclude that the rate of this reaction must be independent of [B], or *zero-order* in B. The rate equation is therefore

$$\text{Rate} = -\frac{\Delta[A]}{\Delta t} = -\frac{\Delta[B]}{\Delta t} = k[A]^1[B]^0 = k[A]$$

PROBLEM 14-5 Complex ions of the transition elements in aqueous solution often react with water. For example, pentamminechlorocobalt(III), $Co(NH_3)_5Cl^{2+}$, reacts with water as follows:

$$Co(NH_3)_5Cl^{2+}(aq) + H_2O(l) \longrightarrow Co(NH_3)_5H_2O^{3+}(aq) + Cl^-(aq)$$

The individual rate equation is first-order in concentration of the complex ion. Use the following data (taken at 52°C) to determine **(a)** the rate constant k in reciprocal seconds and **(b)** the half-life in seconds.

t (min):	0	40.0	100.0	160.0	200.0	270.0
$[Co(NH_3)_5Cl^{2+}]$ (mol L^{-1}):	0.0201	0.0183	0.0153	0.0127	0.0119	0.0096

Solution:
(a) Plot the logarithm of the concentration of the complex ion $\log[Co(NH_3)_5Cl^{2+}]$ as a function of time, as in Figure 14-5. Use the slope of the resulting line to obtain the rate constant. (The time is given in minutes and the answers are asked for in seconds. To convert the time from minutes to seconds, we'll determine the rate constant in min^{-1} first and convert that to s^{-1}.)

$[Co(NH_3)_5Cl^{2+}]$	$\log[Co(NH_3)_5Cl^{2+}]$
0.0201	−1.697
0.0183	−1.738
0.0153	−1.815
0.0127	−1.896
0.0119	−1.924
0.0096	−2.018

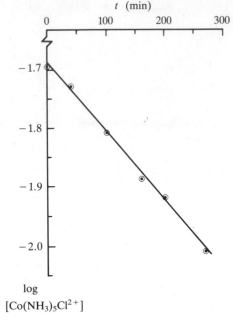

FIGURE 14-5. Problem 14-5

Since

$$\text{slope} = \frac{\log[Co(NH_3)_5Cl^{2+}] - \log[Co(NH_3)_5Cl^{2+}]_0}{t - t_0}$$

we can use any two points on the line to find the slope:

$$\text{slope} = \frac{-2.003 - (-1.706)}{260.0 - 10.0} = -\frac{0.297}{250.0}$$

$$= -1.188 \times 10^{-3}$$

Now we know that the slope is

$$-\frac{k}{2.303} = -1.188 \times 10^{-3} \text{ min}^{-1}$$

so we can find k:

$$k = (2.303)(1.188 \times 10^{-3}) \text{ min}^{-1} = 2.74 \times 10^{-3} \text{ min}^{-1}$$

Converting to reciprocal seconds

$$k = (2.74 \times 10^{-3} \text{ min}^{-1})\left(\frac{1 \text{ min}}{60 \text{ s}}\right) = 4.56 \times 10^{-5} \text{ s}^{-1}$$

(b) Now use the half-life formula to find $t_{1/2}$:

$$t_{1/2} = \frac{0.693}{k} = \frac{0.693}{4.56 \times 10^{-5} \text{ s}^{-1}} = 1.52 \times 10^4 \text{ s}$$

PROBLEM 14-6   The rate constant for the first-order decomposition of N_2O_5 at 320 K is $5.00 \times 10^{-4} \text{ s}^{-1}$. The stoichiometric equation is

$$2N_2O_5(g) \longrightarrow 4NO_2(g) + O_2(g)$$

(a) If you have 6.17×10^{-3} mol of $N_2O_5(g)$ in a 0.500-L container at 320 K, what will be the rate (in $\text{mol L}^{-1}\text{s}^{-1}$) at which N_2O_5 decomposes?

(b) How long will it take for 3.00×10^{-3} mol of N_2O_5 to decompose?

Solution: To do this problem you must understand the difference between rate constant, rate of reaction, and integrated rate equation.

(a) The initial concentration of N_2O_5 is

$$[N_2O_5] = \frac{6.17 \times 10^{-3} \text{ mol}}{0.500 \text{ L}} = 1.23 \times 10^{-2} \text{ mol L}^{-1}$$

Using the rate of N_2O_5 decomposition given in Section 14-2:

$$\text{Rate} = -\frac{\frac{1}{2}\Delta[N_2O_5]}{\Delta t} = k[N_2O_5] \quad \text{or} \quad -\frac{\Delta[N_2O_5]}{\Delta t} = 2k[N_2O_5]$$

$$= 2(5.00 \times 10^{-4} \text{ s}^{-1})(1.23 \times 10^{-2} \text{ mol L}^{-1})$$
$$= 1.23 \times 10^{-5} \text{ mol L}^{-1}\text{s}^{-1}$$

(b) Use the integrated form of the first-order rate equation (14-3):

$$\text{Rate} = \log\frac{[N_2O_5]_0}{[N_2O_5]} = \frac{kt}{2.303}$$

When 3.00×10^{-3} mol of N_2O_5 has decomposed, the remaining concentration will be

$$[N_2O_5] = \frac{(6.17 \times 10^{-3} \text{ mol}) - (3.00 \times 10^{-3} \text{ mol})}{0.500 \text{ L}}$$

$$= 6.34 \times 10^{-3} \text{ mol L}^{-1}$$

The initial concentration $[N_2O_5]_0$ determined in part (a) was 1.23×10^{-2} mol L^{-1}. So substitute the initial and final concentrations into the logarithmic part and the rate constant into the integrated rate equation,

$$\log\left(\frac{1.23 \times 10^{-2}}{6.34 \times 10^{-3}}\right) = 0.2878 = \frac{(5.00 \times 10^{-4} \text{ s}^{-1})t}{2.303}$$

and solve for t:

$$t = \frac{(0.2878)(2.303)}{5.00 \times 10^{-4} \text{ s}^{-1}} = 1.33 \times 10^3 \text{ s} \quad \text{or} \quad 22.2 \text{ min}$$

PROBLEM 14-7 In the hypothetical reaction $B(aq) \rightarrow$ products, of compound B in aqueous solution, the concentration of B at 298 K falls from 1.0 to 0.50 M in 280 s. How long will it take the concentration of B to fall to 0.25 M (a) if the reaction is first-order in B; (b) if the reaction is second-order in B? [*Hint:* Remember that $M = $ mol L^{-1}.]

Solution:
(a) For first-order reactions the half-life of the reactant is a constant quantity. Since [B] has fallen to half its original value (from 1.0 to 0.50 M) in 280 s, it will halve again, from 0.50 M to 0.25 M, in another 280 s; i.e., [B] falls from 1.0 to 0.25 M in 560 s.
(b) If the reaction is second-order, you must use the known t in the integrated form of the second-order rate equation (14-5) to find k:

$$\frac{1}{[B]} = \frac{1}{[B]_0} + kt$$

$$\frac{1}{0.50 \ M} = \frac{1}{1.0 \ M} + k(280 \text{ s})$$

$$k = \frac{(2-1) \ M^{-1}}{280 \text{ s}} = \frac{1}{280} M^{-1}\text{s}^{-1}$$

Now for the second stage you want to find the value of t for which $[B] = 0.25 \ M$ and $[B]_0 = 0.50 \ M$:

$$\left(\frac{1}{0.25 \ M}\right) = \left(\frac{1}{0.50 \ M}\right) + \left(\frac{1}{280} M^{-1}\text{s}^{-1}\right)t$$

$$2.0 \ M^{-1} = \left(\frac{1}{280} M^{-1}\text{s}^{-1}\right)t$$

$$t = \left(280 \times \frac{2.0}{1}\right)\text{s} = 560 \text{ s}$$

So [B] falls from 1.0 to 0.25 M in $(560 + 280) = 840$ s.

PROBLEM 14-8 You find that the rate of a reaction $A + B + C \rightarrow$ products depends on the concentration of OH$^-$ ions. You obtain the following experimental data:

Experiment	Initial rate $-\frac{\Delta[A]}{\Delta t}$ (mol L^{-1}min^{-1})	Initial concentrations (M) [A]	[B]	[C]	[OH$^-$]
1	1×10^{-4}	1×10^{-2}	1×10^{-2}	1×10^{-2}	1×10^{-2}
2	3×10^{-4}	3×10^{-2}	1×10^{-2}	1×10^{-2}	1×10^{-2}
3	9×10^{-4}	1×10^{-2}	3×10^{-2}	3×10^{-2}	1×10^{-2}
4	3×10^{-4}	3×10^{-2}	1×10^{-2}	3×10^{-2}	1×10^{-2}
5	3×10^{-4}	1×10^{-2}	3×10^{-2}	1×10^{-2}	3×10^{-2}

(a) What is the order of the reaction with respect to each component in the rate equation

$$\text{Rate} = -\frac{\Delta[A]}{\Delta t} = k[A]^a[B]^b[C]^c[OH^-]^d$$

(b) What is the value of k?

Solution: There are five sets of data and five unknowns: the exponents a, b, c, d, and the rate constant k. So the problem should be solvable. Even when a rate equation is this complex, you can easily determine the order of the reaction in a component if you compare two data sets in which only one initial concentration is changed. Thus, comparing experiments 1 and 2, where only $[A]$ changes:

$$\text{Rate}_2 = 3 \times 10^{-4}\,\text{mol L}^{-1}\text{min}^{-1} = k(3 \times 10^{-2}\,M)^a(1 \times 10^{-2}\,M)^b(1 \times 10^{-2}\,M)^c(1 \times 10^{-2}\,M)^d$$

$$\text{Rate}_1 = 1 \times 10^{-4}\,\text{mol L}^{-1}\text{min}^{-1} = k(1 \times 10^{-2}\,M)^a(1 \times 10^{-2}\,M)^b(1 \times 10^{-2}\,M)^c(1 \times 10^{-2}\,M)^d$$

Now divide Rate_2 by Rate_1 and $[A]_2$ by $[A]_1$ and set up the equality:

$$\frac{3 \times 10^{-4}\,\text{mol L}^{-1}\text{min}^{-1}}{1 \times 10^{-4}\,\text{mol L}^{-1}\text{min}^{-1}} = \left(\frac{3 \times 10^{-2}\,M}{1 \times 10^{-2}\,M}\right)^a$$

$$3 = 3^a$$

$$a = 1$$

The reaction is first-order in $[A]$.

A similar comparison of Rate_4 with Rate_2, where the rates are the same but $[C]$ has tripled, shows the reaction is zero-order in $[C]$, so $c = 0$. Now compare Rate_3 with Rate_1: The rate has increased 9-fold, $[B]$ has tripled, and $[C]$ has tripled. But you have just shown that the concentration of C has no effect, so you can ignore $[C]$. Since $9 = 3^2$, you know that the order of reaction in $[B] = 2$, so $b = 2$. So far the rate equation is

$$\text{Rate} = k[A][B]^2[OH^-]^d$$

Now compare experiments 3 and 5, where only $[OH^-]$ changes:

$$\frac{9 \times 10^{-4}}{3 \times 10^{-4}} = \left(\frac{1 \times 10^{-2}}{3 \times 10^{-2}}\right)^d$$

$$3 = \left(\frac{1}{3}\right)^d$$

$$d = -1$$

The rate goes down as $[OH^-]$ increases in direct inverse ratio.

You now have the complete rate law:

$$\text{Rate} = k[A][B]^2[OH^-]^{-1} = \frac{k[A][B]^2}{[OH^-]}$$

(b) To find the value of k, use the rate equation just obtained and the data from any experiment. (We'll use experiment 1.)

$$\text{Rate}_1 = 1 \times 10^{-4}\,\text{mol L}^{-1}\,\text{min}^{-1} = k\frac{(1 \times 10^{-2}\,M)(1 \times 10^{-2}\,M)^2}{(1 \times 10^{-2}\,M)}$$

$$= k(1 \times 10^{-4}\,\text{mol}^2\text{L}^{-2})$$

So

$$k = \frac{1 \times 10^{-4}\,\text{mol L}^{-1}\,\text{min}^{-1}}{1 \times 10^{-4}\,\text{mol}^2\text{L}^{-2}} = 1\,\text{L mol}^{-1}\,\text{min}^{-1}$$

PROBLEM 14-9 At $25°C$ in neutral aqueous solution, nitramide (H_2NNO_2) decomposes to $N_2O(g)$ and H_2O:

$$H_2NNO_2 \longrightarrow N_2O + H_2O$$

The following data were obtained for nitramide concentration as a function of time. Determine whether these data fit a first- or a second-order rate equation, and calculate the appropriate value of k.

t (min)	0	60	90	120	150	300	600	1200
$[H_2NNO_2]$ (M)	0.244	0.212	0.199	0.187	0.177	0.139	0.097	0.061

Solution: You'll have to test the linearity of the two integrated rate laws: $\log[A]$ versus t and $1/[A]$ versus t. Make a table of these values and then plot them.

t	$[A]$	$\log[A]$	$\dfrac{1}{[A]}$
0	0.244	-0.613	4.10
60	0.212	-0.674	4.72
90	0.199	-0.701	5.03
120	0.187	-0.728	5.35
150	0.177	-0.752	5.65
300	0.139	-0.857	7.19
600	0.097	-1.01	10.3
1200	0.061	-1.21	16.4

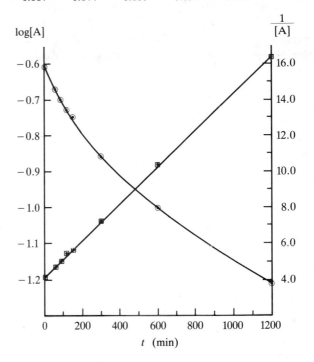

FIGURE 14-6. Problem 14-9

As you can see from Figure 14-6, the $\log[A]$ versus t plot is curved. Therefore the reaction isn't first-order.

note: There's an easier way to spot this—without doing a plot. Remember the half-life concept? The initial concentration has fallen to half its value (from 0.244 M to 0.122 M) in what you'd estimate from the data would be a bit over 300 min. But in a further 300 min it certainly hasn't halved again—as it should if the reaction were first order.

The $1/[A]$ versus t plot is nicely linear over its whole length; therefore the reaction does fit a second-order rate equation.

Since the slope is equal to k, the rate constant is

$$k = \frac{(16.4 - 4.1)\ M^{-1}}{1200\ \text{min}} = 1.02 \times 10^{-2}\ M^{-1}\text{min}^{-1}$$

PROBLEM 14-10 The rate constant for the first-order decomposition of $N_2O_5(g)$ to $NO_2(g)$ and $O_2(g)$ has the value 3.46×10^{-5} s^{-1} at 298.0 K and 4.98×10^{-4} s^{-1} at 318.0 K. Calculate the energy of activation E_a for this reaction in kilojoules.

Solution: Obviously, temperature effects are involved here, so you need the Arrhenius equation (14-6):

$$\log\left(\frac{k_2}{k_1}\right) = \frac{-E_a}{2.303R}\left(\frac{1}{T_2} - \frac{1}{T_1}\right)$$

Call 318 K T_2 and 298 K T_1 and substitute:

$$\log\left(\frac{4.98 \times 10^{-4}\ \text{s}^{-1}}{3.46 \times 10^{-5}\ \text{s}^{-1}}\right) = \left(\frac{-E_a}{(2.303)(8.314\ \text{J K}^{-1}\text{mol}^{-1})}\right)\left(\frac{1}{318.0\ \text{K}} - \frac{1}{298.0\ \text{K}}\right)$$

Since $\dfrac{1}{a} - \dfrac{1}{b}$ can be more easily (and accurately) handled as $\dfrac{b-a}{ab}$, you can write

$$\log 14.4 = 1.158 = \left(\frac{-E_a}{(2.303)(8.314\ \text{J K}^{-1}\text{mol}^{-1})}\right)\left(\frac{298.0 - 318.0}{(298.0)(318.0)}\right)\text{K}^{-1}$$

and so

$$E_a = \left(\frac{-(1.158)(2.303)(8.314)(298.0)(318.0)\ \text{J mol}^{-1}}{-20.0}\right)\left(\frac{1\ \text{kJ}}{10^3\ \text{J}}\right)$$

$$= 105\ \text{kJ mol}^{-1}$$

PROBLEM 14-11 Use the data in Problem 14-10 to predict the rate constant at 338.0 K for the first-order gas-phase decomposition of N_2O_5.

224 College Chemistry

Solution: You know that $k = 4.98 \times 10^{-4}\,\text{s}^{-1}$ at 318.0 K and that $E_a = 105\,\text{kJ mol}^{-1}$. Using the Arrhenius equation (14-6),

$$\log\left(\frac{k_{338}}{k_{318}}\right) = \left(\frac{-E_a}{(2.303)(8.314\,\text{J K}^{-1}\text{mol}^{-1})}\right)\left(\frac{1}{338.0\,\text{K}} - \frac{1}{318.0\,\text{K}}\right)$$

$$= \left(\frac{-105 \times 10^3\,\text{J mol}^{-1}}{(2.303)(8.314\,\text{J K}^{-1}\text{mol}^{-1})}\right)\left(\frac{-20.0}{(338.0)(318.0)}\right)\text{K}^{-1}$$

$$= 1.020$$

Take antilogs of both sides:

$$\frac{k_{338}}{k_{318}} = \text{antilog}(1.020) = 10.5$$

So

$$k_{338} = 10.5 k_{318} = 10.5(4.98 \times 10^{-4}\,\text{s}^{-1}) = 5.23 \times 10^{-3}\,\text{s}^{-1}$$

PROBLEM 14-12 For the reaction $OCl^-(aq) \rightarrow Cl^-(aq) + \frac{1}{2}O_2(g)$, the energy of activation E_a is 47 kJ/mol. The heat of reaction ΔH_r for this exothermic reaction is -51 kJ/mol. Draw a diagram plotting energy against reaction coordinate, clearly identifying E_a, ΔH_r, and the activated complex.

Solution:
See Figure 14-7.

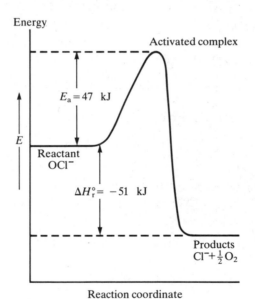

FIGURE 14-7. Problem 14-12

PROBLEM 14-13 The reaction $C_3H_7Br(aq) + OH^-(aq) \rightarrow C_3H_7OH(aq) + Br^-(aq)$ is found to be first-order in $[C_3H_7Br(aq)]$ and first-order in $[OH^-(aq)]$. Which of the following mechanisms is consistent with the rate data?

(a) $C_3H_7Br \longrightarrow C_3H_7^+ + Br^-$ slow
$C_3H_7^+ + OH^- \longrightarrow C_3H_7OH$ fast
(b) $C_3H_7Br + OH^- \longrightarrow C_3H_7OH + Br^-$ slow

Solution: When you're given the elementary reactions, you can write the rate equations directly. Remember: It's the rate of the slow step that determines the overall rate.
In mechanism **(a)**

$$\text{Rate of slow step} = \text{Rate of reaction} = k[C_3H_7Br]$$

There's no $[OH^-]$ term, since OH^- is only involved in the fast step.
In mechanism **(b)**

$$\text{Rate of slow step} = \text{Rate of reaction} = k[C_3H_7Br][OH^-]$$

Experimentally, the reaction is first-order in both C_3H_7Br and OH^-, or Rate $= k[C_3H_7Br][OH^-]$. So you see that, of the two mechanisms proposed, only mechanism (b) is consistent with the rate data.

note: Rate data can never "prove" a mechanism. The best you can say is that the data are "accounted for" by the mechanism.

PROBLEM 14-14 The compound C_4H_9OH reacts with aqueous H^+ and Br^- to produce C_4H_9Br. The rate of the reaction is found to be first-order in $[C_4H_9OH]$, first-order in $[H^+]$, and zero-order in $[Br^-]$. Which of the following mechanisms is consistent with these observations?

(a) $Br^- + C_4H_9OH \longrightarrow C_4H_9Br + OH^-$ slow
   $OH^- + H^+ \longrightarrow H_2O$ fast

(b) $C_4H_9OH + H^+ \rightleftharpoons C_4H_9OH_2^+$ fast equilibrium
   $C_4H_9OH_2^+ \longrightarrow C_4H_9^+ + H_2O$ slow
   $C_4H_9^+ + Br^- \longrightarrow C_4H_9Br$ fast

Solution: You can easily dismiss mechanism (a). The observed rate equation is Rate $= k[C_4H_9OH][H^+]$, but the rate-limiting step in (a)—the slow step—has Rate $= k[C_4H_9OH][Br^-]$, with first-order dependence on $[Br^-]$ and no term in $[H^+]$.

However, mechanism (b) presents a new notion: an equilibrium before the slow step. Now what? You can see that the slow step has a Rate $= k[C_4H_9OH_2^+]$. Now apply what you know about equilibrium to the first step proposed:

$$K = \frac{[C_4H_9OH_2^+]}{[C_4H_9OH][H^+]} \quad \text{or} \quad [C_4H_9OH_2^+] = K[C_4H_9OH][H^+]$$

where K is an equilibrium constant. Now put this value of $[C_4H_9OH_2^+]$ in the rate equation, and it becomes Rate $= kK[C_4H_9OH][H^+]$, which is just the form observed experimentally. (The experimental rate constant kK is the product of the equilibrium constant for the fast equilibrium and the rate constant for the slow step). Therefore mechanism (b) is consistent with the observations.

PROBLEM 14-15 The reaction $2NO(g) + O_2(g) \rightarrow 2NO_2(g)$ is a rare third-order reaction, whose rate equation is Rate $= k[NO]^2[O_2]$. Which of the following mechanisms is consistent with this rate equation? Comment on your answer.

(a) $2NO \rightleftharpoons N_2O_2$ fast equilibrium
   $N_2O_2 + O_2 \longrightarrow 2NO_2$ slow

(b) $NO + O_2 \rightleftharpoons O_2NO$ fast equilibrium
   $O_2NO + NO \longrightarrow 2NO_2$ slow

Solution: Using the idea introduced in Problem 14-14, you can deduce the rate equations for each mechanism.

(a) Rate of slow step $= k[N_2O_2][O_2]$:
   Equilibrium-law expression for the fast-equilibrium step is

$$K = \frac{[N_2O_2]}{[NO]^2}; \quad \text{so} \quad [N_2O_2] = K[NO]^2$$

Substituting into rate equation gives Rate $= kK[NO]^2[O_2]$, which is just like the experimental rate equation.

(b) Rate of slow step $= k[O_2NO][NO]$:
   Equilibrium-law expression for the fast-equilibrium step is

$$K = \frac{[O_2NO]}{[NO][O_2]}; \quad \text{so} \quad [O_2NO] = K[NO][O_2]$$

Substituting into rate equation gives Rate $= kK[NO][O_2][NO] = kK[NO]^2[O_2]$, which is just like the experimental rate equation.

Thus *both* mechanisms are consistent with the rate equation.

comment: The fact that a mechanism gives a predicted rate equation that agrees with experiment is no guarantee that it's correct. Both these mechanisms can't be correct: One (or both) must be wrong—but both agree with the experimental data. You'd need other kinds of experiments to rule one or the other out. For instance, for mechanism (a) you could devise an experiment to see if there's really any evidence for the presence of N_2O_2 in NO. You might try high pressures, which should favor the formation of a dimer.

PROBLEM 14-16 In a popular general chemistry experiment, potassium chlorate ($KClO_3$) is heated in the presence of a trace of manganese dioxide (MnO_2) to yield oxygen:

$$2KClO_3(s) = 2KCl(s) + 3O_2(g)$$

What is the function of the MnO_2 in this reaction?

Solution: The MnO_2 acts as a catalyst to speed up the rate at which O_2 is produced from $KClO_3$. The MnO_2 does not appear in the stoichiometric equation for the reaction, but presumably does combine with $KClO_3$ to produce an unstable intermediate compound that breaks down easily to KCl and O_2, regenerating MnO_2 in the process.

Supplementary Exercises

PROBLEM 14-17 The rate of formation of $FClO_2(g)$ in the reaction $F_2(g) + 2ClO_2(g) \rightarrow 2FClO_2(g)$ at a given temperature is 3.0×10^{-4} mol $L^{-1}s^{-1}$. What is the rate of disappearance of $F_2(g)$?

Answer: 1.5×10^{-4} mol $L^{-1}s^{-1}$

PROBLEM 14-18 The decomposition of HI(g) is second-order: $2HI(g) \rightarrow H_2(g) + I_2(g)$. Write the rate equation for the reaction.

Answer: Rate $= -\dfrac{\frac{1}{2}\Delta[HI]}{\Delta t} = k[HI]^2$

PROBLEM 14-19 The following rate data have been collected at 1000 K for the reaction $2NO(g) + 2H_2(g) \rightarrow N_2(g) + 2H_2O(g)$:

[NO] (mol L^{-1})	[H_2] (mol L^{-1})	NO disappearance (mol $L^{-1}s^{-1}$)
5.0×10^{-3}	2.0×10^{-3}	2.5×10^{-5}
10.0×10^{-3}	2.0×10^{-3}	10.0×10^{-5}
10.0×10^{-3}	4.0×10^{-3}	20.0×10^{-5}

(a) Derive a rate equation for the reaction in terms of NO and H_2 concentrations.
(b) Calculate the value and units of the rate constant. What is the overall order of the reaction?
(c) What would the rate of the reaction be if $[NO] = [H_2] = 3.0 \times 10^{-2}$ M?

Answer: (a) Rate $= -\dfrac{\frac{1}{2}\Delta[NO]}{\Delta t} = -\dfrac{\frac{1}{2}\Delta[H_2]}{\Delta t} = k[NO]^2[H_2]$

(b) $k = 2.5 \times 10^2$ mol^{-2}L^{2}s^{-1}; overall order is 3
(c) 6.8×10^{-3} mol $L^{-1}s^{-1}$

PROBLEM 14-20 The rate constant for the decomposition of cyclobutane (C_4H_8) is 1.23×10^{-4} s^{-1} at 700 K. The reaction $C_4H_8(g) \rightarrow 2C_2H_4(g)$ is first-order.

(a) If you have 3.0×10^{-3} mol of C_4H_8 in a 0.500-L container at 700 K, at what rate (in mol $L^{-1}s^{-1}$) will it be decomposing?
(b) How long will it take for 1.8×10^{-3} mol of the C_4H_8 to decompose?

Answer: (a) 7.4×10^{-7} mol $L^{-1}s^{-1}$ (b) 7.4×10^3 s

PROBLEM 14-21 At 25°C the organic chloride 2-chloro-2-methylpropane (($CH_3)_3CCl$) is hydrolyzed in a mixture of alcohol and water by a reaction that is first-order in ($CH_3)_3CCl$:

$$(CH_3)_3CCl + H_2O \longrightarrow (CH_3)_3COH + HCl$$

The concentrations (mol L^{-1}) at different times (s) are

$[(CH_3)_3CCl]$	6.28×10^{-3}	6.03×10^{-3}	5.53×10^{-3}	5.03×10^{-3}	4.53×10^{-3}	4.28×10^{-3}
t	440	666	1266	1632	2165	2447

Determine **(a)** the rate constant k, **(b)** the concentration of $(CH_3)_3CCl$ at zero time, and **(c)** the half-life in hours.

Answer: **(a)** $1.91 \times 10^{-4}\,s^{-1}$ **(b)** $6.84 \times 10^{-3}\,mol\,L^{-1}$ **(c)** 1.01 h (3630 s)

PROBLEM 14-22   The rate equation for the reaction $2A + B \rightarrow$ products is Rate $= k[A]^2[B]^{-1}$. Which of the following statements about the reaction is/are true?

  (a) Doubling the concentration of A will double the rate.
  (b) Doubling the concentration of B will double the rate.
  (c) Halving the concentration of A will reduce the rate to one-fourth.
  (d) Halving the concentration of B will double the rate.

Answer: **(c)** and **(d)**

PROBLEM 14-23   Calculate the activation energy for a reaction that doubles in rate when the temperature is raised from 298 K to 308 K.

Answer: 52.9 kJ

PROBLEM 14-24   The rate equation for the decomposition of nitramide, $H_2NNO_2 \rightarrow N_2O + H_2O$, is Rate $= k[H_2NNO_2]/[H^+]$. Which of the following mechanisms is consistent with this rate equation?

  (a) $H_2NNO_2 \longrightarrow N_2O + H_2O$ slow
  (b) $H_2NNO_2 + H^+ \rightleftharpoons H_3NNO_2^+$ fast equilibrium
  $\qquad\quad H_3NNO_2^+ \longrightarrow N_2O + H_3O^+$ slow
  $\qquad\qquad\quad H_3O^+ \rightleftharpoons H^+ + H_2O$ fast equilibrium
  (c) $H_2NNO_2 \rightleftharpoons HNNO_2^- + H^+$ fast equilibrium
  $\qquad HNNO_2^- \longrightarrow N_2O + OH^-$ slow
  $\qquad H^+ + OH^- \longrightarrow H_2O$ fast

Answer: **(c)**

PROBLEM 14-25   For the second-order reaction $OH^- + C_2H_5I \rightarrow I^- + C_2H_5OH$, $k = 5.03 \times 10^{-2}\,L\,mol^{-1}s^{-1}$ at 298 K and $k = 6.71\,L\,mol^{-1}s^{-1}$ at 333 K. Use these data to calculate E_a and to estimate k at 273 K.

Answer: $E_a = 115\,kJ$; $k_{273} = 7.18 \times 10^{-4}\,L\,mol^{-1}s^{-1}$

PROBLEM 14-26   In the following reaction sequence, which compound would be described as a catalyst? What is the overall reaction equation?

$$NO + O_3 \longrightarrow NO_2 + O_2$$
$$NO_2 + O \longrightarrow NO + O_2$$

Answer: NO is the catalyst; $O_3 + O \longrightarrow 2O_2$

15 EQUILIBRIUM

THIS CHAPTER IS ABOUT

☑ **The Equilibrium State**
☑ **The Law of Chemical Equilibrium**
☑ **Equilibrium Changes**
☑ **How to Solve Equilibrium Problems**
☑ **Heterogeneous Equilibria**

15-1. The Equilibrium State

Chemical equilibrium is a dynamic process in which simultaneous forward and reverse reactions occur at equal rates in opposite directions, so that the concentration of each species present remains constant at constant temperature. Every reaction is capable of achieving equilibrium, but the equilibrium is not always significant. If the equilibrium of a reaction is significant (i.e., observable), we say that the reaction does not go "to completion," thus indicating the continued presence of reactants in products.

If, for example, one mole of H_2 and one mole of I_2 are heated together at 700 K for a few hours, a reaction occurs: $H_2 + I_2 \xrightarrow{\Delta} 2HI$. An analysis of the resulting gas would reveal a preponderant amount of HI, but it would also indicate a measurable amount of I_2 and H_2. If the reaction were continued at the same temperature and pressure for a few more hours, a second analysis would show that I_2 and H_2 were still present—in the same concentrations that were found in the first analysis. Moreover, if two moles of pure HI were heated at 700 K for a few hours, the reverse reaction would occur: $2HI \xrightarrow{\Delta} H_2 + I_2$. But an analysis of the resulting gas mixture would show the same concentrations of H_2 and I_2 that were found in the analyses for the forward reaction. Obviously, the reaction

$$H_2 + I_2 \underset{}{\overset{\Delta}{\rightleftharpoons}} 2HI$$

is an example of chemical equilibrium. After a few hours, H_2 and I_2 react to give HI and HI breaks down to give H_2 and I_2 at exactly the same rate.

note: Equilibrium reactions can occur in homogeneous (one phase) or heterogeneous (two or more phases) systems. Sections 15-2 through 15-4 deal with homogeneous systems, and Section 15-5 deals with the variations in the rules occasioned by heterogeneity.

15-2. The Law of Chemical Equilibrium

The **law of chemical equilibrium** states that for the general reaction

$$aA + bB + \cdots + \rightleftharpoons pP + qQ + \cdots +$$

at constant temperature, the concentrations of reactants and products at equilibrium are related by the equation

LAW OF CHEMICAL EQUILIBRIUM
$$K = \frac{[P]^p[Q]^q \cdots}{[A]^a[B]^b \cdots} \tag{15-1}$$

where K is the **equilibrium constant** for the reaction. Since the value of K is determined by reactant and final product concentration, K is independent of the number of intermediate steps in the reaction mechanism.

EXAMPLE 15-1: Write an expression for the equilibrium constant for each of the following homogeneous reactions:

(a) $H_2 + I_2 \rightleftharpoons 2HI$ (b) $2O_3 \rightleftharpoons 3O_2$ (c) $6CH_2O \rightleftharpoons C_6H_{12}O_6$

Solution:

(a) $K = \dfrac{[HI]^2}{[H_2][I_2]}$ (b) $K = \dfrac{[O_2]^3}{[O_3]^2}$ (c) $K = \dfrac{[C_6H_{12}O_6]}{[CH_2O]^6}$

note: It is customary to omit the units of K values.

In a homogeneous gas-phase reaction concentration expressions for reactants and products can be replaced with partial pressures, just as in reaction rates. So the law of chemical equilibrium (15-1) can be rewritten as

$$K_p = \frac{p_P^p p_Q^q \cdots}{p_A^a p_B^b \cdots} \tag{15-2}$$

where K_p indicates that K was determined by using partial pressures. (You can use either plain K or K_c to indicate that molar equilibrium concentrations were used.) Pressures in equilibrium expressions are *always* given in atmospheres.

There is, of course, a relationship between pressure and concentration. This relationship is given by the ideal gas law: $pV = nRT$. Thus, for species A,

$$p_A V_A = n_A RT$$

$$p_A = \frac{n_A}{V_A} RT \tag{15-3}$$

$$p_A = [A]RT$$

note: When the number of moles of gaseous reactants is equal to the number of moles of gaseous products, $K_p = K_c$. When the number of moles of reactants does not equal the number of moles of products, $K_p \neq K_c$.

EXAMPLE 15-2: A mixture of H_2, I_2, and HI that had come to equilibrium at 699 K was analyzed and found to contain the following partial pressures of the components: H_2, 0.165 atm; I_2, 0.0978 atm; HI, 0.945 atm. Calculate K_c for the equilibrium

$$H_2(g) + I_2(g) \rightleftharpoons 2HI(g)$$

Solution: K_p is readily given by the equilibrium law as

$$K_p = \frac{p_{HI}^2}{p_{H_2} p_{I_2}} = \frac{(0.945)^2}{(0.165)(0.0978)} = 55.3$$

and since $n_{products} = n_{reactants}$, $K_c = 55.3$.

EXAMPLE 15-3: A mixture of SO_2 and O_2 was allowed to come to equilibrium in a 1.4-L vessel. An analysis of the resulting contents showed 0.42 mol SO_2, 0.28 mol O_2, and 0.35 mol SO_3. Calculate K_c for the reaction

$$2SO_2(g) + O_2(g) \rightleftharpoons 2SO_3(g)$$

Solution: From the equilibrium law

$$K_c = \frac{[SO_3]^2}{[SO_2]^2[O_2]} = \frac{(0.35 \text{ mol}/1.4 \text{ L})^2}{(0.42 \text{ mol}/1.4 \text{ L})^2(0.28 \text{ mol}/1.4 \text{ L})} = 3.5$$

EXAMPLE 15-4: At 27°C and a total pressure of 1.00 atm, the equilibrium mixture of $N_2O_4(g)$ and $NO_2(g)$ contains 19.8% NO_2 by volume. Calculate K_p for the equilibrium

$$N_2O_4(g) \rightleftharpoons 2NO_2(g)$$

Solution: From the equilibrium law

$$K_p = \frac{p_{NO_2}^2}{p_{N_2O_4}}$$

Now we need to find the appropriate pressures for this expression. We know that $\dfrac{\%\text{ by volume}}{100\%}$ is the mole fraction, so

$$\text{mole fraction NO}_2 = \frac{19.8\%}{100\%} = 0.198$$

From Dalton's law, the partial pressure of each gas is (mole fraction) × (total pressure). Therefore,

$$p_{NO_2} = 0.198(1.00 \text{ atm}) = 0.198 \text{ atm}$$

There are only two gases in the mixture, so $p_{NO_2} + p_{N_2O_4} = 1.00$ atm and

$$p_{N_2O_4} = 1.00 \text{ atm} - 0.198 \text{ atm} = 0.802 \text{ atm}$$

Finally, substitute these partial pressures into the equilibrium law:

$$K_p = \frac{(0.198)^2}{0.802} = 0.489$$

15-3. Equilibrium Changes

Equilibrium changes are covered by a general principle—**Le Chatelier's principle:**

- *If an equilibrium system is stressed, the system will change to relieve the stress.*

For our purposes, a "stress" is any change in pressure, concentration, or temperature of a system. There are two ways to relieve stress in a system: The equilibrium can shift, or the equilibrium constant can change.

A. Shifting the equilibrium

- *Pressure changes*: Pressure changes only affect equilibria in gas-phase reactions for which $n_{products} \neq n_{reactants}$. An increase in pressure will shift the equilibrium toward the side having fewer moles of gas, and a decrease in pressure will shift the equilibrium toward the side having more moles of gas. Thus, for example, an increase in pressure for the reaction

$$N_2(g) + 3H_2(g) \rightleftharpoons 2NH_3(g)$$

will favor the formation of $NH_3(g)$.

When $n_{products} = n_{reactants}$ in a gas-phase reaction, changing the pressure has NO effect on the equilibrium.

- *Concentration changes*: An increase in reactant concentration or a decrease in product concentration will shift the equilibrium of a gas- or liquid-phase reaction toward the product side. A decrease in reactant concentration or an increase in product concentration will shift the equilibrium toward the reactant side. For example, for the reaction

$$H_2(g) + I_2(g) \rightleftharpoons 2HI(g)$$

adding H_2 or I_2 (or removing HI) will shift the equilibrium toward the right. Removing H_2 or I_2 (or adding HI) will shift the equilibrium toward the left.

- *note:* Changing pressure or concentration does not affect K. The equilibrium constant *stays* constant when pressure and concentration are changed.

B. Changing the equilibrium constant

- *Temperature changes*: An increase or decrease in temperature changes the value of K. Every equilibrium system involves one endothermic and one exothermic reaction. An increase in the temperature of a system will favor the endothermic reaction. Thus, for the reaction

$$H_2(g) + I_2(g) \rightleftharpoons 2HI(g) \qquad \Delta H_r^\circ = -12.9 \text{ kJ}$$

the forward reaction is exothermic. An increase in the temperature would favor the endothermic (reverse) reaction (i.e., the decomposition of HI), increasing the concentrations of H_2 and I_2 and the value of K would decrease to reflect this shift.

EXAMPLE 15-5: In which direction will the following equilibrium reaction shift if the pressure is increased?

$$2SO_2(g) + O_2(g) \rightleftharpoons 2SO_3(g)$$

Solution: An increase in pressure will cause the reaction to shift toward the side with fewer moles of gas. Three moles of reactants give only two moles of product, so an increase of pressure will produce more SO_3 at equilibrium.

EXAMPLE 15-6: Consider the equilibrium

$$PCl_5(g) \rightleftharpoons PCl_3(g) + Cl_2(g)$$

If some Cl_2 were suddenly injected in the system, what change in the equilibrium partial pressure of PCl_3 would occur?

Solution: The sudden increase of Cl_2 concentration causes the reaction to proceed toward the side that doesn't include Cl_2, i.e., toward PCl_5. The only way this can happen is for some PCl_3 to react with some Cl_2 to give more PCl_5. So the equilibrium partial pressure of PCl_3 must decrease after the injection of Cl_2.

EXAMPLE 15-7: The reaction of PCl_3 with Cl_2 to give PCl_5 is exothermic:

$$PCl_5(g) + \text{heat} \rightleftharpoons PCl_3(g) + Cl_2(g)$$

What effect will raising the temperature have on the equilibrium partial pressure of PCl_5?

Solution: An increase of temperature in an equilibrium system will favor the reaction that is endothermic. Since the reaction of PCl_3 with Cl_2 is exothermic, the reverse reaction (the dissociation of PCl_5) must be endothermic. So if the temperature of this system is raised, the equilibrium partial pressure of PCl_5 will decrease because the endothermic dissociation of PCl_5 will be favored.

15-4. How to Solve Equilibrium Problems

The secret of solving equilibrium problems is organization. If you need to find equilibrium concentrations of reactants or products given K and some initial concentration, you can use the following steps:

Step 1: Balance the equation for the reaction.
Step 2: Write the appropriate equilibrium expression for K.
Step 3: Identify the unknown concentration(s). Sort out the known factors, writing them in terms of the unknown factors. (This step may require some chemical reasoning.)
Step 4: Tabulate the data for initial and equilibrium conditions and the changes in concentration(s).
Step 5: Substitute into the expression for K (in Step 2), and solve for the unknown(s).

> *note:* Step 5 will often give you a quadratic equation to solve, so you'll need to know the general formula: If

$$ax^2 + bx + c = 0$$

then

$$x = \frac{-b \pm \sqrt{b^2 - 4ac}}{2a}$$

EXAMPLE 15-8: For the dissociation of gaseous PCl_5 at 573 K, the value of $K_p = 2.00$. The equilibrium reaction is

$$PCl_5(g) \rightleftharpoons PCl_3(g) + Cl_2(g)$$

If a sample of PCl_5 is heated to 573 K while the total pressure is kept constant at 5.00 atm, what would the partial pressures of reactant and products be at equilibrium?

Solution: Follow the steps prescribed for finding equilibrium concentrations.

Step 1: The equation for the equilibrium reaction is already balanced:

$$PCl_5 \rightleftharpoons PCl_3 + Cl_2$$

Step 2: From the partial-pressure form of the law of chemical equilibrium

$$K_p = \frac{p_{PCl_3} p_{Cl_2}}{p_{PCl_5}} = 2.00$$

Step 3: The problem asks for the equilibrium partial pressures of reactants and products, so those are the unknowns that have to be related to the knowns. The total pressure is given as 5.00 atm, so (by Dalton's law)

$$p_{PCl_5} + p_{PCl_3} + p_{Cl_2} = 5.00 \text{ atm}$$

Then from the stoichiometry of the reaction it is clear that $p_{PCl_3} = p_{Cl_2}$, so we can designate the value of p_{PCl_3} as x and write $p_{PCl_3} + p_{Cl_2} = x + x$.

Step 4: Tabulate the changes using the stoichiometry of the equation:

p (atm)	PCl_5	PCl_3	Cl_2
initial pressures	5	0	0
change in pressure	$-2x$	$+x$	$+x$
equilibrium pressure	$5 - 2x$	x	x

Step 5: Substitute:

$$K_p = \frac{p_{PCl_3} p_{Cl_2}}{p_{PCl_5}} = \frac{(x)(x)}{5.00 - 2x} = 2.00$$

$$2.00 = \frac{x^2}{5.00 - 2x}$$

$$10.00 - 4.00x = x^2$$

So

$$x^2 + 4.00x - 10.00 = 0$$

Now solve the quadratic equation for the unknown x:

$$x = \frac{-b \pm \sqrt{b^2 - 4ac}}{2a}$$

$$= \frac{-4.00 \pm \sqrt{(4.00)^2 - 4(1)(-10.00)}}{2(1)}$$

$$= \frac{-4.00 \pm \sqrt{16.00 + 40.0}}{2}$$

$$= \frac{-4.00 \pm 7.48}{2}$$

$$= -5.74 \quad \text{or} \quad 1.74$$

A negative partial pressure makes no sense, so we ignore it and say that $x = 1.74$. Now from the tabulation in Step 4 we have

p (atm)	PCl$_5$	PCl$_3$	Cl$_2$
equilibrium pressure	5 − 2x	x	x
	5 − 2(1.74) = 1.52	1.74	1.74

15-5. Heterogeneous Equilibria

Many important reactions occur under heterogeneous conditions. Consider the equilibrium

$$CaCO_3(s) \rightleftharpoons CaO(s) + CO_2(g)$$

If you write the equilibrium-law expression in the usual way, you get

$$K = \frac{[CaO(s)][CO_2(g)]}{[CaCO_3(s)]}$$

But the concentration of a pure solid (or pure liquid) is a constant, so the constant terms $[CaCO_3(s)]$ and $[CaO(s)]$ can be dropped from the expression, leaving

$$K_c = [CO_2(g)] \quad \text{or} \quad K_p = p_{CO_2}$$

note 1: An increase in the amount of a solid is also an increase in the volume of a solid, so the ratio of moles to volume (concentration) remains the same. The same is true for pure liquids.

note 2: In reactions that occur in water, H_2O is usually present in vast excess, which allows us to consider $H_2O(l)$ as a pure liquid for all practical purposes.

EXAMPLE 15-9: Write an equilibrium-law expression for each of the following reactions:

(a) $CO_2(g) + C(s) \rightleftharpoons 2CO(g)$
(b) $Fe_2O_3(s) + 3H_2(g) \rightleftharpoons 2Fe(s) + 3H_2O(g)$

Solution:
(a) Because carbon is present as a solid phase, it can be left out of the equilibrium-law expression. So

$$K_p = \frac{p_{CO}^2}{p_{CO_2}} \quad \text{or} \quad K_c = \frac{[CO]^2}{[CO_2]}$$

(b) Both Fe_2O_3 and Fe are solids, so they can be left out. The expression then becomes

$$K_p = \frac{p_{H_2O}^3}{p_{H_2}^3} \quad \text{or} \quad K_c = \frac{[H_2O]^3}{[H_2]^3}$$

SUMMARY

1. When a chemical reaction is at equilibrium, the rate of reaction in one direction is equal to the rate of reaction in the opposite direction.
2. The equilibrium-law expression gives the relationship between the equilibrium concentrations of the components and the equilibrium constant K.
3. The concentrations of gases may be given in units of concentration (moles per liter) or of pressure (atmospheres).
4. The effects of changing concentrations, external pressure, and temperature on an equilibrium mixture can be determined qualitatively by applying Le Chatelier's principle.
5. The equilibrium-law expression can be used to determine the equilibrium constant if the concentrations of the substances in the equilibrium equation are known; if the equilibrium constant is known, the equilibrium-law expression can be used to determine the concentrations of the substances in the equilibrium equation.
6. The concentrations of pure solid and pure liquid phases are not included in the equilibrium-law expression for a heterogeneous equilibrium.

RAISE YOUR GRADES

Can you...?

☑ write the equilibrium-law expression for a gas-phase reaction and a solution reaction
☑ write the equilibrium-law expression for a heterogeneous equilibrium
☑ state Le Chatelier's principle
☑ apply Le Chatelier's principle to a system at equilibrium and tell how external changes will affect the concentrations of the components at equilibrium
☑ use the concentrations or pressures of gases at equilibrium to determine the equilibrium constant
☑ use the equilibrium constant and the initial concentrations or pressures to determine the final concentrations or pressures at equilibrium

SOLVED PROBLEMS

PROBLEM 15-1   Write equilibrium-law expressions for the following equilibria using K_p for gas reactions and K_c for solution reactions:

(a) $2SO_2(g) + O_2(g) \rightleftharpoons 2SO_3(g)$
(b) $2SO_3(g) \rightleftharpoons 2SO_2(g) + O_2(g)$
(c) $CH_3CO_2H(l) + CH_3OH(l) \rightleftharpoons CH_3CO_2CH_3(l) + H_2O(l)$
(d) $HN_3(aq) \rightleftharpoons H^+(aq) + N_3^-(aq)$

Solution:
(a) This is a gas reaction, so we use the partial-pressure form of the equilibrium law (15-2):

$$K_p = \frac{p_{SO_3}^2}{p_{SO_2}^2 p_{O_2}}$$

(b) This reaction is just the reverse of the one in (a), so

$$K_p = \frac{p_{SO_2}^2 p_{O_2}}{p_{SO_3}^2}$$

> *note:* This K_p is the *reciprocal* of the K_p for (a). This is a general rule: *If a reaction is written in reverse, its equilibrium constant becomes the reciprocal of the equilibrium constant for the reaction written in the original direction.*

(c) This is a solution reaction, so the concentration form of the equilibrium law is used:

$$K_c = \frac{[CH_3CO_2CH_3][H_2O]}{[CH_3CO_2H][CH_3OH]}$$

Notice that H_2O is a true product. The reaction does not take place in aqueous solution.
(d) This is another solution reaction: From eq. (15-1)

$$K_c = \frac{[H^+(aq)][N_3^-(aq)]}{[HN_3(aq)]}$$

PROBLEM 15-2   Explain the effects of each of the following changes on the partial pressure of C_2H_6 in the equilibrium mixture resulting from the reaction

$$C_2H_4(g) + H_2(g) \rightleftharpoons C_2H_6(g) \qquad (\Delta H_r^\circ \text{ is negative})$$

(which is exothermic in the left-to-right direction): (a) increasing the total pressure; (b) increasing the temperature; (c) increasing the partial pressure of H_2.

Solution: This question clearly concerns the application of Le Chatelier's principle.

  (a) The external stress is a pressure increase. The system will react so as to produce fewer moles of gas, i.e., toward more C_2H_6. Consequently the partial pressure (concentration) of C_2H_6 at equilibrium will increase.
  (b) Increasing the temperature will favor the reaction direction that is heat-absorbing or endothermic, i.e., the decomposition of C_2H_6 to C_2H_4 and H_2. Thus the partial pressure of C_2H_6 at equilibrium will decrease.
  (c) If the partial pressure of H_2 is increased, the equilibrium will shift to make more H_2 react, and that will produce more C_2H_6. The partial pressure of C_2H_6 at equilibrium will therefore increase.

PROBLEM 15-3 An equilibrium mixture of H_2, I_2, and HI at 764 K is found to have $[H_2] = 2.64\ M$, $[I_2] = 2.31\ M$, and $[HI] = 16.6\ M$. The reaction is

$$2HI(g) \rightleftharpoons H_2(g) + I_2(g)$$

Calculate K for the decomposition of HI.

Solution: The equilibrium-law expression must be written first. Since the data are in concentration terms, write down K_c initially:

$$K_c = \frac{[H_2(g)][I_2(g)]}{[HI(g)]^2} = \frac{(2.64\ M)(2.31\ M)}{(16.6\ M)^2} = 0.0221$$

Since there is no change in the number of moles of gas, $K_c = K_p = K$.

PROBLEM 15-4 A mixture of 2.00 mol of ethanol (C_2H_5OH) and 1.00 mol of acetic acid (CH_3CO_2H) is held at 100.0°C until the following reaction reaches equilibrium:

$$CH_3CO_2H(l) + C_2H_5OH(l) \rightleftharpoons CH_3CO_2C_2H_5(l) + H_2O(l)$$

Analysis of the equilibrium mixture shows that 0.858 mol of $CH_3CO_2C_2H_5$ is present. Calculate K for the reaction as written. (Note that H_2O is a true product.)

Solution: The equation for the reaction is balanced, and so the reaction stoichiometry shows that the amount of H_2O present at equilibrium is

$$(0.858\ \text{mol } CH_3CO_2C_2H_5)\left(\frac{1\ \text{mol } H_2O}{1\ \text{mol } CH_3CO_2C_2H_5}\right) = 0.858\ \text{mol } H_2O$$

Since the stoichiometry shows one mole of C_2H_5OH reacting for each mole of $CH_3CO_2C_2H_5$ produced, the amount of ethanol that reacts to produce 0.858 mol of $CH_3CO_2C_2H_5$ is 0.858 mol. So the amount of ethanol present in the equilibrium mixture is 2.00 mol − 0.858 mol = 1.14 mol. Similarly, the amount of acetic acid present in the equilibrium mixture is 1.00 mol − 0.858 mol = 0.14 mol.
     From the equilibrium law (15-1)

$$K = \frac{[CH_3CO_2C_2H_5][H_2O]}{[CH_3CO_2H][C_2H_5OH]}$$

But we don't have a value for the volume—so we'll just call it V and see what happens:

$$[CH_3CO_2C_2H_5] = 0.858\ \text{mol}/V; \qquad [H_2O] = 0.858\ \text{mol}/V$$
$$[CH_3CO_2H] = 0.14\ \text{mol}/V; \qquad [C_2H_5OH] = 1.14\ \text{mol}/V$$

Substituting, we get

$$K = \frac{(0.858\ \text{mol}/V)(0.858\ \text{mol}/V)}{(0.14\ \text{mol}/V)(1.14\ \text{mol}/V)} = 4.6$$

Notice that the V's cancel out.

PROBLEM 15-5 When a mixture of 3.00 mol of H_2 and 1.00 mol of N_2 is held at 350°C and 30.0 atm and allowed to come to equilibrium, the final mixture contains NH_3 at a partial pressure of 5.34 atm:

$$N_2(g) + 3H_2(g) \rightleftharpoons 2NH_3(g)$$

Calculate K_p for the reaction at 350°C.

Solution: From the equilibrium law (15-2)

$$K_p = \frac{p_{NH_3}^2}{p_{N_2} p_{H_2}^3}$$

The known data include p_{NH_3}, P_{tot}, and the initial molar ratios of N_2 to H_2. You have to calculate p_{N_2} and p_{H_2}. So from Dalton's law

$$P_{tot} = 30.0 \text{ atm} = p_{N_2} + p_{H_2} + p_{NH_3}$$
$$= p_{N_2} + p_{H_2} + 5.34 \text{ atm}$$

Thus

$$p_{N_2} + p_{H_2} = (30.0 - 5.34) \text{ atm} = 24.7 \text{ atm}$$

The stoichiometry of the reaction is $3H_2 : 1N_2$. The starting mixture was (3.00 mol H_2):(1.00 mol N_2). Thus in the equilibrium mixture the molar ratio must be $3H_2 : 1N_2$. By Avogadro's law, then, $p_{H_2} = 3p_{N_2}$. So since $p_{N_2} + p_{H_2} = 24.7$ atm,

$$p_{N_2} + 3p_{N_2} = 4p_{N_2} = 24.7 \text{ atm}$$

and thus

$$p_{N_2} = \frac{24.7 \text{ atm}}{4} = 6.18 \text{ atm} \qquad \text{and} \qquad p_{H_2} = 3p_{N_2} = 18.5 \text{ atm}$$

Hence

$$K_p = \frac{p_{NH_3}^2}{p_{N_2} p_{H_2}^3} = \frac{(5.34)^2}{(6.18)(18.5)^3} = 7.29 \times 10^{-4}$$

PROBLEM 15-6   At high temperatures molecular bromine dissociates into atoms: $Br_2(g) \rightleftharpoons 2Br(g)$. A 0.570-g sample of Br_2 is placed in a 1.00-L quartz flask and heated to 1550°C, after which the total pressure in the flask is found to be 0.750 atm. Calculate K_p for the equilibrium at 1550°C.

Solution: Because this problem requires some chemical reasoning, we'll use the steps for solving equilibrium problems.

Step 1:   The balanced equation is

$$Br_2(g) \rightleftharpoons 2Br(g)$$

Step 2:   Since the problem asks for K_p, the appropriate equilibrium expression is

$$K_p = \frac{p_{Br}^2}{p_{Br_2}}$$

Step 3:   The unknown values needed to find K_p are the respective partial pressures of atomic bromine and molecular bromine, p_{Br} and p_{Br_2}. But we know only the initial weight of Br_2, the volume, and the total equilibrium pressure, so we'll have to work with the only common term among these data—the number of moles. The total pressure at equilibrium is $P_{tot} = 0.750$ atm, which we can use to find the number of moles of gas in the flask at 1550°C by the ideal gas law:

$$n_{e \, tot} = \frac{P_{tot} V}{RT} = \frac{(0.750 \text{ atm})(1.00 \text{ L})}{(0.0821 \text{ L atm mol}^{-1}\text{K}^{-1})(1550 + 273) \text{ K}}$$
$$= 5.01 \times 10^{-3} \text{ mol}$$

From the initial weight of Br_2, we can find the number of moles of gas in the flask initially:

$$n_{i \, tot} = \frac{0.570 \text{ g Br}_2}{2(79.9 \text{ g mol}^{-1})} = 3.57 \times 10^{-3} \text{ mol}$$

Step 4:   The data for initial and equilibrium conditions can now be tabulated in terms of numbers of moles. Let x be the number of moles of Br_2 that dissociate:

	Br_2 ($\times 10^{-3}$ mol)	Br ($\times 10^{-3}$ mol)	Total ($\times 10^{-3}$ mol)
n_i	3.57	0	3.57
change	$-x$	$+2x$	
n_e	$3.57 - x$	$2x$	5.01

So

$$n_{e\,Br_2} + n_{e\,Br} = n_{e\,tot}$$
$$(3.57 \times 10^{-3} - x) + 2x = 5.01 \times 10^{-3} \text{ mol}$$

And thus

$$x = (5.01 \times 10^{-3} \text{ mol}) - (3.57 \times 10^{-3} \text{ mol}) = 1.44 \times 10^{-3} \text{ mol}$$

Substituting into the tabulated data,

	Br_2 ($\times 10^{-3}$ mol)	Br ($\times 10^{-3}$ mol)
n_e:	$3.57 - 1.44 = 2.13$	$2(1.44) = 2.88$

But the problem asks for K_p, so we'll need to convert numbers of moles into partial pressures by the ideal gas law:

$$p_{Br_2} = \frac{n_{Br_2}RT}{V}$$

$$= \frac{(2.13 \times 10^{-3} \text{ mol})(0.0821 \text{ L atm mol}^{-1}K^{-1})(1823\ K)}{1.00 \text{ L}}$$

$$= 3.19 \times 10^{-1} \text{ atm}$$

$$p_{Br} = \frac{n_{Br}RT}{V}$$

$$= \frac{(2.88 \times 10^{-3} \text{ mol})(0.0821 \text{ L atm mol}^{-1}K^{-1})(1823\ K)}{1.00 \text{ L}}$$

$$= 4.31 \times 10^{-1} \text{ atm}$$

Step 5: Substituting into the equilibrium expression,

$$K_p = \frac{(4.31 \times 10^{-1})^2}{3.19 \times 10^{-1}} = 5.82 \times 10^{-1}$$

PROBLEM 15-7 The equilibrium constant K_c for the homogeneous dissociation $N_2O_4(g) \rightleftharpoons 2NO_2(g)$ is 0.211 at 100.0°C. Calculate the equilibrium concentrations of N_2O_4 and NO_2 when 1.00 g of N_2O_4 is heated to 100.0°C in a 325-mL bulb.

Solution:

Step 1: $N_2O_4(g) \rightleftharpoons 2NO_2(g)$

Step 2: $K_c = \dfrac{[NO_2(g)]^2}{[N_2O_4(g)]} = 0.211$

Step 3: We want to find the equilibrium concentrations of N_2O_4 and NO_2. Given the initial weight of N_2O_4, we find the number of moles of N_2O_4 initially:

$$\frac{1.00 \text{ g}}{92.0 \text{ g mol}^{-1}} = 0.0109 \text{ mol}$$

Let x be the number of moles of N_2O_4 that decompose.

Step 4: Tabulating the data in terms of numbers of moles gives us the concentration in terms of x:

	N_2O_4	$2NO_2$
n_i (mol)	0.0109	0
change in n	$-x$	$+2x$
n_e (mol)	$0.0109 - x$	$2x$
concentration (mol L^{-1})	$\dfrac{(0.0109 - x)}{0.325}$	$\dfrac{2x}{0.325}$

Step 5: Substituting into the equilibrium expression, we get a quadratic equation:

$$K_c = 0.211 = \frac{\left(\dfrac{2x}{0.325}\right)^2}{\left(\dfrac{0.0109 - x}{0.325}\right)}$$

$$= \left(\frac{2x}{0.325}\right)^2 \left(\frac{0.325}{0.0109 - x}\right)$$

$$= \frac{4x^2}{(0.325)(0.0109 - x)}$$

$$4x^2 = (0.211)(0.325)(0.0109 - x) = (7.47 \times 10^{-4}) - (6.86 \times 10^{-2})x$$

So

$$4x^2 + (6.86 \times 10^{-2})x - (7.47 \times 10^{-4}) = 0$$

Using the formula for solving quadratic equations, we solve for x:

$$x = \frac{-6.86 \times 10^{-2} \pm \sqrt{(6.86 \times 10^{-2})^2 - 4(4)(-7.47 \times 10^{-4})}}{2(4)}$$

$$= \frac{-6.86 \times 10^{-2} \pm \sqrt{4.71 \times 10^{-3} + 1.20 \times 10^{-2}}}{8}$$

$$= \frac{-6.86 \times 10^{-2} \pm 1.29 \times 10^{-1}}{8}$$

$$= +7.55 \times 10^{-3} \quad \text{or} \quad -2.47 \times 10^{-2} \quad \text{(disregard the negative concentration)}$$

So the concentrations are

$$[NO_2] = \frac{2(7.55 \times 10^{-3})}{0.325} = 4.65 \times 10^{-2} \ M$$

$$[N_2O_4] = \frac{(0.0109 - 7.55 \times 10^{-3})}{0.325} = 1.03 \times 10^{-2} \ M$$

Check: $[NO_2]^2/[N_2O_4] = 0.210 \cong 0.211$, the given value for K_c.

PROBLEM 15-8   For the equilibrium $Cl_2(g) \rightleftharpoons 2Cl(g)$, the equilibrium constant $K_p = 2.45 \times 10^{-7}$ at 1273 K. If a sample of Cl_2 is allowed to dissociate at a total pressure of 1.00 atm, what fraction of it is dissociated into Cl atoms at equilibrium at 1273 K?

Solution: You know that

$$P_{tot} = p_{Cl_2} + p_{Cl} = 1.00 \text{ atm}$$

and that

$$K_p = 2.45 \times 10^{-7} = \frac{p_{Cl}^2}{p_{Cl_2}}$$

You now have two equations in two unknowns: p_{Cl} and p_{Cl_2}. Then you solve for the unknowns by setting

$$p_{Cl_2} = (1.00 \text{ atm}) - p_{Cl}$$

and substituting into the K_p expression:

$$2.45 \times 10^{-7} = \frac{p_{Cl}^2}{(1.00 - p_{Cl})}$$

You could solve the quadratic equation in p_{Cl} directly, but try a little chemical and mathematical reasoning first. Since K_p is a very small constant, you know that there's only a little dissociation of Cl_2 and that p_{Cl} ought to be much less than 1.00 atm. In that case, it may be reasonable to neglect p_{Cl} in comparison with 1.00 atm in the denominator of the K_p expression. If you do that, you get

$$2.45 \times 10^{-7} \cong \frac{p_{Cl}^2}{1.00 \text{ atm}}$$

and so

$$p_{Cl} \cong \sqrt{2.45 \times 10^{-7} \text{ atm}^2} \cong 4.95 \times 10^{-4} \text{ atm}$$

This partial pressure is certainly small compared to 1.00 atm, so your approximation is valid.

Since $p_{Cl} = 4.95 \times 10^{-4}$ atm, the pressure of Cl_2 that dissociates to produce this can be calculated from the stoichiometry of the equation:

$$(4.95 \times 10^{-4} \text{ atm Cl})\left(\frac{1 \text{ atm Cl}_2}{2 \text{ atm Cl}}\right) = 2.47 \times 10^{-4} \text{ atm Cl}_2$$

So the *fraction* of Cl_2 dissociated is

$$\frac{2.47 \times 10^{-4} \text{ atm Cl}_2}{1.00 \text{ atm Cl}_2} = 2.47 \times 10^{-4}$$

PROBLEM 15-9 The Haber process for the synthesis of ammonia takes place at 745 K and $K_p = 2.8 \times 10^{-5}$. The reaction is

$$N_2(g) + 3H_2(g) \rightleftharpoons 2NH_3(g)$$

If a mixture of 6.0 atm of H_2, 2.0 atm of N_2, and 0.10 atm of NH_3 is made at 745 K and allowed to equilibrate, will the partial pressure of NH_3 at equilibrium be greater or less than 0.10 atm?

Solution: It's easy to write an equilibrium expression for this reaction:

$$K_p = \frac{p_{NH_3}^2}{p_{N_2} p_{H_2}^3}$$

But in this expression all the p values are at equilibrium. For problems of this type— in which you do not need a mathematically precise answer and only need to find which *direction* is favored—you can conveniently use nonequilibrium values of p (denoted p') in a **reaction quotient** Q:

REACTION QUOTIENT
$$Q = \frac{(p'_{NH_3})^2}{p'_{N_2}(p'_{H_2})^3}$$

where the values of p' are the pressures at *any* selected moment. Then you can compare Q with K: If $Q > K$, the reverse reaction is favored; if $Q < K$, the forward reaction is favored. Thus, using the initial pressure values for p', you get

$$Q = \frac{(0.10)^2}{(2.0)(6.0)^3} = 2.3 \times 10^{-5}$$

Here $Q < K_p$, so the forward reaction is favored and the partial pressure of NH_3 at equilibrium will be greater than the initial 0.10 atm.

PROBLEM 15-10 Write equilibrium-law expressions (both K_p and K_c) for the following reactions:

(a) $2BaO_2(s) \rightleftharpoons 2BaO(s) + O_2(g)$
(b) $NH_4Cl(s) \rightleftharpoons NH_3(g) + HCl(g)$
(c) $Cr_2O_3(s) + 3CCl_4(g) \rightleftharpoons 2CrCl_3(s) + 3COCl_2(g)$

Solution: Only the concentrations of the gaseous substances will appear in these expressions; no concentration terms involving pure solids or pure liquids are included.

(a) $K_p = p_{O_2}$, $\qquad K_c = [O_2(g)]$
(b) $K_p = p_{NH_3} p_{HCl}$, $\qquad K_c = [NH_3(g)][HCl(g)]$

(c) $K_p = \dfrac{p_{COCl_2}^3}{p_{CCl_4}^3}$, $\qquad K_c = \dfrac{[COCl_2(g)]^3}{[CCl_4(g)]^3}$

PROBLEM 15-11 Write an equilibrium-law expression for the reaction

$$CaCO_3(s) + 2HCl(aq) \rightleftharpoons CaCl_2(aq) + CO_2(g) + H_2O(l)$$

Solution: Only those substances that have (variable) concentrations can be included in the equilibrium-law expression. $CaCO_3(s)$ and $H_2O(l)$ must therefore be excluded because they are pure substances and have constant "concentrations." But HCl(aq) and CaCl$_2$(aq) are not pure substances (being present in solution), so

$$K = \frac{p_{CO_2}[CaCl_2(aq)]}{[HCl(aq)]^2}$$

This last expression may appear somewhat strange, but don't let it bother you. You will come upon problems for which you can write no better expression for K than a mixed expression of this type.

PROBLEM 15-12   A sample of solid ammonium carbamate is heated in a closed container to 298 K and allowed to reach equilibrium. The reaction equation is

$$NH_4CO_2NH_2(s) \rightleftharpoons 2NH_3(g) + CO_2(g)$$

The total pressure of the system is 0.114 atm. Determine the value of K_p.

Solution: The appropriate equilibrium-law expression (15-2) is

$$K_p = p_{NH_3}^2 p_{CO_2}$$

By Dalton's law

$$P_{tot} = p_{NH_3} + p_{CO_2} = 0.114 \text{ atm}$$

From the stoichiometry of the reaction $p_{NH_3} = 2p_{CO_2}$, so

$$P_{tot} = 2p_{CO_2} + p_{CO_2} = 3p_{CO_2} = 0.114 \text{ atm}$$
$$p_{CO_2} = 3.80 \times 10^{-2} \text{ atm}$$

and

$$p_{NH_3} = 2p_{CO_2} = 2(3.80 \times 10^{-2})$$
$$= 7.60 \times 10^{-2} \text{ atm}$$

Thus

$$K_p = (7.60 \times 10^{-2})^2(3.80 \times 10^{-2})$$
$$= 2.19 \times 10^{-4}$$

PROBLEM 15-13   The dissociation of ammonium hydrosulfide

$$NH_4HS(s) \rightleftharpoons NH_3(g) + H_2S(g)$$

has an equilibrium constant $K_c = 1.2 \times 10^{-4}$ at 295 K. If excess solid NH$_4$HS is heated to 295 K in a 1-L flask that contains 0.500 atm of H$_2$S, what would the partial pressure of NH$_3$ be at equilibrium?

Solution:

Step 1: The equilibrium equation is already balanced.
Step 2: The problem asks for p_{NH_3} at equilibrium, so we determine K_c first:

$$K_c = [NH_3(g)][H_2S(g)] = 1.2 \times 10^{-4}$$

then convert to K_p, using the ideal gas law (eq. 15-3):

$$K_p = p_{NH_3}p_{H_2S} = ([NH_3(g)]RT)([H_2S(g)]RT)$$
$$= [NH_3(g)][H_2S(g)](RT)^2$$
$$= K_c(RT)^2$$
$$= (1.2 \times 10^{-4})(0.0821)^2(295)^2$$
$$= 7.0 \times 10^{-2}$$

Step 3: To find p_{NH_3}, let x be the partial pressure of NH$_3$ at equilibrium.
Step 4: Tabulate the data in terms of partial pressure of NH$_3$:

p (atm)	NH$_3$(g)	H$_2$S(g)
initial	0	0.500
change	+x	+x
equilibrium	x	0.500 + x

Step 5: Substitute into the equilibrium-law expression and solve the resulting equation:

$$K_p = p_{NH_3} p_{H_2S}$$
$$7.0 \times 10^{-2} = x(0.500 + x)$$
$$= x^2 + 0.500x$$
$$x^2 + 0.500x - (7.0 \times 10^{-2}) = 0$$

$$x = \frac{-0.500 \pm \sqrt{(0.500)^2 - 4(1)(-7.0 \times 10^{-2})}}{2}$$

$$= \frac{-0.500 \pm 0.73}{2}$$

$$= +0.12 \text{ atm} \quad \text{or} \quad -0.61 \text{ atm}$$

Disregarding the negative pressure, $p_{NH_3} = 0.12$ atm.

PROBLEM 15-14 At 420°C the equilibrium

$$CO(g) + H_2O(g) \rightleftharpoons CO_2(g) + H_2(g)$$

has $K = 10.0$. Calculate the equilibrium concentrations of all components if 1.0 mol of CO and 1.0 mol of H_2O are mixed in a 2.5-L flask at 420°C.

Solution:

Step 1: The equation is already balanced.
Step 2: From the equilibrium law (15-1)

$$K_c = \frac{[CO_2][H_2]}{[CO][H_2O]} = 10.0$$

Because $n_{reactants} = n_{products}$, $K_c = K_p = K$.
Step 3: You are asked to find the equilibrium concentrations of all components. You are given

$$[CO_2]_0 = [H_2O]_0 = \frac{1.0 \text{ mol}}{2.5 \text{ L}} = 0.40 \ M$$

From the stoichiometry of the reaction, the equilibrium concentrations of CO_2 and H_2 will be equal. Let x be the concentration of either product.

Step 4:

Concentration	CO	H₂O	CO₂	H₂
initial	0.40	0.40	0	0
change	$-x$	$-x$	$+x$	$+x$
equilibrium	$(0.40 - x)$	$(0.40 - x)$	x	x

Step 5:

$$K_c = \frac{(x)(x)}{(0.40 - x)(0.40 - x)} = 10.0$$

$$\left(\frac{x}{0.40 - x}\right)^2 = 10.0$$

$$\frac{x}{0.40 - x} = \sqrt{10.0} = 3.162$$

$$x = (0.40 - x)(3.162) = 1.265 - 3.162x$$
$$3.162x + x = 1.265$$
$$4.162x = 1.265$$
$$x = 0.30$$
$$0.40 - x = 0.10$$

So the equilibrium concentrations of CO_2 and H_2 are 0.30 M; and those of CO and H_2O are 0.10 M.

note: If you put these values back into the K expression, you get $K = 9.0$ (not exactly the 10.0 given). That's the trouble with working to only two significant figures—all the data warrant. If you make the smallest possible change in x, increasing it to 0.31 (again, two significant figures), you'll find a "K" of 12, which is even further from 10.0 than the K calculated here. But it is rarely possible to be more precise than this in many equilibrium problems.

PROBLEM 15-15   The decomposition of H_2S at 1300 K

$$2H_2S(g) \rightleftharpoons 2H_2(g) + S_2(g)$$

has an equilibrium constant $K_p = 2.6 \times 10^{-2}$. What percentage of H_2S will be dissociated at equilibrium if 0.10 mol of H_2S is heated in a cylinder at 1300 K at a total pressure of 2.0 atm?

Solution: This is one tough problem! But if you can solve it, you can solve just about any equilibrium problem you're likely to encounter. Use the steps and follow the reasoning.

Step 1: The equation is already balanced:

$$2H_2S(g) \rightleftharpoons 2H_2(g) + S_2(g)$$

Step 2:

$$K_p = \frac{p_{H_2}^2 p_{S_2}}{p_{H_2S}^2} = 2.6 \times 10^{-2}$$

Step 3: You're asked to find the *percentage* of H_2S that will dissociate; you are given P_{tot}. Now remember that the partial pressure of a gas is equal to its mole fraction times the total pressure, so $p_A = X_A P_{tot}$, where X_A is the mole fraction of substance A. Then let $2y$ be the amount of H_2S dissociated.

Step 4: Tabulate your data in terms of moles:

Moles	$2H_2S$	$2H_2$	S_2
initial	0.10	0	0
change	$-2y$	$+2y$	$+y$
equilibrium	$0.10 - 2y$	$2y$	y

Thus the total number of moles is $(0.10 - 2y) + (2y) + (y) = 0.10 + y$. Now use $p_A = X_A P_{tot}$ to get the partial pressure for each species:

$$p_{H_2S} = \left(\frac{0.10 - 2y}{0.10 + y}\right)(2 \text{ atm}) \qquad p_{H_2} = \left(\frac{2y}{0.10 + y}\right)(2 \text{ atm}) \qquad p_{S_2} = \left(\frac{y}{0.10 + y}\right)(2 \text{ atm})$$

Step 5: Using the equilibrium-law expression in Step 2,

$$K_p = \frac{p_{H_2}^2 p_{S_2}}{p_{H_2S}^2}$$

substitute in the partial pressures:

$$2.6 \times 10^{-2} = \frac{\left(\dfrac{2(2y)}{0.10 + y}\right)^2 \left(\dfrac{2y}{0.10 + y}\right)}{\left(\dfrac{2(0.10 - 2y)}{(0.10 + y)}\right)^2}$$

$$= \left(\frac{16y^2}{(0.10 + y)^2}\right)\left(\frac{2y}{0.10 + y}\right)\left(\frac{(0.10 + y)^2}{4(0.10 - 2y)^2}\right)$$

$$= \frac{8y^3}{(0.10 + y)(0.10 - 2y)^2}$$

This is a *cubic* equation with no simple analytic solution.

But you don't have to give up with a cubic equation—approximations can help you solve the problem! Since K isn't very large, you can start by assuming that $y \ll 0.10$ (i.e., y is much smaller than 0.10). If that's so, then

$$\frac{8y^3}{(0.10)(0.10)^2} \cong 2.6 \times 10^{-2}$$

which leads to

$$y = \sqrt[3]{\frac{2.6 \times 10^{-5}}{8}} \cong 1.5 \times 10^{-2}$$

Now 1.5×10^{-2} is 15% of 0.10, so you can't exactly say $y \ll 0.10$. However, this value of y can be plugged back into the cubic as a second trial.

$$\frac{8y^3}{(0.10 + 1.5 \times 10^{-2})[0.10 - 2(1.5 \times 10^{-2})]^2} = 2.6 \times 10^{-2}$$

which gives $y \cong 1.25 \times 10^{-2}$ as a second trial answer. Plug this in for a third iteration and you get

$$\frac{8y^3}{(0.10 + 1.25 \times 10^{-2})[0.10 - 2(1.25 \times 10^{-2})]^2} = 2.6 \times 10^{-2}$$

and $y \cong 1.28 \times 10^{-2}$.

These two last values of y are essentially the same (to two significant figures), so you can conclude that $y = 1.3 \times 10^{-2}$. Thus, $2(1.3 \times 10^{-2}) = 2.6 \times 10^{-2}$ is the number of moles of H_2S that have dissociated. Then the percentage dissociation is

$$\frac{2.6 \times 10^{-2} \text{ mol}}{0.10 \text{ mol}} \times 100\% = 26\%$$

note: This is a very hard problem, and you'll rarely—if ever—be given a problem that requires you to solve a cubic equation. But working—or working *at*—this problem should show you that even in a complicated situation like this, some common sense and a reasonable approximation or two can bring you through to a solution.

Supplementary Exercises

PROBLEM 15-16 Write equilibrium-law expressions for the following, giving both K_p and K_c:

(a) $2NOCl(g) \rightleftharpoons 2NO(g) + Cl_2(g)$
(b) $5B_2H_6(g) + 2BCl_3(g) \rightleftharpoons 6B_2H_5Cl(g)$

Answer: (a) $K_p = \dfrac{p_{NO}^2 p_{Cl_2}}{p_{NOCl}^2}$, $\quad K_c = \dfrac{[NO]^2[Cl_2]}{[NOCl]^2}$ $\qquad$ (b) $K_p = \dfrac{p_{B_2H_5Cl}^6}{p_{B_2H_6}^5 p_{BCl_3}^2}$, $\quad K_c = \dfrac{[B_2H_5Cl]^6}{[B_2H_6]^5[BCl_3]^2}$

PROBLEM 15-17 Write equilibrium-law expressions for the following, giving both K_p and K_c:

(a) $CO_2(g) + C(s) \rightleftharpoons 2CO(g)$
(b) $3Fe(s) + 4H_2O(g) \rightleftharpoons Fe_3O_4(s) + 4H_2(g)$

Answer: (a) $K_p = \dfrac{p_{CO}^2}{p_{CO_2}}$, $\quad K_c = \dfrac{[CO]^2}{[CO_2]}$ $\qquad$ (b) $K_p = \dfrac{p_{H_2}^4}{p_{H_2O}^4}$, $\quad K_c = \dfrac{[H_2]^4}{[H_2O]^4}$

PROBLEM 15-18 Consider the reaction

$$2B_2H_6(g) \rightleftharpoons B_4H_{10}(g) + H_2(g), \qquad \Delta H_r^\circ = -5.8 \text{ kJ}$$

(a) Write the equilibrium-law expression for K_p.
(b) If you have this mixture at equilibrium, how will increasing the total pressure (decreasing the volume) affect the amount of $B_4H_{10}(g)$ in the mixture?
(c) If you have this mixture at equilibrium at 50°C and then change the temperature to 25°C, how will the amount of $B_4H_{10}(g)$ change?

(d) If you have this mixture at equilibrium and then add more H_2 at constant pressure, how will that affect the amount of $B_4H_{10}(g)$ in the mixture?

Answer: **(a)** $K_p = \dfrac{p_{B_4H_{10}}p_{H_2}}{p_{B_2H_6}^2}$ **(b)** no change   **(c)** increase   **(d)** decrease

PROBLEM 15-19   The gases SO_2 and Cl_2 combine to form SO_2Cl_2:

$$SO_2Cl_2(g) \rightleftharpoons SO_2(g) + Cl_2(g)$$

When 6.50×10^{-3} mol of SO_2 and 6.50×10^{-3} mol of Cl_2 were placed in a 0.750-L container, it was found that 5.96×10^{-4} mol of $SO_2Cl_2(g)$ had formed at equilibrium at 375°C.

(a) Calculate the equilibrium constant K_c for this reaction.
(b) For the reaction equation given, $\Delta H_r^\circ = +84$ kJ. At 325°C will K_c be higher or lower than in part **(a)**?
(c) At 325°C will more or less SO_2Cl_2 form?
(d) If the volume of the container had been 1.00 L, would K_c be greater or less than in part **(a)**?
(e) If the volume of the container had been 1.00 L, would more or less SO_2Cl_2 form than in part **(a)**?

Answer: **(a)** 0.138 M **(b)** K_c is lower   **(c)** more SO_2Cl_2 will form   **(d)** K_c would be the same   **(e)** less SO_2Cl_2 would form

PROBLEM 15-20   Nitrosyl bromide NOBr is not a very stable gas and decomposes rapidly:

$$2NOBr(g) \rightleftharpoons 2NO(g) + Br_2(g)$$

When a sample of pure NOBr(g) was placed in a container and brought to equilibrium at 25°C, 34% of the NOBr dissociated, as shown in the reaction equation. The total pressure was found to be 0.25 atm. Calculate K_p.

Answer: $K_p = 9.6 \times 10^{-3}$

PROBLEM 15-21   When 1.000 g of $CaCO_3(s)$ is placed in a 9.97-L container and heated to 1000 K, 1.00×10^{-3} g of $CaCO_3$ decomposes. Calculate the equilibrium constant K_p for the reaction

$$CaCO_3(s) \rightleftharpoons CaO(s) + CO_2(g)$$

Answer: $K_p = 8.2 \times 10^{-5}$

PROBLEM 15-22   When 3.00×10^{-2} mol of pure phosgene ($COCl_2$) was placed in a 1.50-L container and heated to 800 K, it gave the equilibrium mixture

$$COCl_2(g) \rightleftharpoons CO(g) + Cl_2(g)$$

The pressure of CO was found to be 0.497 atm.

(a) Calculate the equilibrium pressures of $COCl_2(g)$ and $Cl_2(g)$.
(b) Calculate the equilibrium constant K_p for the reaction.

Answer: **(a)** $p_{CO} = p_{Cl_2} = 0.497$ atm; $p_{COCl_2} = 0.817$ atm   **(b)** $K_p = 0.302$

PROBLEM 15-23   AgCl and NH_3 gas combine to form a solid compound:

$$Ag(NH_3)Cl(s) \rightleftharpoons AgCl(s) + NH_3(g)$$

At 42°C, $K_p = 0.310$ and at 63°C, $K_p = 0.859$.

(a) Write the equilibrium-law expression for this reaction.
(b) Is the reaction exothermic or endothermic?

Answer: **(a)** $K_p = p_{NH_3}$ or $K_c = [NH_3]$ **(b)** endothermic

PROBLEM 15-24 Derive a *general* relationship between K_p and K_c for a gas reaction in which a total number of a mol of gaseous reactants is converted to a total number of b mol of gaseous products.

Answer: $K_p = K_c(RT)^{b-a}$

PROBLEM 15-25 At 923 K the equilibrium constant K_p for the following reaction is 6.40:

$$2MgCl_2(s) + O_2(g) \rightleftharpoons 2MgO(s) + 2Cl_2(g)$$

A 20.0-g sample of $MgCl_2(s)$ is put into a 1-L quartz flask filled with O_2 at a pressure of 1.00 atm at 293 K. The apparatus is then sealed and heated to 923 K. Calculate the equilibrium partial pressures of Cl_2 and of O_2 at 923 K (assume that the free volume in the flask remains constant at 1.00 L).

Answer: $p_{Cl_2} = 3.17$ atm; $p_{O_2} = 1.57$ atm

16 CHEMICAL THERMODYNAMICS

THIS CHAPTER IS ABOUT

☑ **Chemical Systems**
☑ **The First Law**
☑ **The Second Law and Entropy**
☑ **Free Energy**

Thermochemistry and chemical equilibrium form part of a fairly abstract but extremely useful branch of science—**thermodynamics**. We've already used the two thermodynamic concepts of *enthalpy change* and *equilibrium constants* to explain why chemical reactions occur as they do. In this chapter we'll discuss two additional thermodynamic concepts: *entropy* and *free energy*. Both are of great value in predicting the equilibrium character of a chemical system. These four concepts provide information on reaction spontaneity, energy changes, and reactant/product concentrations—all from a limited amount of thermochemical data.

note: Although chemical thermodynamics is about real chemical systems at equilibrium, most students find it to be a very abstract area of chemistry. The abstractions are rather like the scaffolding erected about a house that is being built: essential during the construction phase, but superfluous after the job is completed. We'll try to keep the scaffolding here to an absolute minimum.

16-1. Chemical Systems

A *system* in thermodynamics is any part of the physical universe in which we have a special interest; everything else in the universe forms the *surroundings*. A system in which no mechanical or thermal transfer (work or heat) can occur between the object(s) of interest and the surroundings is an *isolated* or **adiabatic** system.

The **state** of a system is described by its physical parameters: chemical composition, temperature, pressure, volume, etc. The thermodynamic properties of a system, such as enthalpy (H), are completely determined once the physical parameters are specified. A **change of state** occurs when any one (or more) of the physical quantities is altered. As a result of a change of state, the thermodynamic properties may also change. Changes in thermodynamic properties are independent of the pathway. For example, no matter HOW a system is moved from its initial state ($H_{initial}$) to its final state (H_{final}), the ΔH of the system is the same. (Recall that change is indicated by a capital Greek delta Δ. For example, enthalpy change is $\Delta H = H_{final} - H_{initial}$.)

16-2. The First Law

Heat energy q is the kinetic energy of the particles that make up matter, and temperature (T) measures the intensity with which heat energy is concentrated in a system. The two quantities are related by the equation $q = nC\Delta T$, where C is the heat capacity (see Section 8-2). The change in heat energy of a chemical system that has moved from state A to state B is the enthalpy change ΔH. In an exothermic reaction (where $\Delta H < 0$), the reacting system gives up heat to its surroundings; in an endothermic reaction (where $\Delta H > 0$), the reacting system takes up heat energy from its surroundings.

A well-known expression for the **first law of thermodynamics** states that "the energy of the universe is constant." In symbols we might say that $\Delta E_{system} = -\Delta E_{surroundings}$. Testing the law in this form is impractical, but it does provide a memorable expression of the empirical evidence. Another statement of the first law—one that's easier to handle—is that *the energy of an isolated system is constant*. This doesn't mean that an isolated system can't undergo a change of state: The energy of the system must remain constant, but it can be transformed.

16-3. The Second Law and Entropy

A. Spontaneous changes

You can predict the direction and equilibrium condition of some processes intuitively. For example, if you drop an ice cube into a glass of warm soda, you know the ice cube will melt and the soda will cool down. If you saw a movie scene in which the reverse happened—so that extra ice cubes appeared in the glass—you'd know you were watching a "special effect." But the first law of thermodynamics doesn't even hint that there might be anything odd in that movie scene. Energy is conserved, whichever way the change goes. But still, you know from experience in which direction the equilibrium lies.

Processes like the melting of an ice cube or the explosive reaction of gaseous hydrogen and oxygen to give water, are *spontaneous*: Thus, given the starting conditions, these systems progress to the equilibrium state without an external source of energy or driving force. In contrast, the reverse reaction of the products to the starting materials is *nonspontaneous*: It requires the addition of some external source of energy.

Exothermic reactions are often spontaneous; that is, reactions for which ΔH is negative, so that heat is passed from the system to the surroundings, often proceed spontaneously. On the basis of the first law of thermodynamics, you might expect all exothermic processes to be spontaneous. But they aren't (so much for intuition), so there must be another driving force at work.

B. Entropy

Entropy (S) is a measure of the degree of *randomness* or *disorder* of a system. On a molecular level the particles in all pure substances move more and more vigorously as temperature increases, so the entropies of all pure substances increase as T increases. At 0 K the entropy of any perfect crystalline material is zero. But the entropy of a pure substance, unlike enthalpy, can be obtained at temperatures *above* 0 K by measuring heat capacity changes. Entropy values are expressed in SI units of $J\,mol^{-1}K^{-1}$. *Standard entropy values* ($S°$) are calculated for standard-state conditions at 298 K, unless some other temperature is specified in a subscript.

The **entropy change** (ΔS) is a measure of the increase ($\Delta S > 0$) or decrease ($\Delta S < 0$) in disorder of a system that undergoes a change of state. Entropy change can be calculated from entropy values of the reactants and products. Thus if S_A, S_B are the respective molar entropies of reactants A and B; S_P, S_Q are the molar entropies of products P and Q; and $a, b, \ldots, p, q, \ldots$, are stoichiometric coefficients for the general reaction

$$aA + bB + \cdots \longrightarrow pP + qQ + \cdots$$

then the entropy change is

ENTROPY CHANGE
$$\Delta S° = (pS_P + qS_Q + \cdots) - (aS_A + bS_B + \cdots) \tag{16-1}$$

EXAMPLE 16-1: Use the following $S°$ values (in $J\,mol^{-1}K^{-1}$) to determine $\Delta S°$ for the reaction $2CO(g) + O_2(g) \rightarrow 2CO_2(g)$:

Gas:	CO	O_2	CO_2
$S°$:	197.9	205.0	213.6

Solution: From the equation for entropy change (16-1),

$$\Delta S° = 2S°_{CO_2} - (2S°_{CO} + S°_{O_2})$$
$$= 2(213.6) - 2(197.9) - 205.0$$
$$= -173.6\ J\,mol^{-1}K^{-1}$$

note: When giving the entropy change for a chemical reaction such as this, you can think of the stoichiometric coefficients (2, 1, 2) in the equation as pure numbers, in which case your answer will be in J mol^{-1}K^{-1}. On the other hand, if you think of them as 2 mol, 1 mol, and 2 mol, respectively, the mole units cancel and your answer will be $\Delta S^\circ = -173.6$ J K^{-1}. Both notations are used in textbooks. However, if you are dealing with a specific amount of material, such as 0.1 mol of CO, then your answer would be in J K^{-1}.

You'll need to know how to calculate changes in entropy for the following types of change:

1. *Phase transitions*: constant temperature and 1 atm pressure:

PHASE TRANSITIONS
$$\Delta S = \frac{\Delta H}{T} \quad (T = \text{const})$$
(16-2)

where ΔH is the enthalpy change for the phase transition at the equilibrium temperature.

2. *Change in volume*: constant temperature:

CHANGE IN VOLUME
$$\Delta S = 2.303nR\log\left(\frac{V_f}{V_i}\right)$$
(16-3)

3. *Change in temperature*: constant pressure:

CHANGE IN TEMPERATURE
$$\Delta S = 2.303nC_p\log\left(\frac{T_f}{T_i}\right)$$
(16-4)

where C_p is the molar heat capacity at the stated pressure.

4. *Change in pressure*: ideal gas and constant temperature:

CHANGE IN PRESSURE
$$\Delta S = 2.303nR\log\left(\frac{P_i}{P_f}\right)$$
(16-5)

> *note:* Remember Boyle's law: Pressure and volume are reciprocally proportional at constant temperature.

EXAMPLE 16-2: Calculate the entropy change for 1.00 mol of an ideal gas that expands from 11.2 L to 44.8 L at constant temperature.

Solution: Use the equation for change in volume (16-3):

$$\Delta S = 2.303nR\log\left(\frac{V_f}{V_i}\right)$$

$$= 2.303(1.00 \text{ mol})(8.31 \text{ J mol}^{-1}\text{K}^{-1})\log\left(\frac{44.8 \text{ L}}{11.2 \text{ L}}\right)$$

$$= 19.11(\log 4) = 19.11(0.602)$$
$$= 11.5 \text{ J K}^{-1}$$

EXAMPLE 16-3: The enthalpy change for the transition of liquid water to steam is $\Delta H^\circ_{vap} = 40.8$ kJ mol^{-1} at 100°C. Calculate ΔS° for the process.

Solution: Use the equation for phase transitions (16-2):

$$\Delta S^\circ = \frac{\Delta H^\circ_{vap}}{T}$$

$$= \left(\frac{40.8 \text{ kJ mol}^{-1}}{(100 + 273) \text{ K}}\right)(1000 \text{ J kJ}^{-1})$$

$$= 109 \text{ J mol}^{-1}\text{K}^{-1}$$

Notice that ΔS° is greater than zero. As you might expect, there is an increase in the randomness of the system.

EXAMPLE 16-4: Calculate the entropy change for 1.00 mol of an ideal gas whose pressure decreases from 44.8 atm to 11.2 atm.

Solution: Use the equation for change in pressure (16-5):

$$\Delta S = 2.303nR \log\left(\frac{P_i}{P_f}\right)$$

$$= 2.303(1 \text{ mol})(8.31 \text{ J mol}^{-1}\text{K}^{-1})\log\left(\frac{44.8 \text{ atm}}{11.2 \text{ atm}}\right)$$

$$= 11.5 \text{ J K}^{-1}$$

C. The second law

A well-known expression for the **second law of thermodynamics** is that "the entropy of the universe is constantly increasing." An equivalent statement is "heat energy of itself never flows from a cooler body to a warmer one." Both of these statements are difficult to explore, but for our purposes the most useful statement of the second law is that *entropy increases for a spontaneous change in an isolated system.*

> *note:* We mentioned the third law of thermodynamics in passing: The entropy of a perfect crystal at 0 K is zero.

16-4. Free Energy

Free energy G is a composite function that relates enthalpy and entropy:

FREE ENERGY
$$G = H - TS \tag{16-6}$$

But, just as we can't measure absolute enthalpy, we can't measure free energy: Only *changes* in G are significant, so

CHANGE IN FREE ENERGY
$$\Delta G = \Delta H - T\,\Delta S \tag{16-7}$$

The sign of the free-energy change provides an exact measure of **spontaneity**:

- If $\Delta G < 0$, the reaction or process is *spontaneous.*
- If $\Delta G > 0$, the reaction or process is *nonspontaneous*; the reverse reaction, however, is spontaneous.
- If $\Delta G = 0$, the reaction or process is *at equilibrium.*

We can obtain some valuable insights into the factors that make a reaction spontaneous by looking at Table 16-1. All of the possible combinations of signs of ΔH and ΔS are examined with respect to the sign of ΔG.

TABLE 16-1: Enthalpy, Entropy, and Spontaneity

ΔH	ΔS	ΔG
−	+	− (spontaneous)
+	−	+ (nonspontaneous)
−	−	$\pm$ { (can be +, −, or 0, depending on the actual magnitudes of ΔH and $T\,\Delta S$)
+	+	

A. Standard-state conditions and free energy

The standard-state value $\Delta G°$ is related to ΔG by

$$\Delta G = \Delta G° + 2.303RT \log Q \tag{16-8}$$

where Q is the reaction quotient (see Problem 15-9). This equation leads to some valuable relationships for the various thermodynamic functions.

B. Relationships for thermodynamic functions

1. $\Delta G°$ and K

At equilibrium, $\Delta G = 0$ and $Q = K$ (the equilibrium constant), so $\Delta G°$ is related to the equilibrium constant by

$$\Delta G° = -2.303RT \log K \qquad \text{(16-9)}$$

note: According to standard thermodynamic conventions, values of K_p are given in atm for reactions involving gases and values of K_c are given in mol L^{-1} for solution reactions.

EXAMPLE 16-5: Calculate $\Delta G°$ and the equilibrium constant for the decomposition of solid barium carbonate at 25°C:

$$BaCO_3(s) \rightleftharpoons BaO(s) + CO_2(g)$$

The following are the values of $\Delta G_f°$ for each compound in kJ mol^{-1}: $BaCO_3(s)$ -1138, $BaO(s)$ -525, $CO_2(g)$ -394.

Solution: The relationship discussed in Chapter 8 for $\Delta H_r°$ also applies to $\Delta G_r°$ (see Hess' law):

$$\Delta G_r° = \sum(\Delta G_f° \text{ products}) - \sum(\Delta G_f° \text{ reactants}) \qquad \text{(16-10)}$$

In this case

$$\begin{aligned}
\Delta G_r° &= \Delta G_f°(BaO(s)) + \Delta G_f°(CO_2(g)) - \Delta G_f°(BaCO_3(s)) \\
&= (-525 + (-394) - (-1138)) \text{ kJ mol}^{-1} \\
&= 219 \text{ kJ mol}^{-1}
\end{aligned}$$

To find the equilibrium constant, we use the relationship (16-9):

$$\Delta G° = -2.303RT \log K$$

$$\left(\frac{219 \text{ kJ}}{\text{mol}}\right)\left(\frac{10^3 \text{ J}}{\text{kJ}}\right) = (-2.303)\left(\frac{8.31 \text{ J}}{\text{mol K}}\right)(298 \text{ K}) \log K$$

so that

$$\log K = -\frac{(219)(10^3)}{(2.303)(8.31)(298)} = -38.45$$

and

$$K = 3.5 \times 10^{-39}$$

---

2. $\Delta H°$ and K

A relationship between $\Delta H°$ and K for the standard state can be developed from definition (16-7) and relation (16-9):

$$-2.303RT \log K = \Delta H° - T \Delta S°$$

$$\log K = -\frac{\Delta H°}{2.303RT} + \frac{\Delta S°}{2.303R} \qquad \text{(16-11)}$$

Because $\Delta H°$ and $\Delta S°$ vary only slightly over small shifts in temperature, relation (16-11) can be rewritten to relate $\Delta H°$ and the equilibrium constants for the reaction at two temperatures:

$$\log K_1 - \log K_2 = \log \frac{K_1}{K_2} = \left(\frac{-\Delta H°}{2.303R}\right)\left(\frac{1}{T_1} - \frac{1}{T_2}\right) \qquad \text{(16-12)}$$

or equivalently

$$\log K_2 - \log K_1 = \log \frac{K_2}{K_1} = \left(\frac{-\Delta H°}{2.303R}\right)\left(\frac{1}{T_2} - \frac{1}{T_1}\right)$$

EXAMPLE 16-6: The equilibrium constant for the reaction $H_2(g) + I_2(g) \rightleftharpoons 2HI(g)$ at 340°C is $K = 70.8$; at 460°C, $K = 46.8$. Determine $\Delta H°$ for this temperature range.

Solution: Let $T_1 = 340°C = 613$ K and $T_2 = 460°C = 733$ K. Now to simplify the math, first evaluate $\log(K_1/K_2)$:

$$\log \frac{K_1}{K_2} = \log\left(\frac{70.8}{46.8}\right) = 0.1798$$

Now solve eq. (16-12) for $\Delta H°$ and substitute:

$$0.1798 = \left(\frac{-\Delta H°}{2.303R}\right)\left(\frac{1}{T_1} - \frac{1}{T_2}\right)$$

$$\Delta H° = \frac{-2.303R(0.1798)}{\left(\dfrac{1}{T_1} - \dfrac{1}{T_2}\right)} = \frac{-2.303(8.31 \text{ J mol}^{-1}\text{K}^{-1})(0.1798)}{\left(\dfrac{1}{613} - \dfrac{1}{733}\right)\text{K}^{-1}}$$

$$= -1.29 \times 10^4 \text{ J mol}^{-1} \qquad \text{or} \qquad -12.9 \text{ kJ mol}^{-1}$$

Notice that the reaction is exothermic, and that the equilibrium constant decreases as the temperature increases, as predicted by Le Chatelier's principle.

3. E_a and k

Recall that the Arrhenius equation (see Chapter 14) is

$$\log \frac{k_2}{k_1} = \frac{-\Delta E_a}{2.303R}\left(\frac{1}{T_2} - \frac{1}{T_1}\right)$$

where k is the reaction rate constant and E_a is the energy of activation. The source of this equation should be clear to you now that you've seen eq. (16-12).

SUMMARY

1. The first law of thermodynamics states that the energy of an isolated system is constant.
2. Entropy is a measure of the degree of disorder in a system.
3. The second law of thermodynamics states that entropy increases for a spontaneous change in an isolated system.
4. Free energy is a composite function of enthalpy and entropy: $\Delta G = \Delta H - T\Delta S$. The sign of ΔG is an exact measure of spontaneity. For $\Delta G = 0$, the system is at equilibrium.
5. The change in free energy $\Delta G°$ can be used to find the equilibrium constant for a reaction.

RAISE YOUR GRADES

Can you ...?

☑ determine whether or not a process is spontaneous
☑ determine the entropy change of a system
☑ calculate ΔS for a reaction from the entropies of the reactants and products
☑ calculate the free-energy change from ΔH and ΔS
☑ calculate the free-energy change from the free energies of formation of the reactants and products at constant temperature
☑ calculate the equilibrium constant at a given temperature from the free-energy change at that temperature
☑ use the equilibrium constant and enthalpy change at one temperature to calculate the equilibrium constant at a second temperature

SOLVED PROBLEMS

PROBLEM 16-1   A system at a starting temperature of 300 K is heated until the temperature increases by 53 K. The temperature is then decreased by 135 K. Calculate the final temperature and the net change in temperature.

Solution: The temperature changes are $+53$ K for the first step and -135 K for the second step. The net change in temperature is $\Delta T_{net} = +53 - 135 = -82$ K. The final temperature may be found either by adding the two temperature changes to the initial temperature:

$$300 \text{ K} + 53 \text{ K} + (-135 \text{ K}) = 218 \text{ K}$$

or by adding the net temperature change to the initial temperature:

$$T_f - T_i = \Delta T_{net}$$
$$T_f = T_i + \Delta T_{net} = 300 \text{ K} + (-82 \text{ K})$$
$$= 218 \text{ K}$$

This apparently trivial problem demonstrates an important point: *Temperature is a state function.* You can go from an initial temperature to a final temperature by any route you choose.

PROBLEM 16-2   Are the following processes spontaneous? Why or why not?

(a) A piece of copper and a piece of iron are in contact with each other at exactly the same temperature. Heat flows from the piece of copper to the piece of iron until the temperature of the iron is 10°C higher than that of the copper.

(b) A gas in a large container connected to a smaller container compresses so that all the gas flows into the smaller container, leaving the larger container empty.

Solution:

(a) This process is not spontaneous. We know from experience that heat always flows spontaneously from a hotter body to a colder body, thereby increasing the entropy of the system. A flow of heat from a colder object to a warmer one would *decrease* the randomness of the system, making it more ordered. If both objects are already at the same temperature, they are in the equilibrium state.

(b) This process is not spontaneous. One of the properties of gases is that they tend to fill the entire container. If all of the gas molecules moved to the smaller container, the system would be more ordered than if both containers were filled. Hence the spontaneous process must be the expansion of the gas to fill both containers evenly.

note: Neither of these processes violates the first law of thermodynamics. In both (a) and (b), energy (work) from the surroundings could be applied to the systems to make the stated process occur, but the process would not be spontaneous.

PROBLEM 16-3   Use the standard-state entropies listed here to determine the entropy changes for the following reactions at 298 K:

(a) $2Mg(s) + O_2(g) \longrightarrow 2MgO(s)$
(b) $C_2H_6(g) \longrightarrow H_2(g) + C_2H_4(g)$

Solution: Apply the equation for entropy change (16-1):

$$\Delta S° = (pS_P + qS_Q + \cdots) - (aS_A + bS_B + \cdots)$$

Substance	$S°$ ($J\,mol^{-1}\,K^{-1}$)
$H_2(g)$	131.0
$O_2(g)$	205.0
$Mg(s)$	32.5
$MgO(s)$	26.7
$C_2H_6(g)$	229.5
$C_2H_4(g)$	219.5

(a) $\Delta S° = 2S°_{MgO} - (S°_{O_2} + 2S°_{Mg})$
$= 2(26.7) - 205.0 - 2(32.5)$
$= -216.6 \text{ J mol}^{-1}\text{K}^{-1}$

(b) $\Delta S° = S°_{C_2H_4} + S°_{H_2} - S°_{C_2H_6}$
$= 219.5 + 131.0 - 229.5$
$= 121.0 \text{ J mol}^{-1}\text{K}^{-1}$

Notice how much larger the entropies of gases are than those of solids. The decrease in entropy in reaction (a) is almost entirely due to the fact that O_2 gas reacts to form a solid product.

note: The "mol" in the units $J\,mol^{-1}K^{-1}$ of this problem is a mole of reaction *as written*. If the reaction equation in (a) had been written $Mg(s) + \frac{1}{2}O_2(g) \rightarrow MgO(s)$, then $\Delta S°$ would be equal to $\frac{1}{2}(-216.6\,J\,mol^{-1}K^{-1})$ or $-108.3\,J\,mol^{-1}K^{-1}$, where "mol" again refers to the reaction as written.

PROBLEM 16-4 Calculate the entropy change at 35°C for 0.100 mol of an ideal gas that is compressed from 0.500 atm to 20.0 atm. The temperature remains constant.

Solution: Use the equation for a change in pressure (16-5):

$$\Delta S = 2.30nR\log\left(\frac{P_i}{P_f}\right)$$

$$= 2.30(0.100\,mol)(8.31\,J\,mol^{-1}K^{-1})\log\left(\frac{0.500\,atm}{20.0\,atm}\right)$$

$$= -3.06\,J\,K^{-1}$$

Notice that ΔS is negative. When the gas is compressed from a low pressure to a high pressure, it becomes less random.

PROBLEM 16-5 Given the normal boiling points and standard enthalpies of vaporization, calculate the entropy of vaporization of each liquid listed here:

Liquid	Boiling point (°C)	$\Delta H°_{vap}$ (kJ mol^{-1})
Hexane (C_6H_{14})	68.7	30.8
Ethanol (C_2H_5OH)	78.4	42.4
Toluene ($CH_3C_6H_5$)	110.6	35.2
Carbon tetrachloride (CCl_4)	76.7	32.7

Solution: The type of change here is a phase transition, so we use eq. (16-2):

$$\Delta S° = \frac{\Delta H°_{vap}}{T}$$

Convert the boiling points to degrees Kelvin and the kilojoules to joules:

hexane

$$\Delta S° = \frac{30.8 \times 10^3\,J\,mol^{-1}}{(68.7 + 273.2)\,K} = 90.1\,J\,mol^{-1}K^{-1}$$

ethanol

$$\Delta S° = \frac{42.4 \times 10^3\,J\,mol^{-1}}{(78.4 + 273.2)\,K} = 120.6\,J\,mol^{-1}K^{-1}$$

toluene

$$\Delta S° = \frac{35.2 \times 10^3\,J\,mol^{-1}}{(110.6 + 273.2)\,K} = 91.7\,J\,mol^{-1}K^{-1}$$

carbon tetrachloride

$$\Delta S° = \frac{32.7 \times 10^3\,J\,mol^{-1}}{(76.7 + 273.2)\,K} = 93.5\,J\,mol^{-1}K^{-1}$$

Compare these answers with the entropy of vaporization of water calculated in Example 16-3: $\Delta S° = 109\,J\,mol^{-1}K^{-1}$. This is an illustration of **Trouton's rule**: *The entropy of vaporization of a liquid at its boiling point will be about $90\,J\,mol^{-1}K^{-1}$ if the liquid is not hydrogen-bonded.* Here the two hydrogen-bonded liquids, water and ethanol, have much higher entropies of vaporization than the non–hydrogen-bonded hexane, toluene, and carbon tetrachloride.

PROBLEM 16-6 From the ΔH and ΔS values given decide whether or not these reactions will be spontaneous at 25°C.

Reaction A: $\Delta H = -10.5 \times 10^3$ J mol^{-1}; $\Delta S = +31$ J mol^{-1}K^{-1}
Reaction B: $\Delta H = -11.7 \times 10^3$ J mol^{-1}; $\Delta S = -105$ J mol^{-1}K^{-1}

Solution: In order to decide whether or not a reaction is spontaneous, we need to know the sign of ΔG. So we determine ΔG at 298 K from eq. (16-7):

Reaction A:

$$\Delta G = \Delta H - T\Delta S$$
$$= (-10.5 \times 10^3 \text{ J mol}^{-1}) - (298 \text{ K})(31 \text{ J mol}^{-1}\text{K}^{-1})$$
$$= -19.7 \times 10^3 \text{ J mol}^{-1}$$

Since $\Delta G < 0$, this reaction is spontaneous at 25°C. (We also could have determined the spontaneity by looking at Table 16-1, where it shows that if $\Delta H < 0$ and $\Delta S > 0$, the reaction will be spontaneous.)

Reaction B:

$$\Delta G = \Delta H - T\Delta S$$
$$= (-11.7 \times 10^3 \text{ J mol}^{-1}) - (298 \text{ K})(-105 \text{ J mol}^{-1}\text{K}^{-1})$$
$$= +19.6 \times 10^3 \text{ J mol}^{-1}$$

Here $\Delta G > 0$ and the reaction is not spontaneous. (Note that we cannot use Table 16-1 for this part of the problem because the sign of ΔG depends on the relative magnitudes of ΔH and $T\Delta S$.)

PROBLEM 16-7   Is the following reaction spontaneous at 25°C? Calculate (a) $\Delta G°$ and (b) the equilibrium constant at 25°C:

$$2Fe^{3+}(aq) + Cu(s) \rightleftharpoons Cu^{2+}(aq) + 2Fe^{2+}(aq) \qquad \Delta H° = -15.94 \text{ kJ}; \quad \Delta S° = 229 \text{ J K}^{-1}$$

Solution:
Looking at Table 16-1, you see that this reaction is spontaneous because a reaction in which the enthalpy change is negative and the entropy change is positive will be spontaneous.

(a) Determine $\Delta G°$:

$$\Delta G° = \Delta H° - T\Delta S°$$
$$= (-15.94 \text{ kJ})\left(\frac{1000 \text{ J}}{1 \text{ kJ}}\right) - (298 \text{ K})(229 \text{ J K}^{-1})$$
$$= -84.2 \times 10^3 \text{ J} \quad \text{or} \quad -84.2 \text{ kJ}$$

(b) From relation (16-9)

$$\Delta G° = -2.30RT \log K$$
$$\log K = \frac{-\Delta G°}{2.30\,RT} = \frac{-(-84.2 \times 10^3 \text{ J})}{2.30(8.31 \text{ J K}^{-1})(298 \text{ K})}$$
$$= 14.78$$

So

$$K = \text{antilog}(14.78) = 6.0 \times 10^{14}$$

PROBLEM 16-8   In the system $Br_2(g) \rightarrow Br_2(l)$ at constant pressure, the entropy change is which of the following: (a) exactly 8.314 J mol^{-1}K^{-1}, (b) positive, (c) negative, (d) zero, (e) not measurable.

Solution: This process involves changing a gas into a liquid. Liquids are more ordered than gases; therefore there is a decrease in randomness and a decrease in entropy: $\Delta S < 0$. The correct answer is (c).

Answer (a) is the gas constant, which has the correct units, but is not relevant to this problem as stated. Answer (b) would be correct for the reverse reaction. Answer (d), $\Delta S = 0$, would be very unlikely, especially where a change of state is involved. At the normal boiling point of $Br_2(l)$, ΔG is zero. Answer (e): ΔS could be measured if ΔH_{vap} at the boiling point of $Br_2(l)$ were known.

PROBLEM 16-9   The following are standard entropies in J mol^{-1}K^{-1}: $H_2(g)$ 131, $I_2(s)$ 117, $HI(g)$ 206. The free energy of formation of $HI(g)$ at 298 K is $\Delta G_f° = 1.30$ kJ mol^{-1}.

(a) Calculate $\Delta S°$ and $\Delta H°$ at 298 K for the reaction $H_2(g) + I_2(s) \longrightarrow 2HI(g)$.
(b) Calculate the equilibrium constant K_p at 298 K for the equilibrium $H_2(g) + I_2(s) \rightleftharpoons 2HI(g)$.

Solution:

(a) Use the entropy change equation (16-1) to determine ΔS°:

$$\Delta S^\circ = 2S^\circ_{HI} - (S^\circ_{H_2} + S^\circ_{I_2})$$
$$= 2(206) - 131 - 117$$
$$= 164 \text{ J mol}^{-1}\text{K}^{-1}$$

Then use the Hess' law type of relationship (16-10) to find ΔG°:

$$\Delta G^\circ = 2\Delta G^\circ_f(HI) - [\Delta G^\circ_f(H_2) + \Delta G^\circ_f(I_2)]$$
$$= 2(1.30) - (0 + 0) = 2.60 \text{ kJ mol}^{-1}$$

Now since $\Delta G^\circ = \Delta H^\circ - T\Delta S^\circ$,

$$\Delta H^\circ = \Delta G^\circ + T\Delta S^\circ$$

$$= (2.60\,\text{kJ mol}^{-1}) + (298\text{ K})(164\,\text{J mol}^{-1}\text{K}^{-1})\left(\frac{1\,\text{kJ}}{1000\,\text{J}}\right)$$

$$= 51.5 \text{ kJ mol}^{-1}$$

(b) You calculated ΔG° for this reaction in part (a). Use relation (16-9) to calculate K:

$$\Delta G^\circ = -2.303RT\log K$$

$$\log K = \frac{-\Delta G^\circ}{2.303RT} = \frac{-(2.60\text{ kJ mol}^{-1})(1000\text{ J kJ}^{-1})}{2.303(8.31\text{ J mol}^{-1}\text{K}^{-1})(298\text{ K})}$$

$$= -0.456$$
$$K = \text{antilog}(-0.456) = 0.350$$

Therefore $K_p = \dfrac{p^2_{HI}}{p_{H_2}} = 0.350.$

PROBLEM 16-10   What is the equilibrium constant K_p for the following reaction at 400 K?

$$2NOCl(g) \rightleftharpoons 2NO(g) + Cl_2(g) \qquad \Delta H^\circ = 77.2 \text{ kJ mol}^{-1}; \quad \Delta S^\circ = 122 \text{ J mol}^{-1}\text{K}^{-1} \text{ at 400 K}$$

Solution: Use eq. (16-11) for the relationship between ΔH° and K:

$$-2.30RT\log K = \Delta H - T\Delta S$$

$$\log K = \frac{-\Delta H^\circ}{2.30RT} + \frac{\Delta S^\circ}{2.30R}$$

Since R is in units of $J\,\text{mol}^{-1}\text{K}^{-1}$, we have to change ΔH° to $77.2 \times 10^3\,\text{J mol}^{-1}\text{K}^{-1}$.

$$\log K = \frac{-77.2 \times 10^3\,\text{J mol}^{-1}}{2.30(8.31\,\text{J mol}^{-1}\text{K}^{-1})(400\text{ K})} + \frac{122\,\text{J mol}^{-1}\text{K}^{-1}}{2.30(8.31\,\text{J mol}^{-1}\text{K}^{-1})}$$

$$= -10.10 + 6.38 = -3.72$$
$$K = \text{antilog}(-3.72) = 1.93 \times 10^{-4}$$

Therefore $K_p = \dfrac{p^2_{NO}p_{Cl_2}}{p^2_{NOCl}} = 1.93 \times 10^{-4}.$

Alternate Solution:

$$\Delta G^\circ = \Delta H^\circ - T\Delta S^\circ$$
$$= (7.22 \times 10^4\,\text{J mol}^{-1}) - (400\text{ K})(122\,\text{J mol}^{-1}\text{K}^{-1})$$
$$= 2.84 \times 10^4\,\text{J mol}^{-1}$$

$$\log K = \frac{-\Delta G^\circ}{2.30RT}$$

$$= \frac{-2.84 \times 10^4\,\text{J mol}^{-1}}{(2.30)(8.31\,\text{J mol}^{-1}\text{K}^{-1})(400\text{ K})}$$

$$= -3.71$$
$$K_p = \text{antilog}(-3.71) = 1.95 \times 10^{-4}$$

PROBLEM 16-11   The vapor pressure of diethyl ether is 185.3 torr at 0.0°C and 760.0 torr at 34.6°C. Calculate the enthalpy of vaporization of liquid diethyl ether:

$$(C_2H_5)_2O(l) \longrightarrow (C_2H_5)_2O(g)$$

[*Hint*: Assume that $p_{(C_2H_5)_2O_{2(g)}} = P_{tot}$.]

Solution: The equilibrium constant for the vaporization of diethyl ether is simply the vapor pressure of gaseous diethyl ether: $K = p_{(C_2H_5)_2O(g)}$ (which is also P_{tot} for each temperature). You are given the vapor pressure at two temperatures, so you use eq. (16-12) by substituting P_{T_1} for K_1 and P_{T_2} for K_2 and solving for $\Delta H°$. This form of eq. (16-12) is known as the **Clausius-Clapeyron equation**:

$$\log\left(\frac{K_1}{K_2}\right) = \left(\frac{-\Delta H°_{vap}}{2.303R}\right)\left(\frac{1}{T_1} - \frac{1}{T_2}\right) \qquad \text{(eq. 16-12)}$$

CLAUSIUS-CLAPEYRON EQUATION
$$\log\left(\frac{P_{T_1}}{P_{T_2}}\right) = \left(\frac{-\Delta H°_{vap}}{2.303R}\right)\left(\frac{1}{T_1} - \frac{1}{T_2}\right)$$

$$\Delta H°_{vap} = \left(\frac{-2.303R}{(1/T_1) - (1/T_2)}\right)\log\left(\frac{P_{T_1}}{P_{T_2}}\right)$$

$$= \left(\frac{-(2.303)(8.31\,\text{J mol}^{-1}\text{K}^{-1})}{(1/273.2\,\text{K}) - (1/307.8\,\text{K})}\right)\log\left(\frac{185.3\,\text{torr}}{760.0\,\text{torr}}\right)$$

$$= \left(\frac{-19.14}{4.12 \times 10^{-4}}\right)(-0.6129)$$

$$= 28.5 \times 10^3\,\text{J mol}^{-1}$$

PROBLEM 16-12   How much faster would a reaction proceed at 50°C than at 0°C if the activation energy is 63 kJ?

Solution: We need to find the ratio of the rate constants at the two temperatures, i.e., k_2/k_1. Use the Arrhenius equation:

$$\log\frac{k_1}{k_2} = \left(\frac{-\Delta E_a}{2.303R}\right)\left(\frac{1}{T_1} - \frac{1}{T_2}\right)$$

$$= \left(\frac{(-63\,\text{kJ})(10^3\,\text{J kJ}^{-1})}{2.30(8.31\,\text{J mol}^{-1}\text{K}^{-1})}\right)\left(\frac{1}{273} - \frac{1}{323}\right)$$

$$= -1.87$$

$$\frac{k_1}{k_2} = \text{antilog}(-1.87) = 0.0135$$

$$\frac{k_2}{k_1} = \frac{1}{0.0135} = 74$$

The reaction will proceed roughly 74 times as fast at 50°C as at 0°C.

PROBLEM 16-13   Determine the increase in entropy when 1 mol of neon is mixed with 1 mol of argon at 0°C at constant pressure.

Solution: Noble gases exhibit near-ideal behavior at 273 K (0°C), so you can start with eq. (16-5):

$$\Delta S = 2.303nR\log\left(\frac{P_i}{P_f}\right)$$

From Dalton's law of partial pressures you know that the final pressure is the sum of the partial pressures, and that the partial pressure of each gas in the mixture is proportional to the mole fraction (X) of gas present, so

$$\Delta S_{tot} = \Delta S_{Ar} + \Delta S_{Ne}$$

$$= n_{Ar}(2.303R)\log\left(\frac{P}{X_{Ar}P}\right) + n_{Ne}(2.303R)\log\left(\frac{P}{X_{Ne}P}\right)$$

$$= -2.303R(n_{Ar}\log X_{Ar} + n_{Ne}\log X_{Ne})$$

The mole fraction of each gas after mixing is 0.5. So

$$\Delta S = -2.303(8.31 \text{ J mol}^{-1}\text{K}^{-1})[(1 \text{ mol})\log 0.5 + (1 \text{ mol})\log 0.5]$$
$$= -2.303(8.31)(-0.602)$$
$$= 11.5 \text{ J K}^{-1}$$

Supplementary Exercises

PROBLEM 16-14 In which of the following processes is ΔS less than zero and in which is ΔS greater than zero?

(a) $H_2O(l) \longrightarrow H_2O(g)$
(b) 1.00 mol of gas at 15 atm pressure is compressed until the pressure is 30 atm
(c) $SO_2(g) + Cl_2(g) \longrightarrow Cl_2SO_2(g)$

Answer: (a) $\Delta S > 0$ (b) $\Delta S < 0$ (c) $\Delta S < 0$

PROBLEM 16-15 At 25°C the change in free energy ΔG for the reaction $FeO(s) + H_2(g) \rightarrow Fe(s) + H_2O(l)$ is 7.1 kJ mol^{-1}. The enthalpy change is $\Delta H = -19.3$ kJ mol^{-1}. Calculate the entropy change ΔS in J mol^{-1}K^{-1}.

Answer: -88.6 J mol^{-1}K^{-1}

PROBLEM 16-16 Calculate ΔG at 700 K for the oxidation of NO gas: $2NO(g) + O_2(g) \rightarrow 2NO_2(g)$; $\Delta H = -113.0$ kJ mol^{-1} and $\Delta S = -145$ J mol^{-1}K^{-1}.

Answer: -11.5 kJ mol^{-1}

PROBLEM 16-17 For the reaction $NH_4Cl(s) \rightarrow NH_3(g) + HCl(g)$, $\Delta H = +177$ kJ mol^{-1} and $\Delta S = +285$ J mol^{-1}K^{-1} at 25°C. Calculate ΔG at 25°C. Is this a spontaneous reaction?

Answer: $\Delta G = 92$ kJ mol^{-1}; the reaction is not spontaneous

PROBLEM 16-18 Calculate ΔG at 497°C for the reaction in the previous problem. Assume that ΔH and ΔS remain constant over this temperature change. Is the reaction spontaneous?

Answer: $\Delta G = -42$ kJ mol^{-1}; the reaction is spontaneous

PROBLEM 16-19 At 0°C ice and water are in equilibrium and $\Delta H = 6.00$ kJ mol^{-1} for the process $H_2O(s) \rightarrow H_2O(l)$. Calculate ΔS for the conversion of ice to liquid water.

Answer: $\Delta S = 22.0$ J mol^{-1}K^{-1}

PROBLEM 16-20 Calculate the entropy change for the synthesis of ethanol from ethylene and water at 25°C: $C_2H_4(g) + H_2O(g) \rightarrow C_2H_5OH(g)$. Use these entropies (J mol^{-1}K^{-1}):

Substance:	$H_2O(g)$	$C_2H_4(g)$	$C_2H_5OH(g)$
S at 25°C:	188.7	219.5	282.0

Answer: $\Delta S = -126.2$ J mol^{-1}K^{-1}

PROBLEM 16-21 Calculate ΔG at 25°C for the oxidation of propane by oxygen to form acetone: $C_3H_8(g) + O_2(g) \rightarrow (CH_3)_2CO(l) + H_2O(l)$. Use the following free energies of formation (kJ mol^{-1}):

Substance:	$C_3H_8(g)$	$(CH_3)_2CO(l)$	$H_2O(l)$
ΔG_f at 25°C:	-23.5	-153.6	-237.2

Answer: $\Delta G = -367.3$ kJ mol^{-1}

PROBLEM 16-22   At 395°C the equilibrium constant K_p is 0.0450 atm for the reaction $COCl_2(g) \rightleftharpoons CO(g) + Cl_2(g)$. **(a)** Calculate ΔG at 395°C. **(b)** Using $\Delta H = 113 \text{ kJ mol}^{-1}$, calculate ΔS at 395°C.

Answer: **(a)** 17.2 kJ mol^{-1} **(b)** $143 \text{ J mol}^{-1} \text{K}^{-1}$

PROBLEM 16-23   Use the data given to calculate **(a)** $\Delta G°$ and **(b)** the equilibrium constant K_p at 298 K for the reaction $H(g) + HCl(g) \rightleftharpoons H_2(g) + Cl(g)$.

	ΔH_f at 298 K (kJ mol^{-1})		S at 298 K $(\text{J mol}^{-1}\text{K}^{-1})$
$H(g)$	$+218.0$	$H(g)$	114.6
$HCl(g)$	-92.3	$HCl(g)$	187.0
$Cl(g)$	$+121.4$	$Cl(g)$	165.1
		$H_2(g)$	131.0

Answer: **(a)** $\Delta G° = -2.7 \text{ kJ mol}^{-1}$ **(b)** $K_p = 3.0$

PROBLEM 16-24   At 25°C, $K_p = 0.143$ atm for the reaction $N_2O_4(g) \rightleftharpoons 2NO_2(g)$. The enthalpy change $\Delta H = 58.04 \text{ kJ mol}^{-1}$ at 25°C. Determine K_p at 75°C.

Answer: 4.2 atm

17 ACID–BASE EQUILIBRIA

THIS CHAPTER IS ABOUT

☑ **Acids and Bases: A Reminder**
☑ **Finding the pH in Solutions of Weak Acids**
☑ **Finding the pH in Solutions of Weak Bases**
☑ **Salts of Weak Acids**
☑ **Salts of Weak Bases**
☑ **Buffers and Indicators**
☑ **Titrations**
☑ **Polyprotic Acids**

17-1. Acids and Bases: A Reminder

A. Brønsted-Lowry acids and bases

A *Brønsted-Lowry acid* is a proton donor; the species left after donation of the proton is the *conjugate base* of that acid. A *Brønsted-Lowry base* is a proton acceptor; the species produced by addition of the proton is the *conjugate acid* of that base. (See Chapter 12.)

EXAMPLE 17-1: Identify the Brønsted-Lowry acid, the Brønsted-Lowry base, and their conjugates in the reaction

$$NH_3 + H_2O \rightleftharpoons NH_4^+ + OH^-$$

Solution: The proton acceptor is NH_3, so it is the base; NH_4^+ is its conjugate acid. The proton donor is H_2O, so it is the acid; OH^- is its conjugate base.

> *note:* All of the acids and bases you'll encounter in this chapter will be Brønsted-Lowry in type, and they'll be described simply as "acids" and "bases."

B. Weak acids: K_a and pK_a

The general reaction between an acid HA and water, which acts as base, is

$$HA + H_2O \rightleftharpoons H_3O^+ + A^-$$

The equilibrium constant for this reaction, written according to the equilibrium law (Chapter 15), is

$$K = \frac{[H_3O^+][A^-]}{[HA][H_2O]}$$

In aqueous solutions that are not highly concentrated the concentration term $[H_2O]$ remains virtually constant, because most of the water in the solution is not involved in the reaction with HA.

> *note:* In pure water, $[H_2O] = \dfrac{1000 \text{ g } H_2O}{(1 \text{ L})(18.0 \text{ g mol}^{-1})} = 55.6 \ M$; this concentration will only change by 1 or 2% at most, even if a mole of H_2O reacts completely with a mole of added HA. Typical problems on weak acids and bases involve dilute solutions in which $[H_2O]$ changes are quite insignificant.

TABLE 17-1: Ionization Constants K_a for Monoprotic Acids in H_2O at 25°C

Acid	Formula	K_a
Acetic	CH_3CO_2H	1.75×10^{-5}
Benzoic	$C_6H_5CO_2H$	6.31×10^{-5}
Formic	HCO_2H	1.77×10^{-4}
Hydrocyanic	HCN	4.93×10^{-10}
Hydrofluoric	HF	6.8×10^{-4}
Hypobromous	$HOBr$	2.06×10^{-9}
Hypochlorous	$HOCl$	2.95×10^{-8}
Hypoiodous	HOI	1.0×10^{-10}
Nitrous	HNO_2	4.6×10^{-4}
Phenol	C_6H_5OH	1.28×10^{-10}
Trifluoroacetic	CF_3CO_2H	5.9×10^{-1}

Since $[H_2O]$ is a constant, its value can be included in the value of K, and we can define a new constant

IONIZATION (DISSOCIATION) CONSTANT

$$K_a = \frac{[H_3O^+][A^-]}{[HA]} \qquad (17\text{-}1)$$

where K_a is the *ionization* or *dissociation constant* of the acid. Only if HA is a weak acid—so that there really is some un-ionized HA present at equilibrium—does K_a have a real meaning. If $[HA] = 0$, then K_a becomes infinite, i.e., meaningless. This is the case if HA is a strong acid: For strong acids no K_a value can be determined in water.

There's one more shortcut in writing K_a expressions: $[H_3O^+]$ can be replaced by the simpler $[H^+]$. This means that although we recognize that the proton H^+ is *always* attached to at least one H_2O molecule in aqueous solution, we prefer the equivalent simpler form—just because it's quicker to write it. Summing up the short form of notation for the equilibrium in water:

$$HA \rightleftharpoons H^+ + A^- \qquad \text{(all species are aqueous)}$$

$$K_a = \frac{[H^+][A^-]}{[HA]}$$

The numerical value of K_a expresses the *strength* of the acid HA. The larger K_a is, the stronger the acid. Table 17-1 lists the K_a values of some representative acids.

EXAMPLE 17-2: Arrange the acids HCN, HNO_2, and HF in order of increasing strength.

Solution: According to Table 17-1 the K_a values are HCN 4.9×10^{-10}, HNO_2 4.6×10^{-4}, HF 6.8×10^{-4}. Thus the order of increasing K_a, which is the order of increasing strength, is $HCN < HNO_2 < HF$.

note: Although values of K HAVE units, this is one area of chemistry in which the units are almost always left out.

Because K_a values are often quite small, it's sometimes convenient to express them in the same way that pH is expressed:

$$pH = -\log_{10}[H^+] \qquad \text{and} \qquad pK_a = -\log_{10}K_a$$

The larger pK_a is, the weaker the acid.

EXAMPLE 17-3: Determine pK_a for HCN, for which $K_a = 4.9 \times 10^{-10}$.

Solution:

$$pK_a = -\log K_a$$
$$= -\log(4.9 \times 10^{-10}) = -\log 4.9 - \log 10^{-10} = -0.69 - (-10)$$
$$= 9.31$$

TABLE 17-2: Base Ionization Constants K_b in H_2O at 25°C

Base	Formula	K_b
Ammonia	NH_3	1.79×10^{-5}
Aniline	$C_6H_5NH_2$	4.27×10^{-10}
Benzylamine	$C_6H_5CH_2NH_2$	2.14×10^{-5}
Dimethylamine	$(CH_3)_2NH$	5.40×10^{-4}
Ethylamine	$C_2H_5NH_2$	6.41×10^{-4}
Hydrazine	H_2NNH_2	1.7×10^{-6}
Hydroxylamine	$HONH_2$	1.07×10^{-8}
Methylamine	CH_3NH_2	4.4×10^{-4}
Pyridine	C_5H_5N	1.7×10^{-9}

C. Weak bases: K_b and pK_b

Weak bases give an equilibrium reaction in water in which some hydroxide ion (OH^- is the conjugate base of H_2O, which is a weak acid) and some conjugate acid of the weak base are produced. For example, in a solution of ammonia (a weak base) the following equilibrium is established:

$$NH_3 + H_2O \rightleftharpoons NH_4^+ + OH^-$$

Applying the equilibrium law to this equilibrium, we get

$$K = \frac{[NH_4^+][OH^-]}{[NH_3][H_2O]} \qquad \text{(all species aqueous)}$$

Since $[H_2O]$ is constant, we can define a new, more compact equilibrium constant K_b, the *base ionization* or *dissociation constant*:

BASE IONIZATION (DISSOCIATION) CONSTANT

$$K_b = \frac{[NH_4^+][OH^-]}{[NH_3]} \qquad (17\text{-}2)$$

The larger the value of K_b, the stronger the base. Some K_b values are listed in Table 17-2.

EXAMPLE 17-4: Arrange the following bases in order of increasing strength: ammonia (NH_3), methylamine (CH_3NH_2), pyridine (C_5H_5N).

Solution: Increasing K_b means increasing strength. According to the K_b values listed in Table 17-2, C_5H_5N is the weakest base of the three, NH_3 is in the middle, and CH_3NH_2 is the strongest.

We find pK_b exactly as we find pK_a for acids: p$K_b = -\log_{10} K_b$. Thus NH_3, whose $K_b = 1.79 \times 10^{-5}$, has a p$K_b$ of 4.75.

D. Water: K_w and pK_w

Water can be either an acid or a base; in pure water the following equilibrium exists:

$$H_2O \rightleftharpoons H^+ + OH^-$$

Thus from the equilibrium law

$$K = \frac{[H^+][OH^-]}{[H_2O]}$$

Applying the usual convention of incorporating $[H_2O]$ as a constant into K, we define a new equilibrium constant for water, the *ion product*:

$$K_w = [H^+][OH^-] \qquad (17\text{-}3)$$

At room temperature (25°C), $K_w = 1.0 \times 10^{-14}$. In pure water the stoichiometry of the ionization means that $[H^+] = [OH^-] = 1.0 \times 10^{-7}$. That's how the $[H^+] = 1.0 \times 10^{-7}$ (or pH = 7.00) value arises for pure neutral H_2O.

note: Although the K_w expression was developed for pure water, it applies to EVERY aqueous equilibrium in which H^+ and OH^- are involved. For all aqueous equilibria at room temperature, $[H^+][OH^-] = K_w = 1.0 \times 10^{-14}$.

EXAMPLE 17-5: Calculate $[H^+]$ and pH in a 0.010 *M* solution of KOH.

Solution: The base KOH is strong and completely ionized, so $[OH^-]$ in this solution is 0.010 *M*. Since

$$[H^+][OH^-] = 1.0 \times 10^{-14}$$

$$[H^+] = \frac{1.0 \times 10^{-14}}{[OH^-]} = \frac{1.0 \times 10^{-14}}{0.010}$$

$$= 1.0 \times 10^{-12}$$

$$pH = -\log[H^+] = 12.00$$

Just as $pH = -\log[H^+]$, we can define another useful expression: $pOH = -\log[OH^-]$.

EXAMPLE 17-6: What is the pOH of a 0.010 *M* solution of the strong, fully ionized base $Ba(OH)_2$?

Solution:

$$[OH^-] = (0.010 \ M)\left(\frac{2 \ \text{mol OH}^-}{1 \ \text{mol Ba(OH)}_2}\right) = 0.020 \ M$$

$$pOH = -\log[OH^-]$$
$$= -\log 0.020 = -(-1.70)$$
$$= 1.70$$

By now the symbolism of pK_w should be obvious: It's simply $-\log[K_w] = -\log(1.0 \times 10^{-14}) = 14.00$. Because $[H^+][OH^-] = K_w = 1.0 \times 10^{-14}$ at 25°C, it follows that in ANY aqueous solution at 25°C

$$pH + pOH = pK_w = 14.00 \tag{17-4}$$

EXAMPLE 17-7: Calculate pOH and pH in a 0.010 *M* solution of KOH.

Solution: Since $[OH^-] = 0.010 \ M$,

$$pOH = -\log 0.010 = 2.00$$
$$pH = 14.00 - pOH$$
$$= 14.00 - 2.00 = 12.00$$

note: This method is quicker and easier than the method used in Example 17-5: Learn this approach and use it whenever you can.

17-2. Finding the pH in Solutions of Weak Acids

When a solution of a weak monoprotic acid, such as HF, is made by diluting 0.50 mol of the acid to 1.00 L, the concentration is often described as 0.50 *M*. Although the molarity describes the stoichiometry, it provides no information on the state of ionization of the acid. When we want to calculate the pH of a weak acid solution and we know the K_a value, we can use an approximation method or an exact method.

A. An approximation method

Suppose we want to find the pH of a 0.50 M HF solution, given $K_a = 6.8 \times 10^{-4}$.

First consider all the species present in aqueous HF. There are five of them, according to the (unbalanced) reaction equation:

$$H_2O + HF \longrightarrow H^+ + F^- + OH^-$$

H_2O: The concentration of water stays constant, so we don't worry about it
OH^-: Hydroxide ions are always present in water, so $[OH^-]$ is an unknown
H^+: The concentration of H^+ will give us the pH, so $[H^+]$ is an unknown
F^-: $[F^-]$ is an unknown
HF: $[HF]$ is an unknown

So, of the five species present, we have four unknowns ($[H^+]$, $[OH^-]$, $[F^-]$, and $[HF]$), for which we need four independent equations.

The four independent equations are found as follows: The first two equations are the K expressions for the two equilibria that are established:

$$HF \rightleftharpoons H^+ + F^-: \qquad K_a = \frac{[H^+][F^-]}{[HF]} = 6.8 \times 10^{-4}$$

$$H_2O \rightleftharpoons H^+ + OH^-: \qquad K_w = [H^+][OH^-] = 1.0 \times 10^{-14}$$

The last two equations are the **mass balance equation** (i.e., the stoichiometric equation)

$$[HF] + [F^-] = 0.50\ M$$

and the **charge balance equation**

$$[H^+] = [F^-] + [OH^-]$$

note: All the rules for balancing equations apply to the mass and charge balance. All the F atoms in the original 0.50 mol of HF end up as HF or F^-, so we know that $[HF] + [F^-] = 0.50$. Also, the solution cannot have an excess of positive or negative charge. The only positively charged species is H^+ and the negatively charged species are F^- and OH^-, so we know that $[H^+] = [F^-] + [OH^-]$.

Now that we have the data arranged and equations set up, we can begin the approximation procedure: We *look for terms we can neglect.*

1. Start with the charge balance, $[H^+] = [F^-] + [OH^-]$. We know that HF is an acid, so we can safely predict that $[H^+]$ will be much greater than $[OH^-]$. If we therefore assume that the numerical value of $[OH^-]$ is negligibly small compared to $[H^+]$, we get an approximation

$$[F^-] + [\cancel{OH^-}] \cong [H^+] \cong [F^-] \qquad (1)$$

2. Next think about the mass balance, $[HF] + [F^-] = 0.50$. Since HF is a *weak* acid, it's likely that most of the HF is not ionized, so $[HF] \gg [F^-]$ and we can neglect $[F^-]$ in comparison with $[HF]$.

$$[HF] + [\cancel{F^-}] = 0.50 \cong [HF] \cong 0.50 \qquad (2)$$

3. Now we'll try our two approximations in the K_a equation:

$$K_a = \frac{[H^+][F^-]}{[HF]} = 6.8 \times 10^{-4}$$

$$\cong \frac{[H^+]^2}{[HF]} \cong 6.8 \times 10^{-4} \qquad ([H^+] \cong [F^-])$$

$$\cong \frac{[H^+]^2}{0.50} \cong 6.8 \times 10^{-4} \qquad ([HF] \cong 0.50)$$

so

$$[H^+]^2 \cong 3.4 \times 10^{-4}$$
$$[H^+] \cong \sqrt{3.4 \times 10^{-4}}$$
$$[H^+] \cong 1.8 \times 10^{-2} \; M \tag{3}$$

4. Finally, we check the accuracy of our approximations by calculating the percentage of ionization using the results of approximation (3):

$$\frac{1.8 \times 10^{-2}}{0.50} \times 100\% = 3.6\%$$

A percentage of 3.6% is not very much, so we can consider approximations (1), (2) and (3) a success.

Thus we have a rule:

• Approximations that are valid within 5% are good approximations.

Now we can answer the original question. The pH of the solution is calculated from the H^+ concentration:

$$pH = -\log(1.8 \times 10^{-2}) = 1.74$$

B. An exact method

Suppose we want to find the pH of a 0.02 M solution of HF, given a K_a of 6.8×10^{-4}.

Set up the data:

Unknowns	Equations
$[OH^-]$	$K_a = \dfrac{[H^+][F^-]}{[HF]} = 6.8 \times 10^{-4}$
$[H^+]$	$K_w = [H^+][OH^-] = 1.0 \times 10^{-14}$
$[F^-]$	$[HF] + [F^-] = 0.02$ (mass)
$[HF]$	$[H^+] = [F^-] + [OH^-]$ (charge)

Start the approximation procedure:

1. If $[H^+] \gg [OH^-]$, then $[H^+] \cong [F^-]$ from the charge balance.
2. If $[HF] \gg [F^-]$, then $[HF] \cong 0.02$ from the mass balance.
3. Try the approximations:

$$K_a = \frac{[H^+][F^-]}{[HF]} = 6.8 \times 10^{-4}$$

$$\cong \frac{[H^+]^2}{0.02} \cong 6.8 \times 10^{-4}$$

$$[H^+] \cong \sqrt{1.36 \times 10^{-5}} \cong 3.7 \times 10^{-3}$$

4. Check the percentage of ionization:

$$\frac{3.7 \times 10^{-3}}{0.02} \times 100\% = 19\%$$

According to our rule, 19% is too much—so we check our assumptions.

1. $[H^+] \cong [F^-]$ MUST be reasonable: Because $[H^+] > 10^{-3}$, $[OH^-] < 10^{-11}$ (from the K_w condition). Thus $[OH^-]$ IS negligible.
2. $[HF] \cong 0.02$ is NOT reasonable: $[F^-]$ is NOT negligible compared to $[HF]$, so we have to use the full condition $[HF] + [F^-] = 0.02$. Then, if we keep the $[H^+] \cong [F^-]$ assumption, we have $[HF] + [H^+] = 0.02$, or $[HF] = 0.02 - [H^+]$. Now we at least have all the terms in K_a in the one unknown $[H^+]$.

3. Substitute the revised assumptions into the K_a equation:

$$K_a = \frac{[H^+][F^-]}{[HF]} = 6.8 \times 10^{-4}$$

$$= \frac{[H^+]^2}{0.02 - [H^+]} = 6.8 \times 10^{-4}$$

This is a quadratic equation, so now it's algebra time. The solution to the standard quadratic equation $ax^2 + bx + c = 0$ is

$$x = \frac{-b \pm \sqrt{b^2 - 4ac}}{2a} \qquad\qquad \text{(17-5)}$$

Put the K_a equation into the standard form:

$$[H^+]^2 = (6.8 \times 10^{-4})(0.02 - [H^+])$$

so that

$$[H^+]^2 + (6.8 \times 10^{-4})[H^+] + (-1.4 \times 10^{-5}) = 0$$

with $a = 1$, $b = 6.8 \times 10^{-4}$, and $c = -1.4 \times 10^{-5}$. Consequently

$$[H^+] = \frac{-b \pm \sqrt{b^2 - 4ac}}{2a}$$

$$= \frac{-6.8 \times 10^{-4} \pm \sqrt{(6.8 \times 10^{-4})^2 - 4(1)(-1.4 \times 10^{-5})}}{2}$$

$$= \frac{-6.8 \times 10^{-4} \pm \sqrt{4.62 \times 10^{-7} + 5.6 \times 10^{-5}}}{2}$$

$$= \frac{-6.8 \times 10^{-4} \pm \sqrt{5.64 \times 10^{-5}}}{2}$$

$$= \frac{-6.8 \times 10^{-4} \pm 7.5 \times 10^{-3}}{2}$$

$$= \frac{6.8 \times 10^{-3}}{2} \quad \text{or} \quad \frac{-8.2 \times 10^{-3}}{2}$$

$$= 3.4 \times 10^{-3} \quad \text{or} \quad -4.1 \times 10^{-3}$$

Only one of these answers can have a real chemical meaning, and it's clearly the positive root, 3.4×10^{-3}. A H^+ concentration of -4.1×10^{-3} has no chemical meaning. So

$$pH = -\log(3.4 \times 10^{-3}) = 2.46$$

17-3. Finding the pH in Solutions of Weak Bases

The method for finding the pH of a weak base solution is essentially the same as that for finding the pH of a weak acid solution. In the case of bases, however, $[OH^-] \gg [H^+]$, and so $[H^+]$ can usually be neglected in the charge balance equation. Then an approximation can be tried by neglecting the amount of ionization in the mass balance equation. If the approximated answer indicates that there's more than 5% ionization, a quadratic equation must be used.

EXAMPLE 17-8: Calculate the pH in a 0.10 M solution of NH_3, a weak base, whose $K_b = 1.8 \times 10^{-5}$.

Solution: First write down the reaction equation and then the K_b expression:

$$NH_3 + H_2O \rightleftharpoons NH_4^+ + OH^- \qquad K_b = \frac{[NH_4^+][OH^-]}{[NH_3]} = 1.8 \times 10^{-5}$$

Then write the charge balance and mass balance equations:

$$[NH_4^+] + [H^+] = [OH^-] \quad \text{(charge)}$$
$$[NH_3] + [NH_4^+] = 0.10 \quad \text{(mass)}$$

From the charge balance, make the approximation that since $[OH^-] \gg [H^+]$, then $[NH_4^+] \cong [OH^-]$. From the mass balance, assume that, because NH_3 is a weak base, $[NH_4^+]$ is small and so $[NH_3] \cong 0.10$. Now make the appropriate substitutions into K_b:

$$K_b = \frac{[NH_4^+][OH^-]}{[NH_3]} \cong \frac{[OH^-]^2}{0.10}$$
$$= 1.8 \times 10^{-5}$$

So

$$[OH^-]^2 \cong 1.8 \times 10^{-6}$$
$$[OH^-] \cong 1.3 \times 10^{-3}$$

Now check the approximation:

$$\frac{1.3 \times 10^{-3}}{0.10} \times 100\% = 1.3\%$$

which is well below the 5% range for acceptable approximations. No quadratic is needed for this problem.

Finally, answer the problem by finding the pH:

$$pOH = -\log(1.3 \times 10^{-3}) = 2.89$$
$$pH = 14.00 - pOH$$
$$= 14.00 - 2.89$$
$$= 11.11$$

EXAMPLE 17-9: Calculate $[OH^-]$ in a $1.0 \times 10^{-3}\,M$ solution of the weak base methylamine (CH_3NH_2), whose $K_b = 4.4 \times 10^{-4}$. What percentage of the base has ionized?

Solution: First write out the equilibrium reaction and K_b:

$$CH_3NH_2 + H_2O \rightleftharpoons CH_3NH_3^+ + OH^- \qquad K_b = \frac{[CH_3NH_3^+][OH^-]}{[CH_3NH_2]} = 4.4 \times 10^{-4}$$

Try the usual assumptions:

$$[CH_3NH_3^+] \cong [OH^-] \quad \text{(assuming } [OH^-] \gg [H^+])$$
$$[CH_3NH_2] \cong 1.0 \times 10^{-3} \quad ([CH_3NH_3^+] \text{ negligible?})$$

Then

$$K_b = \frac{[CH_3NH_3^+][OH^-]}{[CH_3NH_2]} \cong \frac{[OH^-]^2}{1.0 \times 10^{-3}} = 4.4 \times 10^{-4}$$

So

$$[OH^-]^2 = 4.4 \times 10^{-7}$$
$$[OH^-] = 6.6 \times 10^{-4}$$

which is

$$\frac{6.6 \times 10^{-4}}{1.0 \times 10^{-3}} \times 100\% = 66\%$$

Drat! The approximation of little ionization is hopeless: $[CH_3NH_3^+]$ is not negligible. We'll have to go to the quadratic, in which

$$[CH_3NH_2] = (1.0 \times 10^{-3}) - [CH_3NH_3^+] = (1.0 \times 10^{-3}) - [OH^-]$$

The neglect of $[H^+]$ is still a good approximation. Now we can make a second trial:

$$K_b = \frac{[CH_3NH_3^+][OH^-]}{[CH_3NH_2]} = \frac{[OH^-]^2}{(1.0 \times 10^{-3}) - [OH^-]}$$

$$= 4.4 \times 10^{-4}$$

$$[OH^-]^2 = (4.4 \times 10^{-7}) - (4.4 \times 10^{-4})[OH^-]$$

So

$$[OH^-]^2 + (4.4 \times 10^{-4})[OH^-] + (-4.4 \times 10^{-7}) = 0$$

Using the standard form,

$$[OH^-] = \frac{-4.4 \times 10^{-4} \pm \sqrt{(4.4 \times 10^{-4})^2 - 4(1)(-4.4 \times 10^{-7})}}{2}$$

$$= \frac{-4.4 \times 10^{-4} \pm 1.40 \times 10^{-3}}{2}$$

$$= 4.8 \times 10^{-4} \quad \text{or} \quad -9.2 \times 10^{-4}$$

Only the positive root can be meaningful here, and so $[OH^-] = 4.8 \times 10^{-4}$ M. The amount of ionization is thus

$$\frac{4.8 \times 10^{-4}}{1.0 \times 10^{-3}} \times 100\% = 48\%$$

which is quite a bit different from the 66% of our first, incorrect, treatment.

note: The quadratic equation should be used when the weak acid or base is not too weak, and when its concentration is low.

17-4. Salts of Weak Acids

The salts of a weak, monoprotic acid HA are ionic compounds that ionize in aqueous solution. According to Brønsted-Lowry definitions, the anion of a weak acid (i.e., the A^- ion) is a base. For example, the salts of the weak acid HF include NaF, KF, etc., whose anion F^- is the conjugate base of HF. (Note that M^+ ions such as Na^+ and K^+ have no acid–base properties in aqueous solutions. They are the cations of strong bases NaOH and KOH, respectively.)

A. Acid–base pair relationship

The ion A^- is a base in water because it reacts with water to produce OH^-, the characteristic basic ion in H_2O. If we apply the equilibrium law to the equilibrium reaction involved, we get

$$A^- + H_2O \rightleftharpoons HA + OH^-$$

$$K = \frac{[HA][OH^-]}{[A^-][H_2O]}$$

Incorporating $[H_2O]$ as a constant, we get

$$K_b = \frac{[HA][OH^-]}{[A^-]}$$

If we multiply this K_b for A^- by K_a for the acid HA, we get

$$K_a \times K_b = \frac{[H^+][A^-]}{[HA]} \times \frac{[HA][OH^-]}{[A^-]}$$

$$= [H^+][OH^-] = K_w \tag{17-6}$$

This relationship always holds for a conjugate acid–base pair in water.

EXAMPLE 17-10: Given that K_a for HF is 6.8×10^{-4}, calculate K_b for F^-.

Solution: We know that $K_a \times K_b = K_w$, so

$$K_b = \frac{K_w}{K_a}$$

We also know that $K_w = 1.0 \times 10^{-14}$; therefore K_b for F^- must be

$$\frac{1.0 \times 10^{-14}}{6.8 \times 10^{-4}} = 1.5 \times 10^{-11}$$

B. pH Calculations

The conjugate base A^- of a weak acid is a weak base. Therefore we can calculate the pH of a salt solution by using the K_b for a conjugate base just as we use K_b for any weak base (Section 17-3): Exactly the same approximations can be applied.

note: Sometimes the basic properties of these conjugate bases are described as being due to "hydrolysis," and the K_b is termed "K_h." We will not follow that usage.

EXAMPLE 17-11: Calculate the pH of a 0.15 M solution of KF. (K_a for HF = 6.8×10^{-4}; K_b for $F^- = 1.5 \times 10^{-11}$.)

Solution: The base reaction is $F^- + H_2O \rightleftharpoons HF + OH^-$, and so

$$K_b = \frac{[HF][OH^-]}{[F^-]}$$

From the stoichiometry of the reaction, $[HF] \cong [OH^-]$. (This neglects any contribution to $[OH^-]$ from ionization of H_2O, which we know will be small because the solution is basic from the F^- reaction. Le Chatelier's principle tells us that the presence of the basic anion will suppress the ionization of H_2O to OH^-.) Then, since F^- is quite a weak base, it's likely that $[F^-] \cong 0.15$ M. At least, that's a fair first approximation. So

$$K_b = \frac{[HF][OH^-]}{[F^-]} \cong \frac{[OH^-]^2}{0.15} \cong 1.5 \times 10^{-11}$$

$$[OH^-]^2 \cong (0.15)(1.5 \times 10^{-11}) \cong 2.2 \times 10^{-12}$$

$$[OH^-] \cong 1.5 \times 10^{-6}$$

The amount of reaction of F^- is only

$$\frac{1.5 \times 10^{-6}}{0.15} \times 100\% = 1 \times 10^{-3}\%$$

The approximation is good. Thus pOH = $-\log(1.5 \times 10^{-6})$ = 5.82 and pH = 8.18, noticeably basic.

note: The salt KF is the stoichiometric product of the "neutralization" of HF by KOH:

$$HF + KOH \longrightarrow KF + H_2O$$

The pH calculation here shows that in the neutralization reaction the *equivalence point*—where one mole of KOH has been added per mole of HF—is going to be noticeably basic.

17-5. Salts of Weak Bases

Salts of weak bases (with anions of strong acids) contain conjugate acids of the weak bases and are therefore acidic. Thus, for example, NH_4^+ is acidic:

$$NH_4^+ \rightleftharpoons NH_3 + H^+ \qquad K_a = \frac{[NH_3][H^+]}{[NH_4^+]}$$

The K_a for the conjugate acid of a weak base whose base ionization constant K_b is known can be derived from the relationship $K_a \times K_b = K_w$.

EXAMPLE 17-12: What is the pH of a 0.10 M solution of NH_4Br? (K_b for $NH_3 = 1.8 \times 10^{-5}$.)

Solution:

$$K_a \text{ for } NH_4^+ = \frac{K_w}{K_b} = \frac{1.0 \times 10^{-14}}{1.8 \times 10^{-5}} = 5.6 \times 10^{-10}$$

Therefore

$$K_a = \frac{[NH_3][H^+]}{[NH_4^+]} = 5.6 \times 10^{-10}$$

From the stoichiometry, assume that $[NH_3] \cong [H^+]$, neglecting the contribution to $[H^+]$ from H_2O ionization. Assume further that, since NH_4^+ is a weak acid, $[NH_4^+] \cong 0.10$. Then

$$K_a = \frac{[H^+]^2}{0.10} \cong 5.6 \times 10^{-10} \quad \text{and} \quad [H^+] = 7.5 \times 10^{-6}$$

The assumptions are good: The extent of reaction of NH_4^+ is only

$$\frac{7.5 \times 10^{-6}}{0.10} \times 100\% = 7.5 \times 10^{-3}\%$$

So pH $= -\log(7.5 \times 10^{-6}) = 5.13$, noticeably acidic.

note: NH_4Br is the product of the 1:1 molar reaction of NH_3 and HBr; thus at the equivalence point in this reaction, the solution will be quite acidic.

17-6. Buffers and Indicators

A. Buffers

Mixtures of weak acids and their salts, or weak bases and their salts, have the property of undergoing only a slight pH change when moderate amounts of strong acids or bases are added to them. Such solutions are called **buffers**, or **buffer solutions**. They are useful in controlling pH in chemical or biochemical processes.

It's quite easy to calculate the pH of a given buffer solution. Consider a weak acid HA and salt Na^+A^- combination for which the concentrations of both salt and acid are known. Then from the K_a expression $K_a = [H^+][A^-]/[HA]$ it follows that

$$[H^+] = \frac{[HA]}{[A^-]} \times K_a \tag{17-7}$$

Practical buffers have fairly large concentrations of HA and A^-, so we can make the approximations that [HA] and $[A^-]$ are simply the stoichiometric concentrations. The tendency for HA to ionize is suppressed by the substantial concentration of A^-, while the tendency for A^- to combine with water to give HA is suppressed by the substantial concentration of HA.

Another useful form of the buffer equation can be obtained by taking logarithms of both sides of eq. (17-7):

$$[H^+] = \frac{[HA]}{[A^-]} \times K_a$$

$$\log[H^+] = \log K_a + \log\left(\frac{[HA]}{[A^-]}\right)$$

$$-\log[H^+] = -\log K_a - \log\left(\frac{[HA]}{[A^-]}\right)$$

$$pH = pK_a - \log\left(\frac{[HA]}{[A^-]}\right)$$

This last equation can be rewritten in a form called the **Henderson-Hasselbach equation**.

HENDERSON-HASSELBACH EQUATION $$pH = pK_a + \log\left(\frac{[A^-]}{[HA]}\right) \qquad (17\text{-}8)$$

EXAMPLE 17-13: What is the pH of a buffer solution that is 0.25 M in HF and 0.10 M in NaF? (K_a for HF $= 6.8 \times 10^{-4}$.)

Solution: We can use the Henderson-Hasselbach equation (17-8) directly. Since $K_a = 6.8 \times 10^{-4}$, we know that $pK_a = -\log(6.8 \times 10^{-4}) = 3.17$, so

$$pH = pK_a + \log\left(\frac{[A^-]}{[HA]}\right)$$

$$= 3.17 + \log\left(\frac{[F^-]}{[HF]}\right)$$

Then if $[A^-]$ and $[HA]$ are the stoichiometric concentrations, $[F^-] = 0.10\ M$ (NaF is completely ionized to Na^+ and F^-) and $[HF] = 0.25\ M$. So

$$pH = 3.17 + \log\left(\frac{0.10}{0.25}\right)$$

$$= 3.17 + \log 0.40 = 3.17 - 0.40$$
$$= 2.77$$

EXAMPLE 17-14: What is the final pH if 5.0 mL of 0.050 M NaOH is added to 50.0 mL of the buffer solution in Example 17-13?

Solution: To answer this we must first consider the reaction that is occurring. Since we're adding a base (OH^-), it will react with the acid present:

$$OH^- + HF \rightleftharpoons H_2O + F^-$$

Then we need to know whether or not this reaction goes to completion. We can show the extent of the reaction by calculating K for it (leaving out $[H_2O]$ as usual):

$$K = \frac{[F^-]}{[OH^-][HF]}$$

which we spot as being $1/K_b$ for F^- because $F^- + H_2O \rightleftharpoons HF + OH^-$ is the reverse of the reaction we're considering. Thus we can use the value we already know—K_a for HF: Since $K_b = K_w/K_a$,

$$K = \frac{1}{K_b} = \frac{K_a}{K_w} = \frac{6.8 \times 10^{-4}}{1.0 \times 10^{-14}} = 6.8 \times 10^{10}$$

That's a large number, so the reaction goes very nearly to completion.

note: This is an important general conclusion: *Weak acids react completely with strong bases; conversely, weak bases react completely with strong acids.*

Now, knowing the stoichiometry and extent of reaction, we can answer the problem. The original 50.00 mL of buffer contained $(50.0\ \text{mL})(0.10\ M\ F^-) = 5.0\ \text{mmol}\ F^-$ and $(50.0\ \text{mL})(0.25\ M\ \text{HF}) = 12.5\ \text{mmol}\ \text{HF}$. The added base contained $(5.0\ \text{mL})(0.050\ M\ OH^-) = 0.25\ \text{mmol}\ OH^-$. This will react with 0.25 mmol HF. So after the addition of the base, the solution contains $(5.0 + 0.25)\ \text{mmol}\ F^- = 5.25\ \text{mmol}\ F^-$ and $(12.5 - 0.25)\ \text{mmol}\ \text{HF} = 12.25\ \text{mmol}\ \text{HF}$. The final solution volume (assuming additivity) is therefore 50.0 mL + 5.0 mL = 55.0 mL. Thus after adding the base, we have

$$[F^-] = \left(\frac{5.25\ \text{mmol}}{55.0\ \text{mL}}\right) = 0.095\ M \qquad [HF] = \left(\frac{12.25\ \text{mmol}}{55.0\ \text{mL}}\right) = 0.22\ M$$

Now we apply the Henderson-Hasselbach equation (17-8):

$$pH = pK_a + \log\left(\frac{[F^-]}{[HF]}\right)$$

$$= 3.17 + \log\left(\frac{0.095}{0.22}\right) = 3.17 + \log(0.43)$$

$$= 3.17 - 0.37$$
$$= 2.80$$

Notice that the pH has gone up from 2.77 to 2.80—a change of only 0.03 units. This very small change demonstrates the fact that a buffer solution does resist major pH change, even when a strong base is added to it.

There are also buffers that resist pH change in the basic region. For these buffers, which are made from weak bases and their salts with strong acids, there are parallel equations to (17-7) and (17-8) that give the hydroxide ion concentration or the pOH:

$$[OH^-] = K_b\left(\frac{[B]}{[BH^+]}\right) \tag{17-9}$$

$$pOH = pK_b + \log\left(\frac{[BH^+]}{[B]}\right) \tag{17-10}$$

EXAMPLE 17-15: Determine the pH of a buffer that is 0.25 M in NH_3 and 0.50 M in NH_4Cl. ($K_b = 1.79 \times 10^{-5}$ and $pK_b = 4.75$ for NH_3.)

Solution: Given that $[BH^+] = 0.50\ M\ NH_4Cl$ and $[B] = 0.25\ M\ NH_3$, we can use eq. (17-10):

$$pOH = pK_b + \log\left(\frac{[BH^+]}{[B]}\right)$$

$$= 4.75 + \log\left(\frac{0.50}{0.25}\right) = 4.75 + 0.30$$

$$= 5.05$$

Thus

$$pH = 14.00 - pOH = 14.00 - 5.05 = 8.95$$

B. Indicators

An **indicator** is a conjugate acid–base pair that is added in very small molar amounts to solutions in order to monitor the pH. The acidic and basic forms of the indicator are of different colors. For the reaction

$$HInd(aq) \qquad Ind^- + H^+ \qquad K_{Ind} = \frac{[H^+][Ind^-]}{[HInd]}$$

where the acidic form (HInd) is one color and the alkaline form (Ind$^-$) is another, the [Ind$^-$]/[HInd] ratio determines which color will be seen. In general, one form of the indicator must be present in less than tenfold excess for the other form to be visible. We can rewrite the K_{Ind} expression as follows:

$$[H^+] = \left(\frac{[HInd]}{[Ind^-]}\right)K_{Ind}$$

This form of the K_{Ind} expression allows us to predict the $[H^+]$ or pH at which colors are visible:

Acid color visible	Alkaline color visible
$[H^+] = \left(\dfrac{10}{1}\right)K_{Ind}$	$[H^+] = \left(\dfrac{1}{10}\right)K_{Ind}$
$pH = -\log 10 - \log K_{Ind}$	$pH = -\log 0.1 - \log K_{Ind}$
$= pK_{Ind} - 1$	$= pK_{Ind} + 1$

The useful pH range of most indicators is around 2 units ($pH = pK_{Ind} \pm 1$). Two well-known indicators are phenolphthalein and methyl red. Phenolphthalein changes from colorless at $pH < 8.3$, to pink at $pH = 8.3–10.0$, and to intense red at $pH > 10.0$. Methyl red is red at $pH < 4.2$, turns orange at $pH = 4.2–6.3$, and is yellow at $pH > 6.3$.

EXAMPLE 17-16: In the titration of acetic acid with sodium hydroxide, the pH at the equivalence point (when exactly one mole of NaOH has been added per mole of acetic acid) is 8.5. Would methyl red or phenolphthalein be the better indicator for this titration?

Solution: Phenolphthalein undergoes its indicator color change in the pH range from 8.3 to 10.0, a range that includes the equivalence point of this titration. However, methyl red undergoes its indicator color change from pH 4.2 to pH 6.3, which does not include the equivalence pH of this titration. Therefore phenolphthalein would be the better indicator for this titration.

17-7. Titrations

The analysis of strong acids or strong bases is usually done by titrating them against their strong opposites: strong acids with strong bases, and strong bases with strong acids. Weak acids and weak bases can also be analyzed by titration against their *strong* opposites. Now we need to recognize that titrations take place in stages, and that each stage requires a different technique for calculating its pH.

A. Titration of strong acids with strong bases

The titration of strong acids (bases) with strong bases (acids) takes place in four stages:

1. *Start*: Strong acid (base) solution alone
2. *Partial conversion to salt*: Acid (base) and salt together in solution uncomplicated by interaction between ions and water
3. *Equivalence point*: Solution of the salt of a strong acid (base) and a strong base (acid), $pH = 7.00$
4. *Beyond the equivalence point*: Dilution of strong base (acid)

EXAMPLE 17-17: A solution of 35.0 mL of 0.0100 M HCl is titrated with 0.0100 M NaOH. Calculate the pH at each of the following stages: (a) before any base is added, (b) when 30.0 mL of base has been added, (c) at the equivalence point, and (d) when 36.0 mL of base has been added.

Solution:
(a) $[H^+] = 1.00 \times 10^{-2}\ M$; $pH = -\log(1.00 \times 10^{-2}) = 2.00$
(b) (0.350 mmol HCl) − (0.300 mmol NaOH) gives 0.050 mmol of H^+ left in 65.0 mL of solution, so

$$[H^+] = \frac{0.050}{65.0} = 7.7 \times 10^{-4}$$

$$pH = -\log(7.7 \times 10^{-4}) = 3.1$$

(c) $pH = 7.0$ at equivalence point

(d) $[OH^-] = \dfrac{0.010 \text{ mmol}}{(35 + 36) \text{ mL}} = 1.4 \times 10^{-4} \, M$

$$pOH = -\log(1.4 \times 10^{-4}) = 3.9$$
$$pH = 14.0 - 3.9 = 10.1$$

B. Titration of weak acids with strong bases and weak bases with strong acids

There are also four stages in the titration of weak acids and bases. The following table describes these stages and lists the section number under which the appropriate technique for calculating pH changes can be found.

Stage	Weak acid vs. strong base	Weak base vs. strong acid
1. Start:	Weak acid solution alone	Weak base solution alone
pH:	Section 17-2	Section 17-3
2. Partial conversion to salt:	Weak acid partially converted to salt; weak acid and salt together form a buffer solution against strong base	Weak base partially converted to salt; weak base and salt together form a buffer solution against strong acid
pH:	Section 17-6	Section 17-6
3. Equivalence point:	Solution of the salt of the weak acid and strong base	Solution of the salt of the weak base and strong acid
pH:	Section 17-4	Section 17-5
4. Beyond equivalence point:	Dilution of strong base; weak acid salt has no effect	Dilution of strong acid; weak base salt has no effect
pH:	Section 17-1	Section 17-1

EXAMPLE 17-18: A solution of 10.0 mL of 0.050 M nitrous acid (HNO$_2$: $K_a = 4.6 \times 10^{-4}$; $pK_a = 3.34$) is titrated with 0.050 M KOH. Calculate the pH **(a)** at the start of the titration, **(b)** after 5.0 mL of KOH has been added, **(c)** at the equivalence point, and **(d)** after 10.5 mL (total) of KOH has been added.

Solution:

(a) At the start (Stage 1) of this titration, the pH of the solution is that of HNO$_2$ alone, so we use the technique for calculating the pH of a weak acid solution. Thus

$$K_a = \frac{[H^+][NO_2^-]}{[HNO_2]} = 4.6 \times 10^{-4}$$

The charge balance, with the normal approximation, is $[H^+] \cong [NO_2^-]$. The mass balance is $[HNO_2] + [NO_2^-] = 0.050 \, M$. If $[NO_2^-]$ is small, we can approximate that $[HNO_2] \cong 0.050 \, M$, and so

$$\frac{[H^+]^2}{0.050} = 4.6 \times 10^{-4} \quad \text{and} \quad [H^+] = 4.8 \times 10^{-3}$$

Then the extent of ionization is

$$\frac{4.8 \times 10^{-3}}{0.050} \times 100\% = 9.6\%$$

This percentage is too large, so we need the quadratic: If $[HNO_2] = 0.050 - [NO_2^-] = 0.050 - [H^+]$, then

$$\frac{[H^+]^2}{0.050 - [H^+]} = 4.6 \times 10^{-4}$$

and

$$[H^+]^2 + (4.6 \times 10^{-4})[H^+] + (-2.3 \times 10^{-5}) = 0$$

so

$$[H^+] = \frac{-4.6 \times 10^{-4} \pm \sqrt{(4.6 \times 10^{-4})^2 - (4)(1)(-2.3 \times 10^{-5})}}{2}$$

Only the positive root makes chemical sense, so $[H^+] = 4.6 \times 10^{-3}$ and pH = $-(\log 4.6 \times 10^{-3}) = 2.34$.

(b) The addition of 5.0 mL of KOH represents Stage 2 of the titration, so we need to consider the titration reaction

$$HNO_2 + OH^- \longrightarrow H_2O + NO_2^-$$

which we know goes to completion, because weak acids react completely with strong bases. Then we use the reaction equation to calculate the stoichiometric concentrations of HNO_2 (HA) and NO_2^- (A$^-$) after the addition of 5.0 mL of KOH. In the initial solution there was (10.0 mL)(0.050 mol/L) = 0.50 mmol of HNO_2. The amount of OH^- added is (5.0 mL)(0.050 mol/L) = 0.25 mmol OH^-, which reacts with 0.25 mmol of HNO_2 to give 0.25 mmol of NO_2^-. The total volume of the solution after the addition of 5.0 mL of KOH is 10.0 mL + 5.0 mL = 15 mL. So the concentrations are

$$[HNO_2] = \frac{(0.50 - 0.25) \text{ mmol}}{15.0 \text{ mL}} = 1.7 \times 10^{-2} \ M$$

$$[NO_2^-] = \frac{0.25 \text{ mmol}}{15.0 \text{ mL}} = 1.7 \times 10^{-2} \ M$$

Now we know that the solution contains both the weak acid and its salt and is therefore a buffer solution. So we use the technique for calculating the pH of a buffer, the Henderson-Hasselbach equation (17-8):

$$pH = pK_a + \log\left(\frac{[A^-]}{[HA]}\right)$$

substituting in the calculated concentrations of NO_2^- and HNO_2:

$$pH = pK_a + \log\left(\frac{1.7 \times 10^{-2}}{1.7 \times 10^{-2}}\right)$$

$$= pK_a + \log 1.00$$

Then, since $\log 1.00 = 0$, we can write

$$pH = pK_a = 3.34$$

note: This is a general result: *At the* **half-equivalence point**, *i.e., when* $[A^-] = [HA]$, *the pH is equal to the* pK_a.

(c) At the equivalence point, or Stage 3 of the titration, exactly 0.50 mmol of KOH has been added to the initial 0.50 mmol of HNO_2. Thus the volume of KOH added is (0.50 mmol)/(0.050 mol/L) = 10.0 mL and the total solution volume must be 10.0 mL + 10.0 mL = 20.0 mL. This solution now contains $K^+NO_2^-$, a salt that is a weak base (the conjugate base of HNO_2), so we must use the technique for calculating the pH of a solution of a salt of a weak acid.

Since for a conjugate acid–base pair, $K_a \times K_b = K_w$ (eq. 17-6),

$$K_b = \frac{K_w}{K_a} = \frac{1.00 \times 10^{-14}}{4.6 \times 10^{-4}} = 2.2 \times 10^{-11}$$

The reaction taking place is

$$NO_2^- + H_2O \rightleftharpoons HNO_2 + OH^-$$

so

$$K_b = \frac{[HNO_2][OH^-]}{[NO_2^-]} = 2.2 \times 10^{-11}$$

Now we need to find the concentration values for the K_b equation. By the stoichiometry, $[HNO_2] = [OH^-]$. Then we know that the $K^+NO_2^-$ concentration is $(0.050\text{ mmol})/(20.0\text{ mL}) = 0.025\ M$, so we assume that little reaction is taking place and thus $[NO_2^-] \cong 0.025\ M$ (an assumption that will have to be checked). Thus

$$K_b = \frac{[OH^-]^2}{0.025} = 2.2 \times 10^{-11} \quad \text{and} \quad [OH^-] = 7.4 \times 10^{-7}$$

The extent of reaction is $(7.4 \times 10^{-7}/0.025) \times 100\% = 3 \times 10^{-3}\%$, so the approximation is good. Thus

$$pOH = -\log(7.4 \times 10^{-7}) = 6.13$$
$$pH = 14.00 - 6.13 = 7.87$$

(d) When a total volume of 10.5 mL of KOH is added, the resulting solution has gone beyond the equivalence point (Stage 4 of the titration). The volume of base OH^- at the equivalence point was 10.0 mL, so the amount of base in excess of one equivalent is $(10.5\text{ mL} - 10.0\text{ mL}) \times (0.050\ M\ OH^-) = 0.025$ mmol OH^-. The total volume of solution at this point is $10.0\text{ mL} + 10.5\text{ mL} = 20.5\text{ mL}$.

Then the calculation of the pH is simple. Since the presence of the weak base NO_2^- will have negligible effects compared to the effect of the strong base OH^-, we can write directly

$$[OH^-] = \frac{0.025\text{ mmol}}{20.5\text{ mL}} = 1.2 \times 10^{-3}\ M$$

$$pOH = -\log(1.2 \times 10^{-3}) = 2.92$$
$$pH = 14.00 - 2.92 = 11.08$$

17-8. Polyprotic Acids

Acids that have more than one ionizable hydrogen per molecule are called **polyprotic acids**, of which some familiar examples are sulfuric acid (H_2SO_4), carbonic acid (H_2CO_3), and phosphoric acid (H_3PO_4). Polyprotic acids ionize in a stepwise fashion—one step for each hydrogen—and for each ionization step a K_a can be written. For example, for H_2CO_3 there are two K_a expressions:

$$\text{(1)} \quad H_2CO_3 \rightleftharpoons H^+ + HCO_3^- \qquad K_{a_1} = \frac{[H^+][HCO_3^-]}{[H_2CO_3]} = 4.3 \times 10^{-7}$$

$$\text{(2)} \quad HCO_3^- \rightleftharpoons H^+ + CO_3^{2-} \qquad K_{a_2} = \frac{[H^+][CO_3^{2-}]}{[HCO_3^-]} = 5.6 \times 10^{-11}$$

For most common polyprotic acids there's a large difference (about 10^4 or 10^5) between successive K_a values (see Table 17-3). That means that simplifying assumptions can usually be made in calculating concentrations for polyprotic acid solutions.

TABLE 17-3: Ionization Constants K_a for Polyprotic Acids in H_2O at 25°C

Species ionizing	Formula	K_a	
Carbonic acid	H_2CO_3	4.3×10^{-7}	(K_{a_1})
hydrogen carbonate	HCO_3^-	5.6×10^{-11}	(K_{a_2})
Phosphoric acid	H_3PO_4	7.52×10^{-3}	(K_{a_1})
dihydrogen phosphate	$H_2PO_4^-$	6.23×10^{-8}	(K_{a_2})
hydrogen phosphate	HPO_4^{2-}	2.2×10^{-13}	(K_{a_3})
Sulfurous acid	H_2SO_3	1.54×10^{-2}	(K_{a_1})
hydrogen sulfite	HSO_3^-	1.02×10^{-7}	(K_{a_2})

EXAMPLE 17-19: The concentration of H_2CO_3 in a saturated solution at 303 K under 1.0 atm of CO_2 is $0.048\ M$. Calculate the pH of this solution and the concentrations of HCO_3^- and CO_3^{2-} present

Solution: We start by ignoring the second ionization step and concentrating on the first ionization step:

$$K_{a_1} = \frac{[H^+][HCO_3^-]}{[H_2CO_3]} = 4.3 \times 10^{-7}$$

Making our usual assumptions on charge and mass balance, we have

$$[H^+] \cong [HCO_3^-] \quad \text{and} \quad [H_2CO_3] \cong 0.048$$

So

$$\frac{[H^+]^2}{0.048} = 4.3 \times 10^{-7} \quad \text{and} \quad [H^+] \cong 1.4 \times 10^{-4}$$

The extent of ionization is $(1.4 \times 10^{-4}/0.048) \times 100\% = 0.3\%$, so the assumptions are valid. Thus we have a preliminary answer of

$$[H^+] = 1.4 \times 10^{-4} = [HCO_3^-] \quad \blacktriangle$$

Now think about the second ionization:

$$K_{a_2} = \frac{[H^+][CO_3^{2-}]}{[HCO_3^-]} = 5.6 \times 10^{-11}$$

HCO_3^- has a low K_{a_2}, so very little of it ionizes. Consequently, $[HCO_3^-] \cong 1.4 \times 10^{-4}$, which is the concentration produced by the first ionization. And $[H^+] \cong 1.4 \times 10^{-4}$, also the concentration produced by the first ionization. So we substitute these approximate values into the K_{a_2} expression:

$$\frac{(1.4 \times 10^{-4})[CO_3^{2-}]}{1.4 \times 10^{-4}} = 5.6 \times 10^{-11}$$

and so

$$[CO_3^{2-}] = 5.6 \times 10^{-11} \quad \blacktriangle$$

The extent of the second ionization is $(5.6 \times 10^{-11}/1.4 \times 10^{-4}) \times 100\% = 4 \times 10^{-5}\%$, which is very small and confirms our approximations.

Now we can put it all together:

$$[H^+] \quad = 1.4 \times 10^{-4}\ M \quad \text{and} \quad pH = -\log(1.4 \times 10^{-4}) = 3.85$$
$$[HCO_3^-] = 1.4 \times 10^{-4}\ M$$
$$[CO_3^{2-}] \ = 5.6 \times 10^{-11}\ M$$

SUMMARY

1. The equilibrium constant K_a (K_b) for a weak acid (base) measures the strength of the acid (base):

$$pK_a = -\log K_a; \qquad pK_b = -\log K_b$$

2. The equilibrium constant K_w is the ion product:

$$K_w = [H^+][OH^-] = 1.0 \times 10^{-14}$$
$$pK_w = 14.00 = pH + pOH$$

3. Weak acid (base) equilibrium problems can be solved by employing the equilibrium-constant expressions and making simplifying approximations for the mass balance and the charge balance.
   (a) Simplifying approximations that lead to $\leqslant 5\%$ of ionization are valid;
   (b) When a simplifying approximation leads to $>5\%$ of ionization, the approximation for extent of ionization should be replaced by an exact quadratic equation (in one unknown).
4. (a) The anion A^- of a weak acid HA is a weak base; its $K_b = K_w/K_a$ of the acid.
   (b) The cation BH^+ of a weak base B is a weak acid; its $K_a = K_w/K_b$ of the base.
5. (a) The equivalence point in the titration of a weak acid with a strong base is basic.
   (b) The equivalence point in the titration of a weak base with a strong acid is acidic.
   (c) The equivalence point in the titration of a strong acid (base) with a strong base (acid) is neutral.

6. Mixtures of weak acids and their salts (or weak bases and their salts) are buffer solutions, which resist pH change when moderate amounts of acid or base are added to them.

7. **(a)** For an acid buffer

$$pH = pK_a + \log\left(\frac{[A^-]}{[HA]}\right)$$

(b) For a basic buffer

$$pOH = pK_b + \log\left(\frac{[BH^+]}{[B]}\right)$$

8. Weak acids (bases) react completely with strong bases (acids).

9. Polyprotic acids, which have more than one hydrogen, ionize in steps—one for each hydrogen. (Successive K_a values usually differ by $\sim 10^5$.)

RAISE YOUR GRADES

Can you ...?

☑ arrange weak acids in order of strength, given the K_a values
☑ arrange weak bases in order of strength, given the K_b values
☑ determine K_a and K_b from pK_a and pK_b, and vice versa
☑ determine pOH from pH, and vice versa

Can you calculate ...?

☑ pH in a given solution of a weak acid or base
☑ K_b for the conjugate base of a weak acid
☑ K_a for the conjugate acid of a weak base
☑ pH in a given solution of the salt of a weak acid or base
☑ pH in a buffer solution before and after the addition of strong acid or base
☑ pH at the various stages of different types of titration
☑ the concentrations present in solutions of polyprotic acids

SOLVED PROBLEMS

PROBLEM 17-1   Identify the Brønsted-Lowry conjugate acid–base pairs in the following equilibria:

(a) $C_2H_5NH_2 + H_2O \rightleftharpoons C_2H_5NH_3^+ + OH^-$
(b) $HN_3 + H_2O \rightleftharpoons H_3O^+ + N_3^-$
(c) $CH_3CO_2^- + H_2O \rightleftharpoons CH_3CO_2H + OH^-$

Solution: Recall the definitions: An acid is a proton donor; a base is a proton acceptor. The answers are therefore

(a) H_2O, acid; OH^-, conjugate base. $C_2H_5NH_2$, base; $C_2H_5NH_3^+$, conjugate acid.
(b) HN_3, acid; N_3^-, conjugate base. H_2O, base; H_3O^+, conjugate acid.
(c) H_2O, acid; OH^-, conjugate base. $CH_3CO_2^-$, base; CH_3CO_2H, conjugate acid.

PROBLEM 17-2   Hypobromous acid (HOBr) is a weak acid. When 0.10 mol of the acid is dissolved in water and diluted to 1.00 L, the resulting solution is found to have a pH of 4.84. Calculate K_a and pK_a for HOBr.

Solution: The ionization reaction and K_a expression of the weak acid HOBr (a proton donor) must be

$$HOBr \rightleftharpoons H^+ + OBr^- \qquad K_a = \frac{[H^+][OBr^-]}{[HOBr]}$$

Tables 17-1, 17-2, and 17-3 are reproduced here for your convenience.

TABLE 17-1: Ionization Constants K_a for Monoprotic Acids in H_2O at 25°C

Acid	Formula	K_a
Acetic	CH_3CO_2H	1.75×10^{-5}
Benzoic	$C_6H_5CO_2H$	6.31×10^{-5}
Formic	HCO_2H	1.77×10^{-4}
Hydrocyanic	HCN	4.93×10^{-10}
Hydrofluoric	HF	6.8×10^{-4}
Hypobromous	$HOBr$	2.06×10^{-9}
Hypochlorous	$HOCl$	2.95×10^{-8}
Hypoiodous	HOI	1.0×10^{-10}
Nitrous	HNO_2	4.6×10^{-4}
Phenol	C_6H_5OH	1.28×10^{-10}
Trifluoroacetic	CF_3CO_2H	5.9×10^{-1}

TABLE 17-2: Base Ionization Constants K_b in H_2O at 25°C

Base	Formula	K_b
Ammonia	NH_3	1.79×10^{-5}
Aniline	$C_6H_5NH_2$	4.27×10^{-10}
Benzylamine	$C_6H_5CH_2NH_2$	2.14×10^{-5}
Dimethylamine	$(CH_3)_2NH$	5.40×10^{-4}
Ethylamine	$C_2H_5NH_2$	6.41×10^{-4}
Hydrazine	H_2NNH_2	1.7×10^{-6}
Hydroxylamine	$HONH_2$	1.07×10^{-8}
Methylamine	CH_3NH_2	4.4×10^{-4}
Pyridine	C_5H_5N	1.7×10^{-9}

TABLE 17-3: Ionization Constants K_a for Polyprotic Acids in H_2O at 25°C

Species ionizing	Formula	K_a	
Carbonic acid	H_2CO_3	4.3×10^{-7}	(K_{a_1})
hydrogen carbonate	HCO_3^-	5.6×10^{-11}	(K_{a_2})
Phosphoric acid	H_3PO_4	7.52×10^{-3}	(K_{a_1})
dihydrogen phosphate	$H_2PO_4^-$	6.23×10^{-8}	(K_{a_2})
hydrogen phosphate	HPO_4^{2-}	2.2×10^{-13}	(K_{a_3})
Sulfurous acid	H_2SO_3	1.54×10^{-2}	(K_{a_1})
hydrogen sulfite	HSO_3^-	1.02×10^{-7}	(K_{a_2})

Having written the reaction equation and the K_a expression, you set up your data for substitution into the equilibrium equation. The pH gives you the value of $[H^+]$ at once:

$$pH = -\log[H^+] = 4.84; \qquad \log[H^+] = -4.84$$
$$[H^+] = \text{antilog}(-4.84) = 1.45 \times 10^{-5}$$

Then by the stoichiometry of the ionization reaction, $[OBr^-] = [H^+]$, so $[H^+] = [OBr^-] = 1.45 \times 10^{-5}$. (Notice that you can neglect the contribution of the ionization of water to $[H^+]$. This contribution will be less than 1.0×10^{-7} because the self-ionization of water to give H^+ is suppressed by the addition of an acid. Thus according to Le Chatelier's principle, this neglect is justifiable.) Next you tackle $[HOBr]$:

$$HOBr = 0.10 \text{ mol (initial amount)} - 1.45 \times 10^{-5} \text{ mol (amount ionized)} \cong 0.10 \text{ mol}$$

Obviously, the amount of ionization is negligible.

Finally, you substitute the values you've calculated into the equilibrium expression to get K_a:

$$K_a = \frac{[H^+][OBr^-]}{[HOBr]} = \frac{(1.45 \times 10^{-5})(1.45 \times 10^{-5})}{0.10} = 2.1 \times 10^{-9} \qquad \text{(units are } M \text{ but are usually left out)}$$

and calculate pK_a:

$$pK_a = -\log K_a = -\log(2.1 \times 10^{-9}) = -(-8.68) = 8.68$$

PROBLEM 17-3   Which is the strongest acid listed in Table 17-3?

Solution: The value of K_a measures the strength of an acid. The largest value of K_a in Table 17-3 is listed for sulfurous acid, H_2SO_3 ($K_a = 1.54 \times 10^{-2}$), which is thus the strongest acid listed.

PROBLEM 17-4   Which is the weakest base listed in Table 17-2?

Solution: The smaller the value of K_b, the weaker the base. The smallest value of K_b in Table 17-2 is listed for aniline, $C_6H_5NH_2$ ($K_b = 4.27 \times 10^{-10}$), which is therefore the weakest base listed.

PROBLEM 17-5   What is the OH^- concentration in a solution of pH 5.50?

Solution: You know that pH + pOH – 14.00, so

$$pOH = 14.00 - pH = 14.00 - 5.50 = 8.50$$

Then

$$pOH = -\log[OH^-] = 8.50$$
$$\log[OH^-] = -8.50$$
$$[OH^-] = \text{antilog}(-8.50) = 3.2 \times 10^{-9} M$$

(Notice how the units creep into the last step; the units of aqueous equilibrium concentration problems in this book are always M.)

PROBLEM 17-6   A solution is prepared by dissolving 0.10 mol of formic acid (HCO_2H) in water and diluting it until the total volume is 500.0 mL. Write down all the equations that would enable you, in principle, to determine the concentrations at equilibrium of HCO_2H, HCO_2^-, H^+, and OH^-. [Hint: Check Table 17-1 for K_a values.]

Solution: The equations are the two equilibrium-constant expressions K_a and K_w, the mass balance, and the charge balance:
   K_a: The reaction equation is $HCO_2H \rightleftharpoons H^+ + HCO_2^-$, so

$$K_a = \frac{[H^+][HCO_2^-]}{[HCO_2H]} = 1.77 \times 10^{-4} \qquad \text{(Table 17-1)} \tag{1}$$

   K_w: The ion product for water is $[H^+][OH^-]$, so

$$K_w = [H^+][OH^-] = 1.00 \times 10^{-14} \tag{2}$$

Mass balance: The stoichiometric concentration of formic acid put into solution is

$$\frac{0.10 \text{ mol}}{500.0 \text{ mL}} \times \frac{10^3 \text{ mL}}{L} = 0.20 M$$

so

$$[HCO_2H] + [HCO_2^-] = 0.20 \tag{3}$$

Charge balance: There are only three charged species, so

$$[H^+] = [HCO_2^-] + [OH^-] \tag{4}$$

You now have four equations to determine four unknowns.

PROBLEM 17-7   Use the equations from Problem 17-6 to calculate the pH of the formic acid solution.

Solution: You'll need to simplify the four equations by making some approximations. First assume that, since the solution will be quite acidic, $[H^+] \gg [OH^-]$. Consequently, in the charge balance equation, $[H^+] \cong [HCO_2^-]$.

Then remember that formic acid is a weak acid and that not much of it will be ionized, so your next approximation is in the mass balance equation. Since $[HCO_2^-]$ is small, $[HCO_2H] \cong 0.20$. Now you substitute into the K_a expression:

$$K_a = \frac{[H^+][HCO_2^-]}{[HCO_2H]} \cong \frac{[H^+]^2}{0.20} = 1.77 \times 10^{-4}$$

and calculate $[H^+]$

$$[H^+]^2 = 3.54 \times 10^{-5}$$
$$[H^+] = 5.95 \times 10^{-3}$$

Check: Certainly the charge balance approximation is justified:

$$[OH^-] = \frac{1.00 \times 10^{-14}}{[H^+]} = \frac{1.00 \times 10^{-14}}{5.95 \times 10^{-3}} = 1.7 \times 10^{-12}$$

so $[OH^-]$ is quite negligible compared with 5.95×10^{-3}. The extent of ionization is

$$\frac{5.95 \times 10^{-3}}{0.20} \times 100\% = 2.98\%$$

This is within the allowable guideline of 5%; so the mass-balance approximation is also acceptable. Now finish the problem by calculating the pH:

$$pH = -\log[H^+] = -\log(5.95 \times 10^{-3}) = 2.23$$

PROBLEM 17-8   Determine the pH of a 0.020 M formic acid solution.

Solution:

$$K_a = \frac{[H^+][HCO_2^-]}{[HCO_2H]} = 1.77 \times 10^{-4}$$

The mass balance equation is $[HCO_2H] + [HCO_2^-] = 0.020$. If you make exactly the same assumptions that you made in Problem 17-7, you'll get $[H^+] \cong [HCO_2^-]$ (from the charge balance) and $[HCO_2H] \cong 0.020$ (from the mass balance). So from K_a

$$\frac{[H^+]^2}{0.020} = 1.77 \times 10^{-4}$$

$$[H^+]^2 = 3.54 \times 10^{-6}$$
$$[H^+] = 1.88 \times 10^{-3}$$

Check: Extent of ionization

$$\frac{1.88 \times 10^{-3}}{0.020} \times 100\% = 9.4\%$$

This is larger than the 5% guideline, so you have to reconsider your approximations. The charge balance approximation is still good, but the mass balance approximation won't work because there's too much ionization.

Obviously, you need an exact method now, so rework the mass balance in one unknown: If $[H^+] \cong [HCO_2^-]$, then

$$[HCO_2H] = 0.020 - [HCO_2^-] = 0.020 - [H^+]$$

and set up a quadratic equation

$$\frac{[H^+]^2}{0.020 - [H^+]} = 1.77 \times 10^{-4}$$

which you rearrange into standard form

$$[H^+]^2 + (1.77 \times 10^{-4})[H^+] + (-3.54 \times 10^{-6}) = 0$$

and solve:

$$[H^+] = \frac{-1.77 \times 10^{-4} \pm \sqrt{(1.77 \times 10^{-4})^2 - 4(1)(-3.54 \times 10^{-6})}}{2}$$

$$= \frac{-1.77 \times 10^{-4} \pm \sqrt{3.133 \times 10^{-8} + 1.416 \times 10^{-5}}}{2}$$

$$= \frac{-1.77 \times 10^{-4} \pm \sqrt{1.419 \times 10^{-5}}}{2}$$

$$= \frac{-1.77 \times 10^{-4} \pm 3.77 \times 10^{-3}}{2}$$

$$= 1.80 \times 10^{-3} \quad \text{or} \quad -1.97 \times 10^{-3}$$

Only the positive root makes chemical sense, so

$$[H^+] = 1.80 \times 10^{-3} \quad \text{and} \quad pH = -\log(1.80 \times 10^{-3}) = 2.74.$$

PROBLEM 17-9 A 25.0-mmol sample of a new monoprotic acid HA was diluted with water to a volume of 100.0 mL. The pH of the solution was found to be 1.96. Calculate K_a for HA.

Solution:

$$K_a = \frac{[H^+][A^-]}{[HA]}$$

You know that the pH gives you the value of $[H^+]$ and that the values of $[A^-]$ and $[HA]$ can be derived from the charge and mass balance equations, so you make the following assumptions and calculations:
   From pH = 1.96, you get $\log[H^+] = -1.96$ and $[H^+] = 1.1 \times 10^{-2}$.
   The charge balance is $[H^+] = [A^-] + [OH^-]$, but in this case you know from the $[H^+]$ what $[OH^-]$ is: It's about 1×10^{-12} and quite negligible. So $[H^+] \cong [A^-]$.
   The mass balance is $[HA] + [A^-] = (25.0 \text{ mmol})/(100.0 \text{ mL}) = 0.250$.
Now use your calculations: If $[A^-] = [H^+] = 1.1 \times 10^{-2}$, then

$$[HA] = 0.250 - 1.1 \times 10^{-2} = 0.239$$

thus

$$K_a = \frac{[1.1 \times 10^{-2}]^2}{0.239} = 5.1 \times 10^{-4}$$

PROBLEM 17-10 Determine the pH of a 0.030 M solution of the weak base pyridine.

Solution: In Table 17-2 you see that pyridine (C_5H_5N) has a K_b of 1.7×10^{-9}. Its ionization reaction and equilibrium expression will be

$$C_5H_5N + H_2O \rightleftharpoons C_5H_5NH^+ + OH^- \qquad K_b = \frac{[C_5H_5NH^+][OH^-]}{[C_5H_5N]} = 1.7 \times 10^{-9}$$

   Charge balance: $[C_5H_5NH^+] + [H^+] = [OH^-]$. Assume $[H^+]$ is negligible compared with $[OH^-]$ (it's a basic solution). Then $[C_5H_5NH^+] \cong [OH^-]$.
   Mass balance: $[C_5H_5N] + [C_5H_5NH^+] = 0.030$. Assuming little ionization, $[C_5H_5N] \cong 0.030$.
Substitute:

$$K_b = \frac{[OH^-]^2}{0.030} = 1.7 \times 10^{-9}; \qquad [OH^-] = 7.1 \times 10^{-6}$$

Check: Extent of ionization: $\dfrac{7.1 \times 10^{-6}}{0.030} \times 100\% = 0.02\%$. The assumptions are valid: $[OH^-] = 7.1 \times 10^{-6}$.

Calculate the pH:

$$pOH = -\log(7.1 \times 10^{-6}) = 5.15$$
$$pH = 14.00 - 5.15 = 8.85$$

PROBLEM 17-11   You are working with the weak base ethylamine. **(a)** At what concentration will ethylamine be 25% ionized? **(b)** What will the solution pH be at that concentration?

Solution: From Table 17-2 you see that ethylamine ($C_2H_5NH_2$) has $K_b = 6.41 \times 10^{-4}$. So

$$K_b = \frac{[C_2H_5NH_3^+][OH^-]}{[C_2H_5NH_2]} = 6.41 \times 10^{-4}$$

(a) Let the concentration be c.

If the ethylamine is 25% ionized to $C_2H_5NH_3^+$, then 75% will be $C_2H_5NH_2$. Therefore

$$\frac{[C_2H_5NH_3^+]}{[C_2H_5NH_2]} = \frac{25}{75} \quad \text{and} \quad [C_2H_5NH_3^+] + [C_2H_5NH_2] = c$$

Thus

$$[C_2H_5NH_3^+] = 0.25c \quad \text{and} \quad [C_2H_5NH_2] = 0.75c$$

From the charge balance condition, $[C_2H_5NH_3^+] + [H^+] = [OH^-]$. But $[H^+]$ will be negligible compared with $[OH^-]$, so

$$[C_2H_5NH_3^+] = [OH^-] = 0.25c$$

Substituting:

$$K_b = \frac{(0.25\cancel{c})(0.25c)}{0.75\cancel{c}} = 6.41 \times 10^{-4} \quad \text{and} \quad c = 7.7 \times 10^{-3}$$

So the required concentration is 7.7×10^{-3} M.

(b) At that concentration, $[OH^-] = 0.25c = 1.93 \times 10^{-3}$. Thus

$$pOH = -\log(1.93 \times 10^{-3}) = 2.72$$
$$pH = 14.00 - 2.72 = 11.28$$

PROBLEM 17-12   Will the ion HCO_2^- be an acid or a base in water? Calculate its ionization constant.

Solution: Since HCO_2^- is an anion (negatively charged), it's most likely to be the conjugate base of a Brønsted-Lowry acid, which you can look up. Sure enough, HCO_2H (formic acid) is listed in Table 17-1 as a monoprotic acid whose K_a is 1.77×10^{-4}. To find the ionization constant, you use eq. (17-6) for a conjugate acid–base pair: $K_aK_b = K_w$. Thus

$$K_b = \frac{K_w}{K_a} = \frac{1.00 \times 10^{-14}}{1.77 \times 10^{-4}} = 5.65 \times 10^{-11}$$

PROBLEM 17-13   List the following ions in order of increasing base strength: $CH_3CO_2^-$, HCO_2^-, $C_6H_5O^-$.

Solution: Equation (17-6) shows the reciprocal relationship between K_a and K_b for conjugate acid–base pairs. The stronger the acid, the weaker the base, and vice versa. The order of acid strengths, or K_a values, from Table 17-1 is therefore $HCO_2H > CH_3CO_2H > C_6H_5OH$. So the order of base strengths of the conjugate bases is just the reverse: $HCO_2^- < CH_3CO_2^- < C_6H_5O^-$.

PROBLEM 17-14   Calculate the pH of a 0.25 M solution of sodium hypochlorite (Na^+OCl^-).

Solution: The sodium ion Na^+ has no acid or base properties in water, but the hypochlorite ion OCl^- is the anion of a weak acid HOCl and is therefore a base. K_a for HOCl (from Table 17-1) $= 2.95 \times 10^{-8}$, and so

$$.K_b \text{ for } OCl^- = \frac{K_w}{K_a \text{ for } HOCl} = \frac{1.00 \times 10^{-14}}{2.95 \times 10^{-8}} = 3.39 \times 10^{-7}$$

Then the reaction and equilibrium equation are

$$OCl^- + H_2O \rightleftharpoons HOCl + OH^- \quad \text{and} \quad K_b = \frac{[HOCl][OH^-]}{[OCl^-]} = 3.39 \times 10^{-7}$$

By the stoichiometry, of the reaction, the charge balance is $[HOCl] \cong [OH^-]$ (assuming a negligible contribution to $[OH^-]$ from H_2O). The mass balance on hypochlorite is $[HOCl] + [OCl^-] = 0.25$; but since the base is weak, you can try the approximation that $[OCl^-] \cong 0.25$. Then

$$\frac{[HOCl][OH^-]}{[OCl^-]} = \frac{[OH^-]^2}{0.25} = 3.39 \times 10^{-7}$$

$$[OH^-]^2 = 8.47 \times 10^{-8}$$

$$[OH^-] = 2.9 \times 10^{-4}$$

Check: The extent of OCl^- reaction is $\dfrac{2.9 \times 10^{-4}}{0.25} \times 100\% = 0.1\%$, way under the 5% guideline, so the approximations are justified.

Since $[OH^-] = 2.9 \times 10^{-4}$,

$$pOH = -(\log 2.9 \times 10^{-4}) = 3.54$$

$$pH = 14.00 - 3.54 = 10.46$$

PROBLEM 17-15 Aniline ($C_6H_5NH_2$), a weak base, forms a salt $C_6H_5NH_3^+Cl^-$ with HCl. Determine the pH of a 0.10 M solution of this salt.

Solution: As the conjugate acid of a weak base, $C_6H_5NH_3^+$ will be a Brønsted-Lowry acid:

$$C_6H_5NH_3^+ \rightleftharpoons C_6H_5NH_2 + H^+$$

$$K_a \text{ for } C_6H_5NH_3^+ = \frac{K_w}{K_b \text{ for } C_6H_5NH_2} = \frac{1.00 \times 10^{-14}}{4.27 \times 10^{-10}} = 2.34 \times 10^{-5} \qquad (K_b \text{ from Table 17-2})$$

By the reaction stoichiometry, $[C_6H_5NH_2] = [H^+]$ (neglecting any small contribution to $[H^+]$ from ionization of H_2O). And since $C_6H_5NH_3^+$ is a weak acid, you can also approximate in the mass balance equation: $[C_6H_5NH_3^+] + [C_6H_5NH_2] = 0.10$ by assuming $[C_6H_5NH_3^+] \cong 0.10$. Therefore

$$K_a = \frac{[C_6H_5NH_3^+][H^+]}{[C_6H_5NH_3^+]} = 2.34 \times 10^{-5}$$

$$\cong \frac{[H^+]^2}{0.10} = 2.34 \times 10^{-5}$$

so

$$[H^+] = \sqrt{2.34 \times 10^{-6}} \qquad \text{and} \qquad [H^+] = 1.5 \times 10^{-3}$$

Check: The extent of reaction is $\dfrac{1.5 \times 10^{-3}}{0.10} \times 100\% = 1.5\%$. This is less than the 5% guideline, so the approximations are valid.

Thus

$$pH = -\log(1.5 \times 10^{-3}) = 2.82$$

PROBLEM 17-16 What is the pH of a buffer solution that is 0.50 M in acetic acid and 0.30 M in sodium acetate?

Solution: The Henderson-Hasselbach equation (17-8) shows that, for a buffer made from a weak acid and its anion, the pH is

$$pH = pK_a + \log\left(\frac{[A^-]}{[HA]}\right)$$

For acetic acid $K_a = 1.75 \times 10^{-5}$, and so $pK_a = -\log K_a = 4.76$. In this buffer solution $[HA]$ is given as 0.50 M acetic acid, and $[A^-]$ is 0.30 M sodium acetate. Thus

$$pH = 4.76 + \log\left(\frac{0.30}{0.50}\right) = 4.76 - 0.22 = 4.54$$

PROBLEM 17-17 To a 100.0-mL portion of the buffer in Problem 17-16, you've added 15.0 mL of 0.10 M HCl. Calculate the pH of the mixture.

Solution: The added HCl (a strong acid) will react completely with acetate ion to convert it into acetic acid:

$$H^+ + CH_3CO_2^- \longrightarrow CH_3CO_2H$$

The 100.0 mL of buffer contains

$$(100.0 \text{ mL})(0.50 \text{ M acetic acid}) = 50.0 \text{ mmol acetic acid}$$

and

$$(100.0 \text{ mL})(0.30 \text{ M acetate}) = 30.0 \text{ mmol acetate ion}$$

The amount of H^+ added is

$$(15.0 \text{ mL})\left(\frac{1 \text{ mol } H^+}{1 \text{ mol HCl}}\right)(0.10 \text{ M HCl}) = 1.50 \text{ mmol } H^+$$

From the reaction equation, 1.50 mmol H^+ will react with 1.50 mmol $CH_3CO_2^-$ to produce 1.50 mmol of CH_3CO_2H. The new total solution volume will be 100.0 mL + 15.0 mL = 115.0 mL. Then the new $CH_3CO_2^-$ concentration will be

$$[CH_3CO_2^-] = \frac{(30.0 \text{ mmol original acetate}) - (1.50 \text{ mmol reacted})}{115.0 \text{ mL}} = 0.248 \text{ M}$$

The new CH_3CO_2H concentration will be

$$[CH_3CO_2H] = \frac{(50.0 \text{ mmol original acetic acid}) + (1.5 \text{ mmol from reaction})}{115.0 \text{ mL}} = 0.448 \text{ M}$$

So

$$pH = pK_a + \log\left(\frac{[A^-]}{[HA]}\right) = 4.76 + \log\left(\frac{0.248}{0.448}\right) = 4.76 - 0.26 = 4.50$$

(Notice that the drop in pH for this substantial addition of acid is only 0.04 units.)

PROBLEM 17-18   Equal volumes of 0.25 M pyridine (C_5H_5N) solution and 0.20 M pyridine hydrochloride ($C_5H_5NH^+Cl^-$) solution are mixed to make 100.0 mL of solution. Calculate the solution pH **(a)** after mixing and **(b)** after 10.0 mL of 0.02 M NaOH solution is added to it.

Solution: Pyridine has $K_b = 1.7 \times 10^{-9}$ (from Table 17-2) and $pK_b = -\log(1.7 \times 10^{-9}) = 8.77$

(a) After the mixing

$$[C_5H_5N] = \frac{0.25 \text{ M}}{2} = 0.125 \text{ M} \quad \text{and} \quad [C_5H_5NH^+] = \frac{0.20 \text{ M}}{2} = 0.10 \text{ M}$$

From eq. (17-10)

$$pOH = pK_b + \log\left(\frac{[BH^+]}{[B]}\right)$$

$$= 8.77 + \log\left(\frac{0.10}{0.125}\right) = 8.77 - 0.10 = 8.67$$

$$pH = 14.00 - 8.67 = 5.33$$

(Notice that this very weak base buffers in the acid region; that's because its conjugate acid is a fairly "strong" weak acid.)

(b) The reaction of Na^+OH^- with $C_5H_5NH^+$ goes to completion:

$$OH^- + C_5H_5NH^+ \dashrightarrow H_2O + C_5H_5N$$

Before the NaOH is added, the buffer solution contained

$$(100.0 \text{ mL})(0.125 \text{ M } C_5H_5N) = 12.5 \text{ mmol } C_5H_5N$$
$$(100.0 \text{ mL})(0.10 \text{ M } C_5H_5NH^+) = 10.0 \text{ mmol } C_5H_5NH^+$$

The amount of OH^- added is (10.0 mL)(0.02 M NaOH) = 0.2 mmol OH^-, which reacts with 0.2 mmol

$C_5H_5NH^+$ to produce 0.2 mmol C_5H_5N. The new total volume = 100.0 mL + 10.0 mL = 110.0 mL. Then

$$[C_5H_5N] = \frac{12.5 \text{ mmol} + 0.2 \text{ mmol}}{110.0 \text{ mL}} = 0.115 \ M$$

$$[C_5H_5NH^+] = \frac{10.0 \text{ mmol} - 0.2 \text{ mmol}}{110.0 \text{ mL}} = 0.089 \ M$$

So

$$pOH = pK_b + \log\left(\frac{[BH^+]}{[B]}\right)$$

$$= 8.77 + \log\left(\frac{0.089}{0.115}\right) = 8.77 - 0.11 = 8.66$$

$$pH = 14.00 - 8.66 = 5.34$$

(Notice that the solution has become just detectably more basic—by only 0.01 pH units—after the addition of the strongly basic NaOH solution.)

PROBLEM 17-19 How many grams of solid sodium acetate ($Na^+CH_3CO_2^-$) would you need to add to 100.0 mL of 0.20 M acetic acid to make a buffer with pH = 5.00? (Assume that the solution volume does not change when the sodium acetate is added.)

Solution: According to eq. (17-8), pH = pK_a + log([A^-]/[HA]). In this problem the final pH is given as 5.00, and pK_a for acetic acid (for which $K_a = 1.75 \times 10^{-5}$) = 4.76. Therefore

$$5.00 = 4.76 + \log\left(\frac{[A^-]}{[HA]}\right)$$

$$\log\left(\frac{[A^-]}{[HA]}\right) = 5.00 - 4.76 = 0.24$$

Thus

$$\frac{[A^-]}{[HA]} = \text{antilog } 0.24 = 1.74 \quad \text{and} \quad [A^-] = 1.74[HA]$$

Now the acetic acid solution is 0.20 M, so you need

$$[A^-] = 1.74(0.20 \ M) = 0.35 \ M$$

and the number of moles of sodium acetate needed to make 100.0 mL of 0.35 M solution is

$$\left(\frac{100.0 \text{ mL}}{10^3 \text{ mL/L}}\right)(0.35 \ M) = 3.5 \times 10^{-2} \text{ mol}$$

Finally, since the molar mass of $Na^+CH_3CO_2^-$ = 82.0 g/mol, the mass of $Na^+CH_3CO_2^-$ needed is

$$(3.5 \times 10^{-2} \text{ mol})(82.0 \text{ g/mol}) = 2.9 \text{ g}$$

PROBLEM 17-20 Calculate the pH at the following points as 10.0 mL of a 0.24 M solution of methylamine (CH_3NH_2) is titrated with 0.17 M HCl: (a) at the start, (b) after 6.0 mL of HCl has been added, (c) at the equivalence point, (d) when an additional 0.3 mL of HCl solution has been added (beyond the equivalence point).

Solution: K_b for CH_3NH_2 is 4.4×10^{-4} (Table 17-2); $pK_b = -\log(4.4 \times 10^{-4}) = 3.36$.
  (a) Stage 1 of the titration: Find the pH of methylamine alone.

The ionization reaction and equilibrium equation are

$$CH_3NH_2 + H_2O \rightleftharpoons CH_3NH_3^+ + OH^- \qquad K_b = \frac{[CH_3NH_3^+][OH^-]}{[CH_3NH_2]}$$

Making the usual approximations for the charge and mass balance, you get

$$[CH_3NH_3^+] \cong [OH^-] \qquad \text{(charge balance)}$$

and

$$[CH_3NH_2] \cong 0.24 \ M \qquad \text{(ionization is slight; approximation to be checked)}$$

Then

$$\frac{[OH^-]^2}{0.24} = 4.4 \times 10^{-4} \qquad \text{and} \qquad [OH^-] = 1.0 \times 10^{-2}$$

Check: $\dfrac{1.0 \times 10^{-2}}{0.24} \times 100\% = 4.3\%$ ionization. This is within the 5% guideline, so the approximation is correct.

Finally,

$$pOH = -\log(1.0 \times 10^{-2}) = 2.00$$
$$pH = 14.00 - 2.00 = 12.00$$

(b) Stage 2 of the titration: Find the pH of the buffer solution.

The reaction between the weak base methylamine and the strong acid HCl goes to completion:

$$CH_3NH_2 + H^+ \longrightarrow CH_3NH_3^+$$

The original methylamine solution contains $(10.0 \ \text{mL})(0.24 \ M) = 2.4$ mmol CH_3NH_2. The added HCl contains $(6.0 \ \text{mL})(0.17 \ M) = 1.0$ mmol H^+, which reacts with 1.0 mmol of CH_3NH_2 to give 1.0 mmol of $CH_3NH_3^+$. The new total volume is $10.0 \ \text{mL} + 6.0 \ \text{mL} = 16.0$ mL. The solution now contains a weak base CH_3NH_2 and its salt $CH_3NH_3^+$, so it is indeed a buffer, and its pH may be calculated by eq. (17-10):

$$pOH = pK_b + \log\left(\frac{[BH^+]}{[B]}\right)$$

Thus if

$$pK_b = 3.36$$

$$[CH_3NH_3^+] = \frac{1.0 \ \text{mmol}}{16.0 \ \text{mL}} = 6.2 \times 10^{-2} \ M$$

$$[CH_3NH_2] = \frac{(2.4 - 1.0) \ \text{mmol}}{16.0 \ \text{mL}} = 8.8 \times 10^{-2} \ M$$

Then

$$pOH = 3.36 + \log\left(\frac{6.2 \times 10^{-2}}{8.8 \times 10^{-2}}\right)$$

$$= 3.36 - 0.15 = 3.21$$
$$pH = 14.00 - 3.21 = 10.79$$

(c) Stage 3 of the titration: Find the pH of the salt solution.

The equivalence point occurs when exactly 2.4 mmol of HCl has been added to the 2.4 mmol of CH_3NH_2.

$$\text{Volume of HCl needed} = \frac{2.4 \ \text{mmol}}{0.17 \ M} = 14 \ \text{mL}$$

$$\text{Volume of solution} = 10.0 \ \text{mL} + 14 \ \text{mL} = 24 \ \text{mL}$$

At this point, you have a solution of the salt $CH_3NH_3^+$, the conjugate acid of the weak base CH_3NH_2. The salt solution is itself a weak acid, and its concentration is

$$[CH_3NH_3^+] = \frac{2.4 \ \text{mmol}}{24 \ \text{mL}} = 0.10 \ M$$

Then

$$K_a \text{ for } CH_3NH_3^+ = \frac{K_w}{K_b \text{ for } CH_3NH_2} = \frac{1.00 \times 10^{-14}}{4.4 \times 10^{-4}}$$

$$= 2.3 \times 10^{-11}$$

and for the ionization reaction $CH_3NH_3^+ \rightleftharpoons H^+ + CH_3NH_2$

$$K_a = \frac{[H^+][CH_3NH_2]}{[CH_3NH_3^+]} = 2.3 \times 10^{-11}$$

Now you can calculate the pH using the approximation procedure: If

$$[H^+] \cong [CH_3NH_2] \quad \text{and} \quad [CH_3NH_3^+] \cong 0.10 \ M$$

then

$$\frac{[H^+]^2}{0.10} = 2.3 \times 10^{-11} \quad \text{and} \quad [H^+] = 1.5 \times 10^{-6}$$

Check: $\frac{1.5 \times 10^{-6}}{0.10} \times 100\% = 1.5 \times 10^{-3} \%$ ionization. The approximation is valid.

Thus

$$pH = -\log(1.5 \times 10^{-6}) = 5.82$$

(**d**) Stage 4 of the titration: Find the pH of the acid solution.

The strong acid HCl is now being added to the weak acid $CH_3NH_3^+$, which will not affect the action of the strong acid. Thus you only have to find the H^+ concentration directly:

$$\text{Excess } H^+ = (0.3 \ mL)(0.17 \ M) = 5 \times 10^{-2} \ mmol$$

$$\text{Total solution volume} = 24 \ mL + 0.3 \ mL = 24.3 \ mL$$

$$[H^+] = \frac{5 \times 10^{-2} \ mmol}{24.3 \ mL} = 2 \times 10^{-3} \ M$$

$$pH = -\log(2 \times 10^{-2}) = 2.7$$

PROBLEM 17-21   Calculate the pH of a 0.15 M solution of phosphoric acid (H_3PO_4) and find the concentrations of the ions $H_2PO_4^-$, HPO_4^{2-}, and PO_4^{3-}.

Solution: Obviously, H_3PO_4 is a polyprotic acid, whose three K_a values you must look up: $K_{a_1} = 7.52 \times 10^{-3}$, $K_{a_2} = 6.23 \times 10^{-8}$, $K_{a_3} = 2.2 \times 10^{-13}$ (from Table 17-3). Then you break the problem into three stages (one for each ionization) and calculate the required concentrations successively, using the values obtained in one stage as data for the next.

First ionization:

$$H_3PO_4 \rightleftharpoons H^+ + H_2PO_4^- \qquad K_{a_1} = \frac{[H^+][H_2PO_4^-]}{[H_3PO_4]} = 7.52 \times 10^{-3}$$

Make the usual approximations for a weak acid solution:
  Charge balance: $[H^+] \cong [H_2PO_4^-]$
  Mass balance: $[H_3PO_4] \cong 0.15 \ M$ (to be checked)
Substitute:

$$\frac{[H^+]^2}{0.15} = 7.52 \times 10^{-3}, \qquad [H^+] = 3.4 \times 10^{-2}$$

Check: $\frac{3.4 \times 10^{-2}}{0.15} \times 100\% = 22\%$ ionization. The mass balance approximation is invalid.

Rework the mass balance in one unknown:

$$[H_3PO_4] = 0.15 - [H_2PO_4^-] = 0.15 - [H^+]$$

And find the positive root for the quadratic:

$$\frac{[H^+]^2}{0.15 - [H^+]} = 7.52 \times 10^{-3}$$

$$[H^+]^2 + (7.52 \times 10^{-3})[H^+] + (-1.13 \times 10^{-3}) = 0$$

$$[H^+] = \frac{-7.52 \times 10^{-3} \pm \sqrt{(7.52 \times 10^{-3})^2 - (4)(1)(-1.13 \times 10^{-3})}}{2}$$

$$= 3.0 \times 10^{-2}$$

Now you can find the pH and $[H_2PO_4^-]$

$$pH = -\log(3.0 \times 10^{-2}) = 1.52 \quad \blacktriangle$$

and

$$[H_2PO_4^-] = [H^+] = 3.0 \times 10^{-2} \ M \quad \blacktriangle$$

Second ionization:

$$H_2PO_4^- \rightleftharpoons H^+ + HPO_4^{2-} \qquad K_{a_2} = \frac{[H^+][HPO_4^{2-}]}{[H_2PO_4^-]} = 6.23 \times 10^{-8}$$

Approximations: Since K_{a_2} is small, very little of the $H_2PO_4^-$ ionizes. So you can use the concentrations calculated in stage 1:

$$[H^+] = 3.0 \times 10^{-2} \quad \text{and} \quad [H_2PO_4^-] = 3.0 \times 10^{-2}$$

Substitute:

$$\frac{(3.0 \times 10^{-2})[HPO_4^{2-}]}{3.0 \times 10^{-2}} = 6.23 \times 10^{-8}$$

$$[HPO_4^{2-}] = 6.23 \times 10^{-8} \ M \quad \blacktriangle$$

Third ionization:

$$HPO_4^{2-} \rightleftharpoons H^+ + PO_4^{3-} \qquad K_{a_3} = \frac{[H^+][PO_4^{3-}]}{[HPO_4^{2-}]} = 2.2 \times 10^{-13}$$

Approximations: K_{a_3} is very small, so the calculations for H^+ and HPO_4^{2-} in stage 2 hold good:

$$[H^+] = 3.0 \times 10^{-2} \quad \text{and} \quad [HPO_4^{2-}] = 6.23 \times 10^{-8}$$

Substitute:

$$\frac{(3.0 \times 10^{-2})[PO_4^{3-}]}{6.23 \times 10^{-8}} = 2.2 \times 10^{-13}$$

$$[PO_4^{3-}] = 2.2 \times 10^{-13}\left(\frac{6.23 \times 10^{-8}}{3.0 \times 10^{-2}}\right) = 4.6 \times 10^{-19} \ M \quad \blacktriangle$$

In summary:

$$pH = 1.52, \quad [H_2PO_4^-] = 3.0 \times 10^{-2} \ M, \quad [HPO_4^{2-}] = 6.23 \times 10^{-8} \ M, \quad [PO_4^{3-}] = 4.6 \times 10^{-19} \ M.$$

PROBLEM 17-22 Calculate the pH of a solution that is 0.25 M in $Na^+H_2PO_4^-$ and 0.19 M in $(Na^+)_2HPO_4^{2-}$.

Solution: Analyze your data first to find out what you're dealing with. You should see that there is a connection between $H_2PO_4^-$ and HPO_4^{2-}. They differ by a proton, so HPO_4^{2-} is the conjugate base of $H_2PO_4^-$, which is a weak acid. And a solution containing a weak acid and its conjugate base is a buffer solution, so you have the clue you need:
   Look up the K_a value for $H_2PO_4^-$ as an acid, and calculate its pK_a (see Table 17-3):

$$K_a = 6.23 \times 10^{-8} \quad \text{and} \quad pK_a = -\log(6.23 \times 10^{-8}) = 7.21$$

Use eq. (17-8) for acid buffers:

$$pH = pK_a + \log\left(\frac{[A^-]}{[HA]}\right)$$

$$= 7.21 + \log\left(\frac{0.19}{0.25}\right) = 7.21 - 0.12$$

$$= 7.09$$

(Phosphate buffers of this kind are widely used in biochemistry.)

Supplementary Exercises

PROBLEM 17-23 The pH of a 0.14 M solution of butyric acid ($C_3H_7CO_2H$, abbreviated HBu) is 2.84. Calculate K_a for HBu.

Answer: 1.5×10^{-5}

PROBLEM 17-24 Find (a) the strongest base and (b) the second strongest base in Table 17-2.

Answer: (a) ethylamine (b) dimethylamine

PROBLEM 17-25 If a solution of acetic acid has a pH of 3.00, what is its concentration?

Answer: $5.7 \times 10^{-2} M$

PROBLEM 17-26 What is the hydrogen ion concentration in a solution for which (a) pOH = 0.10 and (b) pOH = 4.62?

Answer: (a) $1.3 \times 10^{-14} M$ (b) $4.2 \times 10^{-10} M$

PROBLEM 17-27 Determine the pH of a 0.15 M solution of the weak base hydrazine (H_2NNH_2).

Answer: pH = 10.70

PROBLEM 17-28 Hydroxylamine ($HONH_2$) is a weak base. (a) At what concentration will it be 1.0% ionized? (b) What will the pH of the solution in (a) be?

Answer: (a) $1.1 \times 10^{-4} M$ (b) pH = 8.04

PROBLEM 17-29 Calculate (a) the extent of ionization and (b) the pH of a $1.0 \times 10^{-3} M$ solution of benzylamine.

Answer: (a) 14% (b) pH = 10.13

PROBLEM 17-30 Calculate the pK_a of $H_2NNH_3^+$.

Answer: $pK_a = 8.23$

PROBLEM 17-31 Calculate the pH of a 0.16 M solution of hydrazine hydrochloride, $H_2NNH_3^+Cl^-$.

Answer: pH = 4.51

PROBLEM 17-32 When a 0.25 M solution of cyanic acid (HCNO; $K_a = 1.2 \times 10^{-4}$) is titrated with 0.25 M KOH, what is the pH at the equivalence point?

Answer: pH = 8.51

PROBLEM 17-33 (a) What is the pH of a buffer that is 0.50 M in acetic acid and 0.50 M in potassium acetate? (b) When 50.0 mL of the buffer in (a) is mixed with 150.0 mL of water, what is the pH of the mixture?

Answer: (a) pH = 4.76 (b) pH = 4.76

PROBLEM 17-34 A 10.0-mL portion of 0.10 M HCl is added to 55.0 mL of a buffer that is 0.25 M in formic acid and 0.15 M in sodium formate. Calculate the pH of the buffer (a) before the addition of HCl and (b) after its addition.

Answer: (a) pH = 3.53 (b) pH = 3.44

PROBLEM 17-35   Solid sodium hydroxide is added to 1.00 L of 0.100 M acetic acid solution until the pH is 4.38. What mass of NaOH was used?

Answer: 1.2 g

PROBLEM 17-36   (a) What is the pH of a 0.25 M methylamine ($K_b = 4.4 \times 10^{-4}$) solution? (b) What volume of 1.0 M HCl solution must be added to 50.0 mL of 0.25 M methylamine solution to get a buffer solution of pH = 10.50?

Answer: (a) pH = 12.02   (b) 7.3 mL

PROBLEM 17-37   Calculate the pH at the following points in the titration of 25.0 mL of a 5.0×10^{-2} M solution of NH_3 ($K_b = 1.79 \times 10^{-5}$) with a 5.0×10^{-2} M solution of HCl: (a) at the start in the NH_3 solution, (b) after 17.0 mL of the HCl solution has been added, (c) at the equivalence point, (d) after 25.1 mL of the HCl solution has been added.

Answer: (a) 10.98   (b) 8.93   (c) 5.43   (d) 4.0

PROBLEM 17-38   A solution is prepared by mixing 25.0 mL of 0.17 M acetic acid with 25.0 mL of 0.10 M KOH. Calculate the concentrations of all ionic species in the resulting solution.

Answer: $[H^+] = 1.2 \times 10^{-5}$ M, $[OH^-] = 8.2 \times 10^{-10}$ M, $[K^+] = 5.0 \times 10^{-2}$ M, $[CH_3CO_2^-] = 5.0 \times 10^{-2}$ M

PROBLEM 17-39   Calculate the mass of sodium formate (HCO_2Na) you'd need to add to 1.00 L of 0.50 M formic acid to make a buffer solution of pH = 3.95.

Answer: 54 g

PROBLEM 17-40   Calculate the pH at the following points in the titration of a 50.0-mL portion of 0.10 M cacodylic acid solution (a weak acid, $K_a = 6.4 \times 10^{-7}$) with 0.35 M NaOH: (a) at the start in the cacodylic acid solution, (b) after 10.0 mL of NaOH solution has been added, and (c) at the equivalence point.

Answer: (a) 3.60   (b) 6.56   (c) 9.54

PROBLEM 17-41   Lactic acid is a weak acid, $K_a = 1.4 \times 10^{-4}$. Calculate the extent of its ionization (a) in a 1.0 M solution and (b) in a 1.0×10^{-2} M solution.

Answer: (a) 1.2%   (b) 11%

PROBLEM 17-42   Determine the concentrations of all species containing sulfur in a 0.10 M solution of sulfurous acid, a diprotic acid, together with the pH of the solution.

Answer: $[H_2SO_3] = 6.8 \times 10^{-2}$ M, $[HSO_3^-] = 3.2 \times 10^{-2}$ M, $SO_3^{2-} = 1.0 \times 10^{-7}$ M, pH = 1.49

PROBLEM 17-43   Hydrogen telluride (H_2Te) is a diprotic acid, $K_{a_1} = 2.3 \times 10^{-3}$ and $K_{a_2} = 1 \times 10^{-11}$. Calculate the concentrations of HTe^- and Te^{2-} in a 0.10 M solution of H_2Te.

Answer: $[HTe^-] = 1.4 \times 10^{-2}$ M $[Te^{2-}] = 1 \times 10^{-11}$ M

PROBLEM 17-44   What is the pH of a solution that is 0.15 M in $NaHCO_3$ and 0.10 M in Na_2CO_3?

Answer: pH = 10.07

PROBLEM 17-45   A solution of 40.0 mL of 0.0100 M NaOH is titrated with 0.0100 M HCl. Calculate the pH at each of the following stages: (a) before any acid is added, (b) after 37.0 mL of acid has been added, (c) at the equivalence point, and (d) after 42.0 mL of acid has been added.

Answer: (a) 12.00   (b) 10.6   (c) 7.0   (d) 3.6

18 *HETEROGENEOUS SOLUTION EQUILIBRIUM*

THIS CHAPTER IS ABOUT

☑ **Solubility Product**
☑ **The Common Ion Effect**
☑ **pH Effects**
☑ **Precipitation**
☑ **Solubility Rules**

Many salts have low solubilities in water—a property that makes them very useful to chemists. For example, if we want to test an aqueous solution for the presence of sulfate ions (SO_4^{2-}), we add a solution of barium chloride $(BaCl_2)$, which is a strong electrolyte that ionizes fully to Ba^{2+} and Cl^-. If sulfate ions are present in the solution, an easily visible precipitate of the very insoluble barium sulfate $(BaSO_4)$ will form when the Ba^{2+} and SO_4^{2-} ions combine. The net ionic reaction for the precipitation of $BaSO_4$ is

$$Ba^{2+}(aq) + SO_4^{2-}(aq) \longrightarrow BaSO_4(s) \downarrow$$

(The downward-pointing arrow is sometimes used in reaction equations to indicate that a component precipitates.) Obviously, this reaction is a good qualitative test, and it can also be used in the quantitative analysis of sulfate and barium compounds.

Because precipitation reactions are widely used in analytical chemistry, it is important to explore quantitatively the heterogeneous solution equilibria involved in them.

note: Do not confuse the *solubility* of a salt, expressed in moles per liter or grams per 100 mL, with *extent of dissociation* (expressed in percentages). A salt may be sparingly soluble in aqueous solution, but the ions that do get into solution are completely dissociated; that is, a salt may be a strong electrolyte (100% ionization) even though few ions are present in solution.

18-1. Solubility Product

Salts of low solubility are often found in contact with their saturated solutions, since it takes very little of the salt to saturate a solution. In such solutions the salt is usually fully ionized and a dynamic equilibrium is established between the solid salt and its ions in solution. For example, the heterogeneous solution equilibrium for $BaSO_4$ is

$$BaSO_4(s) \rightleftharpoons Ba^{2+}(aq) + SO^{2-}(aq)$$

for which the equilibrium-law expression would be

$$K = \frac{[Ba^{2+}][SO_4^{2-}]}{[BaSO_4]}$$

But applying the convention that the concentrations of pure solids are incorporated into K (because they are constants), we can write another equilibrium-constant expression

$$K_{sp} = [Ba^{2+}][SO_4^{2-}]$$

where K_{sp} is the **solubility product** or **solubility product constant**. Thus, for the general reaction

$$M_pA_q(s) \overset{H_2O}{\rightleftharpoons} pM^{x+} + qA^{y-}$$

the solubility product is

SOLUBILITY PRODUCT $K_{sp} = [M^{x+}]^p [A^{y-}]^q$ (18-1)

where $x+$ and $y-$ are the charges of ions M and A, respectively, and p and q are powers whose numerical values are equal to the coefficients in the general reaction.

EXAMPLE 18-1: The solubility of $BaSO_4$ in water at 293 K is 2.4×10^{-4} g/100 mL. Calculate K_{sp} for $BaSO_4$ at this temperature.

Solution: Since the molar mass of $BaSO_4$ is 233.4 g/mol, the number of moles of $BaSO_4$ that will dissolve per liter is

$$\left(\frac{2.4 \times 10^{-4} \text{ g}}{100 \text{ mL}} \right) \left(\frac{10^3 \text{ mL/L}}{233.4 \text{ g/mol}} \right) = 1.0 \times 10^{-5} \ M$$

so the concentrations of the ions are

$$[Ba^{2+}] = (1.0 \times 10^{-5} \ M \ BaSO_4) \left(\frac{1 \text{ mol } Ba^{2+}}{1 \text{ mol } BaSO_4} \right) = 1.0 \times 10^{-5} \ M$$

$$[SO_4^{2-}] = (1.0 \times 10^{-5} \ M \ BaSO_4) \left(\frac{1 \text{ mol } SO_4^{2-}}{1 \text{ mol } BaSO_4} \right) = 1.0 \times 10^{-5} \ M$$

And since K_{sp} for $BaSO_4$ is the product of the ion concentrations,

$$K_{sp} = [Ba^{2+}][SO_4^{2-}] = (1.0 \times 10^{-5} \ M)(1.0 \times 10^{-5} \ M)$$
$$= 1.0 \times 10^{-10}$$

EXAMPLE 18-2: Exactly 1.00 L of a saturated aqueous solution of CaF_2 was carefully evaporated at 25°C, and a residue of 0.017 g of CaF_2 was obtained. Calculate K_{sp} for CaF_2 at 25°C.

Solution: Since the molar mass of CaF_2 is 78.1 g/mol, the number of moles of CaF_2 dissolved in 1.00 L of solution is

$$\frac{0.017 \text{ g}}{78.1 \text{ g/mol}} = 2.18 \times 10^{-4} \text{ mol}$$

Now write the balanced equation and the solubility product expression:

$$CaF_2(s) \rightleftharpoons Ca^{2+}(aq) + 2F^-(aq) \qquad K_{sp} = [Ca^{2+}][F^-]^2$$

Calculate the ion concentrations:

$$[Ca^{2+}] = (2.18 \times 10^{-4} \ M \ CaF_2) \left(\frac{1 \text{ mol } Ca^{2+}}{1 \text{ mol } CaF_2} \right) = 2.18 \times 10^{-4} \ M$$

$$[F^-] = (2.18 \times 10^{-4} \ M \ CaF_2) \left(\frac{2 \text{ mol } F^-}{1 \text{ mol } CaF_2} \right) = 4.36 \times 10^{-4} \ M$$

and substitute into the K_{sp} expression:

$$K_{sp} = [Ca^{2+}][F^-]^2 = (2.18 \times 10^{-4} \ M)(4.36 \times 10^{-4} \ M)^2$$
$$= 4.1 \times 10^{-11} \qquad \text{(two significant figures)}$$

EXAMPLE 18-3: The value of K_{sp} for $Cr(OH)_3$, which dissolves to give $Cr^{3+}(aq)$ and $OH^-(aq)$, is 1.0×10^{-33}. What is the molar solubility of $Cr(OH)_3$?

Solution: The procedure is simply the reverse of the one used in Examples 18-1 and 18-2. The balanced equation and K_{sp} expression are

$$Cr(OH)_3(s) \rightleftharpoons Cr^{3+}(aq) + 3OH^-(aq) \qquad K_{sp} = [Cr^{3+}][OH^-]^3 = 1.0 \times 10^{-33}$$

Let S be the molar solubility of $Cr(OH)_3$ in moles per liter. Then

$$[Cr^{3+}] = S\left(\frac{1 \text{ mol } Cr^{3+}}{1 \text{ mol } Cr(OH)_3}\right) = S$$

$$[OH^-] = S\left(\frac{3 \text{ mol } OH^-}{1 \text{ mol } Cr(OH)_3}\right) = 3S$$

Then substitute into the K_{sp} expression:

$$[Cr^{3+}][OH^-]^3 = S(3S)^3$$
$$K_{sp} = 27S^4 = 1.0 \times 10^{-33}$$
$$S^4 = 3.70 \times 10^{-35}$$
$$S = (3.70 \times 10^{-35})^{1/4} = 2.5 \times 10^{-9}$$

Thus the molar solubility of $Cr(OH)_3$ is 2.5×10^{-9} M.

18-2. The Common Ion Effect

If an ion of a salt is already present in the solvent solution—i.e., if salt and solution have a common ion—the solubility of the salt is decreased. This phenomenon, called the **common ion effect**, is predictable by Le Chatelier's principle. For example, the sparingly soluble $BaSO_4$ is even less soluble in $NaSO_4$ than it is in H_2O, owing to the presence of the common ion SO_4^{2-} already in solution.

The K_{sp} values of a salt can be used to predict the size of the common ion effect.

EXAMPLE 18-4: Calculate the molar solubility of $BaSO_4$ ($K_{sp} = 1.0 \times 10^{-10}$) in a 0.10 M solution of Na_2SO_4.

Solution: The salt Na_2SO_4 is a strong electrolyte that is completely ionized to $2Na^+$ and SO_4^{2-}, so the contribution of this salt to the final equilibrium $[SO_4^{2-}]$ must be 0.10 M. Now if we let S be the molar solubility of $BaSO_4$, we can say that $[Ba^{2+}] = S$ and that the contribution of $BaSO_4$ to the final $[SO_4^{2-}]$ must also be S. Adding the contributions of $NaSO_4$ and $BaSO_4$ to $[SO_4^{2-}]$, we get $[SO_4^{2-}] = 0.10 + S$. Thus

$$K_{sp} = [Ba^{2+}][SO_4^{2-}] = S(0.10 + S) = 1.0 \times 10^{-10}$$

Solving this quadratic equation could solve the problem, but the data we have do not justify that exact method. So we proceed by approximations: Since the K_{sp} of $BaSO_4$ is small, the value of S must be very small in comparison with 0.10 M. Thus, if $S \ll 0.10$, we can neglect S in the term $(0.10 + S)$ to get

$$S(0.10) = 1.0 \times 10^{-10} \quad \text{and} \quad S = 1.0 \times 10^{-9} \ M$$

note: Compare this solubility with that calculated for $BaSO_4$ in pure water: 1.0×10^{-9} M in Na_2SO_4 vs. 1.0×10^{-5} M in H_2O. The common ion effect has reduced the solubility of $BaSO_4$ by a factor of 10 000.

EXAMPLE 18-5: The K_{sp} value for silver chromate (Ag_2CrO_4) is 2.4×10^{-12}. Calculate the molar solubility (S) of this salt in 5.0×10^{-2} M potassium chromate (K_2CrO_4) solution.

Solution: The balanced equation and solubility product are

$$Ag_2CrO_4 \rightleftharpoons 2Ag^+ + CrO_4^{2-} \qquad K_{sp} = [Ag^+]^2[CrO_4^{2-}]$$

Then $[Ag^+] = 2S$ and $[CrO_4^{2-}] = S$ (from Ag_2CrO_4) + 5.0×10^{-2} (from K_2CrO_4), so

$$K_{sp} = (2S)^2(S + 5.0 \times 10^{-2}) = 2.4 \times 10^{-12}$$

But we don't want (or need) to solve this cubic equation; so we assume that $S \ll 5.0 \times 10^{-2}$ M, in which case

$$(2S)^2(5.0 \times 10^{-2}) = 2.4 \times 10^{-12}$$

and

$$S = \sqrt{\frac{2.4 \times 10^{-12}}{4(5.0 \times 10^{-2})}} = 3.5 \times 10^{-6}$$

Thus we see that S is much less than 5.0×10^{-2} and that our assumption is valid: The molar solubility of Ag_2CrO_4 in 5.0×10^{-2} M K_2CrO_4 is 3.5×10^{-6} M.

EXAMPLE 18-6: Calculate the molar solubility of $Zn(OH)_2$ ($K_{sp} = 2.0 \times 10^{-17}$) in a buffer solution whose pH is 9.00.

Solution: First write the appropriate equations:

$$Zn(OH)_2(s) \rightleftharpoons Zn^{2+}(aq) + 2OH^-(aq) \qquad K_{sp} = [Zn^{2+}][OH^-]^2$$

Now notice that there is a common ion effect here: A buffer fixes $[H^+]$ and $[OH^-]$, and OH^- is one of the ions produced when $Zn(OH)_2$ dissolves. Therefore at pH = 9.00, pOH = 14.00 − 9.00 = 5.00 and $[OH^-] = 1.0 \times 10^{-5}$. Moreover, we don't have to worry about any contribution of $Zn(OH)_2$ to $[OH^-]$ because the buffer holds both $[H^+]$ and $[OH^-]$ constant. Then if we let S be the molar solubility of $Zn(OH)_2$, we get $[Zn^{2+}] = S$ and we can substitute:

$$K_{sp} = S(1.0 \times 10^{-5})^2 = 2.0 \times 10^{-17}$$
$$S = 2.0 \times 10^{-7}$$

Thus the molar solubility of $Zn(OH)_2$ in a buffer solution at pH = 9.00 is 2.0×10^{-7} M.

18-3. pH Effects

The pH of a solution can affect the solubility of a sparingly soluble salt (see Example 18-6). If the anion of the salt is an ion of a weak acid, it follows from Le Chatelier's principle that the addition of acid to the solution will increase the solubility of the salt.

EXAMPLE 18-7: Explain how the addition of some HCl to a saturated solution of CaF_2 in contact with $CaF_2(s)$ changes the solubility of the CaF_2.

Solution: Consider this system in stages. First the K_{sp} equilibrium is established:

$$CaF_2(s) \rightleftharpoons Ca^{2+}(aq) + 2F^-(aq)$$

Now the completely ionized strong acid $H^+(aq) + Cl^-(aq)$ is added and a new equilibrium is established:

$$H^+(aq) + F^-(aq) \rightleftharpoons HF$$

Since HF is a weak acid, most of the F^- in solution combines with H^+ to give un-ionized HF. This means that in the K_{sp} equilibrium reaction the F^- concentration is reduced, and so more CaF_2 has to dissolve to reestablish the K_{sp} equilibrium condition.

18-4. Precipitation

The **ion product** is the product of the concentration of ions in a nonequilibrium or nonsaturated solution. For example, the ion product of any solution containing barium ions and sulfate ions is $[Ba^{2+}][SO_4^{2-}]$. There are three possible relationships between the value of the ion product and the value of K_{sp}:

- If the ion product is *less than* K_{sp}, then all ions are present *in solution*.
- If the ion product is *equal to* K_{sp}, then the solution is *saturated* and is at equilibrium.
- If the ion product is *greater than* K_{sp}, then the solution is *supersaturated* and solid salt will *precipitate* until the ion product equals K_{sp}.

EXAMPLE 18-8: Will a precipitate form if 15 mL of 2.0×10^{-3} M Na_2SO_4 solution is added to 25 mL of 1.0×10^{-3} M $SrCl_2$ solution? (K_{sp} for $SrSO_4 = 3.2 \times 10^{-7}$.) Assume that volumes ($V$) are additive.

Solution: First calculate the ion product $[Sr^{2+}][SO_4^{2-}]$ and then compare this value with K_{sp}. The calculation of the ion concentrations is just a dilution problem. We'll set up the calculations using the subscripts i for initial and f for final:

$$[Sr^{2+}]_f = (1.0 \times 10^{-3} \ M)\left(\frac{V_i}{V_f}\right) = (1.0 \times 10^{-3} \ M)\left(\frac{25 \ \text{mL}}{25 \ \text{mL} + 15 \ \text{mL}}\right) = 6.3 \times 10^{-4} \ M$$

$$[SO_4^{2-}]_f = (2.0 \times 10^{-3} \ M)\left(\frac{V_i}{V_f}\right) = (2.0 \times 10^{-3} \ M)\left(\frac{15 \ \text{mL}}{15 \ \text{mL} + 25 \ \text{mL}}\right) = 7.5 \times 10^{-4} \ M$$

Then the ion product is

$$[Sr^{2+}][SO_4^{2-}] = (6.3 \times 10^{-4})(7.5 \times 10^{-4}) = 4.7 \times 10^{-7}$$

Since the ion product 4.7×10^{-7} is greater than $K_{sp} = 3.2 \times 10^{-7}$, a precipitate of $SrSO_4$ will form.

EXAMPLE 18-9: A very concentrated Na_2CO_3 solution is slowly added to a second solution that is 0.10 M in both Mg^{2+} and Ca^{2+}. Determine **(a)** the composition of the first precipitate and the concentration of CO_3^{2-} when it forms, and **(b)** the composition of the second precipitate and the concentration of Ca^{2+} when it forms. (Neglect any dilution effects on the concentration. The K_{sp} values are 4.8×10^{-9} for $CaCO_3$ and 1.0×10^{-5} for $MgCO_3$.)

Solution: At the start, $[Mg^{2+}] = [Ca^{2+}] = 0.10$ M.
 (a) For $MgCO_3$

$$K_{sp} = [Mg^{2+}][CO_3^{2-}] = 1.0 \times 10^{-5}$$

For $CaCO_3$

$$K_{sp} = [Ca^{2+}][CO_3^{2-}] = 4.8 \times 10^{-9}$$

When the ion product for $CaCO_3$ reaches the K_{sp} value, solid $CaCO_3$ will precipitate. The $[CO_3^{2-}]$ required for this precipitation is much lower than that required for $MgCO_3$, so $CaCO_3$ will precipitate first. The CO_3^{2-} concentration at that point can be found from the K_{sp} for $CaCO_3$:

$$K_{sp} = [Ca^{2+}][CO_3^{2-}] = 4.8 \times 10^{-9}$$
$$(0.10)[CO_3^{2-}] = 4.8 \times 10^{-9}$$
$$[CO_3^{2-}] = 4.8 \times 10^{-8} \ M$$

(b) Now the K_{sp} value for $MgCO_3$ can be used to determine the CO_3^{2-} concentration required to precipitate $MgCO_3$:

$$K_{sp} = [Mg^{2+}][CO_3^{2-}] = 1.0 \times 10^{-5}$$
$$(0.10)[CO_3^{2-}] = 1.0 \times 10^{-5}$$
$$[CO_3^{2-}] = 1.0 \times 10^{-4}$$

The $[Ca^{2+}]$ has now decreased to

$$[Ca^{2+}] = \frac{K_{sp}}{[CO_3^{2-}]} = \frac{4.8 \times 10^{-9}}{1.0 \times 10^{-4}} = 4.8 \times 10^{-5} \ M$$

note: Virtually all of the Ca^{2+} precipitates as $CaCO_3$ before any $MgCO_3$ forms. The concentration of Ca^{2+} falls from 0.10 M to 4.8×10^{-5} M or $\dfrac{4.8 \times 10^{-5}}{0.10} \times 100\% = 0.048\%$ of its initial value. Selective precipitation of this kind, in which one component is virtually quantitatively precipitated before another component, plays an important role in analytical chemistry.

18-5. Solubility Rules

Suppose that a solution of $CaCl_2$ is mixed with a solution of Na_2CO_3 and that a precipitate forms. The ions present are Na^+, Ca^{2+}, Cl^-, and CO_3^{2-}. *Can the composition of the precipitate be predicted?*

There are no theories of solubility to guide us in predicting the composition of mixed solutions. The best we can do is to develop a list of general rules based on observation. Only if we know that $CaCO_3$ is insoluble in water and NaCl is soluble can we predict that $Ca^{2+} + CO_3^{2-} \rightarrow CaCO_3\downarrow$ is the net ionic reaction of the mixture and that the Na^+ and Cl^- remain in solution as spectator ions.

A soluble salt is usually defined as a salt that generates an ionic concentration greater than 0.1 M in water. The following "rules" should be useful in determining which combination of ions will form an insoluble salt at room temperature.

SOLUBLE SALTS

1. All salts of the alkali metals (Li^+, Na^+, K^+, etc.) and NH_4^+
2. All nitrates (NO_3^-), chlorates (ClO_3^-), perchlorates (ClO_4^-), and acetates ($CH_3CO_2^-$)
3. All chlorides (Cl^-), bromides (Br^-), and iodides (I^-), except those in combination with Ag^+, Pb^{2+}, and Hg_2^{2+}
4. All sulfates (SO_4^{2-}), except those in combination with Pb^{2+}, Ca^{2+}, Sr^{2+}, Ba^{2+}, Hg^{2+}, Hg_2^{2+}, and Ag^+

INSOLUBLE SALTS

5. Hydroxides (OH^-), except those in combination with the alkali metals, NH_4^+, Sr^{2+}, and Ba^{2+}
6. Carbonates (CO_3^{2-}), phosphates (PO_4^{3-}), sulfides (S^{2-}), and sulfites (SO_3^{2-}), except in combination with the alkali metals and NH_4^+

EXAMPLE 18-10: Determine the net ionic equation for the reaction that occurs when the following solutions are mixed: (**a**) 0.50 M $BaCl_2$ and 0.50 M K_2SO_4 and (**b**) a solution of 0.50 M $Ba(NO_3)_2$ and 0.50 M NH_4ClO_4.

Solution:
(**a**) The ions present in solution are Ba^{2+}, Cl^-, K^+, and SO_4^-. The overall reaction is

$$BaCl_2 + K_2SO_4 \xrightarrow{H_2O} BaSO_4 + 2KCl$$

From rules 1 and 3, KCl is soluble; from rule 4, $BaSO_4$ is not. So the net ionic equation is

$$Ba^{2+} + SO_4^{2-} \xrightarrow{H_2O} BaSO_4\downarrow$$

(**b**) The ions present in solution are Ba^{2+}, NO_3^-, NH_4^+, and ClO_4^-. Because all salts from these anions and cations are soluble, no precipitate will form and no reaction will occur.

SUMMARY

1. For the general equilibrium $M_pA_q(s) \xrightarrow{H_2O} pM^{x+} + qA^{y-}$, the solubility product K_{sp} is given by $K_{sp} = [M^{x+}]^p[A^{y-}]^q$.
2. A common ion in solution decreases the solubility of a salt that contains that ion.
3. Solution pH can affect the solubility of salts containing ions that are conjugates of weak acids or bases.
4. The ion product is the product of the concentration of ions in a nonequilibrium or nonsaturated solution.
5. If the ion product is less than K_{sp}, all ions are present in solution; if the ion product equals K_{sp}, the solution is saturated and at equilibrium; if the ion product is greater than K_{sp}, the solution is supersaturated and a precipitate will form.

RAISE YOUR GRADES

Can you...?

☑ calculate solubility product constants from solubilities and vice versa
☑ determine the magnitude of common ion effects
☑ determine the effects of pH on solubility
☑ determine whether or not and in what order a precipitate or precipitates will form when two solutions involving salts of known K_{sp} values are mixed
☑ use the general rules for soluble salts to determine whether a given salt is likely to precipitate

SOLVED PROBLEMS

PROBLEM 18-1 Use the following solubilities in water to determine the solubility product constants for AgI and $Cd(OH)_2$: (a) AgI 2.14×10^{-6} g/L, (b) $Cd(OH)_2$ 2.6×10^{-4} g/100 mL.

Solution:
(a) The molar mass of AgI is 234.8 g/mol. The number of moles of AgI that dissolve per liter is

$$\frac{2.14 \times 10^{-6} \text{ g/L}}{234.8 \text{ g/mol}} = 9.11 \times 10^{-9} \text{ mol/L}$$

So the concentrations are

$$[Ag^+] = (9.11 \times 10^{-9} \ M \text{ AgI})\left(\frac{1 \text{ mol Ag}^+}{1 \text{ mol AgI}}\right) = 9.11 \times 10^{-9} \ M$$

$$[I^-] = (9.11 \times 10^{-9} \ M \text{ AgI})\left(\frac{1 \text{ mol I}^-}{1 \text{ mol AgI}}\right) = 9.11 \times 10^{-9} \ M$$

and the solubility product is

$$K_{sp} = [Ag^+][I^-] = (9.11 \times 10^{-9})^2 = 8.30 \times 10^{-17}$$

(b) The molar mass of $Cd(OH)_2 = 146.4$ g/mol. The number of moles of $Cd(OH)_2$ that dissolve per liter is

$$\left(\frac{2.6 \times 10^{-4} \text{ g}}{100.0 \text{ mL}}\right)\left(\frac{10^3 \text{ mL}}{1 \text{ L}}\right)\left(\frac{1}{146.4 \text{ g/mol}}\right) = 1.8 \times 10^{-5} \ M$$

So the concentrations are

$$[Cd^{2+}] = (1.8 \times 10^{-5} \ M \text{ Cd(OH)}_2)\left(\frac{1 \text{ mol Cd}^{2+}}{1 \text{ mol Cd(OH)}_2}\right) = 1.8 \times 10^{-5} \ M$$

$$[OH^-] = (1.8 \times 10^{-5} \ M \text{ Cd(OH)}_2)\left(\frac{2 \text{ mol OH}^-}{1 \text{ mol Cd(OH)}_2}\right) = 3.6 \times 10^{-5} \ M$$

and the solubility product is

$$K_{sp} = [Cd^{2+}][OH^-]^2 = (1.8 \times 10^{-5})(3.6 \times 10^{-5})^2 = 2.3 \times 10^{-14}$$

PROBLEM 18-2 Some 0.050 M KIO_3 solution was saturated with solid $Cu(IO_3)_2$. It was then found that the equilibrium concentration of Cu^{2+} was 5.6×10^{-5} M. Calculate the solubility product of $Cu(IO_3)_2$.

Solution: If $[Cu^{2+}] = 5.6 \times 10^{-5}$ M, then

$$[IO_3^-] = 0.050 \ M + \left(\frac{2 \text{ mol IO}_3^-}{1 \text{ mol Cu}^{2+}}\right)(5.6 \times 10^{-5} \ M \text{ Cu}^{2+})$$

$$= 0.050 \ M + 1.1 \times 10^{-4} \ M = 0.050 \ M$$

The $[IO_3^-]$ that comes from the $Cu(IO_3)_2$ is far less than the 0.050 mol/L already in the solution, so you can neglect it when calculating the solubility product:

$$K_{sp} = [Cu^{2+}][IO_3^-]^2 = (5.6 \times 10^{-5})(0.050)^2 = 1.4 \times 10^{-7}$$

PROBLEM 18-3   The solubility product of MgF_2 is 6.9×10^{-9}. Calculate the molar solubility of MgF_2 in water.

Solution:

$$MgF_2 \rightleftharpoons Mg^{2+} + 2F^- \qquad K_{sp} = [Mg^{2+}][F^-]^2 = 6.9 \times 10^{-9}$$

Let S be the solubility in moles per liter. Then $[Mg^{2+}] = S$ and $[F^-] = 2S$, so

$$K_{sp} = (S)(2S)^2 = 6.9 \times 10^{-9}$$

$$4S^3 = 6.9 \times 10^{-9}$$

$$S^3 = \frac{6.9 \times 10^{-9}}{4} = 1.73 \times 10^{-9}$$

$$S = (1.73 \times 10^{-9})^{1/3} = 1.2 \times 10^{-3}\ M$$

PROBLEM 18-4   Calculate the molar solubility of $AgCl$ in a solution made by dissolving 10.0 g of $CaCl_2$ in 1.00 L of solution ($K_{sp} = 1.6 \times 10^{-10}$ for $AgCl$.)

Solution:

$$AgCl \rightleftharpoons Ag^+ + Cl^- \qquad K_{sp} = [Ag^+][Cl^-]$$

The Cl^- will come from the $CaCl_2$, whose molar mass is 111 g/mol. Thus

$$[Cl^-] = \left(\frac{10.0\ \text{g } CaCl_2/L}{111\ \text{g } CaCl_2/\text{mol}}\right)\left(\frac{2\ \text{mol } Cl^-}{1\ \text{mol } CaCl_2}\right) = 0.180\ M$$

and the solubility of $AgCl$ will be $[Ag^+]$:

$$[Ag^+] = \frac{K_{sp}}{[Cl^-]} = \frac{1.6 \times 10^{-10}}{0.180} = 8.9 \times 10^{-10}\ M$$

PROBLEM 18-5   The solubility product for $PbBr_2$ is 4.6×10^{-6}. The solubility product for $CaCO_3 = 4.7 \times 10^{-9}$. Which of these salts will have *decreased* solubility in 0.10 M HBr solution compared with water?

Solution: HBr is a strong acid; a 0.10 M HBr solution will have $[Br^-] = 0.10\ M$. Increasing the concentration of the common ion Br^- in the equilibrium $PbBr_2 \rightleftharpoons Pb^{2+} + 2Br^-$ would decrease the Pb^{2+} concentration and decrease the solubility of $PbBr_2$.

PROBLEM 18-6   Refer to Problem 18-5. What effect will 0.10 M HBr have on the $CaCO_3$?

Solution: $CaCO_3$ is the salt of the weak acid H_2CO_3. Most of the CO_3^{2-} in solution will combine with the H^+ from the HBr to form H_2CO_3 or HCO_3^-. In the equilibrium $CaCO_3 \rightleftharpoons Ca^{2+} + CO_3^{2-}$ the concentration of CO_3^{2-} will decrease, which means that more $CaCO_3$ will dissolve.

PROBLEM 18-7   A solution is made up of 0.010 M sodium chromate (Na_2CrO_4) and 0.10 M NaCl, to which a concentrated solution of $AgNO_3$ is slowly added. (The volume of the solution does not change much.) For Ag_2CrO_4, $K_{sp} = 1.9 \times 10^{-12}$; for $AgCl$, $K_{sp} = 1.56 \times 10^{-10}$.

   (a) What solid precipitates first, and what is the concentration of Ag^+ when that salt starts to form?
   (b) What is the concentration of Cl^- when Ag_2CrO_4 starts to precipitate?

Solution:
(a) All sodium salts and all nitrates are soluble, so you can expect that Na^+ and NO_3^- will not be involved as precipitates. Since the solubility products of Ag_2CrO_4 and $AgCl$ are given, you can expect one of them to be the first precipitate formed. Calculate the minimum $[Ag^+]$ required to precipitate each insoluble salt:

$$AgCl \rightleftharpoons Ag^+ + Cl^- \qquad K_{sp} = [Ag^+][Cl^-] = 1.56 \times 10^{-10}$$

$$[Ag^+] = \frac{K_{sp}}{[Cl^-]} = \frac{1.56 \times 10^{-10}}{0.10} = 1.6 \times 10^{-9}\ M$$

$$Ag_2CrO_4 \rightleftharpoons 2Ag^+ + CrO_4^{2-} \qquad K_{sp} = [Ag^+]^2[CrO_4^{2-}] = 1.9 \times 10^{-12}$$

$$[Ag^+]^2 = \frac{K_{sp}}{[CrO_4^{2-}]} = \frac{1.9 \times 10^{-12}}{0.010} = 1.9 \times 10^{-10}$$

$$[Ag^+] = (1.9 \times 10^{-10})^{1/2} = 1.4 \times 10^{-5} \ M$$

It takes less Ag^+ to precipitate AgCl, so AgCl precipitates first. The $[Ag^+]$ is 1.6×10^{-9} M when AgCl begins to form.

(b) When the $[Ag^+]$ has increased to 1.4×10^{-5} M, Ag_2CrO_4 will start to precipitate. The $[Cl^-]$ is calculated from

$$[Cl^-] = \frac{K_{sp}}{[Ag^+]} = \frac{1.56 \times 10^{-10}}{1.4 \times 10^{-5}} = 1.1 \times 10^{-5} \ M$$

note: Although the Cl^- has not been completely removed from the solution, the $[Cl^-]$ has been reduced from 0.10 M to 1.1×10^{-5} M while the $[CrO_4^{2-}]$ remains at 0.010 M. Thus, for all practical purposes, the two ions have been separated.

PROBLEM 18-8   When H_2S gas is dissolved in water, it is a weak diprotic acid:

$$H_2S(g) \xrightleftharpoons{H_2O} 2H^+(aq) + S^{2-}(aq)$$

The concentration of undissociated H_2S is constant at about 0.10 M if the temperature is 25°C and the partial pressure of H_2S is 1 atm. The equilibrium expression (ion product) for the reaction is

$$K_{ip} = [H^+]^2[S^{2-}] = 1.2 \times 10^{-22} \qquad ([H_2S] \text{ is constant})$$

Calculate the molar solubility of FeS and PbS in a solution saturated with H_2S and buffered at pH = 3.0. For FeS, $K_{sp} = 4 \times 10^{-19}$; for PbS, $K_{sp} = 7 \times 10^{-29}$.

Solution: You are told that the solution is buffered at pH = 3.0, so you know that the $[H^+]$ is held constant at 1×10^{-3} M. Then you can use the K_{ip} equilibrium expression to determine the $[S^{2-}]$. Once you know $[S^{2-}]$, you can determine the concentrations of Fe^{2+} and Pb^{2+} in the saturated H_2S solution from their respective solubility products. In each case, the solubility will be equal to the concentration of the metal ion.

$$[S^{2-}] = \frac{K_{ip}}{[H^+]^2} = \frac{1.2 \times 10^{-22}}{(1 \times 10^{-3})^2} = 1.2 \times 10^{-16} \ M$$

$$[Fe^{2+}] = \frac{K_{sp}}{[S^{2-}]} = \frac{4 \times 10^{-19}}{1.2 \times 10^{-16}} = 3 \times 10^{-3} \ M$$

$$[Pb^{2+}] = \frac{K_{sp}}{[S^{2-}]} = \frac{7 \times 10^{-29}}{1.2 \times 10^{-16}} = 6 \times 10^{-13} \ M$$

Notice the big difference in the solubilities. By setting the solution at the proper pH, you could remove all of the Pb as PbS and leave the Fe^{2+} in solution.

PROBLEM 18-9   The solubility product for $CaCO_3$ is 4.8×10^{-9}. The second ionization for H_2CO_3 is

$$HCO_3^- \rightleftharpoons H^+ + CO_3^{2-} \qquad K_{a_2} = \frac{[H^+][CO_3^{2-}]}{[HCO_3^-]} = 5.6 \times 10^{-11}$$

What must the pH be to make 1.0×10^{-3} mol of $CaCO_3$ dissolve in 1.00 L of saturated solution? (Assume that HCO_3^- is the only species besides Ca^{2+} and CO_3^{2-} formed in the solution and that $Ca(HCO_3)_2$ is soluble.)

Solution: When the $CaCO_3$ solubility is 1.0×10^{-3} M, the Ca^{2+} concentration must be 1.0×10^{-3}. You can determine the CO_3^{2-} concentration from the solubility product:

$$[CO_3^{2-}] = \frac{K_{sp}}{[Ca^{2+}]} = \frac{4.8 \times 10^{-9}}{1.0 \times 10^{-3}} = 4.8 \times 10^{-6} \ M$$

The equilibrium concentration of CO_3^{2-} is very small compared to the solubility of $CaCO_3$; therefore most of the CO_3^{2-} produced by the $CaCO_3$ must turn into HCO_3^- and $[Ca^{2+}] \cong [HCO_3^-] \cong 1.0 \times 10^{-3}$ M. The H^+

concentration is determined from the HCO_3^- ionization:

$$[H^+] = \frac{K_{a_2}[HCO_3^-]}{[CO_3^{2-}]} = \frac{(5.6 \times 10^{-11})(1.0 \times 10^{-3})}{4.8 \times 10^{-6}} = 1.2 \times 10^{-8}$$

$$pH = -\log[H^+] = -\log(1.2 \times 10^{-8}) = 7.9$$

PROBLEM 18-10 The solubility product of $Mg(OH)_2$ is 1.2×10^{-11}. Calculate the solubility of $Mg(OH)_2$ in pure water and the pH of the resulting solution.

Solution: The solubility S is equal to the Mg^{2+} concentration:

$$S = [Mg^{2+}] \quad \text{and} \quad [OH^-] = 2S$$

Substitute S and $2S$ for $[Mg^{2+}]$ and $[OH^-]$ in the solubility product expression:

$$K_{sp} = [Mg^{2+}][OH^-]^2 = (S)(2S)^2 = 4S^3 = 1.2 \times 10^{-11}$$

$$S^3 = \frac{1.2 \times 10^{-11}}{4} = 3.0 \times 10^{-12}$$

$$S = (3.0 \times 10^{-12})^{1/3} = 1.4 \times 10^{-4} \ M$$

Therefore

$$[OH^-] = 2S = 2(1.4 \times 10^{-4}) = 2.8 \times 10^{-4} \ M$$

$$[H^+] = \frac{K_w}{[OH^-]} = \frac{1.0 \times 10^{-14}}{2.8 \times 10^{-4}} = 3.6 \times 10^{-11}$$

$$pH = -\log[H^+] = -\log(3.6 \times 10^{-11}) = 10.45$$

PROBLEM 18-11 Arrange the carbonates $CaCO_3$ ($K_{sp} = 4.8 \times 10^{-9}$), Ag_2CO_3 ($K_{sp} = 8.2 \times 10^{-12}$), and $MgCO_3$ ($K_{sp} = 1.0 \times 10^{-5}$) in order of decreasing solubility in water. (Neglect hydrolysis of CO_3^{2-}.)

Solution: Determine the solubility of each of these carbonates in water.

$CaCO_3$: $CaCO_3 \rightleftharpoons Ca^{2+} + CO_3^{2-}$

$$S = [Ca^{2+}] = [CO_3^{2-}]$$
$$K_{sp} = [Ca^{2+}][CO_3^{2-}] = (S)(S) = S^2 = 4.8 \times 10^{-9}$$
$$S = (4.8 \times 10^{-9})^{1/2} = 6.9 \times 10^{-5} \ M$$

Ag_2CO_3: $Ag_2CO_3 \rightleftharpoons 2Ag^+ + CO_3^{2-}$

$$2S = [Ag^+]; \qquad S = [CO_3^{2-}]$$
$$K_{sp} = [Ag^+]^2[CO_3^{2-}] = (2S)^2(S) = 4S^3 = 8.2 \times 10^{-12}$$

$$S^3 = \frac{8.2 \times 10^{-12}}{4} = 2.05 \times 10^{-12}$$

$$S = (2.05 \times 10^{-12})^{1/3} = 1.3 \times 10^{-4} \ M$$

$MgCO_3$: $MgCO_3 \rightleftharpoons Mg^{2+} + CO_3^{2-}$

$$S = [Mg^{2+}] = [CO_3^{2-}]$$
$$K_{sp} = [Mg^{2+}][CO_3^{2-}] = (S)(S) = S^2 = 1.0 \times 10^{-5}$$
$$S = (1.0 \times 10^{-5})^{1/2} = 3.2 \times 10^{-3} \ M$$

The order of decreasing solubility is therefore

$$MgCO_3 > Ag_2CO_3 > CaCO_3$$

(Notice that Ag_2CO_3 has the smallest K_{sp}, but it is not the least soluble.)

PROBLEM 18-12 Calculate the solubility in grams per liter of PbI_2 ($K_{sp} = 7.5 \times 10^{-9}$) in 0.050 M $Pb(NO_3)_2$ solution.

Solution:

$$PbI_2 \rightleftharpoons Pb^{2+} + 2I^- \qquad K_{sp} = [Pb^{2+}][I^-]^2 = 7.5 \times 10^{-9}$$

Because $[Pb^{2+}]$ is dominated by the 0.050 M $PbNO_3$, $S \neq [Pb^{2+}]$. But $[I^-] = 2S$ because PbI_2 is the only source of I^- in solution. Thus

$$[I^-]^2 = (2S)^2 = \frac{K_{sp}}{[Pb^{2+}]} = \frac{7.5 \times 10^{-9}}{0.050}$$

$$S^2 = \frac{7.5 \times 10^{-9}}{(4)(0.050)} = 3.75 \times 10^{-8}$$

$$S = \sqrt{3.75 \times 10^{-8}} = 1.9 \times 10^{-4} \ M$$

Converting from moles per liter to grams per liter (the molar mass of PbI_2 is 461.0 g/mol),

$$(1.9 \times 10^{-4} \ \text{mol/L})(461 \ \text{g/mol}) = 8.8 \times 10^{-2} \ \text{g/L}$$

PROBLEM 18-13 Calculate the equilibrium concentrations of Ba^{2+}, Pb^{2+}, and SO_4^{2-} after 0.0300 mol of solid $PbSO_4$ is added to 0.500 L of 0.100 M $Ba(NO_3)_2$ solution. ($K_{sp} = 1.8 \times 10^{-8}$ for $PbSO_4$ and $K_{sp} = 1.0 \times 10^{-10}$ for $BaSO_4$.)

Solution: The reaction that occurs here is an example of **transposition**. The solid $PbSO_4$ is transposed into $BaSO_4$:

$$PbSO_4(s) + Ba^{2+}(aq) \rightleftharpoons BaSO_4(s) + Pb^{2+}(aq)$$

for which the equilibrium constant expression is

$$K = \frac{[Pb^{2+}]}{[Ba^{2+}]} = \frac{[Pb^{2+}][SO_4^{2-}]}{[Ba^{2+}][SO_4^{2-}]} = \frac{K_{sp}(PbSO_4)}{K_{sp}(BaSO_4)} = \frac{1.8 \times 10^{-8}}{1.0 \times 10^{-10}} = 1.8 \times 10^2$$

This transposition will continue either until the ratio $[Pb^{2+}]/[Ba^{2+}] = 180$ or until all of the solid $PbSO_4$ disappears.

In this mixture there are 0.0300 mol of Pb^{2+} and (0.500 L)(0.100 mol/L) = 0.0500 mol Ba^{2+}: Obviously, the ratio $[Pb^{2+}]/[Ba^{2+}]$ must be less than 180. Therefore, all of the $PbSO_4$ will be converted into $BaSO_4$, giving a solution in which the Pb^{2+} concentration is

$$[Pb^{2+}] = \frac{0.0300 \ \text{mol}}{0.500 \ \text{L}} = 0.0600 \ M \quad \blacktriangle$$

Part of the Ba^{2+} in the original solution will be consumed to form $BaSO_4$: $Ba^{2+} + SO_4^{2-} \rightarrow BaSO_4(s)$. The $[Ba^{2+}]$ left in solution will be

$$0.0500 \ \text{mol} \ Ba^{2+} - (0.0300 \ \text{mol} \ PbSO_4)\left(\frac{1 \ \text{mol} \ Ba^{2+}}{1 \ \text{mol} \ PbSO_4}\right) = 0.0200 \ \text{mol} \ Ba^{2+}$$

Since the volume of the solution is still 0.500 L,

$$[Ba^{2+}] = \frac{0.0200 \ \text{mol} \ Ba^{2+}}{0.500 \ \text{L}} = 0.0400 \ M \quad \blacktriangle$$

Of course there is still some Ba^{2+} from solubility of $BaSO_4$, but it is very small:

$$BaSO_4 \rightleftharpoons Ba^{2+} + SO_4^{2-} \qquad K_{sp}(BaSO_4) = [Ba^{2+}][SO_4^{2-}] = 1.0 \times 10^{-10}$$

Now all you need to do is solve for $[SO_4^{2-}]$:

$$[SO_4^{2-}] = \frac{K_{sp}}{[Ba^{2+}]} = \frac{1.0 \times 10^{-10}}{0.0400} = 2.5 \times 10^{-9} \ M \quad \blacktriangle$$

PROBLEM 18-14 If you mixed 25 mL of 0.25 M calcium chloride ($CaCl_2$) solution with 25 mL of 0.25 M lead nitrate ($Pb(NO_3)_2$) solution, would you expect a precipitate to form? If so, what would it be?

Solution: The ions present in the two solutions are Ca^{2+}, Cl^-, Pb^{2+}, and NO_3^-. The solubility rules indicate that all nitrates are soluble (rule 2) and most chlorides are soluble; however, chlorides with Pb^{2+} are not soluble (rule 3). So you could expect to see a precipitate of $PbCl_2$.

PROBLEM 18-15 Give the balanced stoichiometric equation and the net ionic equation for the reaction that would occur if you mixed a solution of K_3PO_4 with one of $SrBr_2$.

Solution: The ions present when the solutions are mixed are K^+, Sr^{2+}, Br^-, and PO_4^{3-}. The solubility rules indicate that most bromides are soluble, and K^+ and Sr^{2+} are not exceptions (rule 3). Therefore, no solid bromide salt will form. Although most phosphates are insoluble, the phosphates of alkali metals are exceptions—so no solid potassium salt will form. But strontium is not an alkali metal, so strontium phosphate ($Sr_3(PO_4)_2$) should be insoluble and would therefore precipitate. Consequently, the stoichiometric equation for the reaction you'd expect is

$$3SrBr_2 + 2K_3PO_4 \longrightarrow Sr_3(PO_4)_2\downarrow + 6KBr$$

for which the net ionic equation is

$$3Sr^{2+}(aq) + 2PO_4^{3-}(aq) \longrightarrow Sr_3(PO_4)_2(s)$$

Supplementary Exercises

PROBLEM 18-16 The solubility of SrF_2 in water is 1.1×10^{-2} g in 100 mL of solution. Calculate the solubility product of SrF_2.

Answer: 2.7×10^{-9}

PROBLEM 18-17 The solubility of $BaCO_3$ in water is 4.6×10^{-3} g in 1.0 L of solution. Calculate the solubility product of $BaCO_3$.

Answer: 5.4×10^{-10}

PROBLEM 18-18 The solubility product of AgCl is 1.56×10^{-10}. Calculate the solubility of AgCl in water in moles per liter.

Answer: 1.25×10^{-5} mol/L

PROBLEM 18-19 The solubility of CaF_2 in 0.100 M $CaCl_2$ solution is 9.9×10^{-6} mol CaF_2 per liter. Calculate the solubility product K_{sp} for CaF_2.

Answer: 3.9×10^{-11}

PROBLEM 18-20 When solid MgF_2 is mixed with 100 mL of 0.0500 M NaF solution, the $[Mg^{2+}]$ is 2.6×10^{-6} M. Calculate the solubility product K_{sp} for MgF_2.

Answer: 6.5×10^{-9}

PROBLEM 18-21 If you start with 500 mL of solution in which $[BaCl_2]$ is 0.010 M and $[SrCl_2]$ is 0.020 M and then add Na_2SO_4, what solid precipitates first: $SrSO_4$ or $BaSO_4$? When the second sulfate starts to precipitate, what is the equilibrium concentration of the first cation? (K_{sp} for $SrSO_4 = 3.2 \times 10^{-7}$ and K_{sp} for $BaSO_4 = 1.0 \times 10^{-10}$.)

Answer: $BaSO_4$ precipitates first. When $SrSO_4$ begins to precipitate, $[Ba^{2+}] = 6.3 \times 10^{-6}$ M

PROBLEM 18-22 The solubility product of AgOH is $K_{sp} = 1.52 \times 10^{-8}$. Calculate the molar solubility of AgOH in pure water and the pH of the resulting solution.

Answer: $S = 1.23 \times 10^{-4}$ M; pH = 10.09

PROBLEM 18-23 Calculate the solubility of $Mg(OH)_2$ in a solution that is buffered at pH = 11.0. The solubility product of $Mg(OH)_2$ is $K_{sp} = 1.2 \times 10^{-11}$.

Answer: 1.2×10^{-5} M

PROBLEM 18-24 What is the solubility of Ag_2CO_3 ($K_{sp} = 8.2 \times 10^{-12}$) in 0.10 M Na_2CO_3 solution?

Answer: 4.5×10^{-6} M

PROBLEM 18-25 Calculate the solubility in moles per liter of MgF_2 in 0.025 M $Mg(NO_3)_2$ solution. ($K_{sp} = 6.9 \times 10^{-9}$ for MgF_2.)

Answer: 2.6×10^{-4} M

PROBLEM 18-26 Calculate the solubility of $BaSO_4$ in a solution made by saturating pure water with $PbSO_4$. ($K_{sp} = 1.0 \times 10^{-10}$ for $BaSO_4$ and $K_{sp} = 1.8 \times 10^{-8}$ for $PbSO_4$.)

Answer: 7.5×10^{-7} M

PROBLEM 18-27 The solubility product of $Ba(IO_3)_2$ is $K_{sp} = 1.3 \times 10^{-9}$: $Ba(IO_3)_2 \rightleftharpoons Ba^{2+} + 2IO_3^-$. (a) Calculate the solubility in moles per liter of $Ba(IO_3)_2$ in water. (b) Calculate the solubility of $Ba(IO_3)_2$ in 0.10 M $NaIO_3$ solution.

Answer: (a) 6.9×10^{-4} M (b) 1.3×10^{-7} M

PROBLEM 18-28 Lanthanum hydroxide is slightly soluble in water: $La(OH)_3 \rightleftharpoons La^{3+} + 3OH^-$. It is found that the solubility of $La(OH)_3$ is 2.0×10^{-13} M in a solution buffered at pH = 12.00. Calculate the solubility product K_{sp} for $La(OH)_3$.

Answer: 2.0×10^{-19}

PROBLEM 18-29 Will any precipitate form if you mix solutions of sodium carbonate (Na_2CO_3) and calcium chloride ($CaCl_2$)? If so, what will the precipitate be?

Answer: Yes; $CaCO_3$

PROBLEM 18-30 Solutions of ammonium acetate ($CH_3CO_2NH_4$) and cadmium sulfate ($CdSO_4$) are mixed. Write the stoichiometric equation and the net ionic equation for any reaction that will occur.

Answer: No reaction occurs

PROBLEM 18-31 A solution of 200 mL of 0.0400 M NaI is mixed with 300 mL of 0.0450 M $Pb(NO_3)_2$. Calculate the concentration of each of the four ions after mixing and the amount of PbI_2 (in grams) that precipitates. (K_{sp} for PbI_2 is 7.5×10^{-9}.)

Answer: $[Na^+] = 0.0160$ M; $[NO_3^-] = 0.0540$ M; $[Pb^{2+}] = 0.0190$ M; $[I^-] = 6.3 \times 10^{-4}$ M; 1.84 g PbI_2

19 ELECTROCHEMISTRY

THIS CHAPTER IS ABOUT

☑ **Electrochemical Fundamentals**
☑ **Electrolytic Cells**
☑ **Galvanic (Voltaic) Cells**
☑ **Standard Electrode Potentials**
☑ **The Nernst Equation**

19-1. Electrochemical Fundamentals

A. Charge and current

The SI unit of charge is the **coulomb** (C). All electrons carry the same amount of negative charge, and all protons carry the same amount of positive charge. The charge carried by an electron, 1.602×10^{-19} C, is equal in magnitude to the charge carried by the proton. The charges on all other charged particles are integral multiples of the charge on the electron.

   Electric current is the rate of flow of charge, measured in **amperes** (A), one of the fundamental SI units. One ampere is the flow of one coulomb per second:

CURRENT   amperes = coulombs per second   or $A = \dfrac{C}{s}$ (19-1a)

   or

CHARGE   coulombs = amperes × seconds   or $C = A \cdot s$ (19-1b)

B. Potential difference

The concept of potential difference is easiest to see in an analogy. You know that a skier has more potential energy at the top of a ski jump than at the bottom. Think of the work done by the ski lift in moving the skier to the top of the jump as increasing the skier's potential energy. Analogously, work is required to transfer electrical charge within an electric field. The **potential difference** is the change in the potential energy of a charge when work is done on it. The SI unit of potential difference is the **volt** (V), which is defined as the potential difference between two points if one joule (newton meter) of work is required to move one coulomb of charge from one point to the other:

**POTENTIAL
DIFFERENCE**   volts = joules per coulomb   or $V = \dfrac{J}{C}$ (19-2)

C. Electromotive force

Electromotive force (emf) is the work per unit charge required to move a charge between two points in an electric field. For our purposes, emf will represent the chemical driving force of any chemical system or cell in which chemical energy can be reversibly transformed into electrical energy.

D. Resistance

The flow of electric current through certain materials generates heat (e.g., the heating elements on an electric range). Such materials offer **resistance** to the flow of current, as a result of which some of the electrical energy is converted to heat energy. The SI unit of resistance is the **ohm** (Ω). A material has a resistance of one ohm if a potential difference of one volt produces a current of one ampere. The relationship is given by **Ohm's law**:

OHM'S LAW   voltage = current × resistance   or $V = A\Omega$ **(19-3)**

E. Conductivity

A material that offers little resistance to the flow of electric current is said to be a good conductor, or to have high **conductivity**. Conductors can be grouped by their properties:

1. *Metallic conductors*: Free electrons carry the electric current. Examples include copper, iron, aluminum, and mercury.
2. *Ionic conductors*: Ions carry the electric current. Both molten ionic compounds and solutions of ionic compounds can be good conductors.
3. *Semiconductors*: Free electrons carry the electric current. Semiconductors have low conductivity, but their properties make them of enormous importance in electronics. Examples include silicon and germanium.
4. *Insulators*: Very little current is carried in materials of extremely low conductivity. Examples include glass, paper, paraffin, and polyethylene.

19-2. Electrolytic Cells

Electrolysis is the term applied to any oxidation–reduction reaction induced by an electric current. For example, the decomposition of H_2O into $H_2(g)$ and $O_2(g)$ can be brought about by an electric current. The decomposition occurs at the **electrodes**—metal conductors used to introduce the current into the water. Oxidation occurs at the **anode**, reduction at the **cathode**.

EXAMPLE 19-1: Identify the anode and cathode in the electrolytic reaction shown in Figure 19-1.

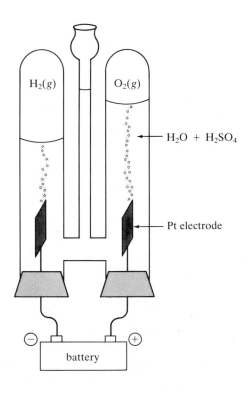

FIGURE 19-1. An electrolytic cell.

Solution: For the decomposition of H_2O, the reaction is

$$2H_2O \longrightarrow 2H_2(g) + O_2(g)$$

Because the oxidation number of hydrogen changes from $+1$ in H_2O to 0 in H_2 (gain of electrons), the reduction half-reaction must occur at the site of hydrogen production. So the platinum electrode on the left is the cathode. Because the oxidation number of oxygen changes from -2 in H_2O to 0 in O_2, the oxidation half-reaction must occur at the site of O_2 production. So the platinum electrode on the right is the anode.

Oxidation–reduction reactions are electroneutral; that is, the positive and negative charges must balance. This principle of electroneutrality holds for electrolysis as well—there can be no build-up of positive or negative charge. Obviously, some mechanism is at work that recycles the electrons through a continuous loop. We'll examine this mechanism and quantify its magnitude in the following sections.

A. Electrode reactions

Metallic magnesium—a metal of low density used in aerospace applications—is produced industrially by electrolysis from molten $MgCl_2$, which consists of Mg^{2+} and Cl^- ions. The Mg^{2+} ions are reduced at the cathode to $Mg(s)$ according to the half-reaction

$$Mg^{2+} + 2e^- \longrightarrow Mg(s)$$

and chloride ions are oxidized at the anode to $Cl_2(g)$ according to the half-reaction

$$2Cl^- \longrightarrow Cl_2 + 2e^-$$

Notice that the overall reaction

$$Mg^{2+} + 2Cl^- \xrightarrow{\text{electrolysis}} Mg + Cl_2$$

is the reverse of the vigorous spontaneous reaction between magnesium metal and chlorine gas to form $MgCl_2$. Thus the input of electrical energy can reverse the direction of spontaneous chemical reactions.

B. Faraday's law

The extent of chemical change produced by electrolysis is given by

- **Faraday's law**: *The amount of chemical change is directly proportional to the charge passing through the electrolytic reaction.*

In other words, the mass of any substance deposited on or liberated from an electrode is directly proportional to the number of coulombs that pass through the electrolytic reaction. The Faraday constant $\mathscr{F}$—the charge carried by one mole of electrons—equals $96\,485$ C, which is usually rounded off to 9.65×10^4 C.

EXAMPLE 19-2: Given that the charge on one electron is $1.602\,189\,2 \times 10^{-19}$ C, calculate Faraday's constant.

Solution: Since Faraday's constant is the charge carried by one mole of electrons, you'll need Avogadro's number for this calculation. And you'll want to use an equally precise value for Avogadro's number, $6.022\,045 \times 10^{23}$ electrons/mole:

$$\mathscr{F} = \left(\frac{1.602\,189\,2 \times 10^{-19}\text{ C}}{e^-}\right)\left(\frac{6.022\,045 \times 10^{23}\,e^-}{\text{mol}}\right)$$

$$= 96\,484.56\text{ C/mol}$$

- Faraday's constant is so useful in chemistry that the quantity of electricity $9.648\,456 \times 10^4$ C has been defined as a unit—the **faraday** (F).

EXAMPLE 19-3: What mass of Mg and of Cl_2 is produced in the electrolysis of molten $MgCl_2$, if a current of 7.50×10^2 A is passed through the molten salt for 1.00 hour?

Solution: First find the charge C from its definition (19-1):

$$C = A \cdot s$$
$$= (7.50 \times 10^2 \text{ A})(1.00 \text{ h})(60 \text{ min/h})(60 \text{ s/min})$$
$$= 2.70 \times 10^6 \text{ C}$$

So the amount of charge passed in 1.00 h at both the cathode and the anode is 2.70×10^6 C.

Now find the mass in each half-reaction. The cathode reaction is

$$Mg^{2+} + 2e^- \longrightarrow Mg$$

In stoichiometric terms, 1 mol of Mg^{2+} ions is reduced by 2 mol of electrons, i.e., by a charge of 2 F, to give metallic Mg. The mass of Mg produced is

$$\left(\frac{1 \text{ mol Mg}}{2 \text{ F}}\right)\left(\frac{24.31 \text{ g Mg}}{1 \text{ mol Mg}}\right)\left(\frac{1 \text{ F}}{9.65 \times 10^4 \text{ C}}\right)(2.70 \times 10^6 \text{ C}) = 3.40 \times 10^2 \text{ g Mg}$$

The anode reaction is

$$2Cl^- \longrightarrow Cl_2 + 2e^-$$

and the mass of Cl_2 produced is

$$\left(\frac{1 \text{ mol Cl}_2}{2 \text{ F}}\right)\left(\frac{2 \times 35.45 \text{ g Cl}_2}{1 \text{ mol Cl}_2}\right)\left(\frac{1 \text{ F}}{9.65 \times 10^4 \text{ C}}\right)(2.70 \times 10^6 \text{ C}) = 9.92 \times 10^2 \text{ g Cl}_2$$

19-3. Galvanic (Voltaic) Cells

Galvanic or **voltaic cells** are devices in which a redox reaction produces an electric current. The oxidizing and reducing agents of a spontaneous oxidation–reduction reaction are maintained in separate **half-cells**, which are connected by a conductor for the transfer of electrons and a salt bridge for the diffusion of ions. Consider, for example, the spontaneous redox reaction

$$Zn(s) + Cu^{2+}(aq) \longrightarrow Zn^{2+}(aq) + Cu(s)$$

separated spatially as shown in Figure 19-2a. Metallic zinc is oxidized at the anode:

$$Zn(s) \longrightarrow Zn^{2+}(aq) + 2e^-$$

This oxidation provides electrons, so the anode is the negative end of the external electrical circuit. The electrons flow to the cathode and reduce Cu^{2+} ions:

$$Cu^{2+} + 2e^- \longrightarrow Cu$$

This reduction requires electrons, so the cathode is the positive end of the external circuit.
note: A shorthand way of writing this galvanic cell is

$$Zn|ZnSO_4(aq)\|CuSO_4(aq)|Cu$$

The single lines show a boundary between two phases (in this case solid and solution); the double line represents the salt bridge.

Conventionally, the anode is shown at the left and the cathode at the right of the array. When no current is being drawn, nothing happens in the cell. When a conductor including, say, a flashlight bulb or a voltmeter is connected by an external circuit to both ends of the array, a spontaneous reaction occurs (see Figure 19-2b). The chemical reaction moves from left to right in the array. That is, Zn metal goes into solution as Zn^{2+} ions, and Cu^{2+} ions move out of solution to be deposited as Cu metal on the cathode. The electrons move in the external circuit in the opposite direction. The electrons leave the Zn metal, move in the external circuit, supply energy to the load (flashlight bulb), and return to the Cu. Thus you can see that a galvanic cell is in effect an *electron pump.*

The voltmeter provides a way to measure the **electromotive force** (emf, denoted E), that is, the electrochemical potential difference between the two half-cells. The total emf is supplied by an individual contribution from each half-cell its **electrode potential**. Because both half-reactions must occur

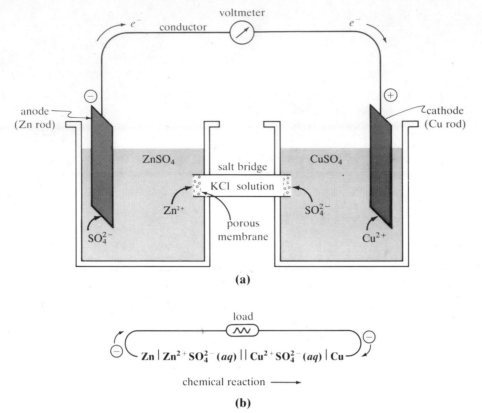

FIGURE 19-2. (a) A galvanic cell; (b) galvanic-cell shorthand.

simultaneously to generate an electrical current, there is no way to measure the electrode potential for a half-reaction. For this reason, chemists have adopted a convention for separating total emf into electrode potentials.

19-4. Standard Electrode Potentials

The emf of a chemical cell varies with concentration and changes of state. As a result, standard conditions have been established for measuring the electrical nature of chemical cells. For our purposes, *standard conditions* are as follows:

- Gases—one standard atmosphere at 298 K
- Liquids and solids—in their normal, pure states at 298 K
- Solutions—at a concentration of exactly 1 M at 298 K

A. The standard hydrogen electrode

The **standard hydrogen electrode** (SHE) has been designated as the reference point for measuring electrode potentials. The reaction is

$$2H^+(aq) + 2e^- \longrightarrow H_2(g)$$

under standard conditions: $[H^+] = 1$ M, $p_{H_2} = 1$ atm, $T = 298$ K. The **standard electrode potential** $E°$ (also called the **standard reduction potential**) for this reaction is equal to 0.0000 V by definition.

B. Measuring standard electrode potentials

Chemists use the SHE to measure the standard reduction potentials for other half-reactions. For example, one half-cell might consist of a pure copper rod suspended in a 1 M $Cu^{2+}SO_4^{2-}$ solution at 25°C, connected to the SHE in the other half-cell by a wire that passes to and from a voltmeter of very high resistance (very little current flow, hence no electrolysis). The magnitude of the voltage,

TABLE 19-1: Standard Reduction Potentials*

Half-reaction	$E°$ (V)	Half-reaction	$E°$ (V)
$Li^+ + e^- \longrightarrow Li$	-3.040	$2H^+ + 2e^- \longrightarrow H_2$	0.000
$K^+ + e^- \longrightarrow K$	-2.931	$Hg_2Cl_2 + 2e^- \longrightarrow 2Hg + 2Cl^-$	0.268
$Na^+ + e^- \longrightarrow Na$	-2.711	$Cu^{2+} + 2e^- \longrightarrow Cu$	0.342
$Mg^{2+} + 2e^- \longrightarrow Mg$	-2.372	$Cu^+ + e^- \longrightarrow Cu$	0.521
$Zn^{2+} + 2e^- \longrightarrow Zn$	-0.762	$I_2 + 2e^- \longrightarrow 2I^-$	0.536
$Cr^{3+} + 3e^- \longrightarrow Cr$	-0.744	$Fe^{3+} + e^- \longrightarrow Fe^{2+}$	0.771
$Fe^{2+} + 2e^- \longrightarrow Fe$	-0.447	$Hg_2^{2+} + 2e^- \longrightarrow 2Hg$	0.797
$Cd^{2+} + 2e^- \longrightarrow Cd$	-0.403	$Ag^+ + e^- \longrightarrow Ag$	0.800
$Ni^{2+} + 2e^- \longrightarrow Ni$	-0.257	$2Hg^{2+} + 2e^- \longrightarrow Hg_2^{2+}$	0.920
$Sn^{2+} + 2e^- \longrightarrow Sn$	-0.138	$Br_2(aq) + 2e^- \longrightarrow 2Br^-$	1.087
$Pb^{2+} + 2e^- \longrightarrow Pb$	-0.126	$O_2 + 4H^+ + 4e^- \longrightarrow 2H_2O$	1.229
$Fe^{3+} + 3e^- \longrightarrow Fe$	-0.037	$Cl_2 + 2e^- \longrightarrow 2Cl^-$	1.358
$2H^+ + 2e^- \longrightarrow H_2$	0.000	$MnO_4^- + 8H^+ + 5e^- \longrightarrow Mn^{2+} + 4H_2O$	1.507
		$Ce^{4+} + e^- \longrightarrow Ce^{3+}$	1.61

* Data taken from *Handbook of Chemistry and Physics* (64th Ed.), CRC Press, Boca Raton FL.

along with the sign, observed for the electrode with respect to the SHE is the standard reduction potential. For the $Cu|Cu^{2+}(aq)$ electrode, $E° = 0.34$ V, which implies that electrons flow from the SHE to the copper electrode (the copper has a greater tendency to gain electrons than the hydrogen ion).

The standard reduction potentials of some important half-reactions are given in Table 19-1. You can predict the outcome of many redox reactions, or the nature of a chemical cell, by using these values.

note: If your textbook lists standard *oxidation* potentials, the magnitude of each $E°$ value will equal that shown in Table 19-1, but the sign and the direction of the reaction will be reversed. For $Cu^{2+} + 2e^- \to Cu$, the standard reduction potential is $+0.342$ V; for $Cu \to Cu^{2+} + 2e^-$, the standard oxidation potential is -0.342 V.

EXAMPLE 19-4: Give the standard electrode potential for

$$2I^- \longrightarrow I_2 + 2e^-$$

Solution: This reaction is written as an oxidation. According to Table 19-1, for the reaction $I_2 + 2e^- \to 2I^-$, $E° = 0.536$ V. Therefore $E°$ for the *reverse* reaction is -0.536 V.

Note the following properties of $E°$:

- If the half-reaction is multiplied by a positive integer, the value of the standard reduction potential is unchanged.
- A positive $E°$ indicates that the species accepting electrons is a stronger oxidizing agent than H^+. The larger the positive $E°$, the stronger the oxidizing agent.
- A negative $E°$ indicates that the species donating electrons—the species on the right-hand side of the half-reaction—is a stronger reducing agent than H_2. The larger the negative $E°$, the stronger the reducing agent.

EXAMPLE 19-5: Arrange MnO_4^-, Ce^{4+}, and Cl_2 in order of increasing strength as oxidizing agents.

Solution: Table 19-1 lists the following $E°$ values in volts: MnO_4^- 1.507, Ce^{4+} 1.61, Cl_2 1.358. Since increasing $E°$ means stronger oxidizing power, the order of increasing oxidizing strength is

$$Cl_2 < MnO_4^- < Ce^{4+}$$

EXAMPLE 19-6: Arrange Cd, Mg, and Zn in order of increasing strength as reducing agents.

Solution: Table 19-1 lists the following $E°$ values in volts: Cd -0.403, Mg -2.372, Zn -0.762. Since a large negative $E°$ value means stronger reducing power, the order of increasing strength as reducing agents is

$$Cd < Zn < Mg$$

C. Calculating standard cell potentials

Balanced oxidation–reduction reaction equations can be obtained by adding half-reactions (see Chapter 13). If you add the appropriate standard electrode potentials for the half-reactions, you can obtain the **standard cell potential** for the overall redox reaction. There are only three simple steps to remember:

1. Arrange the half-reactions so that electrons cancel; you mustn't have any electrons left over in the balanced redox reaction.
2. If you reverse the direction of a half-reaction, reverse the sign of its standard electrode potential.
3. If you multiply a half-reaction by a positive integer, don't change the magnitude of its standard electrode potential.

EXAMPLE 19-7: Calculate the standard cell potential for

$$Zn + Cu^{2+} \longrightarrow Zn^{2+} + Cu$$

Solution: Find the standard reduction potential for each half-reaction from Table 19-1:

$$Cu^{2+} + 2e^- \longrightarrow Cu \qquad E° = 0.342 \text{ V}$$
$$Zn^{2+} + 2e^- \longrightarrow Zn \qquad E° = -0.762 \text{ V}$$

To obtain the given reaction, reverse the Zn^{2+} reduction half-reaction (remember to change the sign of its potential):

$$Zn \longrightarrow Zn^{2+} + 2e^- \qquad E° = 0.762 \text{ V}$$

Now add the reactions and the potentials:

$$
\begin{array}{llll}
Cu^{2+} + 2e^- \longrightarrow & & Cu & E° = 0.342 \text{ V} \\
Zn & \longrightarrow Zn^{2+} + 2e^- & & E° = 0.762 \text{ V} \\
\hline
Zn + Cu^{2+} & \rightleftharpoons Zn^{2+} & + Cu & E° = 1.104 \text{ V}
\end{array}
$$

The standard cell potential is 1.104 V.

EXAMPLE 19-8: Calculate the standard cell potential for

$$2Cr + 3Cd^{2+} \longrightarrow 2Cr^{3+} + 3Cd$$

Solution: From Table 19-1,

$$Cr^{3+} + 3e^- \longrightarrow Cr \qquad E° = -0.744 \text{ V}$$
$$Cd^{2+} + 2e^- \longrightarrow Cd \qquad E° = -0.403 \text{ V}$$

Multiply the Cr^{3+} reduction by 2, but note that this doesn't affect its standard electrode potential:

$$2Cr^{3+} + 6e^- \longrightarrow 2Cr \qquad E° = -0.744 \text{ V}$$

Reverse the Cr^{3+} reduction, changing the sign of the electrode potential:

$$2Cr \longrightarrow 2Cr^{3+} + 6e^- \qquad E° = 0.744 \text{ V}$$

The Cd^{2+} reduction must be multiplied by 3, but its potential is unchanged:

$$3Cd^{2+} + 6e^- \longrightarrow 3Cd \qquad E° = -0.403 \text{ V}$$

Now add the half-reactions (the electrons cancel) and the potentials:

$$2Cr + 3Cd^{2+} \longrightarrow 2Cr^{3+} + 3Cd \qquad E° = 0.744 - 0.403 = 0.341 \text{ V}$$

EXAMPLE 19-9: Calculate the standard cell potential for

$$Fe^{3+} + Ag \longrightarrow Fe^{2+} + Ag^+$$

Solution: From Table 19-1,

$$Fe^{3+} + e^- \longrightarrow Fe^{2+} \qquad E° = 0.771 \text{ V}$$
$$Ag^+ + e^- \longrightarrow Ag \qquad E° = 0.800 \text{ V}$$

Reverse the Ag^+ reduction:

$$Ag \longrightarrow Ag^+ + e^- \qquad E° = -0.800 \text{ V}$$

Now add:

$$Fe^{3+} + Ag \longrightarrow Fe^{2+} + Ag^+ \qquad E° = 0.771 - 0.800 = -0.029 \text{ V}$$

D. Spontaneous and nonspontaneous reactions

A redox reaction that is spontaneous in a thermodynamic sense (see Section 16-4) will have a positive $E°$. If a reaction has a negative $E°$, then it is not thermodynamically spontaneous; the reverse of the reaction will have a positive $E°$ and will be spontaneous.

EXAMPLE 19-10: Is the reaction $Fe^{3+} + Ag \rightarrow Fe^{2+} + Ag^+$ spontaneous? If not, write the spontaneous reaction involving these species and determine its $E°$.

Solution: As we saw in Example 19-9, $E°$ for the reaction $Fe^{3+} + Ag \rightarrow Fe^{2+} + Ag^+$ is negative (-0.029 V), so the reaction is not spontaneous. The reverse reaction, $Fe^{2+} + Ag^+ \rightarrow Fe^{3+} + Ag$, will be spontaneous and its $E°$ will have the same magnitude but opposite sign: $E° = 0.029$ V.

E. $E°$, $\Delta G°$, and K

Measurements of $E°$ are related to the free-energy change $\Delta G°$ of a reaction. For our purposes, the only work performed in a galvanic cell is the electrical work performed by n moles of electrons, each mole having charge $\mathscr{F}$. The total standard free-energy change $\Delta G°$ is given by

$$\Delta G° = -n\mathscr{F}E° \tag{19-4}$$

and $\Delta G°$ is also given by eq. (16-9)

$$\Delta G° = -2.303RT \log K$$

Setting these two equations equal to each other, we get

$$-n\mathscr{F}E° = -2.303RT \log K$$

which gives us

$$E° = \frac{2.303RT \log K}{n\mathscr{F}} \tag{19-5}$$

where the gas constant $R = 8.314 \text{ J mol}^{-1}\text{K}^{-1}$, T is the Kelvin temperature, n is the number of moles of electrons transferred, the Faraday constant $\mathscr{F} = 9.65 \times 10^4 \text{ J V}^{-1}\text{mol}^{-1}$, and K is the equilibrium constant for the reaction. For $T = 298$ K, eq. (19-5) reduces to

$$E° = \frac{0.0592}{n} \log K \qquad \text{or} \qquad \log K = \frac{nE°}{0.0592} \tag{19-6}$$

EXAMPLE 19-11: Calculate the equilibrium constant at 298 K for the reaction

$$Fe^{2+} + Ag^+ \longrightarrow Fe^{3+} + Ag$$

Solution: We know from Example 19-10 that $E°$ for this reaction is 0.029 V, and we saw in Example 19-9 that $n = 1$ for this reaction. So from eq. (19-6)

$$\log K = \frac{nE°}{0.0592}$$

$$= \frac{(1)(0.029)}{0.0592} = 0.49$$

$$K = \text{antilog}(0.49) = 3.1$$

That means that in this reaction

$$K = \frac{[Fe^{3+}]}{[Fe^{2+}][Ag^+]} = 3.1$$

(By convention, the concentration of any solid species, such as Ag, is not included in the equilibrium-law expression.)

EXAMPLE 19-12: Calculate $\Delta G°$ for $Fe^{2+} + Cd \rightarrow Fe + Cd^{2+}$. What conclusions can you draw from your answer?

Solution: First find $E°$ for the reaction. From Table 19-1,

$$Fe^{2+} + 2e^- \longrightarrow Fe \qquad E° = -0.447 \text{ V}$$
$$Cd^{2+} + 2e^- \longrightarrow Cd \qquad E° = -0.403 \text{ V}$$

Reverse the second reaction and its electrode potential:

$$Cd \longrightarrow Cd^{2+} + 2e^- \qquad E° = 0.403 \text{ V}$$

and add the two half-reactions:

$$Fe^{2+} + Cd \longrightarrow Fe + Cd^{2+} \qquad E° = -0.447 + 0.403 = -0.044 \text{ V}$$

Now use the relationship (19-4) between $\Delta G°$ and $E°$:

$$\Delta G° = -n\mathscr{F}E°$$

From the half-reactions you see that $n = 2$ (two moles of electrons are transferred); therefore substitution gives you

$$\Delta G° = -(2)(96\,500 \text{ J V}^{-1}\text{mol}^{-1})(-0.044 \text{ V})$$
$$= 8.5 \times 10^3 \text{ J mol}^{-1}$$

Finally, you can conclude that, because $\Delta G°$ is positive, the reaction as written is nonspontaneous. The spontaneous reverse reaction, $Fe + Cd^{2+} \rightarrow Fe^{2+} + Cd$, has a $\Delta G°$ value of -8.5×10^3 J mol^{-1}.

19-5. The Nernst Equation

The cell potential of galvanic cells operating under nonstandard conditions (concentrations other than 1 *M*, temperatures other than 298 K) can be markedly different from standard cell potentials. The **Nernst equation** allows you to calculate cell potentials under *any* conditions. For the general cell reaction

$$aA + bB + \cdots \longrightarrow pP + qQ + \cdots$$

the cell potential E is given by

NERNST EQUATION

$$E = E° - \left(\frac{2.303RT}{n\mathscr{F}}\right)\log\left(\frac{[P]^p[Q]^q \cdots}{[A]^a[B]^b \cdots}\right) \qquad (19\text{-}7)$$

(By convention, the concentrations of solids and pure liquids are not included.) For cells operating at 298 K, eq. (19-7) reduces to

$$E = E° - \left(\frac{0.0592}{n}\right)\log Q \qquad \textbf{(19-8)}$$

where Q is the reaction quotient, which represents the concentration expression (see Problem 15-9).

EXAMPLE 19-13: The concentrations of Cd^{2+} and Fe^{2+} in a galvanic cell reaction ($Fe^{2+} + Cd \rightarrow Fe + Cd^{2+}$) operating at 298 K are 0.010 M and 2.00 M, respectively. Determine the cell potential.

Solution: From Example 9-12, $E°$ for the reaction is -0.044 V and $n = 2$. Because the cell is operating at 298 K, you can use the simplified version of the Nernst equation (19-8):

$$E = E° - \left(\frac{0.0592}{n}\right)\log\left(\frac{[Cd^{2+}]}{[Fe^{2+}]}\right)$$

$$= -0.044 - \left(\frac{0.0592}{2}\right)\log\left(\frac{0.010}{2.00}\right)$$

$$= -0.044 - \left(\frac{0.0592}{2}\right)\log 0.005$$

$$= -0.044 - \left(\frac{0.0592}{2}\right)(-2.30)$$

$$= -0.044 + 0.068$$
$$= 0.024 \text{ V}$$

Because the cell potential is positive, the reaction under nonstandard conditions is spontaneous, in contrast to the nonspontaneous reaction under standard conditions ($E° = -0.044$ V). This outcome is in the direction expected from Le Chatelier's principle.

SUMMARY

1. (a) Charge (in coulombs) = amperes × seconds $\quad C = A \cdot s$

   (b) Current (in amperes) = coulombs per second $\quad A = \dfrac{C}{s}$

   (c) Potential difference (in volts) = joules per coulomb $\quad V = \dfrac{J}{C}$

   (d) Voltage (in volts) = current × resistance $\quad V = A\Omega$

2. In an electrolytic cell, oxidation occurs at the anode and reduction occurs at the cathode.
3. Faraday's law states that the amount of chemical change is proportional to the charge passed through an electrolytic cell.
4. Galvanic cells are devices in which a redox reaction produces electric current (redox half-reactions can be physically separated at the electrodes).
5. The cell potential is the sum of two electrode potentials.
6. The standard hydrogen electrode (SHE) has a potential arbitrarily called zero.
7. Standard electrode potentials are used to measure the strength of redox reagents.
8. Electrode potentials can be combined algebraically to give cell potentials.
9. Reversing a redox reaction changes the sign of its potential.
10. A spontaneous cell reaction has a positive potential.
11. $E°$ is related to $\Delta G°$, K, and Q.

RAISE YOUR GRADES
Can you...?

☑ recognize anode and cathode in electrochemical cells
☑ use Faraday's law to quantitate electrolytic reactions
☑ recognize the half-reactions that make up a galvanic cell
☑ use shorthand notation to describe galvanic cells
☑ combine electrode potentials to get cell potentials
☑ use electrode potentials to understand strengths of redox reagents
☑ use electrode potentials to predict spontaneous reactions

Can you calculate...?

☑ $\Delta G°$ from $E°$
☑ K from $E°$
☑ $E°$ under nonstandard conditions

SOLVED PROBLEMS

PROBLEM 19-1   If a current of 0.111 A flows through a wire for 3.60×10^3 s, how many coulombs is that? How many electrons have flowed through the wire?

Solution: If the charge (C) in coulombs is amperes times seconds (eq. 19-1), then

$$C = (0.111 \text{ A})(3.60 \times 10^3 \text{ s}) = 4.00 \times 10^2 \text{ C}$$

Since one electron carries 1.60×10^{-19} C, the number of electrons is

$$\frac{4.00 \times 10^2 \text{ C}}{1.60 \times 10^{-19} \text{ C}/e^-} = 2.50 \times 10^{21} \ e^-$$

PROBLEM 19-2   Calculate the energy in joules that one electron gains when it moves through a potential difference of exactly 1 V.

Solution: Since volts = joules/coulomb (eq. 19-2), you can solve for energy by rearranging: joules = (volts) × (coulombs). Then since the charge on an electron is 1.602×10^{-19} C, you get

$$\text{energy (in joules)} = (1 \text{ V})(1.602 \times 10^{-19} \text{ C})$$
$$= 1.602 \times 10^{-19} \text{ J}$$

This amount of energy is called an **electron-volt** (eV). It is often used for energy measurements on the atomic or molecular level.

PROBLEM 19-3   If an electric current is passed through a solution of $CuCl_2$, the following reactions occur at the electrodes:

$$2Cl^- \longrightarrow Cl_2 + 2e^- \quad \text{and} \quad Cu^{2+} + 2e^- \longrightarrow Cu$$

Which electrode is the anode and which is the cathode?

Solution: At the first electrode, Cl^- is oxidized to Cl_2, so it is the anode. At the second electrode, Cu^{2+} is reduced to Cu, so it is the cathode.

PROBLEM 19-4   Calculate the mass in grams of metallic silver produced in the electrolysis of a $AgNO_3$ solution when a current of 0.155 A is passed through the solution for 0.750 h.

TABLE 19-1: Standard Reduction Potentials*

Half-reaction	$E°$ (V)	Half-reaction	$E°$ (V)
$Li^+ + e^- \longrightarrow Li$	−3.040	$2H^+ + 2e \longrightarrow H_2$	0.000
$K^+ + e^- \longrightarrow K$	−2.931	$Hg_2Cl_2 + 2e^- \longrightarrow 2Hg + 2Cl^-$	0.268
$Na^+ + e^- \longrightarrow Na$	−2.711	$Cu^{2+} + 2e^- \longrightarrow Cu$	0.342
$Mg^{2+} + 2e^- \longrightarrow Mg$	−2.372	$Cu^+ + e^- \longrightarrow Cu$	0.521
$Zn^{2+} + 2e^- \longrightarrow Zn$	−0.762	$I_2 + 2e^- \longrightarrow 2I^-$	0.536
$Cr^{3+} + 3e^- \longrightarrow Cr$	−0.744	$Fe^{3+} + e^- \longrightarrow Fe^{2+}$	0.771
$Fe^{2+} + 2e^- \longrightarrow Fe$	−0.447	$Hg_2^{2+} + 2e^- \longrightarrow 2Hg$	0.797
$Cd^{2+} + 2e^- \longrightarrow Cd$	−0.403	$Ag^+ + e^- \longrightarrow Ag$	0.800
$Ni^{2+} + 2e^- \longrightarrow Ni$	−0.257	$2Hg^{2+} + 2e^- \longrightarrow Hg_2^{2+}$	0.920
$Sn^{2+} + 2e^- \longrightarrow Sn$	−0.138	$Br_2(aq) + 2e^- \longrightarrow 2Br^-$	1.087
$Pb^{2+} + 2e^- \longrightarrow Pb$	−0.126	$O_2 + 4H^+ + 4e^- \longrightarrow 2H_2O$	1.229
$Fe^{3+} + 3e^- \longrightarrow Fe$	−0.037	$Cl_2 + 2e^- \longrightarrow 2Cl^-$	1.358
$2H^+ + 2e^- \longrightarrow H_2$	0.000	$MnO_4^- + 8H^+ + 5e^- \longrightarrow Mn^{2+} + 4H_2O$	1.507
		$Ce^{4+} + e^- \longrightarrow Ce^{3+}$	1.61

* Data taken from *Handbook of Chemistry and Physics* (64th Ed.), CRC Press, Boca Raton FL.

Solution: The Ag is formed at the cathode by the half-reaction $Ag^+ + e^- \rightarrow Ag$. The amount of charge ($C$) passed is

$$C = A \cdot s$$
$$= (0.155 \text{ A})(0.750 \text{ h})(60 \text{ min/h})(60 \text{ s/min})$$
$$= 418.5 \text{ C}$$

Then, remembering that Faraday's constant is 9.65×10^4 C, you can calculate the mass of Ag produced:

$$\left(\frac{1 \text{ mol Ag}}{1 \text{ F}}\right)\left(\frac{107.9 \text{ g Ag}}{1 \text{ mol Ag}}\right)\left(\frac{1 \text{ F}}{9.65 \times 10^4 \text{ C}}\right)(418.5 \text{ C}) = 0.468 \text{ g Ag}$$

PROBLEM 19-5 Give the shorthand notation for a galvanic cell that consists of a piece of platinum (inert electrode) dipping into a solution containing Fe^{2+} and Fe^{3+} for the anode, and a piece of silver dipping into a solution of $AgNO_3$ for the cathode.

Solution: Usually the anode is placed on the left-hand side. Use a single line to show the boundary between two phases and a double line to represent a salt bridge. Separate two ions in the same solution by a comma, showing first in the anode the ion that *loses* electrons—and first in the cathode the ion that *gains* electrons:

$$Pt \mid Fe^{2+}, Fe^{3+} \parallel Ag^+ \mid Ag$$

PROBLEM 19-6 Arrange Cl_2, Mg^{2+}, Cu^{2+}, and O_2 in order of increasing strength as oxidizing agents. [Table 19-1 is reproduced here for your convenience.]

Solution: Find the electrode potentials (in volts) in Table 19-1: Cl_2 1.36, Mg^{2+} −2.37, Cu^{2+} 0.34, O_2 1.23. Since increasing $E°$ (more positive) means a stronger oxidizing agent, the order of increasing oxidizing strength is

$$Mg^{2+} < Cu^{2+} < O_2 < Cl_2$$

PROBLEM 19-7 Arrange Pb, Cd^{2+}, Fe^{2+}, and Cl^- in order of increasing strength as reducing agents.

Solution: The electrode potentials found in Table 19-1 are

$$Pb^{2+} + 2e^- \longrightarrow Pb \qquad E° = -0.126 \text{ V}$$
$$Cd^{2+} + 2e^- \longrightarrow Cd \qquad E° = -0.403 \text{ V}$$
$$Fe^{3+} + e^- \longrightarrow Fe^{2+} \qquad E° = 0.771 \text{ V}$$
$$Fe^{2+} + 2e^- \longrightarrow Fe \qquad E° = -0.447 \text{ V}$$
$$Cl_2 + 2e^- \longrightarrow 2Cl^- \qquad E° = 1.358 \text{ V}$$

The Cd species given in the problem is Cd^{2+} not Cd. But Cd^{2+} is not a reducing agent—or at least it is not listed as a reductant in Table 19-1. Likewise, Fe^{2+} is the proposed reducing agent, not Fe. The half-reactions that are *relevant*, listed in order of decreasing electrode potentials, are therefore

$$Cl_2 + 2e^- \longrightarrow 2Cl^- \qquad E° = \quad 1.358 \text{ V}$$
$$Fe^{3+} + e^- \longrightarrow Fe^{2+} \qquad E° = \quad 0.771 \text{ V}$$
$$Pb^{2+} + 2e^- \longrightarrow Pb \qquad E° = -0.126 \text{ V}$$

So the order of increasing strength as a reducing agent is $Cl^- < Fe^{2+} < Pb$, and Cd^{2+} is out of the running.

PROBLEM 19-8 Calculate the standard cell potentials for the reactions (a) $Cl_2 + 2Br^- \rightarrow 2Cl^- + Br_2$ and (b) $Fe + 2Ag^+ \rightarrow 2Ag + Fe^{2+}$.

Solution:
(a) Find the standard reduction potentials for the half-reactions in Table 19-1.

$$Cl_2 + 2e^- \longrightarrow 2Cl^- \qquad E° = 1.358 \text{ V}$$
$$Br_2 + 2e^- \longrightarrow 2Br^- \qquad E° = 1.087 \text{ V}$$

Reverse the Br_2 half-reaction; change the sign of its electrode potential and add it to that of the Cl_2 half-reaction:

$$Cl_2 + 2Br^- \longrightarrow 2Cl^- + Br_2 \qquad E° = 1.358 + (-1.087) = 0.271 \text{ V}$$

(b)
$$Ag^+ + e^- \longrightarrow Ag \qquad E° = \quad 0.800 \text{ V}$$
$$Fe^{2+} + 2e^- \longrightarrow Fe \qquad E° = -0.447 \text{ V}$$

Reverse the Fe^{2+} half-reaction; change the sign of its $E°$ and add it to $E°$ for the Ag^+ half-reaction.

$$E° = 0.800 + (+0.447) = 1.247 \text{ V}$$

PROBLEM 19-9 Calculate the cell potential for the reaction $Cr + Cd^{2+} \rightarrow Cr^{3+} + Cd$ when the Cd^{2+} concentration is 2.0 M and the Cr^{3+} concentration is 0.10 M.

Solution: First you need to balance the equation. Write the two reduction half-reactions:

$$Cd^{2+} + 2e^- \longrightarrow Cd \qquad E° = -0.403 \text{ V}$$
$$Cr^{3+} + 3e^- \longrightarrow Cr \qquad E° = -0.744 \text{ V}$$

Reverse the Cr^{3+} reduction half-reaction and multiply it by 2. Multiply the Cd^{2+} half-reaction by 3. Then add the two half-reactions and electrode potentials:

$$3Cd^{2+} \quad + 6e^- \longrightarrow 3Cd \qquad\qquad E° = -0.403 \text{ V}$$
$$\underline{\qquad 2Cr \qquad \longrightarrow \qquad 2Cr^{3+} + 6e^- \qquad E° = \quad 0.744 \text{ V}}$$
$$3Cd^{2+} + 2Cr \qquad \longrightarrow 3Cd + 2Cr^{3+} \qquad E° = \quad 0.341 \text{ V}$$

So $E° = +0.341$ V, which you can now use in the simplified Nernst equation (19-8):

$$E = E° - \left(\frac{0.0592}{n}\right)\log\left(\frac{[Cr^{3+}]^2}{[Cd^{2+}]^3}\right)$$

$$= 0.341 - \left(\frac{0.0592}{6}\right)\log\frac{(0.10)^2}{(2.0)^3}$$

$$= 0.341 - \left(\frac{0.0592}{6}\right)\log(1.25 \times 10^{-3}) = 0.341 - (-0.029)$$

$$= 0.370 \text{ V}$$

PROBLEM 19-10 Which of the following is the strongest reducing agent: (a) Br^-, (b) $H_2(g)$, (c) Zn^{2+}, (d) Fe^{3+}, or (e) Fe^{2+}?

Solution: Look up these species in Table 19-1. Notice that Zn^{2+} and Fe^{3+} are in the oxidized form (on the left side of the half-reaction equations). Neither of them will be the strongest reducing agent. Of the remaining three, find the lowest or most negative $E°$ in the table:

$$Br_2 + 2e^- \longrightarrow 2Br^- \qquad E° = 1.087 \text{ V}$$
$$2H^+ + 2e^- \longrightarrow H_2(g) \qquad E° = 0.000 \text{ V}$$
$$Fe^{3+} + e^- \longrightarrow Fe^{2+} \qquad E° = 0.771 \text{ V}$$

The species having the lowest $E°$ is H^+; therefore, $H_2(g)$ is the strongest reducing agent. The correct answer is (b).

PROBLEM 19-11 Which one of the following is a stronger oxidizing agent than Fe^{3+}: **(a)** Cu^{2+}, **(b)** Br^-, **(c)** Hg_2^{2+}, **(d)** Zn, or **(e)** Zn^{2+}?

Solution: Look up these species in Table 19-1. Both Br^- and Zn appear on the right-hand sides of the half-reactions, so they are not oxidizing agents. Of the remaining species, find the one that has a larger or more positive $E°$:

$$Cu^{2+} + 2e^- \longrightarrow Cu \qquad E° = 0.342 \text{ V}$$
$$Hg_2^{2+} + 2e^- \longrightarrow 2Hg \qquad E° = 0.797 \text{ V}$$
$$Zn^{2+} + 2e^- \longrightarrow Zn \qquad E° = -0.762 \text{ V}$$
$$Fe^{3+} + e^- \longrightarrow Fe^{2+} \qquad E° = 0.771 \text{ V}$$

The only $E°$ greater than 0.771 is 0.797 V. The correct answer is **(c)** Hg_2^{2+}.

PROBLEM 19-12 The "*dry cell*" or *LeClanché cell* uses the following reaction to produce electricity:

$$2MnO_2(s) + 2H^+ + Zn(s) \longrightarrow Mn_2O_3(s) + H_2O + Zn^{2+}$$

The voltage for this cell is 1.561 V. The Zn^{2+} and H^+ concentrations are each 0.10 M. Use $E°$ for the Zn^{2+} reduction potential to calculate the reduction potential for the cathode $E_c°$.

Solution: Use the simplified Nernst equation (19-8) to calculate $E°$ for the cell:

$$E = E° - \left(\frac{0.0592}{n}\right)\log\left(\frac{[Zn^{2+}]}{[H^+]^2}\right)$$

$$E° = E + \left(\frac{0.0592}{n}\right)\log\left(\frac{[Zn^{2+}]}{[H^+]^2}\right)$$

$$= 1.561 + \left(\frac{0.0592}{2}\right)\log\frac{(0.10)}{(0.10)^2}$$

$$= 1.561 + \left(\frac{0.0592}{2}\right)\log 10 = 1.561 + 0.0296$$

$$= 1.591 \text{ V}$$

At the anode

$$Zn \longrightarrow Zn^{2+} + 2e^- \qquad E_a° = -(-0.762) = 0.762$$

Therefore, for the cathode reaction

$$2MnO_2(s) + 2H^+ + 2e^- \longrightarrow Mn_2O_3(s) + H_2O$$

$E_c°$ can be found by

$$E° = 1.591 = E_c° + E_a°$$
$$E_c° = 1.591 - 0.762$$
$$= 0.829 \text{ V}$$

PROBLEM 19-13 Will Ce^{4+} oxidize H_2O to O_2 in acidic solution?

Solution: The electrode potentials that you need from Table 19-1 are

$$O_2 + 4e^- + 4H^+ \longrightarrow 2H_2O \qquad E° = 1.229 \text{ V}$$
$$Ce^{4+} + e^- \longrightarrow Ce^{3+} \qquad E° = 1.61 \text{ V}$$

Reverse the O_2 half-reaction:

$$2H_2O \longrightarrow O_2 + 4e^- + 4H^+ \qquad E° = -1.229 \text{ V}$$

Multiply the Ce^{4+} half-reaction by 4:

$$4Ce^{4+} + 4e^- \longrightarrow 4Ce^{3+} \qquad E° = 1.61 \text{ V}$$

Add the two half-reactions and electrode potentials

$$2H_2O + 4Ce^{4+} \longrightarrow O_2 + 4Ce^{3+} + 4H^+ \qquad E° = -1.229 + 1.61 = 0.38 \text{ V}$$

Now notice that $E° > 0$. The reaction is spontaneous, and so Ce^{4+} will oxidize H_2O to O_2. Of course, we've assumed here that the H^+ and Ce^{4+} concentrations are both 1 M. You can make and use solutions of Ce^{4+} salts in water because this reaction is very slow.

PROBLEM 19-14   Would you expect a redox reaction to occur between metallic tin and mercury(I) chloride (Hg_2Cl_2) under standard conditions? If so, what reaction would take place?

Solution: The relevant reduction potentials listed in Table 19-1 are

$$Sn^{2+} + 2e^- \longrightarrow Sn \qquad\qquad E° = -0.138 \text{ V}$$
$$Hg_2Cl_2 + 2e^- \longrightarrow 2Hg + 2Cl^- \qquad E° = +0.268 \text{ V}$$

Since the reagent in this problem is metallic tin, you have to consider the oxidation reactions of Sn, for which the potential is

$$Sn \longrightarrow Sn^{2+} + 2e^- \qquad E° = -(-0.138) = +0.138 \text{ V}$$

Adding the Hg_2Cl_2 reduction half-reaction to the Sn oxidation half-reaction, you get

$$Sn + Hg_2Cl_2 + \cancel{2e^-} \longrightarrow Sn^{2+} + 2Hg + 2Cl^- + \cancel{2e^-} \qquad E° = +0.138 + 0.268 = +0.406 \text{ V}$$

Since $E°$ for the whole reaction is positive, this is the spontaneous reaction that you could expect to occur: Mercury(I) is reduced to metallic mercury by metallic tin, the tin being oxidized to tin(II).

PROBLEM 19-15   The standard reduction potential for the deuterium ion D^+ to $D_2(g)$ is $E° = -0.003$ V.

(a)   Which is the more powerful reducing agent, $D_2(g)$ or $H_2(g)$?
(b)   A solution containing D^+ and H^+ is in equilibrium under a mixture of gases in which the partial pressures of D_2 and H_2 are each 1.00 atm. Calculate the $[D^+]/[H^+]$ ratio in this solution.

Solution: The relevant reduction potentials are

$$D^+ + e^- \longrightarrow \tfrac{1}{2}D_2 \qquad E° = -0.003 \text{ V}$$
$$H^+ + e^- \longrightarrow \tfrac{1}{2}H_2 \qquad E° = 0.000 \text{ V}$$

(a)   The spontaneous reaction, for which $E° > 0$, is

$$H^+ + \cancel{e^-} + \tfrac{1}{2}D_2 \longrightarrow D^+ + \cancel{e^-} + \tfrac{1}{2}H_2 \qquad E° = -(-0.003) + 0.000 = 0.003 \text{ V}$$

So D_2 is just a little more powerful as a reducing agent than H_2. (In fact, as you already know, *any* half-reaction having a negative $E°$ involves a reducing agent more powerful than H_2.)

(b)   For a reaction at equilibrium, E must be exactly 0.000 V. So we can use the simplified Nernst equation (19-8)

$$E = E° - \left(\frac{0.059}{n}\right)\log\left(\frac{p_{H_2}^{1/2}[D^+]}{p_{D_2}^{1/2}[H^+]}\right)$$

$$0 = 0.003 - \left(\frac{0.059}{1}\right)\log\left(\frac{(1.00)^{1/2}[D^+]}{(1.00)^{1/2}[H^+]}\right)$$

$$= 0.003 - 0.059 \log\left(\frac{[D^+]}{[H^+]}\right)$$

Therefore

$$\log\left(\frac{[D^+]}{[H^+]}\right) = \frac{-0.003}{-0.059} = 0.05$$

and

$$\frac{[D^+]}{[H^+]} = 1.1$$

PROBLEM 19-16   What reactions, if any, will occur spontaneously if (a) an iron wire and (b) a silver wire are dipped separately into 1.0 M $CuSO_4$ solution? [*Hint*: The SO_4^{2-} ions are not involved in any reaction.]

Solution:

(a) The relevant half-reactions and reduction potentials are

$$Fe^{2+} + 2e^- \longrightarrow Fe \qquad E° = -0.447$$
$$Cu^{2+} + 2e^- \longrightarrow Cu \qquad E° = 0.342$$

Since the reagents are Cu^{2+} and Fe, you must reverse the first reaction:

$$Fe \longrightarrow Fe^{2+} + 2e^- \qquad E° = +0.447$$

Adding:

$$Fe + Cu^{2+} + \cancel{2e^-} \longrightarrow Fe^{2+} + Cu + \cancel{2e^-} \qquad E° = 0.447 + 0.342 = 0.789 \text{ V}$$

Since the $E°$ is positive, this is a spontaneous reaction: Iron wire will reduce Cu^{2+} to Cu.

(b) The half-reactions and reduction potentials are

$$Ag^+ + e^- \longrightarrow Ag \qquad E° = 0.800 \text{ V}$$
$$Cu^{2+} + 2e^- \longrightarrow Cu \qquad E° = 0.342 \text{ V}$$

Reverse and double the Ag reaction:

$$2Ag \longrightarrow 2Ag^+ + 2e^- \qquad E° = -0.800 \text{ V}$$

Adding:

$$2Ag + Cu^{2+} + \cancel{2e^-} \longrightarrow 2Ag^+ + Cu + \cancel{2e^-} \qquad E° = -0.800 + 0.342 = -0.458 \text{ V}$$

Since $E°$ is negative for this reaction, it is not spontaneous—no reaction will occur.

PROBLEM 19-17   For the oxidation of ammonia

$$4NH_3 + 3O_2 \longrightarrow 2N_2 + 6H_2O \qquad \Delta G° = -1356 \text{ kJ}$$

This reaction can be run in a galvanic cell (or *fuel cell*). Calculate the standard cell potential.

Solution: Equation (19-4) relates $\Delta G°$ to $E°$:

$$\Delta G° = -n\mathscr{F}E°$$

where $\mathscr{F} = 96\,500 \text{ J V}^{-1}\text{mol}^{-1}$. Now you have to find the value of n, and the only way to do that is to write the half-reactions:

oxidation: $4NH_3 \longrightarrow 2N_2 + 12H^+ + 12e^-$

reduction: $3O_2 + 12H^+ + 12e^- \longrightarrow 6H_2O$

So $n = 12$.

   Now you can substitute:

$$E° = \frac{-\Delta G°}{n\mathscr{F}} = \frac{-(-1356 \text{ kJ})(10^3 \text{ J kJ}^{-1})}{12(96\,500 \text{ J V}^{-1})}$$

$$= 1.17 \text{ V}$$

PROBLEM 19-18   Calculate K_{sp} for AgBr(s), given that

$$Ag^+(aq) + e^- \longrightarrow Ag(s) \qquad E° = 0.7996 \text{ V}$$
$$AgBr(s) + e^- \longrightarrow Ag(s) + Br^-(aq) \qquad E° = 0.0713 \text{ V}$$

Solution: You know that K_{sp} for AgBr(s) is the equilibrium constant for the reaction $AgBr(s) \rightleftharpoons Ag^+(aq) + Br^-(aq)$. Thus, if you know $E°$ for this reaction, you can calculate K_{sp} from eq. (19-6):

$$\log K = \frac{nE°}{0.0592}$$

So combine the two equations that are given as data to get the desired equilibrium equation—reverse the first equation and add the second:

$$Ag(s) \longrightarrow Ag^+(aq) + e^- \qquad E° = -0.7996 \text{ V}$$
$$AgBr(s) + e^- \longrightarrow Ag(s) + Br^-(aq) \qquad E° = +0.0713 \text{ V}$$

to get

$$AgBr(s) + \cancel{Ag(s)} + \cancel{e^-} \longrightarrow Ag^+(aq) + Br^-(aq) + \cancel{Ag(s)} + \cancel{e^-}$$
$$E^\circ = -0.7996 + 0.0713 = -0.7283 \text{ V}$$

Then substitute (note that $n = 1$):

$$\log K_{sp} = \frac{nE^\circ}{0.0592} = \frac{(1)(-0.7283)}{0.0592} = -12.3$$

$$K_{sp} = \text{antilog}(-12.3) = 5 \times 10^{-13}$$

PROBLEM 19-19 A galvanic cell is made by combining a standard $Pb|Pb^{2+}$ half-cell with a standard hydrogen electrode (SHE).

(a) What is the spontaneous cell reaction for this combination?
(b) What is the value of E° for this cell?
(c) Sulfide S^{2-} is added to the electrolyte in the $Pb|Pb^{2+}$ half-cell until $[S^{2-}] = 1.0 \times 10^{-2}$ M. PbS ($K_{sp} = 1.0 \times 10^{-29}$) precipitates. Calculate the new cell potential.

Solution:
(a) The half-reactions and reduction potentials are

$$2H^+ + 2e^- \longrightarrow H_2 \qquad E^\circ = 0.000 \text{ V}$$
$$Pb^{2+} + 2e^- \longrightarrow Pb \qquad E^\circ = -0.126 \text{ V}$$

so the spontaneous cell reaction is obtained by reversing the second half-reaction and adding:

$$2H^+ + Pb + \cancel{2e^-} \longrightarrow H_2 + Pb^{2+} + \cancel{2e^-}$$

(b) The E° of the cell is obtained by adding the value of the E° for the H^+ reaction to the E° value of the reversed Pb^{2+} reaction:

$$E^\circ = 0.000 + 0.126 = 0.126 \text{ V}$$

(c) The addition of S^{2-} changes the $[Pb^{2+}]$ enormously:

$$K_{sp} = [Pb^{2+}][S^{2-}] = 1.0 \times 10^{-29}$$

$$[Pb^{2+}] = \frac{1.0 \times 10^{-29}}{[S^{2-}]} = \frac{1.0 \times 10^{-29}}{1.0 \times 10^{-2}}$$

$$= 1.0 \times 10^{-27}$$

Applying the simplified Nernst equation (19-8) to this cell, you get

$$E = E^\circ - \left(\frac{0.0592}{2}\right)\log[Pb^{2+}] \qquad \text{(no change in SHE)}$$

so

$$E = 0.126 - \left(\frac{0.0592}{2}\right)\log(1.0 \times 10^{-27})$$

$$= 0.126 + 0.799 = 0.925 \text{ V}$$

(Notice that this is an example of Le Chatelier's principle in action: The addition of S^{2-} removes Pb^{2+} from the reaction mixture, thereby causing more Pb to be oxidized to Pb^{2+}.)

PROBLEM 19-20 Calculate K_{eq} for the reaction $Cu(s) + Cu^{2+}(aq) \rightleftharpoons 2Cu^+(aq)$.

Solution: If you can determine E° for this reaction, you can easily find K_{eq} by using eq. (19-6), which relates E° and K. Thus from Table 19-1

$$Cu^+(aq) + e^- \longrightarrow Cu(s) \qquad E^\circ = 0.521 \text{ V}$$
$$Cu^{2+}(aq) + 2e^- \longrightarrow Cu(s) \qquad E^\circ = 0.342 \text{ V}$$

Double the first equation and reverse it; then add:

$$Cu^{2+}(aq) + \cancel{2e^-} + \cancel{2Cu(s)} \longrightarrow \cancel{Cu(s)} + 2Cu^+(aq) + \cancel{2e^-} \qquad E^\circ = -0.521 + 0.342 = -0.179 \text{ V}$$

Now substitute into eq. (19-6):

$$\log K_{eq} = \frac{nE^\circ}{0.0592} = \frac{2(-0.179)}{0.0592} = -6.05$$

$$K_{eq} = \text{antilog}(-6.05) = 9 \times 10^{-7}$$

PROBLEM 19-21 A standard hydrogen electrode and a silver wire are dipped into a saturated aqueous solution of silver oxalate ($Ag_2C_2O_4$) containing solid $Ag_2C_2O_4$ at 25°C. The measured potential between the wire and the electrode is 0.589 V, the wire being positive. Calculate K_{sp} for $Ag_2C_2O_4$.

Solution: From Table 19-1,

$$2H^+ + 2e^- \longrightarrow H_2 \qquad E^\circ = 0.000 \text{ V}$$
$$Ag^+ + e^- \longrightarrow Ag \qquad E^\circ = 0.800 \text{ V}$$

The spontaneous reaction would therefore be

$$2Ag^+ + H_2 \longrightarrow 2H^+ + 2Ag \qquad E^\circ = 0.800 \text{ V}$$

Applying the simplified Nernst equation (19-8) gives

$$E = E^\circ - \left(\frac{0.059}{2}\right)\log\left(\frac{1}{[Ag^+]^2}\right)$$

since the SHE is in its standard condition. Thus, if the measured potential is 0.589 V, then

$$0.589 = 0.800 - \left(\frac{0.059}{2}\right)\log\left(\frac{1}{[Ag^+]^2}\right)$$

$$\log\left(\frac{1}{[Ag^+]^2}\right) = 7.13$$

$$[Ag^+] = 2.7 \times 10^{-4} \text{ M}$$

In saturated $Ag_2C_2O_4$ in equilibrium with solid $Ag_2C_2O_4$, the K_{sp} equilibrium must be

$$Ag_2C_2O_4(s) \longrightarrow 2Ag^+ + C_2O_4^{2-} \qquad K_{sp} = [Ag^+]^2[C_2O_4^{2-}]$$

so you use the stoichiometry $[C_2O_4^{2-}] = \frac{1}{2}[Ag^+]$ (i.e., 2 mol of Ag^+ must dissolve for every 1 mol of $C_2O_4^{2+}$) to get

$$K_{sp} = [Ag^+]^2(\tfrac{1}{2})[Ag^+]$$
$$= \tfrac{1}{2}[Ag^+]^3$$
$$= \tfrac{1}{2}(2.7 \times 10^{-4})^3$$
$$= 9.8 \times 10^{-12}$$

Supplementary Exercises

PROBLEM 19-22 Calculate the energy in joules gained by 1.00 mol of electrons that move through a 1.00-V potential difference.

Answer: 9.65×10^4 J

PROBLEM 19-23 Calculate the current in amperes when 6.02×10^{21} electrons flow through an electrical circuit for 2.00×10^3 s.

Answer: 0.482 A

PROBLEM 19-24   Arrange Sn, Cu, Fe, and Br^- in order of increasing strength as reducing agents.

Answer: $Br^- < Cu < Sn < Fe$

PROBLEM 19-25   Calculate $E°$ and the equilibrium constant at 25°C for the equilibrium $Fe^{3+} + Ag \rightleftharpoons Ag^+ + Fe^{2+}$.

Answer: $E° = -0.029$ V; $K_{eq} = 0.32$

PROBLEM 19-26   Calculate the cell potential for the following reaction when the Fe^{2+} concentration is 0.10 M and the I^- concentration is 0.020 M:

$$I_2(s) + Fe(s) \longrightarrow Fe^{2+}(aq) + 2I^-(aq)$$

Answer: $+1.113$ V

PROBLEM 19-27   Which of the following is the strongest oxidizing agent: Fe^{3+}, Ag^+, Cl^-, or Cd^{2+}?

Answer: Ag^+

PROBLEM 19-28   What is the spontaneous reaction under standard conditions involving Ni, Cu, Ni^{2+}, and Cu^{2+}? Calculate the value of its equilibrium constant.

Answer: $Cu^{2+} + Ni \longrightarrow Cu + Ni^{2+}$; $K_{eq} = 1.6 \times 10^{20}$

PROBLEM 19-29   The same number of coulombs of electric current flows successively through two electrolytic cells, the first containing $AgNO_3$ and the second $SnCl_2$. If 2.00 g of silver is deposited on the cathode of the first cell, what mass of tin is deposited on the cathode of the second cell?

Answer: 1.10 g

PROBLEM 19-30   Which of the following reagents could oxidize water to oxygen (O_2) under standard conditions: H^+, Cl_2, MnO_4^-, Fe^{3+}, Hg_2^{2+}?

Answer: Cl_2 and MnO_4^-

PROBLEM 19-31   Calculate the cell potential E for the following reaction when the Fe^{2+} concentration at the $Fe^{2+}|Fe$ electrode is 2.00 M, the pH in the hydrogen electrode electrolyte is 3.75, and the H_2 is at a pressure of 55. torr.

$$Fe + 2H^+ \longrightarrow Fe^{2+} + H_2$$

Answer: 0.250 V

PROBLEM 19-32   The following reaction is allowed to proceed to equilibrium in the presence of excess metallic tin and lead. Calculate the equilibrium ratio $[Sn^{2+}]/[Pb^{2+}]$:

$$Pb^{2+} + Sn \Longleftrightarrow Pb + Sn^{2+}$$

Answer: 2.6:1

PROBLEM 19-33   Which of the following metals will *not* spontaneously produce hydrogen gas on reaction with 1.00 M hydrochloric acid: Mg, Zn, Cr, Cu, Cd, Ag?

Answer: Cu and Ag

PROBLEM 19-34   A spacecraft fuel cell uses the oxidation of methane

$$CH_4(g) + 2O_2(g) \longrightarrow CO_2(g) + 2H_2O(l) \qquad \Delta G° = -818 \text{ kJ/mol}$$

to produce electricity. Calculate the standard potential for such a fuel cell.

Answer: 1.06 V

PROBLEM 19-35   Use data from Table 19-1 to calculate K_{sp} for Hg_2Cl_2.

Answer: $K_{sp} = 1.3 \times 10^{-18}$

20 RADIOCHEMISTRY

THIS CHAPTER IS ABOUT

☑ **Radioactivity**
☑ **Nuclear Reactions**
☑ **Rates of Nuclear Reactions**
☑ **Mass/Energy Equivalence**
☑ **Fusion and Fission**

Heretofore we've been concerned exclusively with the distribution of electrons in bonds, structures, and reactions—but there's more to atoms than electrons. The nucleus of an atom is also capable of undergoing processes that change the character of the atom itself. The study of such changes is the province of **radiochemistry** (or *nuclear chemistry*).

20-1. Radioactivity

Radioactivity is the spontaneous emission of highly energetic particles (see Table 20-1) or electromagnetic radiation—or both—from unstable nuclei. There are both natural and artificial (man-made) substances whose nuclei are unstable; these substances are radioactive reactants that spontaneously *decay*, producing particles or radiation as products. You should note that—

1. The rate of decay and the intensity of the radiation are proportional to the number of radioactive nuclei and are minimally affected by chemical composition, phase, temperature, or pressure.
2. Isotopes or **nuclides** of a radioactive element may decay by different pathways. [Recall that atoms having the same number of protons (and thus the same atomic number Z) but different numbers of neutrons (Z minus the atomic number A) are isotopes of the same element.]
3. The more common products of radioactive decay are the alpha particle α (a helium nucleus), the negative beta particle β^- (an electron), and gamma radiation γ (high-energy short-wavelength electromagnetic radiation).

TABLE 20-1:
Common Particles Observed in
Nuclear Reactions

Name	Symbols	Charge	Rest mass (amu)
Electron (beta minus)	β^-, $_{-1}^{0}e$	-1	5.5×10^{-4}
Positron (beta plus)	β^+, $_{+1}^{0}e$	$+1$	5.5×10^{-4}
Neutron	$_{0}^{1}n$	0	$1.008\,67$
Proton (hydrogen nucleus)	$_{1}^{1}p$, $_{1}^{1}H$	$+1$	$1.007\,83$
Alpha particle (helium nucleus)	$_{2}^{4}\alpha$, $_{2}^{4}He$	$+2$	$4.001\,50$

note: In radiochemical notation the mass number and atomic number of an atom (or particle) are written as parts of the symbol for the atom: The mass number A is written as a superscript and the atomic number Z as a subscript to the left of the symbol for the atom. Thus, for example, the symbols $^{52}_{24}Cr$ and $^{55}_{24}Cr$ show clearly that they represent two isotopes of chromium. There are also various symbols for particles (see Table 20-1). For instance, the α particle—which is a helium nucleus—can be written $^{4}_{2}He^{2+}$, or $^{4}_{2}He$, or $^{4}_{2}\alpha$. Finally, because charge calculations are made on the basis of atomic number, the usual charge superscript (to the right of the atomic symbol) is often omitted; thus $^{4}_{2}He^{2+}$ is usually written as $^{4}_{2}He$ (you can ignore the orbital electrons for once).

20-2. Nuclear Reactions

A. Balancing nuclear equations

A balanced equation for a nuclear reaction must satisfy two conservation principles:

1. *Charge is conserved*—the products must have the same net charge as the reactants.
2. *Mass number is conserved*—the total number of protons plus neutrons in the products must equal the total number of protons plus neutrons in the reactants.

One of the possible results of radioactive decay is **transmutation**—in which one element is transformed into another. For example, $^{55}_{24}Cr$ decays by emitting a β^- particle, thus producing the isotope $^{55}_{25}Mn$. The balanced nuclear reaction equation is

$$^{55}_{24}Cr \longrightarrow {}^{55}_{25}Mn + {}^{0}_{-1}e$$

The subscripts, which give atomic numbers and thus the charges, are in balance: $24 = 25 + (-1)$. (Notice that the new Z determines what the new isotope is.) The superscripts, which give the mass numbers, are also in balance: $55 = 55 + 0$.

EXAMPLE 20-1: Write a balanced nuclear reaction equation for the α decay of $^{226}_{88}Ra$, i.e., the reaction in which $^{226}_{88}Ra$ emits an α particle.

Solution: The α particle is $^{4}_{2}He$, so the desired reaction is

$$^{226}_{88}Ra \longrightarrow {}^{4}_{2}He + {}^{222}_{86}X$$

The mass numbers balance: $226 = 4 + 222$
The atomic numbers balance: $88 = 2 + 86$

Thus element X has $Z = 86$, which must be radon. The complete, balanced reaction equation is

$$^{226}_{88}Ra \longrightarrow {}^{4}_{2}He + {}^{222}_{86}Rn$$

You'll find it helpful to remember the following general decay reactions: They occur repeatedly in radiochemical problems:

Name	General reaction	Example
α Decay	$^{A}_{Z}X \longrightarrow {}^{A-4}_{Z-2}Y + {}^{4}_{2}\alpha$	$^{226}_{88}Ra \longrightarrow {}^{222}_{86}Rn + {}^{4}_{2}\alpha$
β Decay		
β^- emission	$^{A}_{Z}X \longrightarrow {}_{Z+1}{}^{A}Y + {}^{0}_{-1}\beta$	$^{55}_{24}Cr \longrightarrow {}^{55}_{25}Mn + {}^{0}_{-1}\beta$
β^+ emission	$^{A}_{Z}X \longrightarrow {}_{Z-1}{}^{A}Y + {}^{0}_{+1}\beta$	$^{13}_{7}N \longrightarrow {}^{13}_{6}C + {}^{0}_{+1}\beta$
Electron capture	$^{A}_{Z}X + {}^{0}_{-1}\beta \longrightarrow {}_{Z-1}{}^{A}Y$	$^{73}_{33}As + {}^{0}_{-1}\beta \longrightarrow {}^{73}_{32}Ge$

In general, radionuclide decay by particle emission leaves a nucleus in an excited state. The nucleus emits γ radiation when it returns to the ground state.

EXAMPLE 20-2: Carbon-14 decays by emitting a β^- particle. Determine the other product of the decay reaction.

Solution: The balanced decay equation is

$$^{14}_{6}C \longrightarrow {}^{0}_{-1}e + {}^{14}_{7}X \qquad \begin{array}{l}(A: 14 = 0 + 14)\\(Z: 6 = -1 + 7)\end{array}$$

Since $Z = 7$ for the other product, it must be ^{14}N.

B. Artificial radioactivity

Radioactive elements can be made artificially in reactors by bombarding target nuclei with streams of high-energy protons, neutrons, or α particles. For example, when ^{14}N is bombarded with α particles, the following reaction occurs:

$$^{14}_{7}N + {}^{4}_{2}He \longrightarrow {}^{17}_{8}O + {}^{1}_{1}p$$

note: A shorthand way of writing the same reaction is $^{14}N(\alpha, p)^{17}O$. The target nucleus is written first, then, within parentheses, the bombarding particle and the product particle, and finally the product nucleus.

EXAMPLE 20-3: Write out in full, with all super- and subscripts, the following reaction equations, and describe each in words.

(a) $^{9}Be(\alpha, n)^{12}C$ (b) $^{2}H(^{2}H, n)^{3}He$ (c) $^{7}Li(p, \alpha)^{4}He$

Solution:
(a) $^{9}_{4}Be + {}^{4}_{2}He \longrightarrow {}^{1}_{0}n + {}^{12}_{6}C$: Beryllium-9 is bombarded with an α particle to give a neutron and carbon-12.
(b) $^{2}_{1}H + {}^{2}_{1}H \longrightarrow {}^{1}_{0}n + {}^{3}_{2}He$: Hydrogen-2 (deuterium) is bombarded by deuterium to give a neutron and helium-3.
(c) $^{7}_{3}Li + {}^{1}_{1}p \longrightarrow {}^{4}_{2}He + {}^{4}_{2}He$: Lithium-7 is bombarded by a proton to give two helium-4 nuclei (or two α particles).

20-3. Rates of Nuclear Reactions

The decay of all radionuclides follows a first-order rate law. If N_R is the number of radioactive nuclei at time t, then

DECAY RATE $$\text{decay rate} = kN_R \qquad (20\text{-}1)$$

where k is the first-order rate constant. Recall (from Chapter 14) that the half-life of a first-order reaction is given by

HALF-LIFE $$t_{1/2} = \frac{0.693}{k} \qquad (20\text{-}2)$$

For nuclear reactions, the half-life is the time required for one-half of the nuclei in a sample to decay. Half-lives range from fractions of a second,

$$^{214}_{84}Po \longrightarrow {}^{4}_{2}\alpha + {}^{210}_{82}Pb \qquad t_{1/2} = 1.6 \times 10^{-4} \text{ s}$$

to billions of years,

$$^{238}_{92}U \longrightarrow {}^{4}_{2}\alpha + {}^{234}_{90}Th \qquad t_{1/2} = 4.49 \times 10^{9} \text{ yr}$$

The rates of radionuclide decay, unlike the rates of other chemical processes, are not affected by pressure or temperature.

EXAMPLE 20-4: How many nuclei of $^{238}_{92}U$ ($t_{1/2} = 4.49 \times 10^{9}$ yr) decay per second in a 1.0-g sample of the pure isotope?

Solution: First calculate the number of nuclei in the sample. A 1.0-g sample of pure $^{238}_{92}U$ contains $(1.0 \text{ g})/(238 \text{ g mol}^{-1}) = 4.2 \times 10^{-3}$ mol of $^{238}_{92}U$ and $(4.2 \times 10^{-3} \text{ mol})(6.02 \times 10^{23} \text{ nuclei mol}^{-1}) = 2.52 \times 10^{21}$ nuclei.

Then make sure that the given data are in consistent units. The half-life for $^{238}_{92}U$ is given as 4.49×10^9 years. Since the question asks for decay rate per second, convert the half-life to seconds:

$$t_{1/2} = (4.49 \times 10^9 \text{ yr})\left(\frac{365.25 \text{ days}}{1 \text{ yr}}\right)\left(\frac{24 \text{ h}}{1 \text{ day}}\right)\left(\frac{60 \text{ min}}{1 \text{ h}}\right)\left(\frac{60 \text{ s}}{1 \text{ min}}\right)$$

$$= 1.42 \times 10^{17} \text{ s}$$

Now use the half-life formula (20-2) to find k:

$$k = \frac{0.693}{t_{1/2}} = \frac{0.693}{1.42 \times 10^{17} \text{ s}} = 4.88 \times 10^{-18} \text{ s}^{-1}$$

Finally, use the decay rate formula (20-1):

$$\text{decay rate} = kN_R = (4.88 \times 10^{-18} \text{ s}^{-1})(2.52 \times 10^{21} \text{ nuclei})$$
$$= 1.2 \times 10^4 \text{ nuclei s}^{-1}$$

A. Measuring radionuclide decay

The integrated form of the first-order rate law can be used to determine the quantity of a radionuclide A that decays in a specific time period:

**RATE OF DECAY
(INTEGRATED FORM)**
$$\log\left(\frac{[A_0]}{[A]}\right) = \frac{kt}{2.303} \qquad (20\text{-}3)$$

where $[A_0]$ is the initial quantity of radionuclide, and $[A]$ is the quantity remaining at time t. Both $[A]$ and $[A_0]$ must be in the same units: Concentration units (moles per liter) are good, but you can use any unit that is proportional to the number of nuclei present—including grams of isotope and counts per minute on a Geiger counter. If the half-life of the radionuclide is known, eq. (20-3) can be rewritten as

$$\log\left(\frac{[A_0]}{[A]}\right) = \frac{0.693t}{2.303t_{1/2}} \qquad (20\text{-}4)$$

EXAMPLE 20-5: If a sample of a radioisotope that has a half-life of 3.00 days contains 1.00×10^{13} atoms, how many atoms will remain after (a) 6.00 days? (b) after 5.35 days?

Solution:

(a) For a time of 6.00 days, simple arithmetic will suffice. Half of 6.00 days is 3.00 days, which just happens to be the half-life. Thus, after 3.00 days the number of atoms would be one-half the original number:

$$\frac{1.00 \times 10^{13} \text{ atoms}}{2} = 5.00 \times 10^{12} \text{ atoms}$$

After *another* 3.00 days the number of atoms would be one-half of one-half, or

$$\frac{5.00 \times 10^{12} \text{ atoms}}{2} = 2.50 \times 10^{12} \text{ atoms}$$

(b) A time of 5.35 days is more complicated and requires the use of the simplified rate equation in integrated form (20-4):

$$\log\left(\frac{[A_0]}{[A]}\right) = \frac{0.693t}{2.303t_{1/2}}$$

$$\log[A_0] - \log[A] = \frac{0.693t}{2.303t_{1/2}}$$

$$\log[A] = \log[A_0] - \frac{0.693t}{2.303t_{1/2}}$$

$$= \log(1.00 \times 10^{13}) - \frac{0.693(5.35)}{2.303(3.00)}$$

$$= 13 - 0.537 = 12.463$$

$$[A] = \text{antilog}(12.463) = 2.90 \times 10^{12} \text{ atoms}$$

B. Radioactivity as a clock

Because the rate of radioactive decay isn't affected by changes in chemical state, the measurement of radioactivity in some objects can reveal something about their ages. For example, many rocks contain uranium, which decays by a chain of reactions into lead. The ratio of lead to uranium in such a rock can be used to make an estimate of how long ago the rock solidified.

Archaeologists have established the dates of many objects by carbon-14 dating. The isotope ^{14}C is produced naturally high in the atmosphere where cosmic-ray neutrons collide with ^{14}N nuclei:

$$^{14}_{7}N + ^{1}_{0}n \longrightarrow ^{14}_{6}C + ^{1}_{1}H$$

The ^{14}C eventually becomes atmospheric $^{14}CO_2$, which diffuses down to earth to be taken up by living organisms—all of which maintain an equilibrium proportion of ^{14}C in the carbon compounds of their bodies while they are alive and exchanging CO_2 with the atmosphere. But once an organism dies, it no longer exchanges CO_2 with the atmosphere. The amount of ^{14}C present in a dead organism is a fixed proportion, which decays at a known rate, thus decreasing its radioactivity. The half-life of ^{14}C is 5730 years, and it decays by β^- emission:

$$^{14}_{6}C \longrightarrow _{-1}^{0}e + ^{14}_{7}N$$

"Fresh" carbon in organic materials—carbon that is in equilibrium with atmospheric $^{14}CO_2$—has a decay rate of 15.3 nuclear disintegrations per minute (dpm) per gram of carbon.

EXAMPLE 20-6: An oak staff found in a cave in central France was believed to have been cut and used by prehistoric man. Upon analysis, a ^{14}C decay rate of 4.5 disintegrations per minute per gram of carbon was established. When was this staff cut from the oak tree?

Solution: Solve the decay-rate equation (20-4) for t and substitute:

$$\log \frac{[A_0]}{[A]} = \frac{0.693t}{2.303t_{1/2}}$$

$$t = \frac{2.303t_{1/2}\log([A_0]/[A])}{0.693} = \frac{2.303(5730 \text{ yr})\log(15.3/4.5)}{0.693}$$

$$= 1.01 \times 10^4 \text{ yr}$$

The staff was cut from an oak tree 10 100 years ago.

20-4. Mass/Energy Equivalence

A. The equivalence equation

On a molar level, the mass changes that occur in most chemical reactions are insignificant. This is not the case at the nuclear level, where small changes in mass are responsible for large changes in energy. The conservation of mass and energy remains in effect, however, and interconversions of mass and energy obey Einstein's **mass/energy equivalence equation**

MASS/ENERGY EQUIVALENCE $\qquad\qquad E = mc^2 \qquad\qquad$ (20-5)

where E is energy (in joules, or $kg\,m^2\,s^{-2}$), m is mass (in kilograms), and c is the velocity of light ($c = 3.00 \times 10^8 \text{ m s}^{-1}$).

EXAMPLE 20-7: Calculate the energy equivalence in joules of (a) 1.00 amu and (b) 1.00 g.

Solution: Use the mass/energy equivalence equation (20-5).

(a) Since there are 6.02×10^{23} (Avogadro's number) amu per gram,

$$E = mc^2$$

$$E = (1.00 \text{ amu})\left(\frac{1.00 \text{ g}}{6.02 \times 10^{23} \text{ amu}}\right)\left(\frac{1 \text{ kg}}{10^3 \text{ g}}\right)(3.00 \times 10^8 \text{ m s}^{-1})^2$$

$$= 1.50 \times 10^{-10} \text{ kg m}^2 \text{ s}^{-1}$$
$$= 1.50 \times 10^{-10} \text{ J}$$

(b) $$E = mc^2$$

$$= (1.00 \text{ g})\left(\frac{1 \text{ kg}}{10^3 \text{ g}}\right)(3.00 \times 10^8 \text{ m s}^{-1})^2$$

$$= 9.00 \times 10^{13} \text{ kg m}^2 \text{ s}^{-2}$$
$$= 9.00 \times 10^{13} \text{ J}$$

note: Memorize these equivalences: They will come in handy.

B. Nuclear binding energy

We must now abandon our simplistic model of the nucleus as a core of protons and neutrons of fixed mass, encircled by a cloud of massless electrons. It turns out that the mass of any nucleus is less than the sum of the masses of the component nucleons, i.e., the protons and neutrons. (This difference is sometimes called the *mass defect*.) Splitting a nucleus into component nucleons requires the input of mass, as energy, in accordance with the mass/energy equivalence equation. Think of this energy, called the **nuclear binding energy**, as the energy that would be released if we could somehow make a stable nucleus from individual protons and neutrons. The *average* binding energy—the **binding energy per nucleon**—can be used to compare the stability of nuclei.

EXAMPLE 20-8: Calculate (a) the nuclear binding energy and (b) the binding energy per nucleon of the ^{12}C nucleus. Use the particle masses listed in Table 20-1.

Solution:

(a) The mass of the ^{12}C nucleus is, by definition, exactly 12 amu; the nucleus contains 6 protons and 6 neutrons. From Table 20-1, the mass of a proton is 1.007 83 amu and the mass of a neutron is 1.008 67 amu, so

$$
\begin{array}{rr}
\text{mass of 6 protons} = 6(1.007\,83) = & 6.046\,98 \text{ amu} \\
\text{plus mass of 6 neutrons} = 6(1.008\,67) = & +6.052\,02 \text{ amu} \\
\hline
\text{mass of nucleons} = & 12.099\,00 \text{ amu} \\
\text{minus mass of } ^{12}\text{C nucleus} = & -12.000\,00 \text{ amu} \\
\hline
\text{mass converted to nuclear binding energy} = & 0.099\,00 \text{ amu}
\end{array}
$$

Recall from Example 20-7 that 1 amu $= 1.50 \times 10^{-10}$ J, so the nuclear binding energy is

$$(0.099\,00)(1.50 \times 10^{-10}) = 1.49 \times 10^{-11} \text{ J}$$

(b) Since there are 12 nucleons in ^{12}C, the binding energy per nucleon is

$$\frac{1.49 \times 10^{-11}}{12} = 1.24 \times 10^{-12} \text{ J/nucleon}$$

note: Energies are often expressed in millions of electron volts (MeV). The conversion factors are

$$1 \text{ MeV} = 1.602\,189 \times 10^{-13} \text{ J}$$
$$1 \text{ amu} = 931.5017 \text{ MeV}$$

Thus the nuclear binding energy and binding energy calculated in Example 20-8 for ^{12}C are

$$\text{nuclear binding energy} = (0.099\,00\ \text{amu})(931.5017\ \text{MeV amu}^{-1}) = 92.22\ \text{MeV}$$
$$\text{binding energy per nucleon} = (92.22\ \text{MeV})/12 = 7.68\ \text{MeV}$$

C. Mass/energy changes in decay reactions

The total mass of the products of a nuclear decay reaction is less than the mass of the decaying nucleus. The mass difference is supplied to the products as *kinetic energy*.

EXAMPLE 20-9: Calculate the maximum kinetic energy in joules of an electron produced by the β^- decay of $^{20}_{9}$F (actual atomic mass = 19.999 99 amu) to produce $^{20}_{10}$Ne (actual atomic mass = 19.992 44 amu).

Solution: Write the reaction equation:

$$^{20}_{9}\text{F} \longrightarrow _{-1}^{0}e + ^{20}_{10}\text{Ne}$$

Calculate the total mass of the products:

$$\begin{aligned}
\text{mass of } ^{20}_{10}\text{Ne} &= 19.992\,44 \\
\text{plus mass of } _{-1}^{0}e &= \underline{\quad 0.000\,55} \\
&\ \ 19.992\,99 \ \text{amu}
\end{aligned}$$

The mass difference between reactant and product is $19.999\,99 - 19.992\,99 = 0.007\,00$ amu. Then, since 1 amu is 1.50×10^{-10} J, the maximum kinetic energy of any product electron is

$$(0.007\,00\ \text{amu})(1.50 \times 10^{-10}\ \text{J amu}^{-1}) = 1.05 \times 10^{-12}\ \text{J}$$

20-5. Fusion and Fission

If we were to graph the binding energy per nucleon of each element against its mass number, we'd discover that the elements with the highest binding energy per nucleon are those whose mass numbers fall between 40 and 100. Furthermore, it's reasonable to assume that elements having the highest binding energies would be the most stable—and, for the most part, they are. Therefore, there must be some process(es) whereby elements of smaller or larger mass numbers can release energy and be converted into elements in the middle range. Two such processes—both of which release enormous energy—do in fact exist: fusion and fission.

A. Fusion

Nuclear fusion is the *combination* of two light nuclei to produce a heavier element. In practice, very high temperatures and high concentrations of reactants are required.

EXAMPLE 20-10: Calculate the energy released if 1.00 mg of deuterium (^{2}D, nuclear mass = 2.014 102 amu) is converted into helium (^{4}He, nuclear mass = 4.002 603 amu) by the fusion reaction

$$^{2}_{1}\text{D} + ^{2}_{1}\text{D} \longrightarrow ^{4}_{2}\text{He}$$

Solution: The mass change is $2(2.014\,102) - 4.002\,603 = 0.025\,601$ amu. The energy liberated for two ^{2}D nuclei fusing is $(0.025\,601\ \text{amu})(1.50 \times 10^{-10}\ \text{J amu}^{-1}) = 3.84 \times 10^{-12}$ J. The number of ^{2}D nuclei in 1.00 mg of deuterium is

$$(1.00\ \text{mg})(10^{-3}\ \text{g mg}^{-1})\left(\frac{6.02 \times 10^{23}\ \text{nuclei}}{2.014\,102\ \text{g}}\right) = 2.99 \times 10^{20}\ \text{nuclei}$$

The energy liberated from 1.00 mg of ^{2}D is

$$(2.99 \times 10^{20}\ \text{nuclei})\left(\frac{3.84 \times 10^{-12}\ \text{J}}{2\ \text{nuclei}}\right) = 5.74 \times 10^{8}\ \text{J}$$

That's a lot of energy from a reaction involving just one milligram of deuterium. And notice that the product is innocuous and stable helium gas.

B. Fission

Nuclear fission is the *splitting* of a heavy nucleus into two lighter nuclei, either spontaneously or after bombardment with certain nucleons. For example, when ^{235}U reacts with a neutron, it undergoes spontaneous fission—and many different fission reactions occur. One such reaction is

$$^{235}_{92}U + ^{1}_{0}n \longrightarrow ^{136}_{54}Xe + ^{90}_{38}Sr + 10^{1}_{0}n$$

The energy released in this reaction is about 2.2×10^{-11} J/nucleus. Note that one neutron initiates this reaction and that ten neutrons are produced by it. These ten neutrons can go on to induce fission in other ^{235}U nuclei, starting a self-perpetuating nuclear **chain reaction** if the mass of ^{235}U is sufficient.

SUMMARY

1. Radioactivity is the spontaneous breakdown of the nucleus.
2. In radioactive decay, the most common products are α and β particles and γ radiation.
3. In nuclear reactions charge and mass/energy are conserved.
4. Nuclear decay reactions follow first-order rate laws.
5. Radioactive decay can be used in dating objects.
6. The mass/energy equivalence equation is $E = mc^2$.
7. The binding energy of nucleons in a nucleus can be determined from the nuclear mass defect.
8. Fusion is the process of combining light nuclei to make heavier ones.
9. Fission is the splitting up of a heavy nucleus into lighter nuclei.

RAISE YOUR GRADES

Can you ... ?

☑ balance nuclear equations
☑ determine $t_{1/2}$ from k and vice versa
☑ use decay rate equations
☑ calculate the age of organic materials from carbon-14 data
☑ do mass/energy interconversions
☑ calculate nuclear binding energies
☑ determine energy yields in fission or fusion reactions

SOLVED PROBLEMS

PROBLEM 20-1   Complete the following nuclear equations:

(a) $^{14}_{7}N + ^{4}_{2}He \longrightarrow ^{1}_{1}H + ?$

(b) $^{6}_{3}Li + ^{1}_{0}n \longrightarrow ^{4}_{2}He + ?$

(c) $^{234}_{90}Th \longrightarrow ^{0}_{-1}e + ?$

Solution: These equations must balance in mass (superscripts) and nuclear charge (subscripts). The product is determined by the new charge (Z):

(a) $^{14}_{7}\text{N} + ^{4}_{2}\text{He} \longrightarrow ^{1}_{1}\text{H} + ^{17}_{8}\text{X}$ $\qquad Z$ of $X = 8$

$\quad\;\; ^{14}_{7}\text{N} + ^{4}_{2}\text{He} \longrightarrow ^{1}_{1}\text{H} + ^{17}_{8}\text{O}$ $\qquad X = \text{O}$

(b) $^{6}_{3}\text{Li} + ^{1}_{0}n \longrightarrow ^{4}_{2}\text{He} + ^{3}_{1}\text{X}$ $\qquad Z$ of $X = 1$

$\quad\;\; ^{6}_{3}\text{Li} + ^{1}_{0}n \longrightarrow ^{4}_{2}\text{He} + ^{3}_{1}\text{H}$ $\qquad X = \text{H}$

(c) $^{234}_{90}\text{Th} \longrightarrow ^{0}_{-1}e + ^{234}_{91}\text{X}$ $\qquad Z$ of $X = 91$

$\quad\;\; ^{234}_{90}\text{Th} \longrightarrow ^{0}_{-1}e + ^{234}_{91}\text{Pa}$ $\qquad X = \text{Pa}$

PROBLEM 20-2 Identify the intermediates (X and Y) and the final product (Q) of this nuclear-decay series:

$$^{238}_{92}\text{U} \longrightarrow X + ^{4}_{2}\text{He}$$
$$X \longrightarrow Y + ^{0}_{-1}e$$
$$Y \longrightarrow Q + ^{4}_{2}\text{He}$$

Solution: Balance the mass and charge and find the intermediates and product in the periodic chart:

$$^{238}_{92}\text{U} \longrightarrow ^{234}_{90}\text{X} + ^{4}_{2}\text{He} \qquad Z \text{ of } X = 90$$
$$^{234}_{90}\text{X} \longrightarrow ^{234}_{91}\text{Y} + ^{0}_{-1}e \qquad Z \text{ of } Y = 91$$
$$^{234}_{91}\text{Y} \longrightarrow ^{230}_{89}\text{Q} + ^{4}_{2}\text{He} \qquad Z \text{ of } Q = 89$$

So X is $^{234}_{90}\text{Th}$, Y is $^{234}_{91}\text{Pa}$, and Q is $^{230}_{89}\text{Ac}$.

PROBLEM 20-3 Complete and write out in full the following nuclear reactions:

(a) $^{23}_{11}\text{Na}(^{2}_{1}\text{H}, ^{1}_{1}p)?$

(b) $^{10}_{5}\text{B}(^{1}_{0}n, ?)^{7}_{3}\text{Li}$

(c) $^{15}_{7}\text{N}(?, ^{4}_{2}\text{He})^{12}_{6}\text{C}$

Solution: The first species given is the target nucleus; then, within the parentheses, the bombarding particle is given, followed by the ejected particle; the final species is the product.

(a) $^{23}_{11}\text{Na} + ^{2}_{1}\text{H} \longrightarrow ^{1}_{1}p + ^{24}_{11}\text{Na}$ $\qquad$ or $\qquad ^{23}_{11}\text{Na} + ^{2}_{1}\text{H} \longrightarrow ^{1}_{1}\text{H} + ^{24}_{11}\text{Na}$

(b) $^{10}_{5}\text{B} + ^{1}_{0}n \longrightarrow ^{4}_{2}\text{He} + ^{7}_{3}\text{Li}$ $\qquad$ or $\qquad ^{10}_{5}\text{B} + ^{1}_{0}n \longrightarrow ^{4}_{2}\alpha + ^{7}_{3}\text{Li}$

(c) $^{15}_{7}\text{N} + ^{1}_{1}p \longrightarrow ^{4}_{2}\text{He} + ^{12}_{6}\text{C}$ $\qquad$ or $\qquad ^{15}_{7}\text{N} + ^{1}_{1}\text{H} \longrightarrow ^{4}_{2}\alpha + ^{12}_{6}\text{C}$

PROBLEM 20-4 Write balanced equations for the following nuclear reactions: (a) $^{243}_{94}\text{Pu}$ emitting a β particle; (b) $^{236}_{94}\text{Pu}$ emitting an α particle; (c) bombarding $^{209}_{83}\text{Bi}$ with high-energy α particles to give two neutrons and an isotope of astatine.

Solution:

(a) $\qquad ^{243}_{94}\text{Pu} \longrightarrow \beta^{-} + ?$

Write the β particle as an electron, and add a nucleus that will make the mass numbers and atomic numbers balance:

$$^{243}_{94}\text{Pu} \longrightarrow ^{0}_{-1}e + ^{243}_{95}\text{Am}$$

(b) $\qquad ^{236}_{94}\text{Pu} \longrightarrow \alpha + ?$

An α particle is a $^{4}_{2}\text{He}$ nucleus; add a nucleus that will make the mass numbers and atomic numbers balance:

$$^{236}_{94}\text{Pu} \longrightarrow ^{4}_{2}\text{He} + ^{232}_{92}\text{U}$$

(c) Assume that only one α particle ($^{4}_{2}\text{He}$) is captured. Each neutron is $^{1}_{0}n$, and astatine has atomic number 85. All you need to determine is the mass number of the astatine isotope that will make the mass numbers balance:

$$^{209}_{83}\text{Bi} + ^{4}_{2}\text{He} \longrightarrow 2^{1}_{0}n + ^{211}_{85}\text{At}$$

PROBLEM 20-5 The half-life of sulfur-35 ($^{35}_{16}\text{S}$) is 87.1 days. If a sample of radioactive sulfur contains 2.0×10^{17} atoms of $^{35}_{16}\text{S}$, how many atoms will it contain after 70.0 days?

Solution: Given the half-life $t_{1/2} = 87.1$ days, the time $t = 70.0$ days, and an initial quantity $[A_0] = 2.0 \times 10^{17}$, you can use the simplified decay-rate equation (20-4):

$$\log\left(\frac{[A_0]}{[A]}\right) = \frac{0.693t}{2.303t_{1/2}} = \log[A_0] - \log[A]$$

$$\log[A] = (\log 2.0 \times 10^{17}) - \frac{(0.693)(70.0)}{(2.303)(87.1)}$$

$$= (\log 2.0 \times 10^{17}) - 0.242$$
$$= 17.301 - 0.242$$
$$= 17.059$$
$$[A] = \text{antilog}(17.059) = 1.1 \times 10^{17} \text{ atoms}$$

PROBLEM 20-6 The isotope of $^{210}_{84}$Po emits α particles and has a half-life of 138.4 days. How many nuclei would disintegrate per second in a 1.0-μg sample of $^{210}_{84}$Po?

Solution: Calculate the number of nuclei there are in 1.0 μg:

$$\left(\frac{1.0\ \mu g}{10^6\ \mu g\,g^{-1}}\right)\left(\frac{1\ mol}{210\ g}\right)\left(\frac{6.02 \times 10^{23}\ nuclei}{1\ mol}\right) = 2.87 \times 10^{15} \text{ nuclei}$$

Now find the rate constant k, which must be expressed in seconds. Use the half-life formula (20-2): $t_{1/2} = 0.693/k$, or

$$k = \frac{0.693}{t_{1/2}} = \left(\frac{0.693}{138.4\ days}\right)\left(\frac{1\ day}{24\ h}\right)\left(\frac{1\ h}{60\ min}\right)\left(\frac{1\ min}{60\ sec}\right)$$

$$= 5.80 \times 10^{-8}\ s^{-1}$$

Finally, use the decay rate formula (20-1)

$$\begin{aligned}
\text{decay rate} &= kN_R \\
&= (5.80 \times 10^{-8}\ s^{-1})(2.87 \times 10^{15}\ nuclei) \\
&= 1.7 \times 10^8 \text{ nuclei s}^{-1}
\end{aligned}$$

PROBLEM 20-7 The **curie** is a unit of radioactivity equal to 3.70×10^{10} disintegrations per second. What is the activity in curies of a 0.10-g sample of $^{237}_{93}$Np, which has a half-life of 2.2×10^6 yr?

Solution: Since the curie is given in disintegrations per second, you must first determine the number of disintegrations per second of the sample, as in Problem 20-5:

$$(0.10\ g)\left(\frac{1\ mol}{237\ g}\right)\left(\frac{6.02 \times 10^{23}\ nuclei}{1\ mol}\right) = 2.54 \times 10^{20} \text{ nuclei}$$

Then find the decay rate constant (in s^{-1}) by the half-life formula (20-2):

$$k = \frac{0.693}{t_{1/2}} = \left(\frac{0.693}{2.2 \times 10^6\ yr}\right)\left(\frac{1\ yr}{365.25\ days}\right)\left(\frac{1\,day}{24\ h}\right)\left(\frac{1\ h}{60\ min}\right)\left(\frac{1\ min}{60\ s}\right)$$

$$= 9.98 \times 10^{-15}\ s^{-1}$$

Now use the decay rate formula (20-1):

$$\begin{aligned}
\text{decay rate} &= kN_R \\
&= (9.98 \times 10^{-15}\ s^{-1})(2.54 \times 10^{20}\ nuclei) \\
&= 2.5 \times 10^6 \text{ nuclei s}^{-1}
\end{aligned}$$

and convert to curies:

$$\text{No. of curies} = \frac{2.5 \times 10^6\ s^{-1}}{3.70 \times 10^{10}\ s^{-1}\,curie^{-1}}$$

$$= 6.8 \times 10^{-5} \text{ curies}$$

PROBLEM 20-8 Tritium $^3_1$H is a radioactive species that undergoes β^- emission ($t_{1/2} = 12.3$ yr). In ordinary H_2O the ratio of $^1_1$H to $^3_1$H is 1.0×10^{17}. How many disintegrations per minute occur for the tritium in 1.00 kg of H_2O?

Solution: Find k in min^{-1} (eq. 20-2):

$$k = \left(\frac{0.693}{12.3}\right)\left(\frac{1}{365.25}\right)\left(\frac{1}{24}\right)\left(\frac{1}{60}\right) = 1.07 \times 10^{-7}\ min^{-1}$$

The number of tritium atoms in 1.00 kg of H_2O is

$$(1.00 \text{ kg})\left(\frac{10^3 \text{ g}}{1 \text{ kg}}\right)\left(\frac{1 \text{ mol } H_2O}{18.0 \text{ g } H_2O}\right)\left(\frac{6.02 \times 10^{23} \text{ molec. } H_2O}{1 \text{ mol}}\right)\left(\frac{2 \text{ atoms H}}{1 \text{ molec. } H_2O}\right)\left(\frac{1 \text{ atom } {}_1^3H}{1.0 \times 10^{17} \text{ atoms } {}_1^1H}\right)$$

$$= 6.69 \times 10^8 \text{ atoms } {}_1^3H$$

So the number of disintegrations (by eq. 20-1) is

$$kN_R = (1.07 \times 10^{-7} \text{ min}^{-1})(6.69 \times 10^8 \text{ atoms } {}_1^3H)$$
$$= 72 \text{ atoms } {}_1^3H \text{ min}^{-1}$$

PROBLEM 20-9 How much mass is converted into energy each second in a nuclear power plant rated at 1250 megawatts (MW; $1 \text{ W} = 1 \text{ J s}^{-1}$)?

Solution: Because this problem involves mass/energy conversion, you know you'll use the mass/energy equivalence formula $E = mc^2$ (20-5), but first convert 1250 MW to $J s^{-1}$:

$$(1250 \text{ MW})\left(\frac{10^6 \text{ W}}{1 \text{ MW}}\right)\left(\frac{1 \text{ J s}^{-1}}{1 \text{ W}}\right) = 1.25 \times 10^9 \text{ J s}^{-1}$$

then use eq. (20-5):

$$m = \frac{E}{c^2} = \frac{1.25 \times 10^9 \text{ J s}^{-1}}{(3.00 \times 10^8 \text{ m s}^{-1})^2}$$

$$= 1.39 \times 10^{-8} \text{ kg}$$

note: If you use standard SI base units (J, m, s, etc.) in calculating problems like this one, your answers will also be in base units—as in this case, the answer is in the standard base unit kg.

PROBLEM 20-10 Calculate the binding energy per nucleon in joules for (a) the ${}_{10}^{20}$Ne nucleus (19.992 44 amu) and (b) the ${}_{26}^{56}$Fe nucleus (55.935 amu).

Solution: The binding energy per nucleon can be found from the mass defect—the difference between the actual nuclear mass and the sum of the masses of the protons and neutrons in the nucleus.

(a) The nucleus ${}_{10}^{20}$Ne is made up of 10 protons (which give it the charge of $+10$) and 10 neutrons. Their combined masses would be (see Table 20-1)

$$
\begin{array}{rr}
10p \times 1.007\,83 \text{ amu}/p = & 10.0783 \text{ amu} \\
10n \times 1.008\,67 \text{ amu}/n = & +10.0867 \text{ amu} \\ \hline
\text{Total} = & 20.165 \text{ amu}
\end{array}
$$

Subtracting:

$$
\begin{array}{rr}
& 20.165 \text{ amu} \\
\text{minus actual mass} = & -19.992 \text{ amu} \\ \hline
\text{Mass defect} = & 0.173 \text{ amu}
\end{array}
$$

Converting amu to J (1 amu = 1.50×10^{-7} J; see Example 20-7):

$$(0.173 \text{ amu})(1.50 \times 10^{-10} \text{ J amu}^{-1}) = 2.60 \times 10^{-11} \text{ J}$$

Thus the binding energy per nucleon is

$$\frac{2.60 \times 10^{-11} \text{ J}}{20 \text{ nucleons}} = 1.30 \times 10^{-12} \text{ J/nucleon}$$

(b) ${}_{26}^{56}$Fe: $A = \sim 56$, so

$$
\begin{array}{rr}
26p \times 1.007\,83 \text{ amu}/p = & 26.2036 \text{ amu} \\
30n \times 1.008\,67 \text{ amu}/n = & +30.2601 \text{ amu} \\ \hline
\text{Total} = & 56.4637 \text{ amu} \\
\text{minus actual mass} = & -55.935 \text{ amu} \\ \hline
\text{Mass defect} = & 0.529 \text{ amu}
\end{array}
$$

So the binding energy per nucleon is

$$\frac{(0.529 \text{ amu})(1.50 \times 10^{-10} \text{ J amu}^{-1})}{56 \text{ nucleons}} = 1.42 \times 10^{-12} \text{ J/nucleon}$$

[Notice how the binding energy per nucleon increases from ^{12}C (see Example 20-8) to ^{20}Ne to ^{56}Fe.]

PROBLEM 20-11   Calculate the maximum kinetic energy (KE) expected for electrons produced by the β^- decay of $^{10}_4\text{Be}$. ($^{10}_4\text{Be}$ = 10.013 534 amu, $^{10}_5\text{B}$ = 10.012 939 amu.)

Solution: Write the decay reaction:

$$^{10}_4\text{Be} \longrightarrow \, _{-1}^{\,0}e + \, ^{10}_5\text{B}$$

Find the mass to be converted to energy:

$$
\begin{array}{lrl}
\text{Total mass of products:} & ^{10}_5\text{B} = & 10.012\,939 \text{ amu} \\
& _{-1}^{\,0}e = & +\ 0.000\,55 \text{ amu} \\
\cline{2-3}
& \text{Total} = & 10.013\,49 \text{ amu} \\[4pt]
\text{Mass of reactant:} & ^{10}_4\text{Be} = & 10.013\,53 \text{ amu} \\
\text{minus mass of products} & = & -10.013\,49 \text{ amu} \\
\cline{2-3}
\text{Mass converted to } KE & = & 0.000\,04 \text{ amu}
\end{array}
$$

This kinetic energy will be shared between the $^{10}_3\text{B}$ nucleus and the electron, so the maximum kinetic energy of the electron is

$$KE = (0.000\,04 \text{ amu})(1.50 \times 10^{-10} \text{ J amu}^{-1})$$
$$= 6 \times 10^{-15} \text{ J}$$

PROBLEM 20-12   Determine the energy produced if 1.00 g of ^6_3Li is used in the fission reaction

$$^6_3\text{Li} + \, ^2_1\text{H} \longrightarrow 2^4_2\text{He}$$

(^6_3Li = 6.015 125 amu; ^2_1H = 2.014 102 amu; ^4_2He = 4.002 603 amu.)

Solution:

$$
\begin{array}{lrl}
\text{Total mass of reactants:} & ^6_3\text{Li} = & 6.015\,125 \text{ amu} \\
& ^2_1\text{H} = & +2.014\,102 \text{ amu} \\
\cline{2-3}
& \text{Total} = & 8.029\,227 \text{ amu} \\[4pt]
\text{Total mass of product:} & 2^4_2\text{He} = & 2(4.002\,603) \text{ amu} \\
& = & 8.005\,206 \text{ amu} \\[4pt]
\text{Mass of reactants} & = & 8.029\,227 \text{ amu} \\
\text{minus mass of products} & = & -8.005\,206 \text{ amu} \\
\cline{2-3}
\text{Mass converted to energy} & = & 0.024\,021 \text{ amu}
\end{array}
$$

so if the mass converted to energy for one ^6_3Li nucleus is 0.024 021 amu, then for 1.00 g of ^6_3Li, the energy would be

$$E = \left(\frac{0.024\,021 \text{ amu}}{1 \text{ nucleus}}\right)\left(\frac{1.50 \times 10^{-10} \text{ J}}{1 \text{ amu}}\right)\left(\frac{1.00 \text{ g} \, ^6_3\text{Li}}{6.015\,125 \text{ g mol}^{-1} \text{ Li}}\right)\left(\frac{6.02 \times 10^{23} \text{ nuclei}}{1 \text{ mol}}\right)$$

$$= 3.61 \times 10^{11} \text{ J}$$

PROBLEM 20-13   One fission reaction that occurs when neutrons react with uranium-235 is

$$^{235}_{92}\text{U} + \, ^1_0n \longrightarrow \, ^{140}_{56}\text{Ba} + \, ^{90}_{36}\text{Kr} + 6^1_0n$$

(a) Is this reaction exothermic? **(b)** Could this reaction be classified as a chain reaction? **(c)** Determine the energy (in megajoules) produced or absorbed if 1.00 mol of ^{235}U were to undergo this reaction. ($^{235}_{92}\text{U}$ = 235.043 92 amu; $^{140}_{56}\text{Ba}$ = 139.910 57 amu; $^{90}_{36}\text{Kr}$ = 89.919 72 amu.)

Solution:
(a)
$$
\begin{array}{lrl}
\text{Mass of reactants:} & ^{235}_{92}\text{U} = & 235.043\,92 \\
& ^1_0n = & 1.008\,67 \\
\cline{2-3}
& \text{Total} = & 236.052\,59 \text{ amu} \\[4pt]
\text{Mass of products:} & ^{140}_{56}\text{Ba} = & 139.910\,57 \\
& ^{90}_{36}\text{Kr} = & 89.919\,72 \\
6^1_0n = 6(1.008\,67) = & & 6.052\,02 \\
\cline{2-3}
& \text{Total} = & 235.882\,31 \text{ amu}
\end{array}
$$

Since the mass of the products is less than the mass of the reactants, mass has been converted into energy; therefore, this reaction is energy-yielding, or exothermic.
(b) Because one neutron is absorbed and six neutrons are released, the reaction can be classified as a chain reaction.

(c) The amount of mass converted into energy is

$$\left(\frac{236.05259 \text{ amu} - 235.88231 \text{ amu}}{1 \text{ nucleus}}\right)\left(\frac{6.02 \times 10^{23} \text{ nuclei}}{1 \text{ mol}}\right)\left(\frac{1.50 \times 10^{-10} \text{ J}}{1 \text{ amu}}\right)\left(\frac{1 \text{ MJ}}{10^6 \text{ J}}\right) = 1.54 \times 10^7 \text{ MJ/mol}$$

PROBLEM 20-14   Carbon from the twigs of a living sequoia gives a count of 15.3 disintegrations per minute (dpm) per gram of carbon because of its ^{14}C content ($t_{1/2} = 5730$ yr). Carbon from the wood of a fallen sequoia gives a count of 12.9 dpm per gram of carbon. How long ago did the sequoia fall?

Solution: You're given $t_{1/2}$ for ^{14}C, so you can use the integrated form of the rate of decay (eq. 20-4):

$$\log\left(\frac{[A_0]}{[A]}\right) = \frac{0.693t}{2.303t_{1/2}}$$

Set $[A_0] = 15.3$ dpm and $[A] = 12.9$ dpm, substitute, and find t:

$$\log\left(\frac{15.3}{12.9}\right) = \frac{0.693t}{(2.303)(5730)}$$

$$t = \frac{(2.303)(5730 \text{ yr})\log(12.9/15.3)}{0.693}$$

$$= 1.41 \times 10^3 \text{ yr}$$

So the sequoia fell 1410 years ago.

PROBLEM 20-15   The half-life of the isotope ^{29}S is 0.19 s. If you started with a sample that contained 1.0×10^7 nuclei of $^{29}_{16}$S, how long would it take until you had only 1.0×10 nuclei? What would happen to the sample in the following second?

Solution: As in Problem 20-14, the appropriate equation is (20-4), where $[A_0] = 1.0 \times 10^7$, $[A] = 1.0 \times 10$, and $t_{1/2} = 0.19$ s:

$$t = \frac{(2.303t_{1/2})\log([A_0]/[A])}{0.693}$$

$$= \frac{(2.303)(0.19 \text{ s})\log(1.0 \times 10^7/1.0 \times 10^1)}{0.693} = 3.8 \text{ s}$$

So after only 3.8 seconds, you'd be down to 10 nuclei. In the next second—which is more than 5 half-lives—you'd have the following sequence:

after 0.19 s:	10 nuclei $\longrightarrow$ 5	
after 2(0.19 s) = 0.38 s:	5 nuclei $\longrightarrow$ 2 or 3	
after 3(0.19 s) = 0.57 s:	2 or 3 nuclei $\longrightarrow$ 1 or 2	
after 4(0.19 s) = 0.76 s:	1 or 2 nuclei $\longrightarrow$ 0 or 1	
after 5(0.19 s) = 0.95 s:	0 or 1 nucleus $\longrightarrow$ 0 (or 1?)	

So you'd either have one nucleus or no nuclei in the second following 3.8 s.

note: Decay is a statistical process, occurring randomly, and half-life is a *statistical average*. Thus if you start with one ^{29}S nucleus, the chances are 50% (0.5) that it will disintegrate within 0.19 s (one half-life); the chances are 25% ($0.5 \times 0.5 = 0.25$) that it will survive two half-lives (2×0.19 s); the chances are 12.5% ($0.5 \times 0.5 \times 0.5$) that it will survive three half-lives (3×0.19 s); and so on.

PROBLEM 20-16   A sample containing a single radioactive isotope was isolated and placed in a radioactivity detector, where it was found to give 1365 counts per minute (cpm). After 3.00 hours its activity under identical conditions was 832 cpm. Calculate the half-life of the isotope.

Solution: The activity in counts per minute is proportional to the number of nuclei present. Let $[A_0] = 1365$ be the activity at the start and $[A] = 832$ be the activity after $t = 3.00$ h. Use the integrated equation (20-4) for a first-order decay reaction:

$$\log\left(\frac{[A_0]}{[A]}\right) = \frac{0.693t}{2.303t_{1/2}}$$

Solve for $t_{1/2}$:

$$t_{1/2} = \frac{0.693t}{(2.303)\log([A_0]/[A])}$$

$$= \frac{(0.693)(3.00 \text{ h})}{(2.303)\log(1365/832)}$$

$$= 4.20 \text{ h}$$

PROBLEM 20-17   The radioactive cobalt isotope $^{60}_{27}\text{Co}$ ($t_{1/2} = 5.24$ yr) is used in cancer therapy. A sample containing 1.00×10^{15} nuclei of $^{60}_{27}\text{Co}$ is used in a hospital.

(a)   At the end of 3.00 years, how many nuclei will remain?
(b)   What will the rate of decay in disintegrations per second be at the end of 3.00 years?

Solution:
(a)   You are given the half-life $t_{1/2} = 5.24$ yr, the time $t = 3.00$ yr, and the number of nuclei at the start $[A_0] = 1.00 \times 10^{15}$, so you use eq. (20-4) and solve for $[A]$:

$$\log\left(\frac{[A_0]}{[A]}\right) = \frac{0.693t}{2.303 t_{1/2}}$$

$$\log[A_0] - \log[A] = \frac{0.693t}{2.303 t_{1/2}}$$

$$\log[A] = \log[A_0] - \frac{0.693t}{2.303 t_{1/2}}$$

$$= \log(1.00 \times 10^{15}) - \frac{(0.693)(3.00 \text{ yr})}{(2.303)(5.24)}$$

$$= 14.828$$

$$[A] = \text{antilog}(14.828) = 6.73 \times 10^{14} \text{ nuclei}$$

(b)   You'll need the decay-rate formula (20-1) and the half-life formula (20-2):

$$\text{decay rate} = kN_R \quad \text{and} \quad k = \frac{0.693}{t_{1/2}}$$

In part (a) you calculated that $[A] = 6.73 \times 10^{14}$, which is the number of nuclei N_R. So convert the half-life $t_{1/2} = 5.24$ yr to $(5.24)(365.25)(24)(60)(60) = 1.65 \times 10^8$ s, combine the two equations, and solve:

$$\text{decay rate} = \left(\frac{0.693}{t_{1/2}}\right) N_R$$

$$= \left(\frac{0.693}{1.65 \times 10^8 \text{ s}}\right)(6.73 \times 10^{14})$$

$$= 2.83 \times 10^6 \text{ s}^{-1}$$

Supplementary Exercises

PROBLEM 20-18   Fill in the blanks in the following equations:

(a) $^{27}_{13}\text{Al} + ^{1}_{1}\text{H} \longrightarrow ^{26}_{13}\text{Al} + \underline{\quad ? \quad}$
(b) $^{27}_{13}\text{Al} + \underline{\quad ? \quad} \longrightarrow ^{24}_{12}\text{Mg} + ^{4}_{2}\text{He}$
(c) $^{3}_{1}\text{H} + \underline{\quad ? \quad} \longrightarrow ^{4}_{2}\text{He} + ^{1}_{0}n$

(d) $^{239}_{92}U \longrightarrow _{-1}^{0}e + \underline{\quad ? \quad} \longrightarrow _{-1}^{0}e + \underline{\quad ? \quad}$

(e) $^{9}_{4}Be + ^{4}_{2}He \longrightarrow ^{1}_{0}n + \underline{\quad ? \quad}$

Answer: **(a)** $^{2}_{1}H$ **(b)** $^{1}_{1}H$ **(c)** $^{2}_{1}H$ **(d)** $^{239}_{93}Np$; $^{239}_{94}Pu$ **(e)** $^{12}_{6}C$

PROBLEM 20-19 The isotope $^{77}_{32}Ge$ undergoes β decay with a half-life of 11.3 h. A sample of $^{77}_{32}Ge$ contains 1.0×10^6 atoms. How long will it take before the sample has only 1.0×10^3 atoms of $^{77}_{32}Ge$?

Answer: 113 h

PROBLEM 20-20 The isotope $^{232}_{90}Th$ undergoes α decay and has a half-life of 1.41×10^{10} yr. **(a)** What is the other decay product? **(b)** How many disintegrations per second (dps) would be produced in a sample containing 1.0 g of $^{232}_{90}Th$?

Answer: **(a)** $^{228}_{88}Ra$ **(b)** 4.0×10^3 dps

PROBLEM 20-21 A 1.5-mg sample of a protein labeled with radioactive $^{131}_{53}I$, which undergoes β decay and has a half-life of 8.07 days, has an initial decay rate of 5.64×10^4 disintegrations per minute (dpm). **(a)** What mass of $^{131}_{53}I$ is present in the sample? **(b)** What will the decay rate of the sample be after 14.0 days?

Answer: **(a)** 2.06×10^{-13} g **(b)** 1.70×10^4 dpm

PROBLEM 20-22 The curie is a measure of radioactivity equal to 3.70×10^{10} dps. What mass of $^{226}_{88}Ra$, which has a half-life of 1622 years, has an activity equal to 1.0 curie?

Answer: 1.0 g (This is not a coincidence: The curie was originally defined in terms of a 1.0-g mass of radium)

PROBLEM 20-23 A radioactive target, freshly prepared in a particle accelerator, gave 4.2×10^3 dps. After 20.8 min the count rate had fallen to 730 dps. What is the half-life in seconds of the species responsible for the observed activity?

Answer: 4.9×10^2 s

PROBLEM 20-24 Some intact bottles of soda water were found in a wrecked ship. If the water in the bottles gives a count of 12.2 dpm per kilogram of H_2O owing to the tritium it contains, what is the maximum length of time that the ship has been at the bottom of the ocean? (In natural water in equilibrium with the atmosphere, tritium gives rise to 70.2 dpm per kg H_2O; $t_{1/2}$ of ^3_1H = 12.3 yr.)

Answer: 31 yr

PROBLEM 20-25 The enthalpy of combustion of most hydrocarbon fuels is about -55 MJ/kg. How much mass in grams would need to be converted into energy to provide the equivalent of the combustion of 1.0 kg of a hydrocarbon?

Answer: 6.1×10^{-7} g

PROBLEM 20-26 What is the binding energy per nucleon of the ^5Li nucleus whose mass is 5.0125 amu?

Answer: 8.5×10^{-13} J

PROBLEM 20-27 What is the maximum kinetic energy of the electrons produced in the β decay of ^6_2He (mass = 6.018 88 amu) to ^6_3Li (mass = 6.015 12 amu)?

Answer: 4.82×10^{-13} J

PROBLEM 20-28 How much energy in megajoules per second would be produced in a fusion power plant using the reaction $^3_2He + ^3_1H \rightarrow ^4_2He + ^2_1H$ if it consumes 1.0 g of ^3_1H per second? (^3_2He = 3.016 03 amu; ^3_1H = 3.016 05 amu; ^4_2He = 4.002 60 amu; ^2_1H = 2.0140 amu.)

Answer: 4.6×10^5 MJ/s (or MW)

PROBLEM 20-29   You are the curator of the Megalithic Museum of Natural History. An acquaintance of yours has offered to sell you—cheaply—an ancient mummy case alleged to have come from the tomb of a pharaoh who flourished ca. 1000 BC. Before buying the case, you determine that its wood gives a ^{14}C count of 13.7 dpm per gram of carbon. Is your acquaintance your friend? (See Problem 20-14 for pertinent data.)

Answer: No! The wood in the case must have come from a tree that was felled only 910 years ago, so your acquaintance is probably lying

PROBLEM 20-30   The isotope $^{210}_{84}Po$ is an α particle emitter; its half-life is 138.4 days.

(a) Complete the equation $^{210}_{84}Po \longrightarrow \alpha + ?$

(b) If you start with 2.87×10^{14} atoms of $^{210}_{84}Po$, how many atoms will be left after 365 days?

Answer: (a) $^{206}_{82}Pb$ (b) 4.62×10^{13} atoms

21 COORDINATION CHEMISTRY

THIS CHAPTER IS ABOUT

☑ **Coordinate Covalence**
☑ **Ligand Classification**
☑ **Geometry of Complexes**
☑ **Bonding in Complexes**
☑ **Stability of Complexes**

21-1. Coordinate Covalence

A. Coordinate covalent bonds and complexes

If both electrons in a single covalent bond are contributed by only one of the atoms, the result is a *coordinate covalent bond* (see Chapter 10). By this definition, any reaction — i.e., **coordination**— between a Lewis acid (an electron-pair acceptor) and a Lewis base (an electron-pair donor) results in the formation of a coordinate covalent bond (see Chapter 12). The bonding electrons come from the Lewis base, as in the reaction of BF_3 with NH_3:

$$H_3N: + BF_3 \longrightarrow H_3N:BF_3$$

Any compound that results from the reaction of a Lewis acid with a Lewis base may be called a **coordination compound**.

A **complex** is a coordination compound formed when a *metallic* Lewis acid is coordinated to a Lewis base molecule or ion; a **complex ion** is a charged complex. While all metal ions form complexes, the ions of the transition metals form a very large number of complexes. In fact, most of the chemistry of the transition-metal ions is complex-ion chemistry, or coordination chemistry.

Notation for complexes

In the notation for a complex the symbols for the Lewis acid and base species appear within brackets, and the charge for a complex ion is written outside the brackets as a superscript.

EXAMPLE 21-1: Describe in words the complex ions whose formulas are as follows: (a) $[Ag(NH_3)_2]^+$, (b) $[Cu(NH_3)_4]^{2+}$, (c) $[Fe(H_2O)_6]^{2+}$, (d) $[FeCl_4]^-$, (e) $[PtCl_6]^{2-}$.

Solution:

Complex ion	Description of complex
(a) $[Ag(NH_3)_2]^+$	Two NH_3 (base) molecules bound to one Ag^+ (acid) ion; total charge is $2(0) + 1 = +1$
(b) $[Cu(NH_3)_4]^{2+}$	Four NH_3 (base) molecules bound to one Cu^{2+} (acid) ion; total charge is $4(0) + 2 = +2$
(c) $[Fe(H_2O)_6]^{2+}$	Six H_2O (base) molecules bound to one Fe^{2+} (acid) ion; total charge is $6(0) + 2 = +2$
(d) $[FeCl_4]^-$	Four Cl^- (base) ions bound to one Fe^{3+} (acid) ion; total charge is $4(-1) + 3 = -1$
(e) $[PtCl_6]^{2-}$	Six Cl^- (base) ions bound to one Pt^{4+} (acid) ion; total charge is $6(-1) + 4 = -2$

B. Coordination number

The molecules or ions that contain the donor atoms (i.e., the Lewis bases bound to the metal ion) in a complex are called **ligands,** each of which is *coordinated to* the metal atom. The number of nonmetallic ligands bound to the central metal atom or ion in a complex is the **coordination number** of the metal in that complex.

EXAMPLE 21-2: What are the coordination numbers of the metals in the complex ions described in Example 21-1?

Solution:

	Complex ion	Metal	Number of ligands	Coordination number
(a)	$[Ag(NH_3)_2]^+$	Ag^+	$2\ NH_3$	2
(b)	$[Cu(NH_3)_4]^{2+}$	Cu^{2+}	$4\ NH_3$	4
(c)	$[Fe(H_2O)_6]^{2+}$	Fe^{2+}	$6\ H_2O$	6
(d)	$[FeCl_4]^-$	Fe^{3+}	$4\ Cl^-$	4
(e)	$[PtCl_6]^{2-}$	Pt^{4+}	$6\ Cl^-$	6

C. Oxidation state

The oxidation number (or state) of an atom in a compound is the charge that the atom appears to have when combined with other atoms (see Chapter 13). For complex ions the most important oxidation state is usually that of the central metal atom. To find that oxidation state, first separate the complex ion into its constituent ligands and a bare metal ion; then, using the rules given in Chapter 13, determine any charge on the ligands. The apparent opposite charge on the metal ion is its oxidation state.

EXAMPLE 21-3: Determine the coordination numbers and oxidation states of the central metal ions in $[CuBr_4]^{2-}$, $[Ni(NH_3)_4]^{2+}$, $[PtCl_6]^{2-}$, $[Pt(NH_3)_2Br_2]$, and $[Co(NH_3)_4Cl_2]^+$.

Solution:

Complex	Ligands	Charge on complex	Metal ion	Coordination number	Oxidation number
$[CuBr_4]^{2-}$	$4Br^-$	$4(-1) + 2 = -2$	Cu^{2+}	4	$+2$
$[Ni(NH_3)_4]^{2+}$	$4NH_3$	$4(0) + 2 = +2$	Ni^{2+}	4	$+2$
$[PtCl_6]^{2-}$	$6Cl^-$	$6(-1) + 4 = -2$	Pt^{4+}	6	$+4$
$[Pt(NH_3)_2Br_2]$	$2Br^-$ $2NH_3$	$2(-1) + 2(0) + 2 = 0$	Pt^{2+}	4	$+2$
$[Co(NH_3)_4Cl_2]^+$	$4NH_3$ $2Cl^-$	$4(0) + 2(-1) + 3 = +1$	Co^{3+}	6	$+3$

D. Transition-metal ions

The metals of the first transition series are those whose atomic numbers range from 21 through 30 (Sc through Zn). The neutral atoms of these elements have an $[Ar]4s^2 3d^n$ electronic configuration, where $n = 1-10$. When the ions of these metals are formed, the $4s$ electrons ionize first; therefore, the ions have an electronic configuration of the type $[Ar]3d^m$ (with no $4s$ electrons).

  note: One shorthand notation for transition-metal ion configuration uses only the d^m portion of the electronic configuration. Thus, for example, $[Ar]3d^3$ would be called a d^3 ion.

EXAMPLE 21-4: What are the electronic configurations of the ions Ti^{4+}, Ti^{3+}, Fe^{3+}, Co^{3+}, Ni^{2+}, and Sc^{3+}?

Solution:

Ion	Z	Neutral atom configuration	Electrons lost	Ion configuration
Ti^{4+}	22	$[Ar]4s^2 3d^2$	$4s^2 3d^2$	$[Ar]$; noble-gas configuration
Ti^{3+}	22	$[Ar]4s^2 3d^2$	$4s^2 3d^1$	$[Ar]3d^1$; d^1 ion
Fe^{3+}	26	$[Ar]4s^2 3d^6$	$4s^2 3d^1$	$[Ar]3d^5$; d^5 ion
Co^{3+}	27	$[Ar]4s^2 3d^7$	$4s^2 3d^1$	$[Ar]3d^6$; d^6 ion
Ni^{2+}	28	$[Ar]4s^2 3d^8$	$4s^2$	$[Ar]3d^8$; d^8 ion
Sc^{3+}	21	$[Ar]4s^2 3d^1$	$4s^2 3d^1$	$[Ar]$; noble-gas configuration

note: The transition metals in the d series can present a number of different oxidation states, with the exception of scandium, which exhibits an oxidation state of $+3$ in almost all of its compounds.

21-2. Ligand Classification

Ligands can be classified by the number of Lewis base donor sites present on each molecule of the ligand.

1. **Unidentate ligands:** One donor site per molecule.
2. **Bidentate ligands:** Two donor sites per molecule.
3. **Polydentate ligands:** More than two donor sites per molecule.

A single bidentate or polydentate ligand may coordinate to a metal ion to form a ring, a process called **chelation**. Some important chelating ligands are listed in Table 21-1.

TABLE 21-1: Chelating (ring-forming) Ligands

Ligand name	Formula	Classification	Abbreviation
Oxalate ion (Oxolato complex)		bidentate	ox
1,2-Diaminoethane (Ethylenediamine)	$H_2NCH_2CH_2NH_2$	bidentate	en
Ethylenediaminetetraacetic acid anion		polydentate (up to 6 donor sites)	edta

21-3. Geometry of Complexes

The geometry of coordination compounds is generally governed by the same principle that governs VSEPR geometry—the maintenance of maximum distance between substituents on a central atom (see Section 10-5B). Ligands (L) on a central metal (M) tend to be arranged in four ideal geometries, depending (at least in part) on the number of ligands: Two-coordinate (ML_2) complexes are linear; four-coordinate (ML_4) complexes may be square-planar or tetrahedral; six-coordinate (ML_6) complexes tend to be octahedral. (See Figure 21-1; cf. Table 10-4.)

One aspect of the geometric arrangement of ligands in fixed positions about a central atom is **isomerism**, in which compounds having the same chemical formula have different structures (and consequently different properties). It is possible for complexes to form geometric isomers or optical isomers.

A. Geometric isomers

Geometric isomers in coordination compounds have the same chemical formula, but differ in the spatial arrangement of their ligands. Square-planar and octahedral complexes can form geometric isomers.

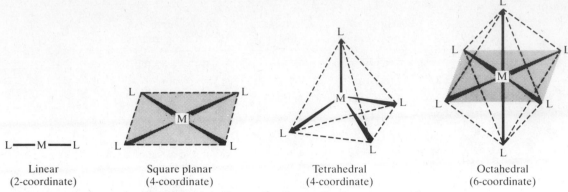

FIGURE 21-1. Geometry of coordination compounds.

1. Square-planar complexes

Consider, for example, the two forms of $[PtCl_2(NH_3)_2]$, a square-planar complex:

Both structures represent the same formula (and geometry); but in the first form the identical ligands are in *adjacent* or **cis** positions, and in the second the identical ligands are in *opposite* or **trans** positions. These two molecules—*cis*-dichlorodiammineplatinum(II) and *trans*-dichlorodiammineplatinum(II)—are examples of **cis–trans isomerism**.

EXAMPLE 21-5: Sketch the cis and trans isomers of $[Pt(NH_3)Cl_2I]^-$, a square-planar complex.

Solution: The differentiation is made on the basis of the position of the chlorine atoms:

In the cis isomer the Cl atoms are adjacent to each other, and in the trans isomer the Cl atoms are opposite each other.

note: You might have drawn

But this sketch is just another view of the cis isomer. Notice that the chlorine atoms are adjacent and that this structure can be superimposed on that of the cis isomer.

- *If one structure can be perfectly superimposed on another, the two structures are the same.*

(Superimposability is much easier to see in three-dimensional models.)

---

2. Octahedral complexes

An important consequence of octahedral geometry is that there is one, and only one, octahedral complex of the general type ML_5X, where L is one kind of ligand and X is another (see Figure 21-2a). There are only two complexes of the type ML_4X_2 (Figure 21-2b): If the two X ligands are in adjacent positions, the isomer is cis; if the two X ligands are in opposition, the isomer is trans. Chelating ligands can span only adjacent positions in octahedral complexes.

note: Recall from geometry that the triangular faces of a regular octahedron are congruent equilateral triangles. (Sketches of octahedrons tend to obscure this fact.) Models can be of real help in understanding isomerism in octahedral complexes.

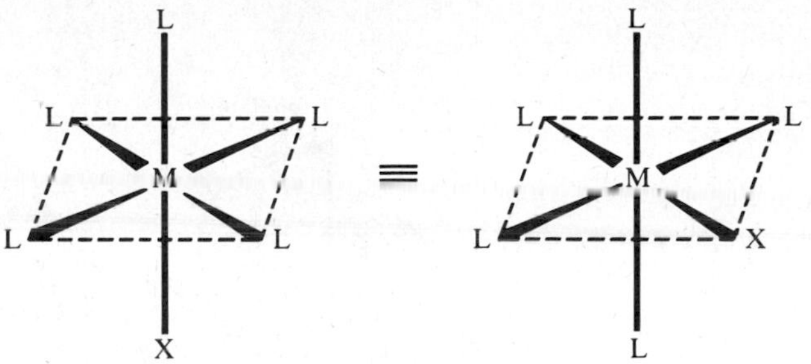

(a) ML$_5$X

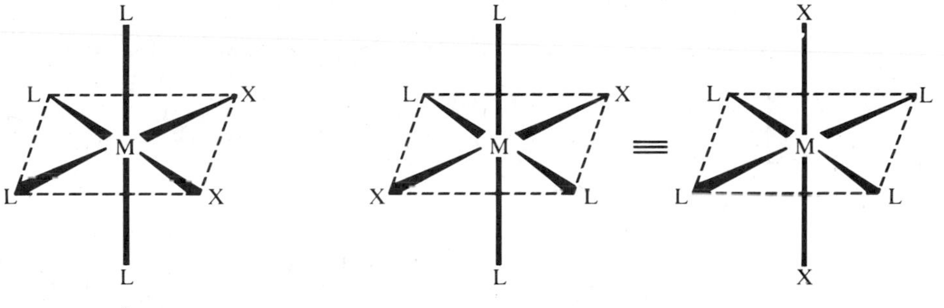

cis

trans

(b) ML$_4$X$_2$

FIGURE 21-2. Octahedral geometry. (**a**) Only one octahedral ML$_5$X compound exists; (**b**) Isomers of ML$_4$X$_2$.

EXAMPLE 21-6: Sketch the isomers of $[Co(NH_3)_4BrCl]^+$, an octahedral complex of the ML$_4$X$_2$ type.

Solution: The differentiation is based on the position of the chlorine and bromine atoms:

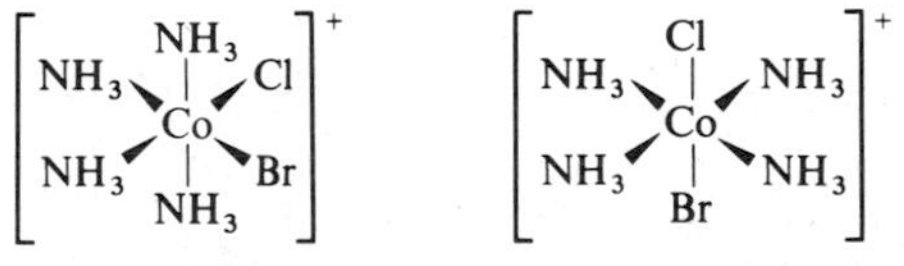

cis   trans

EXAMPLE 21-7: Sketch the geometry of $[Fe(ox)(H_2O)_4]$, an octahedral complex.

Solution: See Table 21-1 for the structure of ox, the oxalate ion. The ox group, a chelating ligand, can span only adjacent positions to make a ring so there's only one possible geometry:

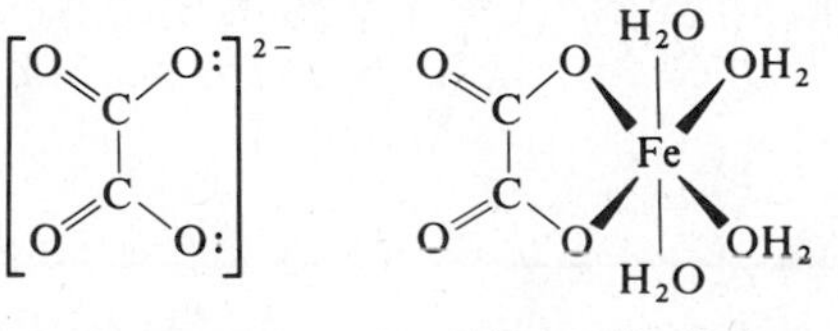

ox $[Fe(ox)(H_2O)_4]$

B. Optical isomers

Optical isomers, called **enantiomers,** are paired species whose molecular formulas are identical and whose structures are mirror images. The compounds represented by these structures rotate plane-polarized light in opposite directions. The D-isomer rotates plane-polarized light to the right (clockwise); the L-isomer rotates it to the left (counterclockwise). Enantiomers are *not* identical, just as a left and right hand are not identical. (Once again, models will be of great value to you in deciding if two isomers are enantiomers.)

All six-coordinate complexes with three chelate rings have optically active enantiomers (see Figure 21-3).

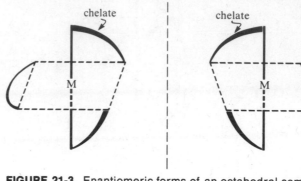

FIGURE 21-3. Enantiomeric forms of an octahedral complex with three chelate rings.

EXAMPLE 21-8: Sketch the enantiomers of $[Co(en)_3]^{3+}$, an optically active octahedral complex.

Solution: The chelating ligand en can span only adjacent positions. For simplicity, use a curved line for en:

21-4. Bonding in Complexes

Many transition-metal complex ions are paramagnetic owing to the presence of unpaired electrons in the complex; they are also colored. These properties can be explained by two models for bonding in complexes: *valence bond theory* and *crystal field theory*.

A. Valence bond theory

Valence bond theory employs the concept of **hybridization**—the overlap or blending of s, p, and d orbitals—to explain bond formation. According to valence bond theory, coordinate covalent bonds in complexes are formed when electron pairs from the ligands occupy hybrid orbitals of equal energy. **Hybrid orbitals** are mathematical combinations of s, p, and d orbitals. For example, the overlap of one s and one p orbital generates TWO sp hybrid orbitals; one s and two p orbitals generate THREE sp^2 hybrid orbitals; one s and three p orbitals generate FOUR sp^3 hybrid orbitals. Orbital shape reflects the contribution of each atomic orbital to the hybrid. (See your textbook for suitable diagrams.)

The geometry of the complex is used to establish the hybrid orbital designation as follows:

Coordination number	Geometry	Type of hybrid
4	tetrahedral	sp^3
4	square planar	dsp^2
6	octahedral	d^2sp^3 or sp^3d^2

The donated electrons from the ligands are placed in the appropriate hybrid orbital and metal d orbital electrons are placed in nonhybridized d orbitals. Although this sounds complicated, it's a simple matter of keeping track of electrons. (See Sections 9-3, 9-4, and 10-6 for a review of orbitals and electronic configurations.)

EXAMPLE 21-9: Determine the valence bond configuration of the following complexes: **(a)** $[FeCl_4]^-$, a paramagnetic tetrahedral complex; **(b)** $[PdBr_4]^{2-}$, a diamagnetic planar complex; **(c)** $[Co(NH_3)_6]^{3+}$, a diamagnetic octahedral complex; **(d)** $[CoF_6]^{3-}$, a paramagnetic octahedral complex.

Solution:

(a) You are told that the complex ion $[FeCl_4]^-$ is tetrahedral, so you choose sp^3 hybrid orbitals. The complex ion is paramagnetic, so there will be unpaired electrons. First determine the oxidation state of the metal ion and its configuration:

(Total charge) − (ligand contribution) = oxidation number of metal ion

$$-1 \quad - \quad (-4) \quad = +3$$

Neutral Fe has 26 electrons and has the electronic configuration $[Ar]4s^2 3d^6$; so the ion Fe^{3+} is $[Ar]3d^5$, a d^5 ion. Now set up a chart to show the electronic configuration. (Remember: You're dealing *only* with the orbitals of the metal ion: The electrons from the ligands will fill the next available orbitals to form hybrid orbitals.)

Electronic configuration
	3d	4s	4p

Fe^{3+} ($[Ar]3d^5$): 3d = [↑][↑][↑][↑][↑]; 4s = [ ]; 4p = [ ][ ][ ]

+ 4Cl⁻: 4s = [↑↓]; 4p = [↑↓][↑↓][↑↓]

$[FeCl_4]^-$: 3d = [↑][↑][↑][↑][↑]; 4s = [↑↓]; 4p = [↑↓][↑↓][↑↓]

four sp^3 hybrid orbitals

(b) $[PdBr_4]^{2-}$ is planar, so you'll use dsp^2 hybrid orbitals—four equal-energy hybrid orbitals obtained from the blending of one 4d orbital, the 5s orbital, and two of the 5p orbitals. Because the complex is diamagnetic, the electrons in the remaining 4d metal orbitals must be paired. The oxidation state of the palladium is $-2 - (-4) = +2$. Neutral Pd has 46 electrons with the electronic configuration $[Kr]4d^{10}$, which is an irregularity in the regular orbital-filling order (see the periodic table or Table 9-2). Therefore, the ion Pd^{2+} is $[Kr]4d^8$, a d^8 ion.

Electronic configuration
	4d	5s	5p

Pd^{2+} ($[Kr]4d^8$): 4d = [↑↓][↑↓][↑↓][↑][↑]; 5s = [ ]; 5p = [ ][ ][ ]

+ 4Br⁻: 4d = [↑↓]; 5s = [↑↓]; 5p = [↑↓][↑↓]

$[PdBr_4]^{2-}$: 4d = [↑↓][↑↓][↑↓][↑↓][↑↓]; 5s = [↑↓]; 5p = [↑↓][↑↓][ ]

four dsp^2 hybrid orbitals

(c) The complex ion $[Co(NH_3)_6]^{3+}$ is octahedral, so you'll use either d^2sp^3 or sp^3d^2 hybrid orbitals. The difference between these choices is governed by the principal quantum number of the d orbital hybrid:

- In d^2sp^3 hybridization the $(n-1)d$ orbitals are filled and the complex is called an **inner-orbital** complex.
- In sp^3d^2 hybridization the $(n-1)d$ orbitals are only partly filled while ligand electrons enter the nd orbitals. The complex is called an **outer-orbital** complex.

For $[Co(NH_3)_6]^{3+}$, the oxidation state of cobalt is $+3$ (the NH_3 molecules are neutral). Neutral Co has 27 electrons and has the electronic configuration $[Ar]4s^2 3d^7$; so Co^{3+} is $[Ar]3d^6$, a d^6 ion. The complex is diamagnetic, so the d orbital electrons must be paired. Then you can see that d^2sp^3 hybridization, which has six hybrid orbitals—two 3d, one 4s, and three 4p orbitals—will nicely hold the 12 electrons contributed by the ligands.

Electronic configuration
	3d	4s	4p

Co^{3+} ($[Ar]3d^6$): 3d = [↑↓][↑][↑][↑][↑]; 4s = [ ]; 4p = [ ][ ][ ]

+ 6NH₃: 3d = [↑↓][↑↓]; 4s = [↑↓]; 4p = [↑↓][↑↓][↑↓]

$[Co(NH_3)_6]^{3+}$: 3d = [↑↓][↑↓][↑↓][↑↓][↑↓]; 4s = [↑↓]; 4p = [↑↓][↑↓][↑↓]

six d^2sp^3 hybrid orbitals

(d) For $[CoF_6]^{3-}$, the oxidation state of cobalt is $-3 - (-6) = +3$. As in part **(c)**, the electronic configuration of Co^{3+} is $[Ar]3d^6$. You are told that the complex ion is paramagnetic, so the electrons in the 3d metal orbitals must remain unpaired. The 12 ligand electrons cannot be accommodated by d^2sp^3 hybridization, so you use sp^3d^2.

	Electronic configuration			
	3d	4s	4p	4d

$Co^{3+}([Ar]3d^6)$ $\boxed{↑↓}\boxed{↑}\boxed{↑}\boxed{↑}\boxed{↑}$ $\square$ $\square\square\square$ $\square\square\square\square\square$

+ 6F⁻ $↑↓$ $↑↓$ $↑↓$ $↑↓$ $↑↓$ $↑↓$

$[CoF_6]^{3-}$ $\boxed{↑↓}\boxed{↑}\boxed{↑}\boxed{↑}\boxed{↑}$ $\boxed{↑↓}$ $\boxed{↑↓}\boxed{↑↓}\boxed{↑↓}$ $\boxed{↑↓}\boxed{↑↓}\square\square\square$

$\underbrace{\hspace{5cm}}$
six sp^3d^2 hybrid orbitals

B. Crystal field theory

Crystal field theory is concerned with the electrostatic forces in a complex. These forces arise primarily from the spatial arrangement of the *d* orbitals of the metal ion in relation to the ligands. Consider, for example, the shapes of the five *d* orbitals shown in Figure 21-4. Two *d* orbitals d_{z^2} and $d_{x^2-y^2}$ are directed along the coordinate axes, while three *d* orbitals d_{xy}, d_{yz}, and d_{xz} are directed along the spaces between the coordinate axes. In a free metal ion, the energy of all five *d* orbitals is identical.

Now consider a six-coordinate octahedral complex, in which six electron–pair-bearing ligands approach a metal ion along the (x, y, z) coordinate axes. The electrons of the metal ion in the d_{z^2} and $d_{x^2-y^2}$ orbitals, called the e_g set, are electrostatically repelled by the electrons of the approaching ligands, increasing the energy level of the e_g set and making the set less stable. At the same time, the energy level of the d_{xy}, d_{yz}, and d_{xz} orbitals, called the t_{2g} set, decreases, making the t_{2g} set more stable. The five *d* orbitals have been "split" into two sets whose energy levels are not equal. The energy gap between the two sets is called the **crystal field splitting**. Its magnitude depends on the nature of the metal ion and the ligands, but the energy gained by the e_g set always equals the energy lost by the t_{2g} set.

In the formation of a tetrahedral complex, the exact reverse holds: The t_{2g} set of orbitals becomes less stable and of higher energy than the e_g set, and the energy gained by the t_{2g} set equals the energy lost by the e_g set.

In determining electronic configuration by crystal field theory, only the *d* electrons of the metal ion are considered. If the crystal field splitting is large, the *d* electrons will pair in the lower-energy orbitals, giving rise to a **low-spin** (or *spin-paired*) **complex**. If the crystal field splitting is small, the

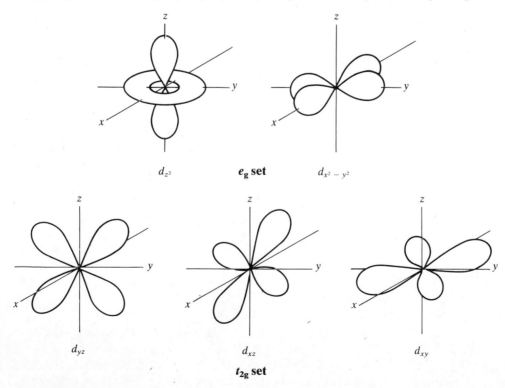

d_{z^2} **e_g set** $d_{x^2-y^2}$

d_{yz} d_{xz} d_{xy}

t_{2g} set

FIGURE 21-4. *d* Orbitals. e_g set: orbitals lying *along* the axes; t_{2g} set: orbitals lying in the spaces *between* the axes.

stability of pairing is outweighed by the energy cost, so the d electrons remain unpaired, giving rise to a **high-spin complex**. In practice

- Tetrahedral complexes are high-spin.
- Octahedral complexes can be either low- or high-spin, depending on the metal ion and the ligands (see Problem 21-9).

EXAMPLE 21-10: Use crystal field theory to explain the magnetic properties of (a) $[FeCl_4]^-$, (b) $[Co(NH_3)_6]^{3+}$, (c) $[CoF_6]^{3-}$. See Example 21-9 for a discussion of the electronic configuration of each metal ion.

Solution:

(a) $[FeCl_4]^-$ is a tetrahedral complex with a d^5 metal ion. Because tetrahedral complexes are high-spin, the orbital occupancy is $(e_g)^2(t_{2g})^3$ (i.e., two electrons in the e_g set and three electrons in the t_{2g} set) and there are no paired electrons:

$$E \uparrow \quad \boxed{\uparrow}\boxed{\uparrow}\boxed{\uparrow} \; t_{2g} \atop \boxed{\uparrow}\boxed{\uparrow} \; e_g \; \} \text{ small crystal field splitting}$$

This configuration agrees with the paramagnetic nature of the complex.

(b) $[Co(NH_3)_6]^{3+}$ is an octahedral complex with a d^6 ion. Because it is diamagnetic, all electrons must be paired in a low-spin complex with a large crystal field splitting. The orbital occupancy is $(t_{2g})^6(e_g)^0$:

$$E \uparrow \quad \boxed{\ }\boxed{\ } \; e_g \atop \boxed{\uparrow\downarrow}\boxed{\uparrow\downarrow}\boxed{\uparrow\downarrow} \; t_{2g} \; \} \text{ large crystal field splitting}$$

(c) $[CoF_6]^{3-}$ is also an octahedral complex with a d^6 ion. Because it is paramagnetic, it must be a high-spin complex. The orbital occupancy is $(t_{2g})^4(e_g)^2$:

$$E \uparrow \quad \boxed{\uparrow}\boxed{\uparrow} \; e_g \atop \boxed{\uparrow\downarrow}\boxed{\uparrow}\boxed{\uparrow} \; t_{2g} \; \} \text{ small crystal field splitting}$$

The sixth electron has no choice; it has to pair, an energetically unfavorable process for an electron. It naturally pairs in the lowest-energy orbital available.

Although it's beyond the scope of this treatment, you should appreciate that the crystal field splitting phenomenon can also explain the colors of transition-metal complexes (see Problem 21-10), something that the valence bond theory cannot easily do.

note: Molecular orbital theory can also be used to explain the physical properties of complexes. This theory incorporates crystal field splitting and the t_{2g} and e_g orbitals at energy levels intermediate to those of the bonding (σ) and antibonding (σ^*) molecular orbitals. Thus for an octahedral complex, the t_{2g} crystal field orbitals are referred to as *nonbonding molecular orbitals*, while the e_g orbitals are split into two sets: e_g orbitals, equivalent to the σ_d *bonding molecular orbitals*, and e_g^* orbitals, equivalent to the σ_d^* *antibonding molecular orbitals*.

21-5. Stability of Complexes

The coordination of a ligand to a metal is an equilibrium process. For example, for the reaction

$$Cu^{2+}(aq) + 4NH_3(aq) \rightleftharpoons [Cu(NH_3)_4]^{2+}(aq)$$

you can write an equilibrium-constant expression

$$K_s = \frac{[[Cu(NH_3)_4]^{2+}]}{[Cu^{2+}][NH_3]^4}$$

where K_s is the **stability constant** of the complex. The reaction of Cu^{2+} with four NH_3 ligands is a four-step process, and equilibrium constants can be written for each step, so that $K_s = K_1K_2K_3K_4$.

note: Your text may use K_d, the **dissociation constant**, for a complex. The two equilibrium constants are reciprocals: $K_d = 1/K_s$.

EXAMPLE 21-11: The stability constant K_s for $[Cu(NH_3)_4]^{2+}$ is 1.4×10^{13}. Calculate the concentration of free $Cu^{2+}(aq)$ in a solution made from 1.0 mL of 0.10 M $Cu^{2+}(aq)$ and 10.0 mL of 1.00 M NH_3.

Solution: Apply the techniques discussed in Chapters 17 and 18 to complexes. Write the reaction equation and K_s expression:

$$Cu^{2+} + 4NH_3 \rightleftharpoons [Cu(NH_3)_4]^{2+} \qquad K_s = \frac{[[Cu(NH_3)_4]^{2+}]}{[Cu^{2+}][NH_3]^4} = 1.4 \times 10^{13}$$

You start with (1.0 mL)(0.10 mmol/mL) = 0.10 mmol of Cu^{2+} and (10.0 mL)(1.00 mmol/mL) = 10.00 mmol of NH_3. Clearly, Cu^{2+} is the limiting reagent, so the amount of NH_3 used in the reaction is

$$(0.10 \text{ mmol } Cu^{2+})\left(\frac{4 \text{ mmol } NH_3}{1 \text{ mmol } Cu^{2+}}\right) = 0.40 \text{ mmol } NH_3$$

Because ammonia is present in vast excess and the complex is quite stable (large K_s), nearly all of the Cu^{2+} will be converted to product; so

$$[[Cu(NH_3)_4]^{2+}] \cong \frac{0.10 \text{ mmol}}{1.0 \text{ mL} + 10.0 \text{ mL}} = 9.1 \times 10^{-3} \text{ M}$$

The concentration of NH_3 at equilibrium is

$$[NH_3] = \frac{(10.0 - 0.40) \text{ mmol}}{(1.0 + 10.0) \text{ mL}} = 0.87 \text{ M}$$

Thus if

$$K_s = \frac{[[Cu(NH_3)_4]^{2+}]}{[Cu^{2+}][NH_3]^4} = 1.4 \times 10^{13}$$

then

$$[Cu^{2+}] = \frac{[[Cu(NH_3)_4]^{2+}]}{K_s[NH_3]^4} = \frac{9.1 \times 10^{-3}}{(1.4 \times 10^{13})(0.87)^4} = \frac{9.1 \times 10^{-3}}{8.02 \times 10^{12}}$$

$$= 1.1 \times 10^{-15} \text{ M}$$

Notice that the assumption that virtually all the Cu^{2+} would be converted to product is justified.

SUMMARY

1. Coordination is a Lewis acid–base reaction.
2. In a Lewis acid–base complex of a metal ion, the bases are termed ligands.
3. The number of ligands bound to a metal is called its coordination number.
4. When transition-metal atoms ionize, their outer s electrons are lost first.
5. Ligands are classified by the number of donor sites they have: unidentate, bidentate, polydentate.
6. Four-coordinate complexes can be tetrahedral or planar; six-coordinate complexes are usually octahedral.
7. Many complexes exist in isomeric or enantiomeric (optical isomeric) forms.
8. Transition-metal complexes are often paramagnetic and colored.
9. Valence bond theory uses hybrid orbitals to explain structures and magnetic properties of complexes: tetrahedral, sp^3; planar, dsp^2; octahedral, d^2sp^3 (inner-orbital) or sp^3d^2 (outer-orbital).
10. Crystal field theory uses electrostatic splitting of d orbitals to explain structures and magnetic properties of complexes: Tetrahedral complexes are high-spin; octahedral complexes may be high- or low-spin, depending on the size of the d-orbital splitting.
11. The binding of ligands to metals is an equilibrium process, which can be described quantitatively by K_s, the stability constant.

RAISE YOUR GRADES

Can you ...?

☑ pick out the Lewis acids and bases making up a complex
☑ derive the coordination number and oxidation state of the metal atom in a complex
☑ deduce the electronic configuration of a transition-metal ion
☑ pick out the donor site(s) of a ligand
☑ sketch the correct isomers and/or enantiomers for 4- and 6-coordinate complexes
☑ explain the structures and shapes of complexes by using either valence bond or crystal field theory
☑ use K_s to calculate equilibrium concentrations in solutions of complexes

SOLVED PROBLEMS

PROBLEM 21-1 Identify the Lewis acid and Lewis base components of the following complexes: **(a)** $[HgBr_4]^{2-}$, **(b)** $[Ni(H_2O)_6]^{2+}$, **(c)** $[PdCl_2(NH_3)_2]$, **(d)** $[Al(OH)_3]$, **(e)** $[Ag(CN)_2]^-$, **(f)** $[Cr(CO)_6]$.

Solution: In each case the Lewis base will be an electron-pair donor: either a negative ion or a neutral molecule with a lone pair of electrons. The Lewis acid, or electron-pair acceptor, will be a metal ion (or, occasionally, a neutral metal atom).

	Lewis acid	Lewis bases	Resulting complex
(a)	Hg^{2+}	$4\ Br^-$	$[HgBr_4]^{2-}$
(b)	Ni^{2+}	$6\ H_2O$	$[Ni(H_2O)_6]^{2+}$
(c)	Pd^{2+}	$2\ Cl^-$ and $2\ NH_3$	$[PdCl_2(NH_3)_2]$
(d)	Al^{3+}	$3\ OH^-$	$[Al(OH)_3]$
(e)	Ag^+	$2\ CN^-$	$[Ag(CN)_2]^-$
(f)	Cr	$6\ CO$	$[Cr(CO)_6]$

note: Example **(d)** shows that a simple metal hydroxide can be considered and treated as a complex. Example **(f)** shows that a neutral metal atom—Cr, which can also be written Cr(0), stressing the zero charge on the metal atom—can form a complex.

PROBLEM 21-2 What are the oxidation states and coordination numbers of the metal ions in the complexes in Problem 21-1?

Solution: The oxidation state is given simply by the charge on the metal ion that is acting as a Lewis acid. The coordination number is the number of ligands bound to the metal ion. Inspection of the formulas analyzed in Problem 21-1 then leads to the following answers:

Complex	Lewis acid	Oxidation state	Coordination number (number of ligands)
(a) $[HgBr_4]^{2-}$	Hg^{2+}	$+2$	4
(b) $[Ni(H_2O)_6]^{2+}$	Ni^{2+}	$+2$	6
(c) $[PdCl_2(NH_3)_2]$	Pd^{2+}	$+2$	4
(d) $[Al(OH)_3]$	Al^{3+}	$+3$	3
(e) $[Ag(CN)_2]^-$	Ag^+	$+1$	2
(f) $[Cr(CO)_6]$	Cr	0	6

PROBLEM 21-3   What are the electronic configurations of the ions (a) V^{3+}, (b) Cr^{2+}, (c) Mn^{2+}, (d) Co^{2+}, (e) Cu^{+}?

Solution:
(a) Neutral V has 23 electrons (because $Z = 23$ for V) and its electronic configuration is $[Ar]4s^23d^3$. The three electrons lost in forming V^{3+} are the two $4s$ electrons (which invariably ionize first in this transition series) and one of the $3d$ electrons. So V^{3+} has the electronic configuration $[Ar]3d^2$ and is a d^2 ion.
(b) Neutral Cr has 24 electrons with configuration $[Ar]4s^23d^4$; Cr^{2+} has two fewer electrons and is therefore $[Ar]3d^4$.
(c) Neutral Mn has 25 electrons with configuration $[Ar]4s^23d^5$; Mn^{2+} is therefore $[Ar]3d^5$.
(d) Neutral Co has 27 electrons with configuration $[Ar]4s^23d^7$; Co^{2+} is therefore $[Ar]3d^7$.
(e) Neutral Cu has 29 electrons with configuration $[Ar]4s^13d^{10}$, so Cu^{+} is $[Ar]3d^{10}$.

PROBLEM 21-4   The glycinate anion ($H_2NCH_2CO_2^-$) is a good bidentate chelating ligand. Draw its valence bond structure and identify the probable donor sites in this ion.

Solution: The valence bond structure, with all the electrons shown, is

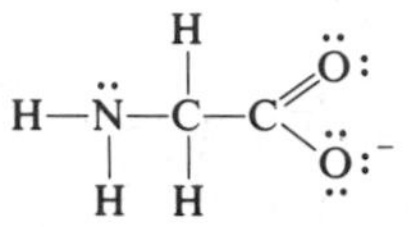

Since the ion is a bidentate chelating agent, it must contain two donor sites that can donate an electron pair to a Lewis acid. The most probable donor sites are therefore $-\ddot{N}-$, which has a lone pair, and $-\ddot{O}{:}^-$, the anionic oxygen atom.

PROBLEM 21-5   Sketch all the isomers of the 4-coordinate complexes (a) $[NiBr_2Cl_2]^{2-}$ (tetrahedral) and (b) $[Pt(Cl)(NO_2)(NH_3)_2]$ (planar).

Solution:
(a) For a tetrahedral complex $[NiBr_2Cl_2]^{2-}$, only one form can exist:

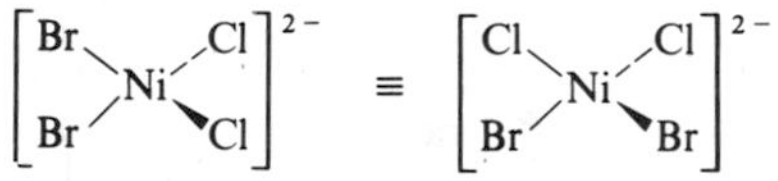

Even if you draw it differently, it's still the same compound. (Build the models if you aren't convinced!)
(b) For the planar complex, there are two isomers: one in which the two NH_3 groups are cis and one in which they are trans:

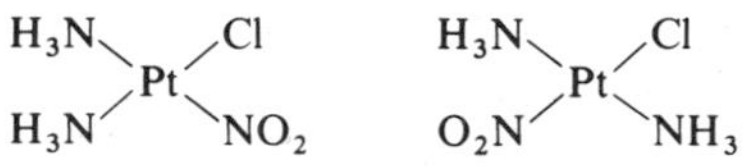

cis   trans

PROBLEM 21-6   How many isomers of the octahedral complex $[Cr(H_2O)_3Cl_3]$ are there? Sketch them.

Solution: There are only two isomers (you'll probably believe this only if you work with models):

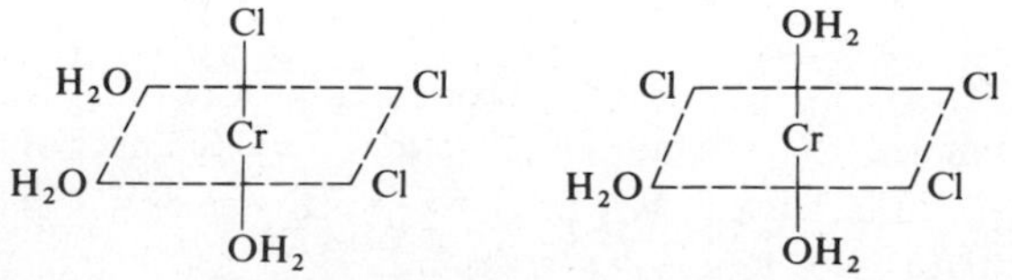

PROBLEM 21-7   Draw the enantiomers of the octahedral complex ion $[cis\text{-}Co(en)_2Cl_2]^+$.

Solution: The chelating ligand en ($H_2\ddot{N}CH_2CH_2\ddot{N}H_2$) (see Table 21-1) can only span adjacent cis positions. Consequently the cis in the description of the ion must refer to the two Cl groups. A brief trial gives the following enantiomers:

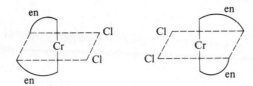

PROBLEM 21-8 Use valence bond theory to explain the shapes and magnetic properties of the complexes (a) $[Ni(CO)_4]$: tetrahedral, diamagnetic; (b) $[Fe(H_2O)_6]^{2+}$: octahedral, 4 unpaired electrons; (c) $[Fe(CN)_6]^{4-}$: octahedral, diamagnetic.

Solution:

(a) $[Ni(CO)_4]$: The four ligands are neutral CO molecules; thus the Ni is in the zero oxidation state and has the electronic configuration $[Ar]4s^2 3d^8$, which shows 10 electrons to be accommodated. The tetrahedral shape of $Ni(CO)_4$ means that there is sp^3 hybridization of the $4s$ and $4p$ orbitals to accept the 8 electrons from the four CO ligands. The 10 electrons from Ni are completely paired in the $3d$ orbital, accounting for the diamagnetism of the complex. The box diagram is

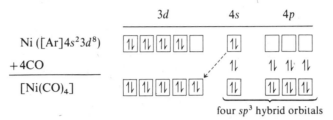

(b) $[Fe(H_2O)_6]^{2+}$. The six ligands are neutral H_2O molecules, so the Fe is in the $+2$ oxidation state and Fe^{2+} has the electronic configuration $[Ar]3d^6$ (i.e., a d^6 ion). The octahedral shape means that the hybridization is d^2sp^3 or sp^3d^2. If it's d^2sp^3 (inner-orbital), only three of the metal $3d$ orbitals will be left to be occupied by the six d electrons from the metal. The inner-orbital box diagram would look like this:

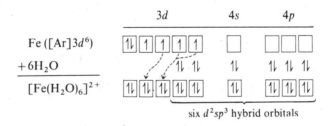

If it's sp^3d^2 (outer-orbital), the diagram would look like this:

	3d	4s	4p	4d
Fe	⬆⬇ ⬆ ⬆ ⬆ ⬆	☐	☐ ☐ ☐	☐ ☐ ☐ ☐ ☐
+6H₂O		⬆⬇	⬆⬇ ⬆⬇ ⬆⬇	⬆⬇ ⬆⬇
$[Fe(H_2O)_6]^{2+}$	⬆⬇ ⬆ ⬆ ⬆ ⬆	⬆⬇	⬆⬇ ⬆⬇ ⬆⬇	⬆⬇ ⬆⬇ ☐ ☐ ☐

six sp^3d^2 hybrid orbitals

Since the complex has four unpaired electrons, it must be outer-orbital in type.

(c) $[Fe(CN)_6]^{4-}$: The six ligands are CN^- ions, and so the Fe is again in the $+2$ oxidation state ($Fe^{2+} + 6CN^- \rightarrow [Fe(CN)_6]^{4-}$). Thus again this is a d^6 ion. From the discussion in part (b), because the complex is diamagnetic, it must be an inner-orbital complex with completely filled $3d$ orbitals.

PROBLEM 21-9 Use crystal field theory to explain the shapes and magnetic properties of the complexes discussed in Problem 21-8.

Solution:

(a) $Ni(CO)_4$. In crystal field theory all the valence electrons of the metal are placed in metal d orbitals. For Ni(0) there are 10 valence electrons ($4s^2 3d^8$) to be placed in the five vacant $3d$ orbitals of the complex. Thus all the $3d$

orbitals are occupied by electron pairs and the complex is diamagnetic. In the high-spin tetrahedral field the orbital occupancy is $(e_g)^4(t_{2g})^6$:

$$E \uparrow \quad \begin{array}{c} t_{2g} \quad \boxed{1\!\!\downarrow}\boxed{1\!\!\downarrow}\boxed{1\!\!\downarrow} \\[6pt] e_g \quad \boxed{1\!\!\downarrow}\boxed{1\!\!\downarrow} \end{array} \Bigg\} \begin{array}{l} \text{crystal field splitting} \\ \text{in tetrahedral field} \end{array}$$

(b) $[Fe(H_2O)_6]^{2+}$: The ion Fe^{2+} is a d^6 ion, and the complex is octahedral, so there are two possibilities, which depend on the nature of the ligand. If the ligands produce a **strong field**, the splitting will be large and the electrons will pair in the lower-lying orbitals (the t_{2g} set in an octahedral complex) to give a *low-spin complex*. If the ligands produce only a **weak field**, the splitting will be small and both the lower and upper orbitals will be occupied. The two cases are as follows:

$$\begin{array}{l} \text{small} \\ \text{splitting} \end{array} \left\{ \begin{array}{c} \boxed{1}\boxed{1} \\[6pt] \boxed{1\!\!\downarrow}\boxed{1}\boxed{1} \end{array} \right. \begin{array}{c} e_g \\[6pt] E \uparrow \quad t_{2g} \end{array} \qquad E \uparrow \begin{array}{c} \boxed{}\boxed{} \\[6pt] \boxed{1\!\!\downarrow}\boxed{1\!\!\downarrow}\boxed{1\!\!\downarrow} \end{array} \left. \begin{array}{l} \text{large} \\ \text{splitting} \end{array} \right.$$

$$\begin{array}{ll} \text{Weak-field ligand} & \qquad\qquad \text{Strong-field ligand} \\ \text{(high-spin complex)} & \qquad\qquad \text{(low-spin complex)} \end{array}$$

Since $[Fe(H_2O)_6]^{2+}$ is paramagnetic, having four unpaired electrons, it must be a weak-field case. The orbital occupancy is $(t_{2g})^4(e_g)^2$.

(c) $[Fe(CN)_6]^{4-}$: Again, Fe^{2+} is a d^6 ion. As the complex is diamagnetic, the arguments given in **(b)** lead you to conclude that its ligands are strong-field ligands, for which the splitting will be large. The orbital occupancy will be $(t_{2g})^6(e_g)^0$.

PROBLEM 21-10 Use a crystal field argument to explain why Cu^+ forms many colorless compounds, while compounds of Cu^{2+} are almost always colored (they're usually blue or blue-green).

Solution: Color in a chemical compound is the result of the absorption of visible light by the compound. The mechanism of the absorption is the excitation of electrons from lower occupied energy levels to higher unoccupied energy levels. The energy gap between these levels must fall in a rather narrow range if *visible* light is to be absorbed.

The Cu atom has the electronic configuration $[Ar]4s^1 3d^{10}$, and so Cu^+ has the configuration $[Ar]3d^{10}$ (not $[Ar]4s^1 3d^9$, because of the extra stability of a filled d shell). If one of the $3d$ electrons is to be excited to a vacant higher energy level, that level must be a $4s$ level—and the energy gap from $3d$ to $4s$ is rather large. This means that Cu^+ compounds absorb electromagnetic radiation in the high-energy ultraviolet region and thus do not appear colored.

On the other hand, Cu^{2+} has the configuration $[Ar]3d^9$; in its compounds the d levels are split by the crystal fields thus:

$$\begin{array}{ccc} \boxed{1\!\!\downarrow}\boxed{1} \quad e_g & & \boxed{1\!\!\downarrow}\boxed{1\!\!\downarrow} \\[4pt] d^9 \text{ ground state} & \xrightarrow{\;hv\;} & d^9 \text{ excited state} \\[4pt] \boxed{1\!\!\downarrow}\boxed{1\!\!\downarrow}\boxed{1\!\!\downarrow} \quad t_{2g} & & \boxed{1\!\!\downarrow}\boxed{1\!\!\downarrow}\boxed{1} \end{array}$$

The lowest-energy excitation of a d^9 ion is a transition of an electron from the lower to the higher of the split d levels to produce the excited state. The energy difference between the split d levels is much smaller than that between $3d$ and $4s$. The absorption of light in the d^9 ion is thus in the visible region (actually in the red/yellow), and so Cu^{2+} compounds appear blue or blue-green.

PROBLEM 21-11 The stability constant for the silver/thiosulfate complex $[Ag(S_2O_3)_2]^{3-}$ formed from Ag^+ and $S_2O_3^{2-}$ is 1.0×10^{13}. What is the concentration of free Ag^+ in a solution made by adding 0.25 mmol of $AgNO_3$ to 10.0 mL of 0.50 M sodium thiosulfate ($Na_2S_2O_3$) solution? (Neglect any changes in volume.)

Solution: The complex-forming reaction is

$$Ag^+ + 2S_2O_3^{2-} \rightleftharpoons [Ag(S_2O_3)_2]^{3-} \qquad K_s = \frac{[[Ag(S_2O_3)_2]^{3-}]}{[Ag^+][S_2O_3^{2-}]^2} = 1.0 \times 10^{13}$$

Since K_s is so large, we can assume that $[Ag^+]$ is small and that almost all of the Ag^+ is converted to complex. At the start the solution contained (10.0 mL)(0.50 M $S_2O_3^{2-}$) = 5.0 mmol $S_2O_3^{2-}$. By the reaction stoichiometry, the added

0.25 mmol of Ag^+ will react with 0.50 mmol of $S_2O_3^{2-}$ to produce (very nearly) 0.25 mmol of $[Ag(S_2O_3)_2]^{3-}$. So

$$[[Ag(S_2O_3)_2]^{3-}] = \frac{0.25 \text{ mmol}}{10.0 \text{ mL}} = 2.5 \times 10^{-2} \text{ } M$$

$$[S_2O_3^{2-}] = \frac{5.0 \text{ mmol} - 0.50 \text{ mmol reacted}}{10.0 \text{ mL}} = 0.45 \text{ } M$$

Thus

$$[Ag^+] = \frac{[[Ag(S_2O_3)_2]^{3-}]}{K_s[S_2O_3^{2-}]^2} = \frac{2.5 \times 10^{-2}}{(1.0 \times 10^{13})(0.45)^2} = 1.2 \times 10^{-14} \text{ } M$$

The assumption that very little Ag^+ remains is fully confirmed.

Supplementary Exercises

PROBLEM 21-12   What are **(a)** the oxidation states and **(b)** the coordination numbers of the metal ions in the complexes **(1)** $[Cu(SCH_2C_6H_5)_3]^-$, **(2)** $[Mo(CN)_8]^{4-}$, **(3)** $[Mg(H_2O)_4]^{2+}$, **(4)** $[Ni(CN)_4]^{4-}$?

Answer:

	(1)	(2)	(3)	(4)
(a)	+2	+4	+2	0
(b)	3	8	4	4

PROBLEM 21-13   Give the electronic configurations of the metal ions **(a)** Mo^{4+}, **(b)** Mg^{2+}, **(c)** Ti^{3+}, **(d)** Ni^{3+}, and **(e)** Pd^{4+}

Answer: **(a)** $[Kr]4d^2$ **(b)** $[Ne]$ **(c)** $[Ar]3d^1$ **(d)** $[Ar]3d^7$ **(e)** $[Kr]4d^6$

PROBLEM 21-14   How many isomers are there for the four-coordinate complexes **(a)** $[Ni(CO)_2(PF_3)_2]$ (tetrahedral), **(b)** $[Ni(CO)(PF_3)_3]$ (tetrahedral), **(c)** $[PtCl_2(PCl_3)_2]$ (planar), **(d)** $[Pt(Cl)(I)(PCl_3)_2]$ (planar)?

Answer: **(a)** 1   **(b)** 1   **(c)** 2   **(d)** 2

PROBLEM 21-15   How many isomers are there for the octahedral complexes **(a)** $[Co(NH_3)_5Br]^{2+}$, **(b)** $[Co(NH_3)_4Br_2]^+$, **(c)** $[Co(NH_3)_3Br_3]$?

Answer: **(a)** 1   **(b)** 2   **(c)** 2

PROBLEM 21-16   Which of the following octahedral complexes should exist in enantiomeric forms?

   (a) $[cis\text{-}Co(NH_3)_4Br_2]^+$
   (b) $[cis\text{-}Co(en)_2Br_2]^+$ (en = chelating $H_2\ddot{N}CH_2CH_2\ddot{N}H_2$)
   (c) $[trans\text{-}Co(NH_3)_4Br_2]^+$
   (d) $[Cr(ox)_3]^{3-}$ (ox = chelating $^-O_2CCO_2^-$)
   (e) $[Cr(ox)Cl_4]^{3-}$

Answer: **(b)** and **(d)**

PROBLEM 21-17   The complex $[Ni(CN)_4]^{2-}$ is diamagnetic. Use the valence bond theory to deduce whether it is planar or tetrahedral.

Answer: square-planar

PROBLEM 21-18 The octahedral complex ion $[CoF_6]^{4-}$ has three unpaired electrons. In terms of valence bond theory, is the complex an inner-orbital or outer-orbital complex?

Answer: outer-orbital

PROBLEM 21-19 The octahedral complex ion $[Co(CN)_6]^{4-}$ has one unpaired electron. In terms of crystal field theory, is the crystal field splitting in this ion large or small?

Answer: large

PROBLEM 21-20 The stability constant K_s for the formation of $[AgBr_4]^{3-}$ from Ag^+ and Br^- is 5.5×10^8. Exactly 1.00 mmol of $AgNO_3$ is added to 100.0 mL of 0.10 M NaBr. Calculate the resulting concentration of free Ag^+.

Answer: $1.4 \times 10^{-6} M$

FINAL EXAM
(Chapters 12–21)

Standard Reduction Potentials*

Half-reaction	$E°$ (V)	Half-reaction	$E°$ (V)
$Li^+ + e^- \longrightarrow Li$	−3.040	$2H^+ + 2e^- \longrightarrow H_2$	0.000
$K^+ + e^- \longrightarrow K$	−2.931	$Hg_2Cl_2 + 2e^- \longrightarrow 2Hg + 2Cl^-$	0.268
$Na^+ + e^- \longrightarrow Na$	−2.711	$Cu^{2+} + 2e^- \longrightarrow Cu$	0.342
$Mg^{2+} + 2e^- \longrightarrow Mg$	−2.372	$Cu^+ + e^- \longrightarrow Cu$	0.521
$Zn^{2+} + 2e^- \longrightarrow Zn$	−0.762	$I_2 + 2e^- \longrightarrow 2I^-$	0.536
$Cr^{3+} + 3e^- \longrightarrow Cr$	−0.744	$Fe^{3+} + e^- \longrightarrow Fe^{2+}$	0.771
$Fe^{2+} + 2e^- \longrightarrow Fe$	−0.447	$Hg_2^{2+} + 2e^- \longrightarrow 2Hg$	0.797
$Cd^{2+} + 2e^- \longrightarrow Cd$	−0.403	$Ag^+ + e^- \longrightarrow Ag$	0.800
$Ni^{2+} + 2e^- \longrightarrow Ni$	−0.257	$2Hg^{2+} + 2e^- \longrightarrow Hg_2^{2+}$	0.920
$Sn^{2+} + 2e^- \longrightarrow Sn$	−0.138	$Br_2(aq) + 2e^- \longrightarrow 2Br^-$	1.087
$Pb^{2+} + 2e^- \longrightarrow Pb$	−0.126	$O_2 + 4H^+ + 4e^- \longrightarrow 2H_2O$	1.229
$Fe^{3+} + 3e^- \longrightarrow Fe$	−0.037	$Cl_2 + 2e^- \longrightarrow 2Cl^-$	1.358
$2H^+ + 2e^- \longrightarrow H_2$	0.000	$MnO_4^- + 8H^+ + 5e^- \longrightarrow Mn^{2+} + 4H_2O$	1.507
		$Ce^{4+} + e^- \longrightarrow Ce^{3+}$	1.61

* Data taken from *Handbook of Chemistry and Physics* (64th Ed.), CRC Press, Boca Raton FL.

1. Identify the Lewis acids and bases in the following reactions:

$$Al^{3+} + 6H_2O \longrightarrow Al(H_2O)_6^{3+} \quad \text{and} \quad NH_3 + BCl_3 \longrightarrow NH_3BCl_3$$

2. Give balanced net ionic equations for the following: (a) sodium formate HCO_2Na reacts with HCl solution to form the conjugate acid of HCO_2^-; (b) KN_3 is a weak base in H_2O.

3. Calculate the pH of a solution in which $[OH]^- = 1.83 \times 10^{-4}\,M$.

4. XeF_4 reacts with H_2O to form XeO_3, Xe, and HF. Write the balanced equation for this reaction and give the oxidation number of each Xe involved.

5. For the reaction

$$CuCl(s) + HClO(aq) \longrightarrow Cl^- + Cu^{2+}$$

in acidic solution, write the balanced oxidation and reduction half-reactions and a balanced redox equation.

6. For the reaction

$$MnO_4^- + SO_3^{2-} \longrightarrow MnO_4^{2-} + SO_4^{2-}$$

in basic solution, write the balanced oxidation and reduction half-reactions and a balanced redox equation.

7. For the reaction

$$Br^- + HSO_4^- \longrightarrow SO_2 + Br_2$$

in acidic solution, write the balanced net ionic equation.

8. The decomposition of hypochlorite ion

$$OCl^-(aq) \longrightarrow Cl^-(aq) + \tfrac{1}{2}O_2(g)$$

is first-order at 30°C and constant pH ($k = 7.0 \times 10^{-4}\,s^{-1}$). Determine (a) the rate law; (b) the time at which the OCl^- concentration will be $0.70 \times 10^{-4}\,M$, if the initial concentration is

$1.40 \times 10^{-4}\,M$; **(c)** the time at which OCl^- concentration will be $0.140 \times 10^{-4}\,M$, if the initial concentration is $1.40 \times 10^{-4}\,M$.

9. The reaction for the decomposition of $(CH_3)_2CO$,

$$(CH_3)_2CO(g) \longrightarrow CH_4(g) + H_2C_2O(g)$$

is second-order. Write its rate equation in integrated form.

10. The decomposition of N_2O_5

$$2N_2O_5(g) \longrightarrow 4NO_2(g) + O_2(g)$$

is first-order ($E_a = 105\,kJ\,mol^{-1}$). If the rate constant is $7.90 \times 10^{-7}\,s^{-1}$ at 0°C, determine the rate constant at 27°C.

11. The rate constant for the reaction

$$CO(g) + NO_2(g) \longrightarrow CO_2(g) + NO(g)$$

was found to be $1.6 \times 10^{-3}\,L\,mol^{-1}s^{-1}$ at 540 K and $6.3 \times 10^{-1}\,L\,mol^{-1}s^{-1}$ at 675 K. Calculate the activation energy for this reaction.

12. For the first-order reaction

$$C_4H_8(g) \longrightarrow 2C_2H_4(g)$$

at 700 K, the rate constant is $1.23 \times 10^{-4}\,s^{-1}$. If 3.00×10^{-3} mol of $C_4H_8(g)$ is placed in an otherwise empty 0.500-L container at 700 K, determine **(a)** the initial rate of decomposition of C_4H_8; **(b)** the amount of C_4H_8 remaining after 195 s; **(c)** the total pressure at 195 s.

13. For the equilibrium

$$COCl_2(g) \rightleftharpoons CO(g) + Cl_2(g)$$

(a) write the equilibrium constant expression. **(b)** If additional Cl_2 is added, which gas(es) will show a decrease in partial pressure? Explain your answer.

14. The stability constant for $Cu(NH_3)_4^{2+}$ is $K_s = 5.0 \times 10^{13}$. If the Cu^{2+} concentration is $1.5 \times 10^{-12}\,M$ and the $Cu(NH_3)_4^{2+}$ concentration is $0.200\,M$ at equilibrium, what is the NH_3 concentration?

15. For the equilibrium

$$2NOCl(g) \rightleftharpoons 2NO(g) + Cl_2(g)$$

$K_p = 1.7 \times 10^{-2}$ at 240°C. If 0.0133 mol of pure NOCl is heated to 240°, so that a partial pressure for Cl_2 of 0.0357 atm is obtained at equilibrium, what will the equilibrium partial pressures of NO and NOCl be?

16. For the equilibrium

$$2HI(g) \rightleftharpoons H_2(g) + I_2(g)$$

the equilibrium constant is 1.98×10^{-2} at 448°C. Calculate the concentration of I_2 at equilibrium when 0.0134 mol of pure HI in a 2.68-L container is heated to 448°C.

17. **(a)** Is S greater than, equal to, or less than zero for $N_2(g)$ in its standard state at 298 K? **(b)** Is ΔS greater than, equal to, or less than zero for the process $Br_2(g) \rightarrow Br_2(l)$ at constant pressure? Explain your answers.

18. Calculate ΔH_r°, ΔG_r°, and ΔS° for the reaction

$$FeO(s) + H_2(g) \longrightarrow Fe(s) + H_2O(l)$$

at 298 K, using the following standard enthalpies and free energies of formation (in $kJ\,mol^{-1}$):

	ΔH_f°	ΔG_f°
FeO(s)	−267	−244
$H_2O(l)$	−286	−237

19. For the reaction

$$2Ag(s) + Cl_2(g) \longrightarrow 2AgCl(s)$$

$\Delta G_r^\circ = -219.5 \text{ kJ mol}^{-1}$ at 25°C and $\Delta H_f^\circ(AgCl) = -127.04 \text{ kJ mol}^{-1}$. Calculate the standard entropy change.

20. For the equilibrium

$$H_2(g) + I_2(s) \rightleftharpoons 2HI(g)$$

at 25°C, calculate the equilibrium constant, ΔS_r°, and ΔH_r°. The following data may be needed: $\Delta G_f^\circ(HI) = 1.30 \text{ kJ mol}^{-1}$;

Substance:	$H_2(g)$	$I_2(s)$	$HI(g)$
S° $(J mol^{-1}K^{-1})$:	131	117	206

21. Benzoic acid $(C_6H_5CO_2H)$ is a weak acid $(K_a = 6.5 \times 10^{-5})$. Calculate the $[H^+]$ and pH of a solution made by adding 60.0 mL of 0.0500 M benzoic acid to 20.0 mL of 0.150 M KOH solution.

22. Adding 2.50 g of $NaNO_2$ to 150.0 mL of 0.180 M HNO_2 (a weak acid) gives a solution whose pH is 3.56. Calculate K_a for HNO_2. (Assume that the volume is unchanged.)

23. Consider the following reaction:

$$Sn^{2+} + KMnO_4 \longrightarrow Mn^{2+} + Sn^{4+}$$

If 25.00 mL of acidic Sn^{2+} solution required 15.83 mL of 0.1753 M $KMnO_4$ solution for complete titration, calculate the molarity of the Sn^{2+} solution.

24. For the equilibrium

$$NH_3 + H_2O \rightleftharpoons NH_4^+ + OH^- \qquad K_b = 1.77 \times 10^{-5}$$

When 175 mL of 0.100 M HCl is mixed with 225 mL of 0.170 M NH_3, the final volume is 400.0 mL. Calculate the final concentrations of NH_4^+, H^+, and Cl^-.

25. When excess solid MgF_2 is mixed with 100 mL of 0.0500 M NaF solution, the Mg^{2+} concentration is found to be 2.6×10^{-6} M. Calculate K_{sp} for MgF_2.

26. $PbBr_2$ and $CaCO_3$ are not very soluble in water. Explain why the solubility of $PbBr_2$ is less in 0.10 M HBr than in water, while $CaCO_3$ exhibits the opposite behavior.

27. Calculate the solubility (S) of AgOH $(K_{sp} = 1.5 \times 10^{-8})$ in a solution buffered at a pH of 9.00.

28. A steady 0.337-A current passes through a solution of indium nitrate $[In(NO_3)_3]$. At the anode, 7.58×10^{-3} mol of O_2 forms; at the cathode, In^{3+} is reduced to indium metal. How much In is formed in grams?

29. How long will it take to produce 7.30×10^3 g of Cl_2 from the electrolysis of NaCl solution, using a current of 3.33 A?

30. Use standard reduction potentials to predict the probable reactions in the following situations:

(a) Silver metal is placed in 1.0 M H^+ solution;
(b) Chromium metal is placed in 1.0 M H^+ solution;
(c) Chlorine gas is added to a 0.5 M NaBr solution;
(d) Liquid bromine is added to a 0.5 M NaCl solution;
(e) Copper metal is placed in 1.00 M $Pb(NO_3)_2$ solution;
(f) Lead is placed in 1.00 M $Cu(NO_3)_2$ solution;
(g) A 1.00 M $Cu(NO_3)_2$ solution is mixed with a 1.00 M $Pb(NO_3)_2$ solution.

31. For the reaction

$$2Ag(s) + Cl_2(g) \longrightarrow 2AgCl(s)$$

calculate $\Delta G°$ at 25°C using standard reduction potentials:

$$Cl_2(g) + 2e^- \longrightarrow 2Cl^-(aq) \qquad\qquad E° = 1.358\ V$$
$$AgCl(s) + e^- \longrightarrow Ag(s) + Cl^-(aq) \qquad E° = 0.2222\ V$$

32. For the reaction

$$2Hg(l) + Br_2(aq) \longrightarrow Hg_2^{2+} + 2Br^-$$

calculate $E°$ and K at 25°C.

33. Consider the reaction

$$2Ag^+ + Pd \longrightarrow Pd^{2+} + 2Ag$$

Calculate the equilibrium concentration of Pd^{2+} when a piece of Pd is dipped into a 0.20 M $AgNO_3$ solution at 25°C. (For $Pd^{2+} + 2e^- \rightarrow Pd$, $E° = 0.987\ V$.)

34. For the galvanic cell reaction

$$Ag(s) + Cl^- + Fe^{3+} \longrightarrow AgCl(s) + Fe^{2+}$$

at 25°C, calculate the cell voltage when the Cl^- concentration is 1.00×10^{-3} M and the Fe^{2+} and Fe^{3+} concentrations are 0.100 M. (For $Ag(s) + Cl^- \rightarrow AgCl(s) + e^-$, $E° = -0.222\ V$.)

35. Write balanced nuclear reaction equations for (**a**) strontium-90 decay by β^- emission, (**b**) nitrogen-15 bombardment by protons to generate carbon-12, and (**c**) chlorine-32 decay by positron emission.

36. Write a balanced nuclear equation for the α decay of polonium-210, $t_{1/2} = 138.4$ days. If you start with 1.00×10^{15} atoms, how many will be left after 365 days?

37. How many atoms of $^{35}_{16}S$ ($t_{1/2} = 87.1$ days) will be left after 71.0 days if the initial sample contains 1.00×10^{18} atoms?

38. Describe the complex paramagnetic ion $[CoF_6]^{3-}$. Write the net ionic reaction for the formation of the complex, and determine its coordination number and the oxidation number of the central metal atom. Discuss the geometry of the complex, and use a valence-bond model to explain its electronic configuration.

39. Discuss the geometry of $[Pt(NH_3)_2Br_2]$.

40. Mixing $AgNO_3$ and $NaCN$ solutions gives a solution with $[CN^-] = 0.047$ M and $[Ag(CN)_2^-] = 3.0 \times 10^{-3}$ M ($K_s = 1.0 \times 10^{21}$). Calculate the concentration of uncomplexed Ag^+ in the solution.

Solutions to Final Exam

1. Lewis acids can accept an electron pair, so Al^{3+} and BCl_3 are Lewis acids; Lewis bases can donate an electron pair, so H_2O and NH_3 are Lewis bases.

2.   (a) $HCO_2^- + H^+ \longrightarrow HCO_2H$ (b) $N_3^- + H_2O \longrightarrow HN_3 + OH^-$

3.
$$[OH^-] = 1.83 \times 10^{-4} \, M$$
$$pOH = -\log[OH] = -\log(1.83 \times 10^{-4})$$
$$= 3.738$$
$$pH = 14.000 - pOH = 14.000 - 3.738$$
$$= 10.262$$

4.
$$3XeF_4 + 6H_2O \longrightarrow 2XeO_3 + Xe + 12HF$$

XeF_4: The apparent charge on the central Xe atom in XeF_4, a compound that does not obey the octet rule, is $+4$, so the oxidation number of Xe in XeF_4 is $+4$

XeO_3: The oxidation number of each O in XeO_3 is -2. The molecule is neutral, so the sum of the oxidation numbers must equal zero. The oxidation number of Xe in XeO_3 is $+6$ because $+6 + 3(-2) = 0$.

Xe: The oxidation number of neutral elements is always 0, so the oxidation number of Xe is 0.

5. *Step 1.* Calculate oxidation numbers:

Species	Oxidation number
CuCl	$+1$ for Cu
HClO	$+1$ for Cl
Cl^-	-1
Cu^{2+}	$+2$

Chlorine in HClO has gained electrons, so it is being reduced; copper has lost electrons, so it is being oxidized. Write the half-reactions and balance, if necessary:

Oxidation half-reaction	**Reduction half-reaction**
$CuCl \longrightarrow Cl^- + Cu^{2+}$	$HClO \longrightarrow Cl^-$

Step 2. All atoms are balanced except H and O.

Step 3. Add H_2O to balance O; then add H^+ to balance H:

$$CuCl \longrightarrow Cl^- + Cu^{2+} \qquad\qquad HClO + H^+ \longrightarrow Cl^- + H_2O$$
(no H's or O's to balance)

Step 4. Add electrons to balance the charges:

$$CuCl \longrightarrow Cl^- + Cu^{2+} + e^- \qquad\qquad HClO + H^+ + 2e^- \longrightarrow Cl^- + H_2O$$

Step 5. Balance the electrons in the two half-reactions:

$$2CuCl \longrightarrow 2Cl^- + 2Cu^{2+} + 2e^- \qquad\qquad HClO + H^+ + 2e^- \longrightarrow Cl^- + H_2O$$

Step 6. Add the two half-reactions and cancel where appropriate:

$$2CuCl + HClO + H^+ \longrightarrow 3Cl^- + 2Cu^{2+} + H_2O$$

Check. Charge balance: $+1 = -3 + 4$

A quick check shows that all atoms and charges are balanced.

6. *Step 1.*

Species	Oxidation number
MnO_4^-	+7 for Mn
SO_3^{2-}	+4 for S
MnO_4^{2-}	+6 for Mn
SO_4^{2-}	+6 for S

reduction

oxidation

Oxidation half-reaction	**Reduction half-reaction**
$SO_3^{2-} \longrightarrow SO_4^{2-}$	$MnO_4^- \longrightarrow MnO_4^{2-}$

Step 2. All atoms are balanced except H and O.

Step 3. Add OH^- and H_2O to balance O:

$$SO_3^{2-} + 2OH^- \longrightarrow SO_4^{2-} + H_2O \qquad | \qquad MnO_4^- \longrightarrow MnO_4^{2-}$$

Step 4. Add electrons to balance the charges:

$$SO_3^{2-} + 2OH^- \longrightarrow SO_4^{2-} + H_2O + 2e^- \qquad | \qquad MnO_4^- + e^- \longrightarrow MnO_4^{2-}$$

Step 5. Balance the electrons in the two half-reactions:

$$SO_3^{2-} + 2OH \longrightarrow SO_4^{2-} + H_2O + 2e^- \qquad | \qquad 2MnO_4^- + 2e^- \longrightarrow 2MnO_4^{2-}$$

Step 6. Add the two half-reactions and cancel where appropriate:

$$2MnO_4^- + SO_3^{2-} + 2OH^- \longrightarrow 2MnO_4^{2-} + SO_4^{2-} + H_2O$$

Check. Charge balance: $-2 + (-2) + (-2) = -4 + (-2)$

7. Determine oxidation numbers:

$$\overset{-1}{Br^-} + \overset{+6}{HSO_4^-} \longrightarrow \overset{+4}{SO_2} + \overset{0}{Br_2}$$

For Br in $Br^- \rightarrow Br_2$, there is an increase of 1 in oxidation number.
For S in $HSO_4^- \rightarrow SO_2$, there is a decrease of 2 in oxidation number. Balance oxidation-number change:

$$2Br^- + HSO_4^- \longrightarrow SO_2 + Br_2$$

Balance the remaining atoms with H^+ and H_2O:

$$2Br^- + HSO_4^- + 3H^+ \longrightarrow SO_2 + Br_2 + 2H_2O$$
$$-2 + (-1) + 3 = 0$$

8. (a) $\text{Rate} = -\dfrac{\Delta[OCl^-]}{\Delta t} = k[OCl^-] = (7.0 \times 10^{-4} \text{ s}^{-1})[OCl^-]$

(b) The final concentration is one-half of the initial concentration, so find $t_{1/2}$:

$$t_{1/2} = \frac{0.693}{k} = \frac{0.693}{7.0 \times 10^{-4} \text{ s}^{-1}} = 9.9 \times 10^2 \text{ s}$$

(c) Use the integrated form of the rate equation and solve for t:

$$\log[OCl^-] = \log[OCl^-]_0 - \frac{kt}{2.303}$$

$$t = -\frac{2.303(\log[OCl^-] - \log[OCl^-]_0)}{k}$$

$$= -\frac{2.303(-4.85 - (-3.85))}{7.0 \times 10^{-4} \text{ s}^{-1}}$$

$$= 3.3 \times 10^3 \text{ s}$$

9. For $\text{Rate} = k[A]^2 = k[(CH_3)_2CO]^2$, the integrated form of the second-order rate equation is

$$\frac{1}{[(CH_3)_2CO]} = \frac{1}{[(CH_3)_2CO]_0} + kt$$

10. Use the Arrhenius equation:

$$\log\left(\frac{k_2}{k_1}\right) = -\left(\frac{E_a}{2.303R}\right)\left(\frac{1}{T_2} - \frac{1}{T_1}\right) = \left(\frac{E_a}{2.303R}\right)\left(\frac{T_2 - T_1}{T_1 T_2}\right)$$

$$= \frac{(105 \text{ kJ mol}^{-1})(10^3 \text{ J kJ}^{-1})}{2.303(8.314 \text{ J mol}^{-1}\text{K}^{-1})}\left(\frac{27}{273(300)} \text{ K}^{-1}\right)$$

$$= 1.808$$

$$\frac{k_2}{k_1} = \text{antilog}(1.808) = 64.3$$

$$k_2 = 64.3k_1 = 64.3(7.90 \times 10^{-7} \text{ s}^{-1})$$
$$= 5.08 \times 10^{-5} \text{ s}^{-1}$$

11. Use the Arrhenius equation:

$$\log\left(\frac{k_2}{k_1}\right) = \left(\frac{E_a}{2.303R}\right)\left(\frac{T_2 - T_1}{T_1 T_2}\right)$$

$$E_a = (2.303R)\left(\log\frac{k_2}{k_1}\right)\left(\frac{T_1 T_2}{T_2 - T_1}\right)$$

$$= 2.303(8.31 \text{ J mol}^{-1}\text{K}^{-1})\left(\log\frac{6.3 \times 10^{-1}}{1.6 \times 10^{-3}}\right)\left(\frac{540(675)}{675 - 540} \text{ K}^{-1}\right)$$

$$= 1.34 \times 10^5 \text{ J mol}^{-1} = 134 \text{ kJ mol}^{-1}$$

12. (a) Use the general rate equation:

$$\text{Rate} = k[C_4H_8]$$

$$= (1.23 \times 10^{-4} \text{ s}^{-1})\left(\frac{3.00 \times 10^{-3} \text{ mol}}{0.500 \text{ L}}\right)$$

$$= 7.38 \times 10^{-7} \text{ mol L}^{-1}\text{s}^{-1}$$

(b) Use the integrated rate law:

$$\log\frac{[C_4H_8]_0}{[C_4H_8]} = \frac{kt}{2.303} = \frac{(1.23 \times 10^{-4} \text{ s}^{-1})(195 \text{ s})}{2.303}$$

$$= 1.04 \times 10^{-2}$$

$$\frac{[C_4H_8]_0}{[C_4H_8]} = \text{antilog}(1.04 \times 10^{-2}) = 1.02$$

$$[C_4H_8] = \frac{[C_4H_8]_0}{1.02} = \frac{3.00 \times 10^{-3} \text{ mol}}{(0.500 \text{ L})(1.02)} = 5.88 \times 10^{-3} \text{ mol L}^{-1}$$

The amount of C_4H_8 equals $(5.88 \times 10^{-3} \text{ mol L}^{-1})(0.500 \text{ L}) = 2.94 \times 10^{-3}$ mol.

(c) Use the stoichiometry of the reaction to find the total moles of gas:

$$n_T = n_{C_4H_8} + n_{C_2H_4}$$
$$n_T = 2.94 \times 10^{-3} \text{ mol} + 2(3.00 \times 10^{-3} - 2.94 \times 10^{-3}) \text{ mol}$$
$$= 3.06 \times 10^{-3} \text{ mol}$$

Now use the ideal gas law:

$$P = \frac{nRT}{V} = \frac{(3.06 \times 10^{-3} \text{ mol})(0.0821 \text{ L atm mol}^{-1}\text{K}^{-1})(700 \text{ K})}{0.500 \text{ L}}$$

$$= 0.352 \text{ atm}$$

13. (a) $K_c = \dfrac{[CO][Cl_2]}{[COCl_2]}$ or $K_p = \dfrac{p_{CO}p_{Cl_2}}{p_{COCl_2}}$

(b) By Le Chatelier's principle, the system will react to minimize the stress. The partial pressure of CO will decrease if more Cl_2 is added. The extra Cl_2 will react with CO to form more $COCl_2$ (thus decreasing the concentration and partial pressure of CO).

14. Balance the equation:

$$Cu^{2+} + 4NH_3 \rightleftharpoons Cu(NH_3)_4^{2+}$$

Write the equilibrium expression for K_s and solve for $[NH_3]$:

$$K_s = \frac{[Cu(NH_3)_4^{2+}]}{[Cu^{2+}][NH_3]^4}$$

$$[NH_3] = \left(\frac{[Cu(NH_3)_4^{2+}]}{[Cu^{2+}]K_s}\right)^{1/4}$$

$$= \left(\frac{0.200}{(1.5 \times 10^{-12})(5.0 \times 10^{13})}\right)^{1/4}$$

$$= 0.23\ M$$

15.
$$K_p = \frac{p_{NO}^2 p_{Cl_2}}{p_{NOCl}^2} = 1.7 \times 10^{-2}$$

From the stoichiometry, $p_{NO} = 2p_{Cl_2} = 2(0.0357) = 7.14 \times 10^{-2}$ atm. Thus

$$p_{NOCl}^2 = \frac{p_{NO}^2 p_{Cl_2}}{K_p} = \frac{(7.14 \times 10^{-2})^2(3.57 \times 10^{-2})}{1.7 \times 10^{-2}} = 1.07 \times 10^{-2}$$

and

$$p_{NOCl} = \sqrt{1.07 \times 10^{-2}} = 1.0 \times 10^{-1}\ \text{atm} \qquad \text{(2 sig. figs.)}$$

16. Because $n_{products} = n_{reactants}$, $K_p = K_c$, and so

$$K_c = \frac{[H_2][I_2]}{[HI]^2} = 1.98 \times 10^{-2}$$

Write all the concentrations in terms of one unknown, $[I_2]$. First $[H_2] = [I_2]$, since you started with pure HI. Then from the stoichiometry

$$[HI] = \left(\frac{0.0134\ \text{mol}}{2.68\ \text{L}}\right) - [I_2]\left(\frac{2\ \text{mol HI}}{1\ \text{mol I}_2}\right) = (5.00 \times 10^{-3}) - 2[I_2]$$

Substitute into the expression for K_c:

$$\frac{[I_2]^2}{((5.00 \times 10^{-3}) - 2[I_2])^2} = 1.98 \times 10^{-2}$$

$$\frac{[I_2]}{(5.00 \times 10^{-3}) - 2[I_2]} = \sqrt{1.98 \times 10^{-2}} = 0.141$$

$$[I_2] = 0.141(5.00 \times 10^{-3} - 2[I_2])$$

$$[I_2] + 0.282[I_2] = 7.04 \times 10^{-4}$$

$$[I_2] = \frac{7.04 \times 10^{-4}}{1.28} = 5.5 \times 10^{-4}\ \text{mol L}^{-1}$$

17. (a) The entropy of a pure, perfect crystalline substance is zero at 0 K. At 298 K, the entropy of N_2 must be greater than zero.
(b) Gases have higher entropies than their corresponding liquids, so $\Delta S < 0$.

18. Since $\Delta H_r^\circ = \sum \Delta H_f^\circ(\text{products}) - \sum \Delta H_f^\circ(\text{reactants})$,

$$\Delta H_r^\circ = \Delta H_f^\circ(H_2O) - \Delta H_f^\circ(FeO) = -286 - (-267) = -19\ \text{kJ mol}^{-1}$$

Similarly

$$\Delta G_r^\circ = \Delta G_f^\circ(H_2O) - \Delta G_f^\circ(FeO) = -237 - (-244) = 7\ \text{kJ mol}^{-1}$$

[*Note*: For elements in standard states ΔH_f° and ΔG_f° are zero.]

Use the expression for free energy to find $\Delta S°$:

$$\Delta G° = \Delta H° - T\Delta S°$$

$$\Delta S° = \frac{\Delta H° - \Delta G°}{T} = \frac{(-19 - 7)\,\text{kJ mol}^{-1}}{298\,\text{K}}$$

$$= -0.087\,\text{kJ mol}^{-1}\text{K}^{-1}$$
$$= -87\,\text{J mol}^{-1}\text{K}^{-1}$$

19. $\quad \Delta H_r° = 2\Delta H_f°(\text{AgCl}) = 2(-127.04)$
$$= -254.08\,\text{kJ mol}^{-1}$$

$$\Delta S° = \frac{\Delta H° - \Delta G°}{T} = \frac{(-254.08 - (-219.5))\,\text{kJ mol}^{-1}}{298\,\text{K}}$$

$$= -0.116\,\text{kJ mol}^{-1}\text{K}^{-1}$$
$$= -116\,\text{J mol}^{-1}\text{K}^{-1}$$

20. $\quad \Delta G_r° = 2\Delta G_f°(\text{HI})\quad \Delta G_f°(\text{H}_2) - \Delta G_f°(\text{I}_2)$
$$= 2(1.30\,\text{kJ mol}^{-1}) - 0 - 0$$
$$= 2.60 \times 10^3\,\text{J mol}^{-1}$$

Use the relationship between $\Delta G°$ and K:

$$\Delta G_r° = -2.303RT\log K$$

$$\log K = \frac{-\Delta G_r°}{2.303RT} = \frac{-2.60 \times 10^3\,\text{J mol}^{-1}}{2.303(8.314\,\text{J mol}^{-1}\text{K}^{-1})(298\,\text{K})}$$

$$K = \text{antilog}(-4.56 \times 10^{-1}) = 0.350$$

Find $\Delta S_r°$:

$$\Delta S_r° = 2S°(\text{HI}) - S°(\text{H}_2) - S°(\text{I}_2)$$
$$= 2(206) - 131 - 117 = 164\,\text{J mol}^{-1}\text{K}^{-1}$$

Now use the relationship for free-energy change to calculate $\Delta H_r°$:

$$\Delta H_r° = \Delta G_r° + T\Delta S_r°$$
$$= 2.60 \times 10^3\,\text{J mol}^{-1} + (298\,\text{K})(164\,\text{J mol}^{-1}\text{K}^{-1})$$
$$= 51.5\,\text{kJ mol}^{-1}$$

21. Write a balanced net ionic equation (using HBz for benzoic acid):

$$\text{OH}^- + \text{HBz} \longrightarrow \text{H}_2\text{O} + \text{Bz}^-$$

Calculate n for OH^- and HBz:

$$n_{\text{OH}^-} = MV = (0.150\,\text{mol L}^{-1})(0.0200\,\text{L}) = 3.00 \times 10^{-3}\,\text{mol}$$
$$n_{\text{HBz}} = MV = (0.0500\,\text{mol L}^{-1})(0.0600\,\text{L}) = 3.00 \times 10^{-3}\,\text{mol}$$

(This is the equivalence point of the reaction). Now calculate $[\text{Bz}^-]$:

$$[\text{Bz}^-] = \frac{3.00 \times 10^{-3}\,\text{mol}}{0.0800\,\text{L}} = 0.0375\,M$$

Since the salt of a weak acid and a strong base is a weak base, use $K_a \times K_b = K_w$ and write a suitable expression for K_b

$$K_b = \frac{K_w}{K_a} = \frac{[\text{HBz}][\text{OH}^-]}{[\text{Bz}^-]} = \frac{1.0 \times 10^{-14}}{6.5 \times 10^{-5}} = 1.54 \times 10^{-10}$$

and substitute, using $[\text{HBz}] = [\text{OH}^-]$ (from the stoichiometry) and assuming little reaction of Bz^-:

$$K_b = \frac{[\text{OH}^-]^2}{0.0375} = 1.54 \times 10^{-10}$$

$$[\text{OH}^-] = \sqrt{(1.54 \times 10^{-10})(0.0375)} = \sqrt{5.8 \times 10^{-12}} = 2.4 \times 10^{-6}$$

Therefore

$$[H^+] = \frac{K_w}{[OH^-]} = \frac{1.00 \times 10^{-14}}{2.4 \times 10^{-6}} = 4.2 \times 10^{-9}$$

and

$$pH = -\log[H^+] = -\log(4.2 \times 10^{-9}) = 8.38$$

22. Write the balanced equation:

$$HNO_2 \rightleftharpoons H^+ + NO_2^-$$

This is a buffer solution. Determine $[NO_2^-]$:

$$[NO_2^-] = \frac{m}{MV} = \frac{2.50 \text{ g NaNO}_2}{(69.0 \text{ g mol}^{-1} \text{ NaNO}_2)(0.1500 \text{ L})} = 0.242 \text{ mol L}^{-1}$$

Calculate the $[H^+]$:

$$pH = -\log[H^+] = 3.56$$
$$[H^+] = \text{antilog}(-3.56) = 2.75 \times 10^{-4}$$

Write an expression for K_a and substitute:

$$K_a = \frac{[H^+][NO_2^-]}{[HNO_2]} = \frac{(2.75 \times 10^{-4})(0.242)}{0.180} = 3.7 \times 10^{-4}$$

23. Balance the net ionic equation:

$$5Sn^{2+} + 2MnO_4^- + 16H^+ \longrightarrow 2Mn^{2+} + 5Sn^{4+} + 8H_2O$$

Calculate n for MnO_4^- and Sn^{2+} required for titration:

$$n_{MnO_4^-} = (15.83 \text{ mL})(0.1753 \text{ M}) = 2.775 \text{ mmol}$$

$$n_{Sn^{2+}} = (2.775 \text{ mmol MnO}_4^-)\left(\frac{5 \text{ mol Sn}^{2+}}{2 \text{ mol MnO}_4^-}\right) = 6.937 \text{ mmol}$$

Now solve for the molarity of Sn^{2+}:

$$M_{Sn^{2+}} = \frac{6.937 \text{ mmol}}{25.00 \text{ mL}} = 2.775 \times 10^{-1} \text{ mol L}^{-1}$$

24. Write an expression for K_b:

$$K_b = \frac{[NH_4^+][OH^-]}{[NH_3]} = 1.77 \times 10^{-5}$$

The reaction $NH_3 + H^+ \rightarrow NH_4^+$ is a buffer of a strong acid and a weak base, and essentially goes to completion. Now calculate reactant concentrations: Use the stoichiometry of the reaction and K_b.

$$n_{NH_3}(\text{initial}) = (225 \text{ mL})(0.170 \text{ mmol mL}^{-1}) = 38.2 \text{ mmol}$$
$$n_{H^+}(\text{added}) = n_{NH_4^+}(\text{produced}) = (175 \text{ mL})(0.100 \text{ mmol mL}^{-1}) = 17.5 \text{ mmol}$$

Therefore

$$[NH_4^+] = [Cl^-] = \frac{17.5 \text{ mmol}}{400.0 \text{ mL}} = 4.38 \times 10^{-2} \text{ M}$$

Furthermore

$$n_{NH_3}(\text{remaining}) = (38.2 - 17.5) \text{ mmol} = 20.7 \text{ mmol}$$

$$[NH_3] = \frac{20.7 \text{ mmol}}{400.0 \text{ mL}} = 5.18 \times 10^{-2} \text{ M}$$

So

$$[OH^-] = \frac{K_b[NH_3]}{[NH_4^+]} = \frac{(1.77 \times 10^{-5})(5.18 \times 10^{-2})}{4.38 \times 10^{-2}} = 2.09 \times 10^{-5} \, M$$

$$[H^+] = \frac{K_w}{[OH^-]} = \frac{1.00 \times 10^{-14}}{2.09 \times 10^{-5}} = 4.78 \times 10^{-10} \, M$$

25. The equilibrium is

$$MgF_2(s) \rightleftharpoons Mg^{2+} + 2F^-$$

The NaF solution is completely ionized, so

$$[F^-] = 0.0500 + 2(2.6 \times 10^{-6}) \cong 0.0500 \, M$$

Now solve for K_{sp}:

$$K_{sp} = [Mg^{2+}][F^-]^2 = (2.6 \times 10^{-6})(0.0500)^2 = 6.5 \times 10^{-9}$$

26. Consider the equilibria of the two solids:

$$PbBr_2(s) \rightleftharpoons Pb^{2+} + 2Br^- \qquad CaCO_3(s) \rightleftharpoons Ca^{2+} + CO_3^{2-}$$

The reduced solubility of $PbBr_2$ is due to the common ion effect: Br^- in HBr solution suppresses dissolution of $PbBr_2$. The increased solubility of $CaCO_3$ in HBr is caused by the reaction of CO_3^{2-} (the anion of a weak acid) with H^+ supplied by the HBr solution. As CO_3^{2-} is converted to H_2O and CO_2, more $CaCO_3$ dissolves to reestablish equilibrium.

27. The equilibrium is $AgOH \rightleftharpoons Ag^+ + OH^-$, so use the pH to calculate $[OH^-]$:

$$pOH = 14.00 - pH = 14.00 - 9.00 = 5.00$$

$$-\log[OH^-] = pOH = 5.00$$

$$[OH^-] = \text{antilog}(-5.00) = 1.0 \times 10^{-5}$$

Now use K_{sp} to find the solubility S:

$$K_{sp} = [Ag^+][OH^-] = S[OH^-]$$

$$S = \frac{K_{sp}}{[OH^-]} = \frac{1.5 \times 10^{-8}}{1.0 \times 10^{-5}} = 1.5 \times 10^{-3} \, \text{mol L}^{-1}$$

28. Anode (oxidation half-reaction): $\quad 2H_2O \rightarrow O_2 + 4H^+ + 4e^-$
Cathode (reduction half-reaction): $\quad In^{3+} + 3e^- \rightarrow In$

The mass of In is calculated from the stoichiometry:

$$(7.58 \times 10^{-3} \, \text{mol O}_2)\left(\frac{4 \, \text{mol } e^-}{1 \, \text{mol O}_2}\right)\left(\frac{1 \, \text{mol In}}{3 \, \text{mol } e^-}\right)\left(\frac{114.8 \, \text{g In}}{1 \, \text{mol In}}\right) = 1.16 \, \text{g In}$$

29. At the anode, $2Cl^- \rightarrow Cl_2 + 2e^-$. Calculate the moles of e^- required:

$$n_{e^-} = \left(\frac{7.30 \times 10^3 \, \text{g Cl}_2}{70.9 \, \text{g Cl}_2/\text{mol}}\right)\left(\frac{2 \, \text{mol } e^-}{1 \, \text{mol Cl}_2}\right)$$

$$= 2.06 \times 10^2 \, \text{mol}$$

Convert the current to coulombs per second (use Faraday's constant) and determine the time required to send 2.06×10^2 mol of e^- through the NaCl solution:

$$\text{time} = \frac{(\text{mol } e^-)(\mathscr{F})}{\text{current}}$$

$$= \frac{(2.06 \times 10^2 \, \text{mol } e^-)(96\,500 \, \text{A s/mol } e^-)}{3.33 \, \text{A}}$$

$$= 5.97 \times 10^6 \, \text{s} \cong 69 \, \text{days}$$

30. In each case, calculate the standard cell potential using a table of standard reduction potentials: If $E°_{cell} > 0$, the reaction will occur.

(a) $2Ag + 2H^+ \rightarrow H_2 + 2Ag^+$ $E°_{cell} = -0.800 + 0.000 = -0.800$ V   no reaction

(b) $2Cr + 6H^+ \rightarrow 3H_2 + 2Cr^{3+}$ $E°_{cell} = 0 + 0.744 = 0.744$ V   reaction

(c) $2Br^- + Cl_2 \rightarrow Br_2 + 2Cl^-$ $E°_{cell} = 1.358 - 1.087 = 0.271$ V   reaction

(d) $Br_2 + 2Cl^- \rightarrow 2Br^- + Cl_2$ $E°_{cell} = 1.087 - 1.358 = -0.271$ V   no reaction

(e) $Pb^{2+} + Cu \rightarrow Pb + Cu^{2+}$ $E°_{cell} = -0.342 - 0.126 = -0.468$ V   no reaction

(f) $Pb + Cu^{2+} \rightarrow Cu + Pb^{2+}$ $E°_{cell} = 0.342 + 0.126 = 0.468$ V   reaction

(g) $Cu^{2+} + Pb^{2+} \rightarrow ?$ no redox possibilities   no reaction

31. Since $E°_{cell} = +1.358 - 0.2222 = 1.136$ V, then

$$\Delta G° = -n\mathscr{F}E° = -2(96\,500)(1.136)$$
$$= -2.19 \times 10^5 \text{ J} = -2.19 \times 10^2 \text{ kJ}$$

32. Since $E°_{cell} = +1.087 - 0.797 = 0.290$ V, then

$$\log K_{25°C} = \frac{nE°}{0.0592} = \frac{2(0.290)}{0.0592} = 9.80$$

$$K_{25°C} = \text{antilog}(9.80) = 6.3 \times 10^9$$

33. Calculate the cell potential: $E°_{cell} = +0.800 - 0.987 = -0.187$ V. Then use the Nernst equation; at equilibrium, $E = 0$:

$$E°_{25°C} = \left(\frac{0.0592}{2}\right)\log\left(\frac{[Pd^{2+}]}{[Ag^+]^2}\right)$$

$$\log\left(\frac{[Pd^{2+}]}{(0.20)^2}\right) = \frac{2(-0.187)}{0.0592}$$

$$\frac{[Pd^{2+}]}{0.040} = \text{antilog}(-6.32) = 4.8 \times 10^{-7}$$

$$[Pd^{2+}] = (0.040)(4.8 \times 10^{-7}) = 1.9 \times 10^{-8} \text{ } M$$

34. $E°_{cell} = +0.771 - 0.222 = 0.549$ V

Use the Nernst equation:

$$E_{25°C} = E°_{cell} - \left(\frac{0.059}{n}\right)\log\left(\frac{[Fe^{2+}]}{[Cl^-][Fe^{3+}]}\right) = 0.549 - 0.059\log\left(\frac{0.100}{(1.00 \times 10^{-3})(0.100)}\right)$$

$$= 0.549 - 0.177 = 0.372 \text{ V}$$

35. (a) $^{90}_{38}Sr \rightarrow ^{90}_{39}Y + ^{0}_{-1}\beta$

(b) $^{15}_{7}N + ^{1}_{1}H \rightarrow ^{12}_{6}C + ^{4}_{2}He$

(c) $^{32}_{17}Cl \rightarrow ^{32}_{16}S + ^{0}_{+1}\beta$

36. $^{210}_{84}Po \longrightarrow ^{4}_{2}He + ^{206}_{82}Pb$

Use the integrated form of the rate of decay equation, in terms of $t_{1/2}$:

$$\log\left(\frac{[A_0]}{[A]}\right) = \frac{0.693t}{2.303t_{1/2}} = \frac{0.693(365)}{2.303(138.4)} = 0.794$$

$$\frac{[A_0]}{[A]} = \text{antilog}(0.794)$$

$$[A] = \frac{[A_0]}{\text{antilog}(0.794)} = \frac{1.00 \times 10^{15}}{6.22} = 1.61 \times 10^{14} \text{ atoms}$$

37.

$$\log\left(\frac{[A_0]}{[A]}\right) = \frac{0.693t}{2.303t_{1/2}} = \frac{0.693(71.0)}{2.303(87.1)} = 0.245$$

$$\frac{[A_0]}{[A]} = \text{antilog}(0.245)$$

$$[A] = \frac{[A_0]}{\text{antilog}(0.245)} = \frac{1.00 \times 10^{18}}{1.76}$$

$$= 5.68 \times 10^{17} \text{ atoms}$$

38. The $[CoF_6]^{3-}$ ion contains six F^- (base) ions bound to one Co^{3+} (acid) ion. Its formation can be represented as

$$6\,:\ddot{\underset{..}{F}}:^- + Co^{3+} \longrightarrow [Co(:\ddot{\underset{..}{F}}:)_6]^{3-}$$

There are six ligands, so the coordination number is 6; the oxidation number of Co is +3. The complex is octahedral and has no isomers. The electronic configuration of Co^{3+} is $[Ar]3d^6$. Since we know that the complex is paramagnetic, we know that there are unpaired electrons in the metal *d* orbitals and that we use sp^3d^2 outer-orbital hybridization:

	Electronic configuration			
	3*d*	4*s*	4*p*	4*d*
$Co^{3+}([Ar]3d^6)$	[↑↓][↑][↑][↑][↑]	[ ]	[ ][ ][ ]	[ ][ ][ ][ ][ ]
$+ 6F^-$		↑↓	↑↓ ↑↓ ↑↓	↑↓ ↑↓
$[CoF_6]^{3-}$	[↑↓][↑][↑][↑][↑]	[↑↓]	[↑↓][↑↓][↑↓]	[↑↓][↑↓][ ][ ][ ]

six sp^3d^2 hybrid orbitals

39. There are four ligands and the apparent charge on Pt is +2, so the coordination number is 4 and the oxidation number is +2, giving a square-planar complex. There are two geometric isomers:

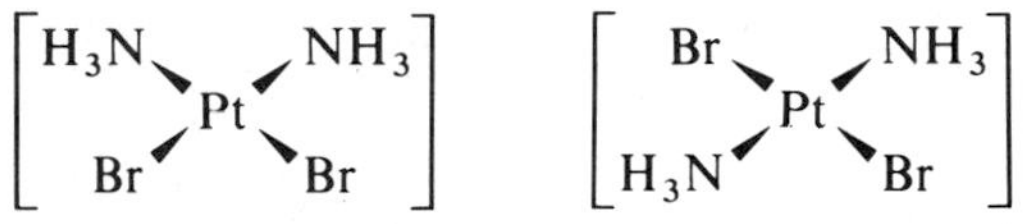

cis trans

40. Write the equilibrium:

$$Ag^+ + 2CN^- \rightleftharpoons Ag(CN)_2^-$$

Isolate $[Ag^+]$ from the K_s expression and substitute:

$$K_s = \frac{[Ag(CN)_2^-]}{[Ag^+][CN^-]^2}$$

$$[Ag^+] = \frac{[Ag(CN)_2^-]}{K_s[CN^-]^2} = \frac{3.0 \times 10^{-3}}{(1.0 \times 10^{21})(0.047)^2}$$

$$= 1.4 \times 10^{-21}\ M$$

INDEX

Page numbers in italics refer to solved problems.

Absolute zero, 54
Acid-base equilibria, 259–76, *277–88*
Acidic solutions, half-reactions in, 193–94
Acid ionization constant, 177–78
Acids and bases
  stoichiometry of, 179–80, *186–87*
  strength of, 176–78, *183–84*
  systems of, 175–76, *181–83*
  titration of, 178–79, *183–84*
Activated complex, 215
Adiabatic system, 246
Amorphous solids, 81, 87–88
Amphiprotism, 176
Amphoterism, 176
Angular momentum quantum number, 108
Anion, 14
Anode, 305–6
Arrhenius equation, 312–14
Arrhenius system, 175, *181*
Artificial radioactivity, 325
Atom, 13
Atomic mass unit, 14
Atomic mass (weight), 15, 19–20, *21*
Atomic number, 13–14, *19*
Atomic orbitals, 107, 109–11
Atomic radius, 131
Atomic spectra, 104
Atoms, electronic configuration of, 111,
  119–20, 121
Avogadro's law, 57, *71*
Avogadro's number, 17, *71*
Azeotropes, 76

Balmer series, 104, *117*
Base ionization constant, 178, 261
Bases. *See* Acids and bases
Base units, 2
Basic solutions, half-reactions in, 194–96
Body-centered cubic structure, 84–85
Bohr model of hydrogen spectrum, 106–7
Boiling point, 75, *77, 79*
  elevation of, 152, 154
Bond dissociation energy, 133–34
Bond energies, 95, 134
Bonding, 124–25, 139–42
Bond length, 131
Bond moment, 130–131
Bond order, in molecular orbitals, 137, *142*
Bond properties, 131–34, 142–44
Boyle's gas law, 55–56, *63*
Brackett series, 106
Brønsted-Lowry system, 175–76, *182*, 259,
  277–79
Buffers, 269–71, *283–85*

Catalysis, 216
Cathode, 305–6
Cation, 14
Cgs (metric) system, 2–3, *7–8*
Change of state, 246
Charge and current, 304, *314*
Charge balance equation, 262, *279–80*
Charles' gas law, 56, *63–64*

Chemical equations, 37
  balancing of, 37–38, *41–42*
  molar interpretation of, 38, *42–44*
  stoichiometry and, 38–39, *46–47*
Chemical equilibrium
  changing constant of, 230–31, *237–39*
  heterogeneous, 233
  law of, 228–30, *234*
  shifting, 230, 235–37
  solving problems of, 231–33, *234–43*
  state of, 228
Chemical kinetics, 208–16, *217–26*
Chemical thermodynamics
  entropy and, 247–49, *252–57*
  first law of, 94, 246–47
  free energy and, 249
  relationships of functions in, 250–51
  second law of, 249
  systems in, 246, *252*
Collision theory, 214
Combined gas law, 56–57, *64, 70*
Common ion effect, 293–94
Complexes
  bonding in, 344–47
  geometry of, 341–44
  stability of, 347–48
Complexion, 339–40
Compounds, 1
Condensed states, 74
Conductivity, 305
Conjugate acids and bases, 175–76
Coordinate covalence, 339–40, *349–50*
Coordinate covalent bonds, 125
Coordination chemistry, 339–48, *349–53*
Covalent bonds, 125, 129
Covalent network solids, 82
Crystal coordination number, 83
Crystal field theory, 346–47, *351–52*
Crystalline solids, 81–82, *88–89*
Crystal structure, 82–86
Cubic closest packing, 83

Dalton, 14
Dalton's law of partial pressures, 55, *65*
De Broglie equation, 107
Decay rate, 325–26, *332–34*
Density
  of cubic structure, 85–86, *89*
  of gases, 58
Derived units, 3
Diffusion, 60–61
Dipole-dipole forces, 74
Dipole moment, 130–31
Distillation, 76
Dynamic equilibrium, 75

Effusion, 61
Electrochemistry
  charge and current in, 304, *314*
  conductivity in, 305
  electrolytic cells, 305–7
  electromotive force in, 304
  galvanic (voltaic) cells, 307–8, *315*, 320

Nernst equation in, 312–13, *320–21*
  potential difference in, 304
  resistance in, 304–5
  standard electrode potentials in, 308–12,
  315–18
Electrode potential, 307–12, *315–18*
Electrode reactions, 306
Electrolysis, 305–6
Electrolytes, 153–54
Electrolytic cells, 305–7
Electromagnetic energy, 103–4
Electromotive force, 304, 307–8
Electron, 13
Electron charge units, 13
Electronegativity, 129–31
Electronic configuration
  of atoms, 111, *119–20, 121*
  of molecules, 136–37
Elements, 1, *7*
Endothermic reactions, 93
Energy, free, 249
Enthalpy, 92, *97–100*
Entropy, 247–49
Equations. *See* Chemical equations
Equilibrium. *See* Chemical equilibrium
Equivalence point (endpoint), 178–79
Exothermic reactions, 93, 215

Face-centered cubic structure, 83
Faraday's law, 306–7
First-order rate equation, 210–11, *218–19*
Formal charge, 127–28
Formula mass (weight), 16, *20–21*
Free energy, 249
Freezing point, depression of, 152, 154,
  158–59
Frequency, 103, *116–17*

Galvanic cells, 307–8
Gases
  Avogadro's law of, 57
  Boyle's law of, 55–56, *63*
  Charles' law of, 56, *63–64*
  combined law of, 56–57, *64, 70*
  composition of, 54–55
  Dalton's law of, 55, *65*
  definition of, 53
  density of, 58
  diffusion of, 60–61
  effusion of, 61
  Graham's law of, 60–61, *68*
  ideal, 61, *64–66*
  ideal law of, 58–60, *64–65*
  kinetic theory of, 60
  molar volume of, 57
  mole fraction of, 54, *62–63, 70*
  pressure of, 53–54
  real, 61
  standard temperature and pressure of,
    STP, 55
  state of, 53–55
  temperature of, 54
  Van der Waals equation of, 61, *69*
  volume of, 53

General rate equation, 209–210
Geometric isomers, 341–42
Geometry of complexes, 341–44
Graham's law, 60–61, *68*

Half-equivalence point, 274
Half-life, 211–12, *219*, 325–26, *333–36*
Half-reactions, 192–95. *See also* Redox
    reactions
Heat capacity, 91–92
Heat energy, 91–92, *96–97*
Heisenberg uncertainty principle, 107
Henderson-Hasselbach equation, 270
Henry's law, 150–51
Hess' law, 93–94
Heterogeneous equilibria, 233
Heterogeneous mixtures, 1
Heterogeneous solution equilibrium, 291–96,
    297–303
Hexagonal closest packing, 83
Homogeneous mixtures, 1
Hybridization, 344–46
Hydrogen spectrum, 106–7

Ideal gas law, 58–60, 64–66
Indicators, 271–72
Inorganic compounds, systematic names for,
    17–18
Intermolecular forces, 74–75
Ionic bonds, 124–25, *138*
Ionic solids, 81
Ionization, degree of, in solutions, 154, *161*
Ionization (dissociation) constant, 260,
    285–86, 287–88
Ionization energies, 111–15, *118, 119*
Ion product, 294–95
Ion product constant, 176–77
Ions, 14

Kinetic theory of gases, 60

Law of chemical equilibrium, 228–30, *234,*
    239–40
Le Chatelier's principle, 230
Lewis formulas, 126–27, *138, 140*
Lewis symbol, 123
Lewis system, 176, *181–182*
Ligand classification, 341, *350*
Limiting reagents, 39–40, *47*
Liquids, 75–76
London forces, 74
Lyman series, 106

Magnetic quantum number, 108
Mass balance equation, 263, *279–80*
Mass calculations from equations, 38–39,
    46–47
Mass/energy equivalence, 327–29
Matter
    classification of, 1
    condensed states of, 74
    properties of, 2
Measurements, 2–3
Metallic bonds, 124
Metallic solids, 81
Metric system, 2–3, *7–8*
Mixtures, 1
Molality of solution, 149–50, *156, 159–60*

Molarity of solution, 149, *156, 159–60,*
    183–84
Molar mass, 17, *21–22, 67,* 149–50, *156, 158,*
    159–60
Molar volume of gases, 57
Mole, 17, *21–22*
Molecular electronic configuration, 136–37
Molecular formula, 16, 25, 28, *33–35*
Molecular mass (weight), 16, *20*
Molecular orbitals, 135–37, *142–43*
Molecular speed, of gases, 60
Molecular structure, 126–29
Molecules, 15
Mole fraction, of gases, 54, *62–63, 70*
Mole fraction, of solution, 148, *156–57*
Moments, dipole, 130–31
Multiple covalent bonds, 125

Nernst equation, 312–13, *320–27*
Net ionic equations, 180
Neutron, 13
Nomenclature, 17
Nonpolar covalent bond, 129
Normality of solution, 150
Nuclear binding energy, 328–29
Nuclear fission, 330
Nuclear fusion, 329–30
Nuclear reactions, 324–27, *331–32*
Nuclei, composition of, 14–15
Nucleus, 13
Nuclides, 323

Octahedral complexes, 342–43, *350–51*
Octet rule, 123–24, 127, *139–40*
Ohm's law, 304–5
Optical isomers, 343–44
Orbitals, 109–11
Osmotic pressure, 152–53
Oxidation numbers, 191–92, 195–96
Oxidation-reduction reactions. *See* Redox
    reactions
Oxidation state, 340
Oxidizing agent, 190

Partial charges, 129
Pascal, 53
Paschen series, 106
Pauli exclusion principle, 110–11
Percentage by mass and volume of solutions,
    148, *155–56*
Percentage composition, 25–26, *30–33*
Percentage yield, 39–40, *47–49*
Percent relative error, 5
Periodic law, 115
Periodic table, 115
pH and pK, 177, *184–85*
    in weak acids and bases, 262–67, *279–83*
Photoelectric effect, 105–6, *117–18*
Photon, 104
Physical properties, 2
Pi molecular orbitals (πMOs), 135–36
Planck's equation, 105, 117
Polar covalent bond, 129
Polarity, 129–31
Polyprotic acids, 275–76
Potential difference, 304
Precipitation, 294–95, *301–2*
Primitive cell, 83

Primitive cubic structure, 83
Principal quantum number, 108
Products, of equations, 37
Proton, 13

Quantities, fundamental and derived, 2
Quantum numbers, 108, *119–20*
Quantum theory of radiation, 104–8

Radioactivity, 323–24, *330–31*
Radiochemistry, 323–30, *331–36*
Radionuclide decay, 326–27
Raoult's law, 151–52, *157–58*
Rate equation, 209–10
Reaction mechanisms, 215–16
Reaction rates
    and concentration effects, 209–13, *218–19,*
        221–24
    definition and expression of, 208–9, *217–18*
    and reaction mechanisms, 214–16
    and temperature effects, 213–14
Reaction rate theory, 215
Reagent, 37
Redox reactions
    analytical uses of, 197, *205*
    balancing, by half-reactions, 192–95,
        199–205
    definitions in, 190, *197–98*
    oxidation numbers for, 191–92, 195–96,
        198–205
Reducing agent, 190
Resistance, 304–5
Resonance, 128–29
Rounding off, 6

Salts, common structures of, 86
Salts of weak acids and bases, 267–69
Saturated solution, 147
Scientific notation, 4, *8*
Second law of thermodynamics, 249
Second-order rate equation, 212–13, *222–23*
Semipermeable membrane, 152–53
Sigma molecular orbitals (σMOs), 134–35
Significant figures, 4–6, *8–9*
Single covalent bonds, 125
SI system, 2–3, *7–8*
Solids, types of, 81–82, 87
Solubility, of solutions, 147, 150–51, *155*
Solubility product, 291–93, *297–300*
Solubility rules, 296
Solute, 147
Solutions
    definitions of, 147, *157*
    colligative properties of, 151–53
    electrolytes as, 153–54
    ionization, degree of, in, 154, *161*
    molality of, 149–50, *156, 159–60*
    molarity of, 149, *156, 159–60*
    mole fraction of, 148, *156–57*
    normality of, 150
    percentage by mass and volume of, 148,
        155–56
    solubility changes in, 150–51
    vapor pressure of, 151–52, *157–58*
Solvent, 147, *157*
Space lattice, 82
Spectator ions, 179–80
Spectroscopy, 104

Spin quantum number, 108
Spontaneous and nonspontaneous reactions, 311
Square-planar complexes, 342
Standard atmosphere, 53
Standard electrode potentials, 308–12
Standard enthalpy of reaction, 94, 134
Stoichiometry, 38–39
   of acids and bases, 179–80, *186–87*
Subatomic particles, 13
Substance, pure, 1, *7*
Surface tension, 76, *79*
Systematic names, 17–18, *21*
Système Internationale d'Unité (SI system), 2–3, *7–8*

Temperature, 91
Thermal equilibrium, 91
Thermodynamics. *See* Chemical thermodynamics

Thermodynamics, first law of, 94, 246–47
Titration, 178–79, 272–75, *285*
Torr, 53
Transition-metal ions, 340–341
Trivial names, 17

Unit cell, 82
Units, systems of, 2–3

Valence, 123–24
Valence-bond formulas, 126–27, *139–42*
Valence bond theory, 344–46
Valence-shell electron-pair repulsion (VSEPR) theory, 131–33, *140–41, 143–44*, 341
Van der Waals equation, 61, *69*
Van der Waals forces, 74
Van't Hoff factor, 153–54
Vapor pressure, 75, *77–78*
Vapor pressure, of solutions, 151–52, 157–58

Viscosity, 76–77, *78–79*
Voltaic cells, 307–8

Wavelength, 103, *116–17, 118–19*
Wave mechanics, 107
Wave number, 103
Weak acids
   pH solutions in, 262–65, *279–81*
   salts of, 267–68
Weak bases, 261
   pH solutions in, 265–67, *281–82*
   salts of, 268–69

X-ray diffraction, 82, *88*

Zeros, significant, 4–5

TEXTBOOK CORRELATION TABLE

The following table shows how the pages of this outline correspond by topic to the pages of five major textbooks on general chemistry. These textbooks are listed below. The topics of this outline and the pages on which they are covered are listed in the first two columns at the left of the table. The corresponding pages of each of the textbooks, designated by author, are listed in the next five columns. To find the pages of your textbook that correspond to the pages of this outline, read across each row and down the column headed by your author's name.

Bailar, Moeller, Kleinberg, Guss, Castellion, and Metz, CHEMISTRY, 2nd ed., 1984, Academic Press

Brady and Holum, FUNDAMENTALS OF CHEMISTRY, 1981, Wiley

Brady and Humiston, GENERAL CHEMISTRY, 3rd ed., 1982, Wiley

Ebbing, GENERAL CHEMISTRY, 1984, Houghton Mifflin

McQuarrie and Rock, GENERAL CHEMISTRY, 1984, Freeman

CHAPTER	THIS OUTLINE	BAILAR et al.	BRADY/ HOLUM	BRADY/ HUMISTON	EBBING	McQUARRIE/ ROCK
1: Matter and Measurement	1–12	6–23 25–27 54–55	5–6 13–18 21 23–24	3–18	7–12 14–18 37–38 A-1, A-2	2–3 25–32 155–156 A-1–A-3
2: Chemical Composition	13–24	38–53 62–64 67–70 74–83 89	62–71 76 143–148 171 254–257	22–23 35–41 46 65–69 132–136	28–34 51–59 63–65	9–23 85–91 95–96
3: Formulas	25–36	64–67 87–89	72–77	42–46	63–65	92–96
4: Equations and Stoichiometry	37–52	71–74 83–86 132–141 143–151 153–155	26–28 86–98	46–55	34–37 60–62 66–71	43–47 96–104
5: Gases	53–73	96–125	32–54	210–239	89–120	155–186
6: Intermolecular Forces and Liquids	74–80	335–341 388–394 396	314–326	247–252 254–256 259–261	42–44 347–357 372–373	473–478 481–484 486–487
7: Solids	81–90	70 397–408	171, 178 326–334	135 245 263–272	283–284 357–371 571–572	126 155–156 481–497
8: Thermochemistry	91–102	21–23 166–187 261–265	10–12 118–119 124–135 230, 485	27–29 356–376	13 312–336	164–165 203–221 228–229 467–468
9: Atomic Structure and the Periodic Table	103–122	197–233 279–288	135–138 160–166 185–207	65–66 70–94 100–102	137–157 164–191	251–279 289–308

CHAPTER	THIS OUTLINE	BAILAR et al.	BRADY/ HOLUM	BRADY/ HUMISTON	EBBING	McQUARRIE/ ROCK
10: Chemical Bonding	123–146	239–266 273–279 289–292 310–335	201–203 211–215 219–245 263 282–289 671–672	108–129 140–167	179 188–191 196–222 236–265	306–308 331–344 365–389 397–414 433–457
11: Solutions	147–162	432–433 446–473 650–652	105–107 346–352 356–376 587	17 199–200 236 317–351	271–272 389–400 404–418	523–547
12: Acids and Bases	175–189	141–144 435–436 638–666	86–87 178–180 373–376 545–570 582–593	176–178 448–460 464–480 491–494	66–69 274–275 570–589 615–619	139–141 647–677 701–710
13: Redox Reactions	190–207	293–298 486–489 749–753	245–254 603–607	130–132 190–192	292–303	781–797
14: Chemical Kinetics	208–227	565–593	453–477	394–417	494–527	601–632
15: Equilibrium	228–245	601–632	505–540	425–443	537–562	565–591
16: Chemical Thermodynamics	246–258	166–171 715–739	482–500 512–515	355–387	656–676	203–204 857–882
17: Acid–Base Equilibria	259–290	638–644 652–659 661–663 676–698	178–180 545–593	464–496	570–589 594–619	647–667 670–675 697–723
18: Heterogeneous Solution Equilibrium	291–303	676–677 680–681 698–705	350–352 527–540 553–556	178–179 501–509	626–648	743–767
19: Electrochemistry	304–322	753–783	559–632	518–554	683–709	805–837
20. Radiochemistry	323–338	347–380	142–154 717–737	67–69 774–794	717–753	943–973
21: Coordination Chemistry	339–354	331–333 1042 1067	232 275 301 693–711	510–511 677–697	856–887	433–446 907–936